STATUTE MILES

0   1200

Miller Cylindrical Projection

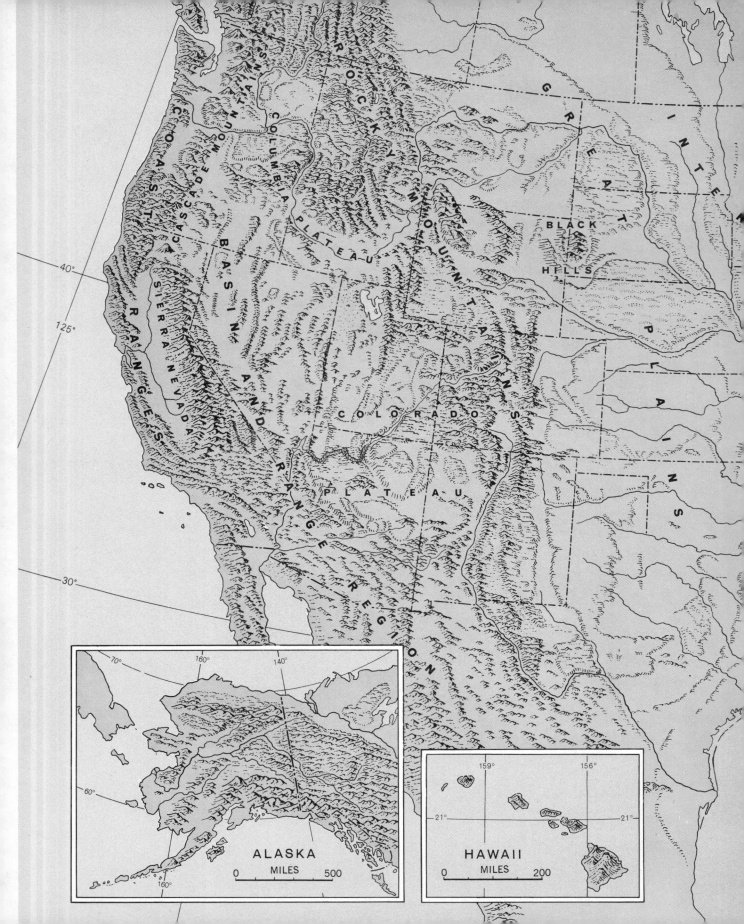

COAST RANGES

SIERRA NEVADA

CASCADE MOUNTAINS

COLUMBIA PLATEAU

ROCKY MOUNTAINS

BASIN AND RANGE REGION

COLORADO PLATEAU

GREAT PLAINS

INTER

BLACK HILLS

40°

125°

30°

ALASKA
MILES
0   500

70°
160°
140°
60°
160°

HAWAII
MILES
0   200

159°
156°
21°   21°

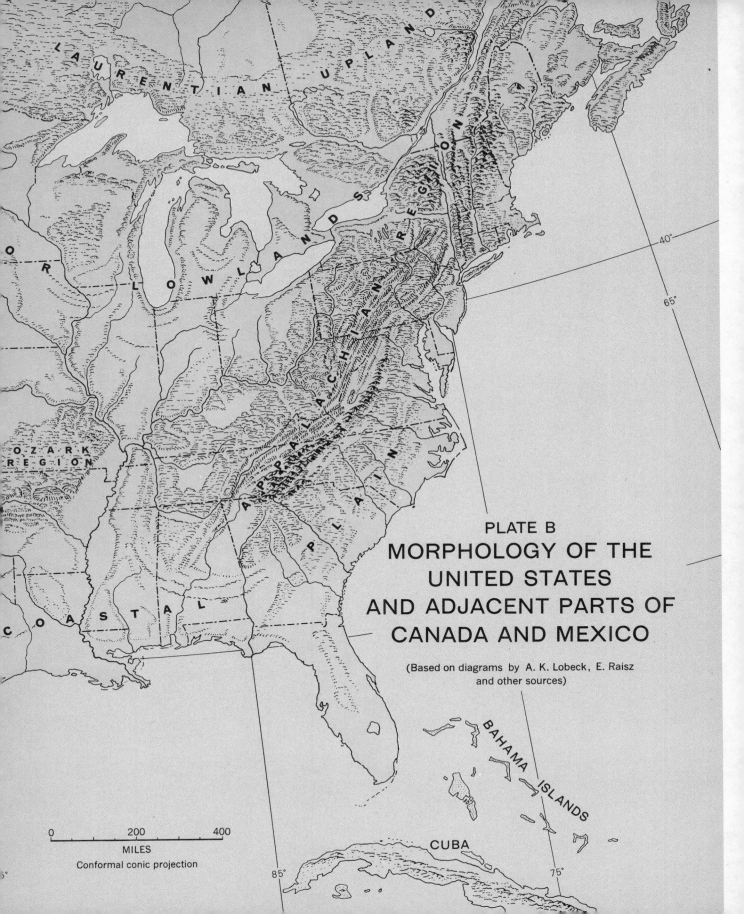

PLATE B

MORPHOLOGY OF THE UNITED STATES AND ADJACENT PARTS OF CANADA AND MEXICO

(Based on diagrams by A. K. Lobeck, E. Raisz and other sources)

LAURENTIAN UPLAND

LOWLANDS

OZARK REGION

APPALACHIAN REGION

APPALACHIAN PLAIN

COASTAL PLAIN

BAHAMA ISLANDS

CUBA

40°

65°

85°

75°

0 200 400
MILES
Conformal conic projection

Physical Geology

Physical Geology

Richard Foster Flint

Brian J. Skinner

Department of Geology and Geophysics, Yale University

John Wiley & Sons, Inc.

New York · London · Sydney · Toronto

Cover Photo:

A lake of basaltic lava on the floor of a caldera at the summit of Kilauea Volcano, Hawaii. Out of the lake flow streams of lava, each a few meters wide (Richard S. Fiske, U.S. Geological Survey).

This book was set in Plantin by York Graphic Services, Inc. It was printed by Murray Printing Co. and bound by Book Press, Inc. The designer was Madelyn Waxman. The drawings were designed and executed by John Balbalis with the assistance of the Wiley Illustration Department. Picture research was done by Marjorie Graham. Joan E. Rosenberg supervised production.

Library of Congress Cataloging in Publication Data:

Flint, Richard Foster, 1902–
  Physical geology.

  Includes bibliographies.
  1. Physical geology. I. Skinner, Brian J.,
1928– joint author. II. Title.
QE28.2.F55   551   73-11429
ISBN 0-471-26440-7

Printed in the United States of America

10 9 8 7 6 5 4 3 2

Preface

During the last few years Earth science has been undergoing a revolution that has three distinct aspects. The first concerns the dynamics of the lithosphere. Recent exploration of ocean floors and of Earth's crust and mantle has revealed crucial new facts that force the rethinking of Earth's dynamics. Many processes are being reexamined; new concepts are replacing older assumptions. A fundamental group of these concepts have coagulated into the dramatic theory of plate tectonics.

The second aspect involves sharply increasing awareness, in industrial nations, of the effect of human activities on environments at the surface of the Earth, accompanied by widespread attempts to analyze such effects and to modify them where necessary. These attempts represent not only scientific concern, but public concern on a broad scale. They are, likewise, closely related to the "energy crisis," which has stimulated renewed appraisal of Earth's sources of energy and of the ways in which energy is used.

The third aspect springs from the start of systematic investigation of the Moon, Mars, and other members of the Solar System. The resulting data bear importantly on the composition and history of Planet Earth itself.

In one way or another, then, the current revolution is profoundly changing our concepts of the way our planet works. The present book, a successor to the long line of Yale textbooks of physical geology, has been written in sharp realization of the revolution, and of the need to make the new facts and ideas intelligible in well-integrated form to students at the beginning level. Eleven of its 21 chapters are entirely new; the others have been rewritten to whatever extent seemed desirable in the light of new information. Rocks and minerals are treated, not as catalogues but as organizations of matter that take part in Earth's cyclical, dynamic processes. In the description of such processes, the concept of the steady-state condition (and its importance for living things) is emphasized repeatedly.

Those chapters (mainly those concerned with external geological processes) which have not been completely rewritten have been made more compact. Indeed the book as a whole is shorter than its immediate predecessor and the number of chapters has been reduced by three. Nevertheless a wider range of topics is covered than in previous editions. Among the new topics three are noteworthy. The sources and forms of energy that drive Earth's dynamic processes are described in Chapter 3; Chapter 19 discusses Moon and planets, summarizing many new recent findings; and Chapter 21 is devoted entirely to the effects of man as a geological agent.

Apart from these things we have continued features of earlier editions that have proved to be well received. Among these are the review summaries at the ends of chapters and the unique scheme of defining technical terms within the body of the text. A term defined in the text appears in **boldface italic** type, and the defining phrase or clause appears in *lightface italics*. The page on which the term is defined is indicated by boldface type in the index. The definition thus occurs in context, in the presence of additional information, and perhaps also in one or more clarifying illustrations. Thus, in some instances, the reader can see how a definition is built and even why some definitions are difficult to frame. In addition, a glossary is included for those who may wish it.

The selected bibliographies at the ends of the chapters include both source material and suggested reading for students. In the captions accompanying illustrations, references to these bibliographies cite only the last name of the author and the date of publication. References to sources not listed in the bibliographies cite full names and dates to lead the reader to the *Bibliography and Index of Geology* published by the Geological Society of America. The few references to titles not listed in that publication are cited in full.

As to units of measurement, we have compromised between the metric and English systems. We abbreviate metric units (10cm, 10km) but spell out English units (50 feet, 15 miles). In general we express in metric units very small or very large dimensions, nearly all measurements of depths and distances at sea, and nearly all values involved in calculations. We use English units in general statements about familiar distances or rates. For temperatures the Celsius scale is used throughout the book.

Our sincere thanks go to Margaret C. H. Flint, who read nearly the entire manuscript and made many suggestions, to Geoffrey W. Smith, Gary D. Webster, and John J. Rusch for many improvements and suggestions, and to co-workers at John Wiley & Sons, Inc., especially D. H. Deneck, and Jeramiah McCarthy. Also we are grateful to the many persons who generously supplied photographs, each of which carries its own credit line.

R. F. F.
B. J. S.

Contents

Physical Geology

"Earthrise" seen from the Moon. (*NASA.*)

Part I Planet Earth

Chapter 1

Introduction to Planet Earth

It rained heavily in New York today. It rained in Kansas City too. Everyone who passed along a street saw the large raindrops crashing down against the wet pavement. Many people crossing the street had to step over a stream of water, in places an inch or more deep, that flowed along the gutter and finally disappeared through an iron grating, falling down to the sewer beneath the street. The stream in the gutter was carrying with it sodden dead leaves, a few twigs, candy wrappers, cigarette ends, and other trash. Only those few people who looked closely saw (Fig. 1.1) that the stream was likewise sweeping much smaller things along the gutter: grains of sand and finer particles that had earlier been part of the soil beneath trees and in grass plots, dust blown in by the wind, sand that had been thrown on the street during icy winter days. Present in the water too, yet invisible because it was dissolved, was some of the salt that had been strewn over the pavement to melt ice and snow.

The runoff of the rain, the gutter streams, and all the bits of things solid and dissolved that they swept along represented movement, activity, and geologic process. The threads of flowing water were lost to view as they plunged into sewers, but the water reappeared where the sewers (Fig. 1.2) poured into a river, large or small, on its way to the ocean. Out in the country, along highways, down slopes, in open fields, wherever the rain was heavy, similar activity was in progress. Water was moving off the land and toward the ocean, carrying with it solid grains that are bits of rock, as well as substances (such as salt from the highways) dissolved from rock. **Rock** is *any naturally formed, firm and coherent*

1

Figure 1.1   Water near the start of a long journey. (*William H. Graham.*)

Figure 1.2   A sewer adds its bit of water to a river. (*John Hendry, Jr. National Audubon Society.*)

aggregate or mass of mineral matter that constitutes part of Earth's crust.

The moving water and the solid particles in it are a very small sample of a great chain of activities that operates over the Earth's whole surface. The water that flows off through gutter and river comes from rain, which comes from clouds, which in turn form from moisture evaporated from the ocean. As rivers empty into the ocean, the traveling water, ocean to land to ocean, completes an endless circle and is ready to begin anew. The bits of rock carried in the flowing water are part of another circle, one that interlocks with the circle traveled by water. The particles of rock came originally from the firm rock that forms all continents and that is continually being broken into particles. Like the water itself, the rock particles are on their way to the ocean, where they will be spread out and deposited, and will eventually form new rock. *The complex group of related processes by which rock is broken down physically and chemically and its products removed* is called **erosion** ("wearing away").

The movement of rock particles in flowing water involves two things: *materials* (substances that are being worked on) and *processes* (*activities* that are expending energy in working on the materials). Rock particles are only one kind of material, and flowing water is only one kind of activity. To get an idea of the other materials and other activities that make the Earth's surface a very busy place, we must take a broader look at both. First, material. Of what substances is the Earth made, and how are they arranged? Second, what active processes are moving these substances around or changing them in other ways?

The Solid Earth—its Layers and Internal Processes

Because we live on Earth's surface, we are most aware of, and most directly affected by, the materials we see here and the activities that work on them. Yet down below is the solid Earth, a realm where very different activities are occurring and where many of the materials we see on the surface were originally fashioned. We rarely think of the deep, internal activities, although unexpected earthquakes, volcanic eruptions, and the sight of strange rocks at the surface remind us that Earth is far from quiet, and that things must be happening down in the depths. The internal processes that work down within the Earth change the surface in ways we can learn to recognize, and they affect all living things, including people.

Core, Mantle, Lithosphere, and Crust. Imaginative speculations about Earth's interior have always figured prominently in mythology, religion, and literature. The sulfurous fumes and eternal fires of Hades were an ancient belief that probably arose in the minds of people who saw volcanic eruptions without understanding their origin. Hot springs, too, were imagined to bring fluids from the center of the Earth, where they had been heated by great internal fires. Sometimes the waters of hot springs were believed to possess magical powers, and we read again and again of people who sought to restore their youth by bathing in spring waters. Because hot springs and volcanoes seemed to need fires to keep them going, people often assumed that a giant inferno raged in Earth's interior (Fig. 1.3). On the other hand, earthquakes did not seem to have any connection with common experience; so they were frequently interpreted as manifestations of the wrath of the gods. Slowly, however, such beliefs changed. Beginning in the Middle Ages and continuing to the present day, clearer ideas and more exact knowledge about Earth's interior were developed and tested, so that today we have a fairly good picture of what Earth's internal structure is like.

Earth is so huge that we cannot sample or examine its depths directly, but by carefully examining its surface we can infer a good deal about what goes on down below. But our greatest understanding of what is happening there comes, not from examining the surface, but from measuring things we cannot see. We have learned much about Earth's composition from indirect measurements. We measure the speed with which earthquake waves pass through Earth's body, we measure irregularities in the paths of orbiting spacecraft, and we measure variations in the force of magnetism from place to place.

Such indirect measurements tell us that the solid Earth consists not of one single material, but of distinct layers, like the skins of an onion. There are three such layers (Fig. 1.4): (1) a massive *core* composed largely of metallic iron, part of it molten, (2) a *mantle* of rather dense rocky matter that encloses the core, and (3) a thin outer zone of somewhat lighter rock. We call this zone the *lithosphere* ("rock sphere"). Core and mantle each have their own nearly uniform composition, but the boundary be-

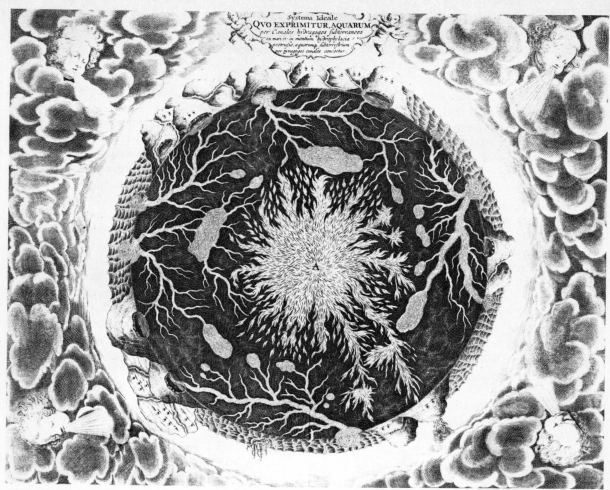

Figure 1.3 In 1678 a German scholar, Athanasius Kircher, depicted the most advanced thinking of his day. Earth's interior was supposed to contain a giant, fiery inferno. The fires heated water that had seeped in from the ocean and gave rise to hot springs. Where tongues of fire came close to Earth's surface (lower right), they started volcanoes. (*From* Mundus Subterraneus, *by Athanasius Kircher, 1678.*)

tween them is distinct. This is because the composition of the metallic core differs sharply from that of the rocky mantle, much as, in a building under construction, the iron beams differ from the concrete. The lithosphere, which of course we can see and touch, is much more varied than the core and mantle appear to be. In particular the outermost 10 kilometers of the lithosphere beneath the ocean and the outermost 20 to 60 kilometers beneath the continents consist of rock materials that differ from those of the deeper lithosphere. These two outermost parts of the lithosphere are called, respectively, *oceanic crust* and *continental crust*, (Fig. 1.5). The boundary

between the lithosphere as a whole and the mantle beneath it is distinct, but it does not reflect a sudden change of composition. The compositions of the base of the lithosphere and the upper part of the mantle are, as far as we can tell, identical. The boundary between them is merely one across which certain physical properties change rapidly, somewhat as they do at the boundary between water and a chunk of ice that floats in it. The lithosphere is rigid, but because the physical properties of materials change at high temperatures and high pressures, the mantle is far from rigid. It is weak and has many properties similar to those of liquids. The boundary between

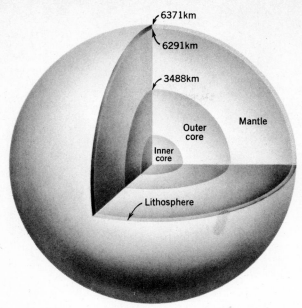

6371km

6291km

3488km

Outer core

Mantle

Inner core

Lithosphere

Figure 1.4 Planet Earth consists of distinct layers, like the skins of an onion. The *core*, with a radius of 3488km (equal to more than half Earth's radius) is composed largely of metallic iron and nickel. The core has an inner solid portion and an outer molten zone. Surrounding the core is the *mantle*, a zone of dense rocky matter about 2800km thick. Above the mantle is the *lithosphere*, a thin outer skin about 80km thick. The lithosphere is a zone of lighter rock that includes our familiar continents as well as the floors of the ocean basins. (*Simplified from E. C. Robertson, U.S. Geol. Survey.*)

lithosphere and mantle is therefore a zone that separates strong, rigid material above from weak, easily deformed material below it. Yet the compositions above and below are essentially the same.

How Solid Is the Solid Earth? Any rock we examine at the surface is obviously solid; so we think of the Earth as a solid. Most people therefore assume the rest of the Earth must be solid too, but this is not strictly true. The familiar picture of a volcano (Fig. 1.6) that pours forth a stream of molten rock (*lava*) reveals that there must be at least some liquid below ground. An internal liquid of this kind, while still beneath the surface, we call *magma*, although after it has come out upon the surface the same liquid is called *lava*. Magma forms because radioactive heat continually tends to raise the temperature of rock below ground, until eventually the rock melts. Yet only a small fraction of Earth's rocky material is molten at any time. Also we can easily see that rocks formed from cooling lavas differ from place to place; so we infer that the composition of magma varies from one place to another. Determining how these variations occur, and why they occur, allows us to infer a great deal about Earth's internal processes.

If a force of any kind disturbs it, magma, like other liquids, tends to move. Consequently, when subjected to uneven pressures, bodies of magma move slowly upward, forcing their way through solid

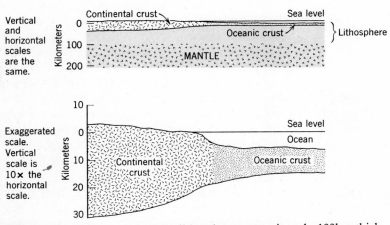

Vertical and horizontal scales are the same.

Continental crust

Sea level

Oceanic crust

Lithosphere

MANTLE

Kilometers 0 / 100 / 200

Exaggerated scale. Vertical scale is 10× the horizontal scale.

Sea level

Ocean

Continental crust

Oceanic crust

Kilometers 10 / 0 / 10 / 20 / 30

Figure 1.5 Cross section of the lithosphere, approximately 100km thick. The lithosphere is capped by a crust of two different kinds. One kind underlies the continents and is 20km to 60km thick; the other kind underlies the oceans and is only 10km thick.

Figure 1.6 Flowing stream of molten rock poured out by Mauna Loa volcano in Hawaii, during an eruption in 1950. The stream flowed many kilometers down the slope of the mountain. Here it is seen cascading over a cliff about 20m high. Light-colored areas are fresh, hot lava; darker areas are cooled crust. (*Courtesy U.S. Air Force.*)

rock, gradually cooling, and in some places reaching the surface, where they emerge as lava.

Because magma obviously exists, then, the "solid Earth" is not as solid as it may seem. Even in places where rock has not melted to form magma, there is evidence that other curious things have happened to it. Figure 1.7 shows rock that has been squeezed. It has actually flowed, as a pat of cold, stiff butter would flow out sidewise if you pressed down on it with your thumb. In contrast, Figure 1.8 shows layered rock that seems to have been bent. We infer that sidewise pressure must have contorted the layers without breaking them. Finally, in Figure 1.9 we see rock that has broken and shattered like glass. It was brittle, and so, when pressed, it responded by shattering. These three photographs show that some of the forces at work within the Earth must be very strong. Such forces continually push bodies of magma around, and also slowly bend and fracture the solid Earth. We cannot see or easily measure the movement of material in the mantle, but after a good deal of study of evidence found in the continental crust, we can infer that great segments of the crust have been thrust up repeatedly to form mountains, and great blocks have been pulled down repeatedly to form basins. The nature of this evidence, the way in which the evidence is obtained, and the way in which the forces arise are topics we are going to discuss in this book.

Movement of Continents; Plate Tectonics.
When we examine the shapes of the bodies of continental crust and their patterns of mountain ranges, we find many puzzling things. For example, the coasts on both sides of the Atlantic have similar shapes, as if they might once have been joined and later torn apart. If we look more closely (Fig. 1.10), we find that we can bring together the areas of continental crust now on opposite sides of the Atlantic. When we do bring them together, we see that there are mountain belts—the familiar Appalachian mountain belt and similar belts in East Greenland, Europe, and Africa. These belts fit together to form one long, continuous chain. It is hard to escape the inference that America, Europe, and Africa were once in contact, sharing a single great chain of mountains, and that these continents were torn apart and somehow separated. This, together with much evidence of other kinds, led as recently as 1960 to a new and compelling idea. This is that huge slabs or plates of lithosphere are being dragged very slowly across the top of the mantle by great internal forces, that tear apart whole continents.

This idea is only a theory, but it answers many questions that earlier had no answers. Also it has suggested new theories about what makes earthquakes happen and about how volcanoes are formed. Eventually it may enable us to predict accurately when earthquakes and volcanic eruptions will occur,

Figure 1.7 Rock showing evidence of extreme deformation, with evidence of flowing movement caused by squeezing. Layers that were originally flat and uniformly thick are now highly contorted, thickened in some places and thinned in others. The rock is limestone, in Aberdeenshire, Scotland. Width of exposure, about 6m. (*Geol. Survey of Great Britain. Crown copyright.*)

Figure 1.8 Strongly bent layers of sandstone, exposed in a cliff about 15m high. The strength of the rock was great enough to resist flowing movement like that seen in Figure 1.7. Pembrokeshire, England. (*Geol. Survey of Great Britain. Crown copyright.*)

0   5 cm

Figure 1.9 Rock that has broken and shattered like glass. The shattered fragments were later cemented together by substances deposited from water slowly circulating through them. The fragments are sandstone, the same kind of rock as that shown in Figure 1.8. Similar rocks therefore respond in different ways to the differing forces within the Earth. West Range, Nevada. (*Yale Peabody Museum.*)

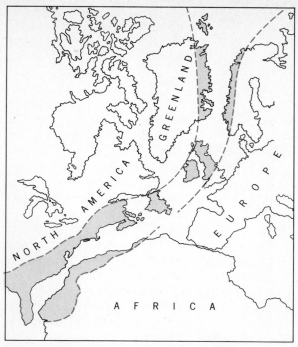

Figure 1.10 The eroded remains of similar mountain belts are found in North America, Greenland, Europe, and North Africa. The mountains were formed between 350 and 470 million years ago. When moved and fitted together, they are seen to form a continuous belt. When reconstructed in this way on a map, the pieces indicate the relative positions of the present continents before they were broken up by movements of the lithosphere. (*Adapted from P. M. Hurley, 1968.*)

and so provide a potential for saving lives in these natural disasters. The theory of movement of lithospheric plates has become so important that it has been given a special title: theory of plate tectonics (Chap. 18). The word **_tectonics,_** *the study of Earth's broad structural features,* is derived from *tekton,* a Greek word meaning a carpenter or builder. The term plate tectonics is apt because we now think the lithosphere itself is divided up into huge plates that are being continuously rebuilt. The zones where building is going on lie along great fractures in the deep ocean floor (Fig. 1.11). Through the fractures magma wells up from the mantle far below. It freezes and forms new lithosphere. The plates of older lithosphere move sideways to accommodate it.

If Earth is continually making new lithosphere, one of two things must be happening. Either the lithosphere as a whole is getting larger, or old lithosphere is somehow being destroyed. The latter process is what is actually happening. In several places in the oceans there are great trenches many kilometers deep, into which moving lithospheric plates tip downward and slowly plunge back into the mantle. Once in the mantle, the plates are heated up, melted, and remixed. The forming of new lithosphere and its destruction beneath the trenches of course require large amounts of energy. The process is a continuous one; lithosphere forms along ocean fractures, moves sideways like a conveyor belt, then returns to the mantle in an endless circle.

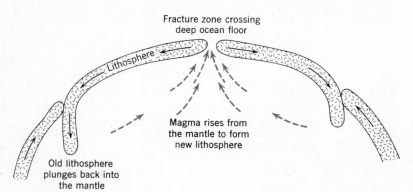

Figure 1.11. Cross section of Earth's outer layers, showing how magma (dotted arrows) moves from the mantle upward into fracture zones in the ocean floors, and cools there to form new lithosphere. To accommodate the new material, the lithosphere (solid arrows) moves away from the fracture zone and eventually plunges slowly down into the mantle again, where it is re-heated, melts, and eventually mixes again with the mantle, to complete a cycle of astonishing proportions.

The Outer Envelopes and Their Activities

The solid Earth, then, consists of a series of layers in which various internal processes are at work. But the outermost part of the solid Earth (which we can think of as a slightly flattened sphere) is surrounded by other layers—layers of water, air, and living things. The water, the air, and the living things are continually moving about and reacting with each other in many ways. In describing these external layers it may be helpful to think of the outer part of Planet Earth as consisting of four spheres: *hydrosphere, atmosphere, regolith,* and *biosphere.* Each is a shell or envelope, wrapping around and to some extent mixing with the sphere next within it. Each consists of characteristic materials. And in each, distinctive activities are at work.

Hydrosphere. The hydrosphere is the "water sphere," embracing the world's oceans, lakes, streams, water under ground, and all snow and ice, including glaciers (Fig. 1.12). A very small part of it exists in the atmosphere in the form of water vapor. Watching a steaming teakettle on the stove, we can see a little bit of the hydrosphere entering the atmosphere around it.

Atmosphere. Although the word *atmosphere* really means "vapor sphere," it would be more logical to label it "air sphere," because it consists of the mix-ture of gases that together we call air. It penetrates into the ground, filling the openings, small and large, that are not already filled with water. The atmosphere is never quiet. Some part of it is always moving, as we are well aware whenever we feel a wind blowing.

Regolith. Where the atmosphere and hydrosphere are in contact with the surface of the land, reactions take place between them. For example, the atmosphere with its content of water vapor acts upon **bedrock,** *the continuous solid rock of the continental crust,* breaking it up mechanically and causing it to decay chemically. As a result, the surface rock, although still a solid, is no longer a continuous solid. It has been subdivided into many pieces, most of them very small particles. This broken-up part of the crust has a separate name, **regolith** ("blanket rock"), because it lies like a blanket draped over (but in places grading into) the continuous solid rock beneath. It is defined as *the blanket of loose, non-cemented rock particles that commonly overlies bed-rock.* We would usually have to use a hammer or even a high-speed drill to collect a sample of bed-rock, but a shovel or pick would ordinarily be enough for sampling regolith.

Not all regolith has been broken up and left in place. Some of it has been moved and set down in a new site, like the sand deposited in our gutter at the end of the rainstorm. It is on its way (with

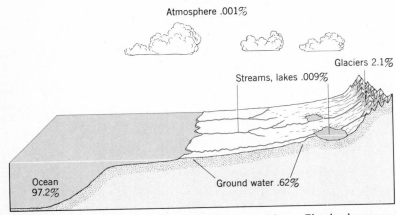

Atmosphere .001%

Glaciers 2.1%

Streams, lakes .009%

Ocean 97.2%

Ground water .62%

Figure 1.12. Distribution of water in the hydrosphere. Clearly the ocean is the great reservoir; water (and ice) elsewhere make up less than 3 per cent of the total.

The sum of the percentages is not 100% but 99.93%. The difference represents uncertainty in measurements and estimates. Although we don't like uncertainty, we have to live with it. (Data from U.S. Geol. Survey.)

stopovers) to the ocean. *Regolith that has been transported* by any of the external processes is called **sediment,** a word that means "settling."

In some places where bare bedrock is exposed at the surface, there is no regolith at all. In other places regolith is 100 meters or more in thickness. But we estimate that on the average its thickness is little more than a meter or two.

Biosphere (the "life sphere"), as its name implies, is *the totality of Earth's organisms,* and, in addition, organic matter that has not yet been completely decomposed. Its composition is distinctive; its chief constituents are compounds of carbon, hydrogen, and oxygen, although it includes small amounts of other chemical elements as well. All these elements are drawn from the other spheres, but they are fixed in patterns peculiar to the biosphere. Whether they form parts of living organisms or dead ones, the

organic compounds remain a part of the biosphere until they have been destroyed by chemical alteration.

The biosphere extends through, or at least into, each of the other three outer envelopes (Fig. 1.13). Although living things manage to exist through an enormous vertical range—nearly 20,000 meters—they are very sparse at great heights and great depths. Most of them are crowded into a narrow zone that extends from a little below sea level to a thousand meters or so above it.

Although very different from the air, water, and rock that make up the rest of Earth's outer part, the biosphere is as much an integral part of Earth as are the other three outer envelopes. All organic matter, without exception, is derived from sources within the Earth. Eventually, in one form or another, it returns to the inorganic Earth and ceases, for a time at least, to be part of the biosphere.

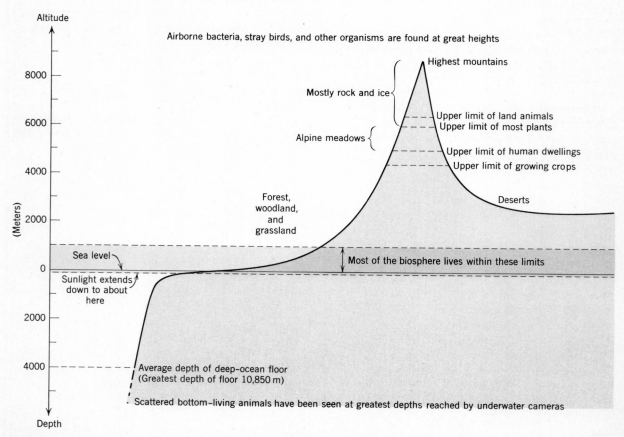

Figure 1.13. Some data on the extent of the biosphere, shown in a schematic way.

External/Internal Interactions

The lithosphere and the outer envelopes interpenetrate each other to some extent, but seen from a distance, the surface or *interface* along which they are in contact looks sharp and distinct. At this interface, exchanges and other activities take place. The busiest of the spheres are atmosphere, hydrosphere, and biosphere, because within them activities are going forward at higher rates than in the solid Earth.

Water and air penetrate the ground as far down as each is able to go. There they promote chemical reactions that cause rock to crumble slowly and become regolith. Driven by energy from the Sun, water and air move as flowing currents. In very cold places water in the air freezes and falls as snow, which in turn forms glaciers. Air currents, water currents, and glaciers (currents of ice) pick up solid particles (recalling the gutter during the rain) and move them long distances. The biosphere penetrates air, water, and regolith so thoroughly that living things occupy every available nook and cranny within the other spheres, filling them all with additional activity. Partly through physical processes, but more importantly through biochemical reactions, organisms help to crumble rock and also to deposit solid substances on the ocean floor.

All these activities, organic as well as inorganic, at or near the surface of the lithosphere constitute *external processes,* in contrast to the *internal processes* that originate down within Earth's solid body. The two kinds of activities influence each other. One part of the crust may be lifted up to form a broad highland and another may be bent down to form a large basin. Rainwater forms streams that flow off the highland and into the basin, carrying sediment with them. Thus the highland continually loses rock material and so becomes lower. The basin, in contrast, receives sediment from the highland and gradually becomes filled with it. The creation of highland and basin are the work of internal processes, but the wasting away of the highland and the filling of the basin with the waste products are the work of external forces as they respond to the opportunity offered by internal ones.

How fast do these things happen? At what rates do parts of the crust go up and down? We noted earlier that plates of the lithosphere can move horizontally at rates of a few centimeters per year. Calculation of the volume of sediment being carried by rivers today tells us that some high mountain ranges are losing by erosion, each year, rock material equal to a layer as much as several centimeters thick. A basin in eastern Europe has been forming, deepening, and filling with sediment during the last four million years, at a rate of one centimeter every forty years—a rate slower than the one quoted above, but still fast enough to have created a basin one kilometer deep. Planet Earth therefore seems to be changing continually. But our planet has existed for billions of years in terms of erosion and deposition, whose rates we can measure. Given such rates, the external processes have had far more than enough time to wear the surfaces of the continents down very nearly to sea level. Yet today Earth's lands are not at sea level; in most places they are well above it. The average altitude of the lands is between 700m and 800m and the highest altitudes exceed 8500m. Evidently internal processes must have lifted up or built up parts of the continents again and again, enabling external processes to continue without pause their work of destruction. The surface of the solid Earth is clearly the scene of a vast conflict, a conflict between lifting up and wearing down, that has been in progress for a very long time.

The Three Kinds of Rock

This lifting up and wearing down of lithosphere, now here and now there, is reflected in the rocks we see at Earth's surface. Examining the rocks, we gradually realize that they have originated in three very different ways. Hence we classify them, according to these different origins, into three families that have been labeled *igneous, sedimentary,* and *metamorphic.* Although we shall treat them in more detail in Chapter 5, we must characterize them here so that we can use the family names in the three chapters that intervene.

Igneous rock is rock that is made by the cooling and solidification of magma or lava.

Most *sedimentary rock,* as its name implies, consists of particles of sediment derived from older rock. The particles were transported by one of Earth's external processes, deposited, and later cemented so as to make firm new rock.

Most *metamorphic rock* consists of igneous or sedimentary rock that became deeply buried and so was subjected to high pressure and great heat, which changed the character of the rock.

Most of the metamorphic- and igneous rocks we see at Earth's surface today have reached their present positions by a combination of lifting up and wearing down. They are lifted up as parts of the lithosphere rise, and as the rocks that lie above them are gradually worn away. In contrast, sedimentary rocks are the rearranged, re-sorted wreckage of older rocks that have been lifted up. Sedimentary rocks, therefore, are wholly the product of Earth's external processes, while the other two families are direct products of internal processes.

Time

Only within the last 200 years have people understood that the form and appearance of Earth's solid surface have been and are being shaped by the continual conflict between the two sets of processes, internal and external. Part of our understanding is the result of thoughtfully watching the processes at work. But the Earth is so large, and the rate of work is so slow by comparison with a human lifespan, that most of the changes accomplished during a few decades are too small to be measured directly. Therefore another part of our understanding, an absolutely essential part, consists of our concept of *time*. Planet Earth has endured long enough to allow the natural processes, working year after year through billions of years, to accomplish the building of mountain ranges and then to destroy them partially or completely.

Our understanding of the changes that take place at the Earth's surface, then, is based on our recognition of two things: (1) how the *processes* do their work, and (2) the vast length of *time* through which our planet has existed, making it possible for the processes, even though most of them work slowly, to accomplish results that are enormous. This concept of great changes brought about cumulatively through a very long time is one that our earlier ancestors did not possess. Only we, in the latest moment of Earth's history, are able to make measurements and observations by which we know, for example, that the Andes and Himalaya mountain chains are actively growing and becoming higher. It is only we who can look back to times when both these great barriers did not exist, to times when the Atlantic Ocean basin had not yet been formed, and to still earlier times when the present flat tundra of northern Canada was crossed by great chains of mountains.

Physical Geology and Historical Geology

Recognizable changes that have occurred in the past are history. Changes induced by people, whether peacefully by law or violently by force, have occurred time and again in the social and political histories of nations. Changes caused by natural processes have occurred in enormous numbers during the history of Planet Earth. The particles of rock poured by streams into the basin mentioned earlier tell us much about what was happening to a certain region of the Earth's surface during a certain length of time. What happened? We can find out for ourselves if we learn what to look for in the sediment that filled the basin. The study of this or of any other portion or feature of the Earth is part of **geology,** *the science of the Earth.*

The study of geology falls into two parts, the parts of our twofold basis for understanding the Earth: *process* and *time.* The first part deals with what is happening on and in the Earth today; we call it *physical geology.* The second deals with what has happened in the past; that is *historical geology.* Although the two studies are closely related, physical geology comes first, because we must first be familiar with processes and their results before we can examine Earth's crust and extract from it information about its past.

Planet Earth is bristling with questions large and small, basic and practical, to which answers are needed. Some questions relate to how the natural processes work, as we observe them today. Others relate to the distribution of natural features, climates, and living things in the past, and so involve the recapture of events in Earth's history.

A third group of questions is of this general kind: "Where can we find, and how can we extract from rock, the mineral substances from which things (such as steel or concrete) can be made, or from which energy (such as nuclear energy) can be derived?" "What is the best location for this bridge, this dam, this skyscraper?" "Will the location provide long life and maximum safety for the structure?"

Finally, questions are rapidly multiplying—in fact, almost overwhelming us—concerning the ways in which human activities affect external processes and natural environments, and especially how environments are affected in undesirable ways. To deal with such questions involves tremendous work, but it offers much opportunity for making new discoveries and for improving the quality of our lives.

After this first look at Earth's composition and at the conflicting activities that are always in progress at its surface, we are ready to look more closely at both. So, in Chapter 2, we examine the "endless circle," the cyclic patterns followed by Earth's external and internal processes—patterns that have existed unchanged through most of Earth's history. Then we see how a sequence of events can be learned from rock layers, and finally how radioactivity has made it possible to measure geologic time.

Summary

1. Water travels continually in a great circle, from ocean to land and back to the ocean.

2. Rock material also moves; it is eroded from high places and is deposited in basins; it becomes deeply buried, melts, and solidifies.

3. The solid Earth is layered. It consists of a core, a mantle, and a lithosphere whose outer part is differentiated into oceanic crust and continental crust.

4. Great plates of lithosphere, some of them carrying continents, appear to be moving sideways. New volcanic rock is being added to them along their trailing edges, while along their leading edges the plates are being melted.

5. Outside the solid Earth are the envelopes we call hydrosphere, atmosphere, regolith, and biosphere.

6. The Earth's active processes can be divided into two groups: internal and external. The two are in continual conflict.

7. Because geologic time has been vastly long, even very slow processes have been able to accomplish enormous changes at the Earth's surface. Indeed Earth's history is the joint result of *processes* and *time*.

8. *Geology is the scientific study of the Earth. Physical geology deals with what is happening today, historical geology with what has happened in the past.*

Chapter 2

Cycles, Geologic Column, and Time

Earth's Cycles

The Cycle Concept

Cycle is a term that describes a sequence of recurring events. The circling of Earth around Sun and of Moon around Earth are cycles. One result of the first is the daily heating and cooling of Earth's surface. One important result of the second is the tide, the recurring rise and fall of the ocean surface. Cycles of one sort or another control almost everything that happens on Earth. Even the growth, development, and death of human beings is part of a cycle—the life cycle; and our human economy involves many kinds of cycles. For instance we "recycle" (actually we *cycle*) glass, a process in which bottles are ground up and melted down to form a liquid, from which new bottles are made. We "recycle" paper, which is pulped and fashioned into new paper. Any single particle of glass or paper substance thus may pass through the same series of states or forms again and again. The operation of Earth's natural processes, both internal and external, and the conflicts between the two groups of processes, involve very distinct cycles. As we shall see in Chapter 3, energy for the internal processes comes from radioactive heat in the Earth's interior, whereas energy for the external ones comes from the Sun (and, through the tides, from the Moon).

Water Cycle

A conspicuous example of a cycle powered by external processes is the **water cycle,** *the cyclic movement of water substance through evaporation, wind trans-*

port, stream flow, percolation, and related processes (Fig. 2.1). Heat from the Sun evaporates ocean water and other surface water. The water vapor enters the atmosphere and moves with the flowing air (the winds). Some of it condenses and is precipitated, as rain or snow, back into the ocean; some is carried over the land and is precipitated. The rain runs off the land as streams and so back into the ocean, completing the cycle. Snow lies on the ground, perhaps through a winter season, before melting and running off. In especially cold areas snow accumulates and forms glaciers. As glaciers it remains on the land longer, through many years or even thousands of years, but sooner or later it too melts or evaporates and so eventually returns to the ocean. Earth's lands, on the average, stand some 700m to 800m above the surface of the ocean. So when solar heat lifts air with its contained water vapor from ocean to atmosphere, and then precipitates water onto land at that altitude, potential energy is created, and the energy enables the water to flow downhill, back to the ocean. We make use of this principle when we build a dam across a stream to form a lake. The dammed-up water conserves and concentrates stream energy for use in generating electric power.

The quantity of water that participates each year in the water cycle has been estimated for the world by comparing measurements of the amounts evaporated and precipitated at many places on land and at sea, as well as the amounts carried through streams, lakes, and glaciers. The estimated total quantity, called the *annual world water budget,* is 410,000 cubic kilometers, enough water to cover an area the size of the State of Illinois to a depth of more than 3km. What happens during a year's time is shown in Figure 2.2. From the numbers given in that figure we can calculate that an amount of water equal to the combined volumes of all the oceans goes through the water cycle once every 3200 years.

The amount of water in the hydrosphere changes very little. Some water is lost for a time when chemical exchange locks it up as part of the composition of a mineral. Sooner or later the locked-up water is freed of its chemical bonds and re-enters the water cycle.

Rock Cycle

Just as water substance travels through the water cycle, rock material travels, in its own ways, through a **rock cycle,** *the cyclic movement of rock material, in the course of which rock is created, destroyed, and altered through the operation of internal and external Earth processes.* The diagram, Figure 2.3, shows the many possible paths of movement from one phase to another. In moving along these paths, any particular particle can travel thousands of kilometers laterally and tens of kilometers downward or upward.

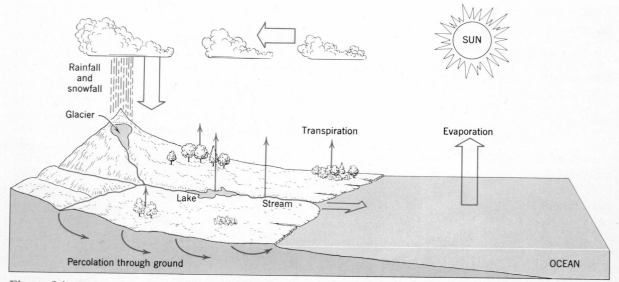

Figure 2.1 The water cycle. All the arrows that point upward represent evaporation and/or transpiration.

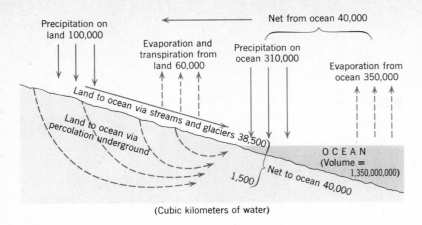

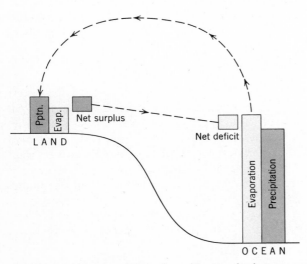

Figure 2.2 Estimated annual world water budget, expressed in cubic kilometers of water. The ocean loses by evaporation more than it gains by direct precipitation, but with the land it is the other way around. The surplus precipitation on the land goes back over or beneath the land surface, to the ocean and so balances the budget. The balance is shown by diagram in *B*, where the two little boxes are of equal size.

None of the numbers shown is precise, because all are estimates calculated from measured samples. (*Modified from data published by R. L. Nace.*)

Rock at Earth's surface is attacked by the complex activities of erosion; it becomes subdivided into loose particles, which are carried away as sediment and deposited. The deposited sediment becomes cemented, usually by substances carried in ground water, and is thereby converted into new sedimentary rock. In places where such rock subsides, it can reach depths at which pressure and heat alter it to metamorphic rock, or depths at which internal heat is great enough to melt it, converting it to magma.

The magma can move upward through the crust, cool, and form a body of igneous rock. When uplift occurs, erosion gradually wears down the surface of the land to a depth so great that the top of the body of igneous rock is uncovered and itself begins to be eroded. Again igneous rock is attacked and broken up, its waste starting once more on its way to the sea. The rock cycle has been completed and has begun again. But, as the arrows in the illustration show, the long circuit can be interrupted by short

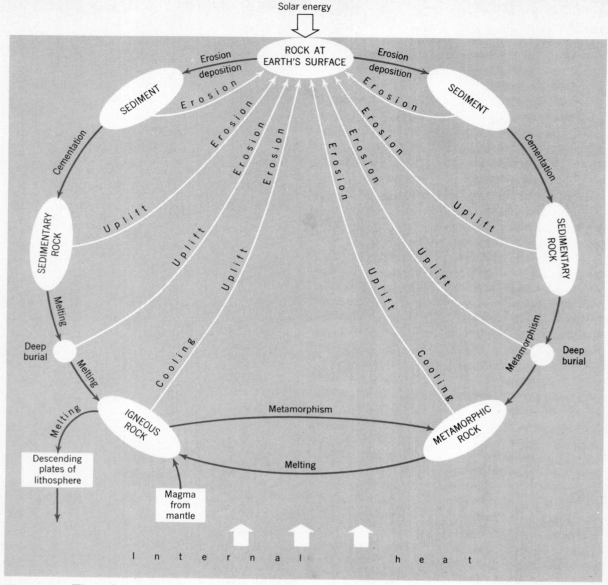

Figure 2.3 The rock cycle. In it rock material can follow any of the arrows from one phase to another. At one time or another it has followed all of them. Phases are labeled in capital letters; paths represent processes, labeled in lower-case letters. The rectangular boxes at lower left show how the cycle of movement of lithosphere plates (Fig. 1.11) interacts with the rock cycle.

circuits of various kinds. Some rock bodies are never melted, never metamorphosed, or never even buried deeply. But whether the circuits are long or short, sooner or later every rock body is exposed at the surface, where it is vulnerable to erosion and is ultimately destroyed. As long as Earth's internal energy is supplied to the crust from below, and as long as solar energy reaches the crust from above, the rock cycle will continue to operate.

At the same time, as erosion gradually breaks down rock and transports waste to the sea or to basins on the land, it sculptures the land itself and greatly alters the landscape. Thus the form of the land goes through a cycle of its own, the erosion cycle, as will be shown in Chapter 8.

Principle of Uniformity

How long has the enormous rock cycle, with its many side paths and short circuits, been rearranging the rock materials of the lithosphere as it is doing today? Apparently for as long as the time that has elapsed since Earth's oldest known rocks were built, a time more than 3 billion years long. This seems a bold statement, yet we can state it with confidence. For all the many bodies of rock in the continents, regardless of their age, belong to the same recognized kinds as those we see being made today, whether by external or internal processes. All consist of the same range of materials, forming the same ranges of patterns. Each of these rock bodies fits into one phase or another of the rock cycle, in which materials are broken up, sorted, transported along the surface, deposited, cemented to form rock, buried deeply, bent and squeezed, melted and forced upward toward Earth's surface, only to be broken up once more.

The similarity of the old rocks to the recently made ones, and the way all of them fit into the many activities in the cycle, forces us to believe that the cycle, operating under the same physical/chemical laws we recognize today, has been going on and on, essentially unchanged, for at least 3 billion years. This belief is not new. It was first reasoned out in 1795, and early in the 19th Century was developed into a principle named the **Principle of Uniformity.** It says that *the external and internal processes we recognize today have been operating unchanged, and at the same set of rates, throughout most of Earth's history.* Recognizing this principle gives us a great ability. We can examine any rock, however old, and compare its characteristics with those of a similar rock that is being formed today in a particular environment. We then know that the old rock was formed in that same sort of environment. The Principle of Uniformity, therefore, provides a first and very long step in understanding Earth's history.

Cycles in the Biosphere

Much of the intense activity that characterizes the biosphere is likewise cyclic. If we tried to sum it all up in a single generalization, we might say that energy arriving from the Sun passes through atmosphere and biosphere, and then, in the form of heat, is radiated from lands and oceans outward again into space. Income is balanced by outgo. If the two did not balance, Earth's surface would cool off or heat up, and would eventually reach a temperature that would be fatal to the biosphere.

The incoming solar energy is the sole source of energy for living things. Through their ability to accomplish the photosynthesis reaction, plants trap that energy and build it into their substance. The substance becomes food for animals, which then are eaten by other animals, the energy in these various foods passing through chains of living things. At every link in every one of these food chains, heat is created and commonly passes into the atmosphere, and so outward into space.

The cycle in which CO_2 is extracted from the atmosphere by land plants, and then, returned to the atmosphere as plant substance, decays or is eaten by animals, is rapid. By means of that cycle alone the entire content of CO_2 in the atmosphere "turns over" once every 4.5 years. So rapid is the turnover that Earth's human population can make an appreciable change in its length, by stepping up the burning of plant substances or by changing the total bulk of vegetation on the land.

The water cycle, of course, is essential to life, which cannot persist without water. Although less fundamental to the biosphere, the much slower rock cycle nevertheless affects organisms because it keeps gradually changing the environments in which they live. It does so as internal forces lift up or depress land areas, in some places submerging lands and in others reclaiming them from the sea. All such events make some environments uninhabitable and others newly attractive. Hence they cause organisms to move from one place to another and also to change slowly by evolution. An organism that lives on the sea floor dies and becomes a **fossil** (*the naturally preserved remains or traces of an animal or a plant*). Preserved in accumulating sediment and later converted to rock, fossils form an ever-growing record of the kinds of plants and animals that lived at the time when the enclosing sediment accumulated. This record, a vital link between the present and the past, leads us directly to what is called the geologic column.

Figure 2.4 Colorado River flows past Desert View Point on its way through the Grand Canyon. (*Joseph Muench.*)

The Geologic Column

In Figure 2.4 we see the Grand Canyon, nearly two kilometers deep, one of the deepest cuts into the Earth's crust made by a river on any continent. This cut, made by the Colorado River, reveals a thick pile of horizontal *strata*. Here we have a word that is used very frequently in geology. A ***stratum,*** of which *strata* is the plural form, is *a distinct layer of sedimentary or igneous rock consisting of material that has been spread out upon the Earth's surface.* The term is used frequently in geology and therefore also in our discussion.

The strata exposed in the Grand Canyon are sedimentary, having been deposited one on top of the other as sediment transported by ancient rivers and spread out on the floor of a shallow sea. The physical characteristics of those particular strata, as well as the fossil animals they contain, indicate that the sedimentary layers were deposited in a basin occupied by an arm of the sea. As they lay buried

for a long time after being deposited, the particles of sediment that composed each stratum became cemented to form sedimentary rock. Later the region was slowly lifted up, the Colorado River system formed, and erosion began as the river started to cut the canyon. The rock waste from the growing canyon and from elsewhere along the river system is the sediment that makes the river muddy today. It has long been accumulating beyond the mouth of the river in the Gulf of California, a rather new arm of the sea (Fig. 2.5).

The modern Gulf of California lies in an area far southwest of the position of that ancient arm of the sea in which the strata now exposed in the walls of the Grand Canyon accumulated. But the sediment now passing through the Canyon on its way to the Gulf consists of the same mineral particles and bits of rock that had been carried by ancient streams into the earlier sea (Fig. 2.6). This whole history, of course, is part of the rock cycle, and part of the water cycle as well. Water has made many round trips between ocean and land while the rock particles are still in the midst of their second trip from land to ocean. How many still-earlier trips those particles made we do not know, but there must have been many.

The lithosphere is a patchwork of rock bodies large and small. Some of these bodies consist of sedimentary strata, remnants of the fillings of former basins, that escaped erosion and dispersal to later basins. Other bodies, after being carried deep down below the surface, have been squeezed and altered to form metamorphic rock. Still other bodies are igneous rock, made by cooling of some earlier rock carried downward and melted at great depth. By examining rocks in detail, we can learn much of their history. As records of the Earth's history, sedimentary strata are useful in a special way because although they occur today as separate, disconnected bodies (Fig. 2.7), they possess characteristics that make it possible to reconnect them on paper, and

(still on paper) to arrange the individual layers in the order, or *sequence*, in which they were deposited originally. Why is it so useful to be able to do this with layers of sedimentary rock? To answer this question we must first say a little more about *sequence*.

Sequence

Toward the end of winter it is often possible to see a layer of old snow that is compact and perhaps also dirty, overlain by fresh, looser, clean snow deposited during the latest snowstorm. Here are two layers, or strata, that were deposited in sequence, one above the other. The dirty layer underneath was deposited first and must therefore be the older of the two. The very simple principle involved here applies also to a whole succession of many snow layers. It applies equally well to layers of sediment and sedimentary rock. Known as the **principle of stratigraphic**

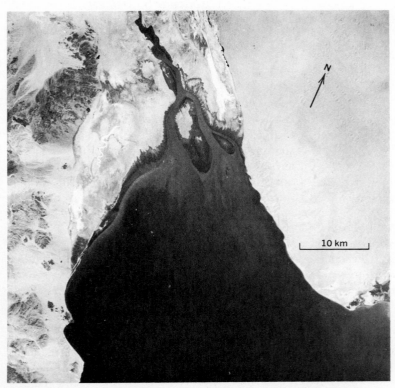

Figure 2.5 The Colorado River empties into the Gulf of California 250km southeast of San Diego, California. Out beyond the delta the water is turbid with silt and clay, brought from up the river and not yet deposited. (*Gemini IV photo, NASA.*)

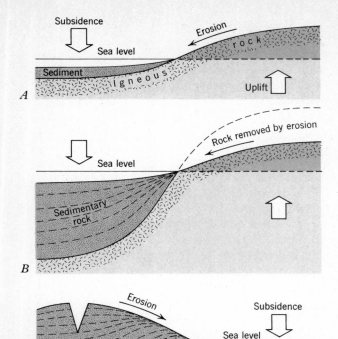

A

B

C

Figure 2.6 The internal processes that cause uplift and subsidence determine the locations of land and shallow sea at any time, and guide the external processes of erosion of rock and deposition of sediment. The bite cut out of the uplifted area in *C* represents a location like that of the Grand Canyon today.

superposition, it says that *in any sequence of strata, not later disturbed, the order in which they were deposited is from bottom to top.*

This principle implies a scale of relative time, by which the age of one stratum, *A,* can be fixed *in relation to* another layer, *B,* according to whether *A* lies beneath or above *B.* The principle does not enable us to fix the age of any stratum in years, as

we reckon time with the use of a calendar; the ages derived from the principle are purely relative.

The principle of stratigraphic superposition was first forcefully presented and widely introduced to science by William Smith, an English civil engineer and land surveyor, shortly before the beginning of the 19th Century. His profession gave him an ideal opportunity to observe not only terrain but the rocks that underlie it. While surveying for the construction of new canals in western England, he observed the sedimentary strata and soon realized that they lie, as he put it, "like slices of bread and butter" in a definite, unvarying sequence. Using the principle of stratigraphic superposition, he became familiar with the physical characteristics of each layer and with the sequence of the layers. By looking at a specimen of sedimentary rock collected from anywhere within a wide region, he could name the layer from which it had come and, of course, the position of the layer in the sequence.

Time Significance of Fossils

In the region where Smith worked, the strata contained abundant fossils of marine invertebrate animals. Smith began to collect the fossils and soon realized that each layer contained distinctive kinds, enabling him to identify it by the sorts of fossil animals it contained, without regard to its physical characteristics. In other words, he recognized that each assemblage of fossils was peculiar to the stratum in which it occurred and thus constituted an identification tag for the stratum. In so doing, Smith discovered what we now call the **law of faunal succession,** which says that *fossil faunas and floras succeed one another in a definite, recognizable order.*

We now believe this relationship between a stratum and its fossils is the effect of the evolution of living things through time. As successive generations of living things gradually change their form,

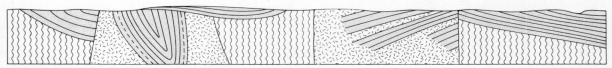

Figure 2.7 A deep slice through continental crust shows disconnected bodies of sedimentary rock ▨ separated in various ways by metamorphic ⟩⟩⟩ and igneous ⟨⟨⟨ rock. Such a slice could be of any length from tens of miles to thousands of miles.

the changes are carried from one part of the world to another through the migration or shifting of populations of organisms as they expand or otherwise change their living areas. The rates at which plants and animals spread or shift seem always to have been comparable with the rates at which organisms evolve. As a result the major evolutionary changes spread through large parts of the world within the generous time intervals represented by the major groups of strata. All this, however, was unknown in Smith's day and did not become clear until after 1859, when Charles Darwin put forth his famous theory of evolution.

Smith's discovery that strata containing similar assemblages of fossils are broadly similar in age, no matter where they occur, was not related to a scientific principle; it was purely practical. Nevertheless, it opened the door to the correlation of sedimentary strata through increasingly wide areas. By *correlation* we mean *determination of equivalence, in geologic age and position, of the sequences of strata found in two or more different areas.* Smith correlated strata, on the dual basis of physical similarity and fossil content, through distances measured in miles and then in tens of miles. But by means of fossils alone it became possible to correlate through hundreds and then thousands of miles (Fig. 2.8).

Figure 2.8 represents a comparatively simple situation, with flat-lying strata, good suites of fossils, and no complications. But returning for a moment to Figure 2.7, we can see in it a situation that is far more complex. Suppose that, on paper, we could unbend and smooth out all the layers of sedimentary rock in the figure, so that they would lie one on top of another in a single pile, in order of decreasing age from bottom to top. The smoothed-out pile would be a very useful standard reference for the strata in that region. If the known layers in all the continents were placed in one single pile, we should have a standard reference for the Earth's entire continental crust.

We do have such a worldwide pile—on paper. It is not complete, but it is being added to and refined continually. We call it the **geologic column** (Table 2.1), *a composite diagram combining in a single column the succession of all known strata, fitted together on the basis of their fossils or of other evidence of relative age.* Because the groups of fossils in each layer of the column record the gradual progress of evolution through time, the column represents not only a succession of layers but also the passage of time. It does not, however, tell us anything about how much time passed between the episodes of deposition of any two given strata. For the actual measurement of time we must look elsewhere.

Time

The question *how much time?* is as important for us as the geologic column itself. It is a question that had to be answered. Many attempts were made to find some way of subdividing the geologic column on a basis of actual years. The earliest attempts consisted of rough estimates, such as the time that would have been needed for the accumulation of all the sedimentary strata in the column, and the time needed for a once-molten Earth to cool to its present temperature. But we know now that all such esti-

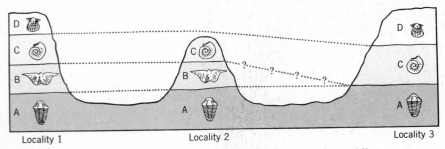

Figure 2.8 Correlation of strata exposed at three localities, many kilometers apart, on a basis of similarity of the groups of fossils they contain. The fossil groups show that at Locality 3, stratum B is missing because C directly overlies A. Was B never deposited there, or was it deposited but later destroyed by erosion, before the deposition of C?

Table 2.1

The Geologic Column, Major Worldwide Subdivisions, Selected Dates,[a] and Events in Evolution. (After R. F. Flint, 1973, The Earth and its history: W. W. Norton & Co., New York.) As the positions of the numbers show, the time divisions are not drawn to a uniform scale. The dates have merely been spotted onto the geologic column.

Uniform Time Scale	Subdivisions Based on **Strata**/*Time*			Radiometric Dates (millions of years ago)	Outstanding Events	
		Systems/*Periods*	**Series**/*Epochs*		In Physical History	In Evolution of Living Things
0				0		
PHANEROZOIC	PHANEROZOIC — CENOZOIC	Quaternary	Recent or Holocene / Pleistocene		Several glacial ages Making of the Great Lakes; Missouri and Ohio Rivers	*Homo sapiens*
				2?		
		Tertiary	Pliocene			Later hominids
				6	Beginning of Colorado River	Primitive hominids
			Miocene			Grasses; grazing mammals
580				22	Creation of mountain ranges and basins in Nevada	
			Oligocene			
				36		
			Eocene		Beginning of volcanic activity at Yellowstone Park	Primitive horses
				58		
			Paleocene		Beginning of making of Rocky Mountains	Spreading of mammals Dinosaurs extinct
				63		
	PHANEROZOIC — MESOZOIC	Cretaceous			Beginning of lower Mississippi River	Flowering plants
				145		Climax of dinosaurs
		Jurassic				Birds
				210		
		Triassic			Beginning of Atlantic Ocean	Conifers, cycads, primitive mammals Dinosaurs
				255	Climax of making of Appalachian Mountains	Mammal-like reptiles
PRECAMBRIAN	PHANEROZOIC — PALEOZOIC	Permian				
				280		
		Pennsylvanian (Upper Carboniferous)				Coal forests, insects, amphibians, reptiles
				320		
		Mississippian (Lower Carboniferous)				
				360		Amphibians
		Devonian				
				415		
		Silurian				Land plants and land animals
				465		
		Ordovician			Beginning of making of Appalachian Mountains	Primitive fishes
				520		
		Cambrian				Marine animals abundant
				580		
	PRECAMBRIAN (Mainly igneous and metamorphic rocks; no worldwide subdivisions.)			1,000		Primitive marine animals Green algae
				2,000		
				3,000	Oldest dated rocks	Bacteria, blue green algae
~4,650	Birth of Planet Earth			~4,650		

(MANY)

[a] Best estimates (after R. L. Armstrong, 1972, unpublished).

mates were far too small. The geologic column represents a much longer time than anyone in the 19th Century supposed, and the errors in the estimates arose because the large amounts of time that elapsed between the periods of deposition of successive strata were unknown. Similarly, the calculated cooling time for a once-molten Earth was in error, because in the 19th Century it was not known that new heat is continually being added to the Earth by natural radioactivity.

To get around this roadblock, what was needed was a way to measure geologic time by some process that runs continuously, that is not influenced by other processes, and that leaves a continuous record without gaps in it. At the end of the 19th Century—in 1896 in fact—the discovery of radioactivity provided the needed method. That discovery opened the door to radiometric dating, a new and reliable means of measuring geologic time.

Radiometric Dating

Natural Radioactivity. As we shall see in Chapter 4, most chemical elements are stable and do not change, but a few are radioactive and are therefore unstable. These unstable elements are continually decaying. It is largely the heat created by this decay that makes Earth's interior very hot.

A radioactive element decays by throwing off—literally shooting out—particles from the nucleus of each of its atoms (App. A). By this process each atom becomes instantly converted into a different, "daughter" atom that has a mass number smaller than that of its parent. The daughter atom can be either an isotope of the same element as its parent or a different element altogether. An example is portrayed in Figure A.3, where we can see one atom of uranium-238 decaying and "bumping down" through a whole series of unstable daughters, finally ending up as one atom of an end product, lead-206, that is not radioactive.

This natural radioactivity of any unstable element follows a definite timetable. Think of a radioactive isotope that decays directly, in one step, to form a stable daughter end product. The number of decaying parent atoms continually decreases while the number of daughter atoms increases. The proportion of atoms that decay during one unit of time is constant. But although the proportion is constant, the

actual number keeps decreasing, because the parent atoms are being used up.

So, in a mineral sample, the overall radioactivity (the sum of the radioactivity of all the parent atoms remaining in the sample) decreases continually (Fig. 2.9). The decrease can be expressed in terms of the **half-life** of the parent, *the time required to reduce the number of parent atoms by one-half.* The time units marked in Figure 2.9 are half-lives. Of course they are of equal length, just as years are. But at the end of each one, the number of atoms that decay (and therefore the combined radioactivity of the sample) has decreased by exactly half.

Age Determination. Through the use of very sensitive instruments, it is possible to measure the amounts of parent atoms and daughter atoms that are present in a rock or mineral sample. If both these amounts are known, we can place the sample at that point on a curve where it belongs, as in Figure 2.9. Then, knowing the half-life of the parent, we can calculate the approximate time in years since decay began—in other words, the age of the sample. Such dates are called *radiometric dates,* because they are determined through measurements of radioactivity.

Radiometric dates are obtained from various isotopes. A particular isotope may be selected because it is present in one of the minerals in the rock we wish to date. Or it may be selected because its half-life is long (if we want to date a very ancient

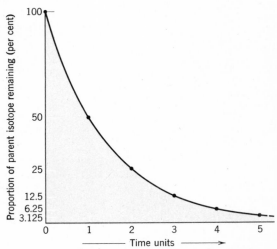

Figure 2.9 Curve showing decay of a radioactive isotope. During each time unit half the atoms remaining in the parent isotope decay into daughter isotopes.

25

rock) or short (if we want to date something young). Table 2.2 lists some of the isotopes commonly used.

The radiometric date of an igneous rock is the date when the mineral that includes the parent isotope crystallized from its parent magma. The date of a metamorphic rock is the date when previously existing rock was metamorphosed, creating the new minerals that contain the parent isotope. Tens of thousands of radiometric dates have been determined.

Dating by Carbon-14. Among the dating methods listed in Table 2.2, the one based on carbon-14 (^{14}C, also known as radiocarbon) is unique for two reasons. The half-life of ^{14}C is short, and the amount of daughter product cannot be measured.

Radiocarbon (^{14}C) is continuously created in the atmosphere through bombardment of nitrogen-14 by neutrons created by cosmic radiation. Carbon-14, with a half-life of 5730 years, decays back to nitrogen-14. The radioactive carbon mixes with ordinary carbon (carbon-12) and diffuses rapidly through the atmosphere, hydrosphere, and biosphere. Because the rates of mixing and exchange are rapid compared with the half-life, the proportion of radiocarbon is nearly constant throughout the system. As long as the production rate remains constant, the radioactivity of natural carbon remains constant because rate of production balances rate of decay.

As long as an organism is alive, it contains this balanced proportion of carbon-14. However, at death the balance is upset, because replenishment by life processes such as feeding, breathing, and photosynthesis ceases. The carbon-14 in organic tissues continually decreases by radioactive decay, at a rate that decreases with time, as can be seen in Figure 2.9. The analysis for the radiocarbon date of a sample involves only a determination of the radioactivity level of the carbon-14 it contains. The daughter product, ^{14}N, cannot be measured, and it is necessary to assume that the rate of production of ^{14}C has been constant throughout the last 50,000 years or so, the

Table 2.2
Some of the Principal Isotopes Used in Radiometric Dating

Isotopes		Half-Life of Parent (Years)	Effective Dating Range (Years)	Minerals and Other Materials That Can Be Dated
Parent	Daughter			
Uranium-238	Lead-206	4.5 billion	10 million to 4.6 billion	Zircon Uraninite Pitchblende
Uranium-235	Lead-207	710 million	10 million to 4.6 billion	Zircon Uraninite Pitchblende
Potassium-40	{ Argon-40 { Calcium-40	1.3 billion	100,000 to 4.6 billion	Muscovite Biotite Hornblende Whole volcanic rock
Rubidium-87	Strontium-87	47 billion	10 million to 4.6 billion	Muscovite Biotite Microcline Whole metamorphic rock
Carbon-14	Nitrogen-14	5,730 ± 30	100 to 50,000	Wood, charcoal, peat, grain, and other plant material Bone, tissue, and other animal material Cloth Shell Stalactites Ground water Ocean water

range of time to which the short half-life of ^{14}C limits the usefulness of the method.

Because the method is based partly on this assumption, the accuracy of carbon-14 dates has been checked against samples whose dates are known independently through historical information. Among these samples are grains of corn, wooden beams, prehistoric clothing, and furniture from ancient Egyptian tombs. In these samples, ^{14}C dates compare fairly well with historical dates. Although none of the historically dated samples is as much as 5000 years old, the annual growth rings of long-lived trees provide a further check on ^{14}C dates of wood from the same trees, that extends back more than 8000 years. Many dates as old as 50,000 years have been calculated without the benefit of these independent checks. Although less accurate than the dates of younger samples, they are still very useful.

Because of its application to organisms (by dating fossil wood, charcoal, peat, bone, and shell material) and its short half-life, radiocarbon has proved to be enormously valuable in establishing dates for prehistoric races of man and for recently extinct animals, and in this way it is of extreme importance in archeology. It is of comparable value in dating the most recent part of geologic history, particularly the latest of the glacial ages. For example, the dates of many samples of wood taken from trees overrun by the advance of the latest of the great ice sheets and buried in the rock debris thus deposited, show that the ice reached its greatest extent in the Ohio-Indiana-Illinois region not more than about 18,000 years ago.

Similarly, radiocarbon dates afford the means for determining rates of movement, such as the rate of advance of the last ice sheet across Ohio, the rates of rise of the sea against the land as glaciers melted throughout the world, and the rates of local uplift of the crust that raised ancient beaches above the sea.

Geologic Time Scale

Through the various methods of radiometric dating, the dates of solidification of many bodies of igneous rock have been determined. Many such bodies have identifiable positions in the geologic column, and through this fact it becomes possible to date approximately a number of the sedimentary layers in the column. We remember that the standard units of the geologic column consist of sedimentary strata containing characteristic fossils, but the typical rocks from which dates (other than radiocarbon dates) are determined are igneous rocks. It is necessary, therefore, to be sure of the time relations between an igneous body that is datable and a sedimentary layer whose fossils closely indicate its position in the column.

Figure 2.10 shows in an idealized manner how apparent ages of sedimentary strata are approximated from the apparent ages of igneous bodies. The age of a stratum is bracketed between bodies of igneous

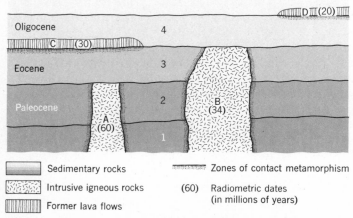

Figure 2.10 Idealized section illustrating application of radiometric dating to the geologic column. For method see text.

rock, the apparent ages of which are known.

In the figure, four series of sedimentary strata, whose geologic ages are known from their fossils, are separated by surfaces of erosion. Related to the strata are two intrusive bodies of igneous rock (A, B) and two sheets of extrusive igneous rock (C, D). From the apparent dates of the igneous bodies and the geologic relations shown, we can draw these inferences as to the ages of the sedimentary strata:

Stratum	Age (millions of years)	That is,
4	<34 <30 >20	age lies between 20 and 30 million years
3	<60 >34 >30	age lies between 34 and 60 million years
2	>60 >34	age of both is more than
1	>60 >34	60 million years

To separate 1 from 2, dates from other localities are needed. Dates from igneous rocks elsewhere could also narrow the possible ages of 3 and 4.

Through this combination of geologic relations and radiometric dating, we are able to fit a scale of time to the geologic column. The scale is being continually refined.

It is a great tribute to the work of geologists during the first half of the 19th Century that the geologic column they established has been fully confirmed by radiometric dating. Comparisons between the numbers column and the names column in Table 2.1 shows this. They show also that the grouping of strata into the successively smaller subdivisions called *systems, series,* and *stages* is matched by the corresponding time units called *periods, epochs,* and *ages.* We can speak of the time units, of course, whether or not we know their dates. We could speak of events that occurred in the Devonian Period (or simply in Devonian time) even if we did not know that the dates of that period fall between 415 and 360 million years ago.

Of course the geologic column is not yet provided with a time scale that is complete. There are gaps. One obvious gap near the top of the column occurs between the "old" limit of carbon-14 dating and the "young" limit of potassium/argon dating. To a great extent this gap is the result of scarcity of datable samples, and in time both it and the other gaps will surely be filled in.

Age of Planet Earth

Table 2.1 shows us that the oldest rocks are the great assemblage of metamorphic and igneous kinds known as Precambrian rocks. Of the many radiometric dates obtained from them, the youngest are around 600 million years, the oldest around 3.5 billion years. The Precambrian unit of the geologic column, then, existed during a *minimum* time equal to 3.5 billion minus 600 million years, or 2.9 billion years—a span nearly 6 times as long as the time elapsed since the Precambrian unit ended.

If some Precambrian rocks are 3.5 billion years old, the beginning of Planet Earth's history must lie still further back in time. This is confirmed by one of the Precambrian rocks, a body of granite in South Africa. Although itself an igneous rock, this ancient granite contains great chunks of quartzite, much as a pudding contains raisins. At an earlier time, before it became enveloped by the granite magma, the quartzite must have been part of a layer of sandstone. Even before that, the sandstone must have been loose sand. Clearly, then, the rock cycle must have been operating in its present manner well before the granite magma solidified. This relationship between quartzite and granite is not unique. Other old igneous bodies that enclose older sedimentary or metamorphic rocks are found in other continents.

Even though 3.5 billion years is the age of the oldest Earth rock dated thus far, other parts of the Solar System are more helpful in fixing the age of Planet Earth. Meteorites—small independent bodies that have "fallen" onto Planet Earth—have been dated by three of the methods listed in Table 2.2 (^{238}U/^{206}Pb, ^{40}K/^{40}Ar, ^{87}Rb/^{87}Sr). According to all three methods, the resulting maximum ages of meteorites are about 4.6 billion years, and "Moon dust" has given a similar maximum age. Because of these similarities it is thought that Planet Earth, and indeed all the other planets and the meteorites in the Solar System, formed about 4.6 billion years ago, with an uncertainty of a few hundred million years.

In its impact on our thinking about the age and history of the Earth, radiometric dating has been perhaps an even more important development for the 20th Century than the construction of the geologic column was for the 19th. The two great developments complement each other in a remarkable way.

Summary

Cycles

1. The water cycle is driven by solar energy. In it, moisture evaporates (chiefly from the ocean), is precipitated, and returns to the sea directly or by longer paths.

2. In the water cycle, a quantity of water equal to the volume of the world ocean is recycled once every 3200 years.

3. The complete rock cycle begins with magma, which solidifies and forms igneous rock. The rock is eroded, creating sediment, which is deposited in layers that become sedimentary rock. Deep burial leads to metamorphism. Rock of all kinds sinks downward to depths at which it can be melted to form new magma.

4. Many cycles operate in the biosphere. A fundamental cycle involves carbon dioxide. This is extracted from the atmosphere by land plants, which decay or are eaten by animals. In both processes, the CO_2 is returned to the atmosphere.

Geologic Column

5. Sedimentary strata are deposited in sequence, oldest at bottom and youngest at top.

6. Major groups of strata contain distinctive assemblages of fossils, by which they can be identified. The strata and their fossils are the basis for the geologic column.

Time

7. Decay of radioactive isotopes of various elements is the basis of radiometric dating.

8. A sedimentary layer is dated by being bracketed between two bodies of igneous rock that have been dated radiometrically.

9. The dates of the bracketed layers establish the correctness of the geologic column.

10. Measurement of carbon-14 activity, mainly in fossil organic matter, yields apparent ages through approximately the last 50,000 years.

11. The age of Planet Earth is obtained through radiometric dating of meteorites. It is about 4.6 million years.

Selected References

Cycles

Garrels, R. M., and Hunt, Cynthia, 1972, The web of life: New York, W. W. Norton.
Kuenen, Ph. H., 1963, Realms of water. Some aspects of its cycle in nature: Rev. ed., New York, John Wiley, Science Editions.

Geologic Column

Harland, W. B., and others, eds., 1964, The Phanerozoic time-scale: Geol. Soc. London, Quart. Jour., v. 120S.

Time

Berry, W. B. N., 1968, Growth of a prehistoric time scale: San Francisco, W. H. Freeman.

Eicher, D. L., 1968, Geologic time: Englewood Cliffs, N. J., Prentice-Hall.

Faul, Henry, 1966, Ages of rocks, planets, and stars: New York, McGraw-Hill.

Libby, W. F., 1961, Radiocarbon dating: Science, v. 133, p. 621–629.

Toulmin, Stephen, and Goodfield, June, 1965, The discovery of time: New York, Harper & Row.

Chapter 3

Energy
and its Sources

Chapters 1 and 2 have described Earth's general makeup, have sketched the cyclic activities represented by the external and internal processes, and have stressed the great importance of *time* in the changes that result. Basically, all these activities represent *energy* working on *materials*. So, in order to examine each activity in more detail, we begin by discussing energy (in Chapter 3) and (in Chapters 4 and 5) minerals and rocks, the materials to which Earth's energy is applied.

In beginning, we ask you to turn a page in this book. You are using energy. Whether walking outside, driving a car, or merely turning on the light—whatever the activity, you are using energy. Activities and energy are so intimately related that we may define **energy** as the *capacity to produce activity*. Energy is vital for our existence, and vital, too, for Earth's existence. Without energy, Earth would be a dead planet.

Energy appears in many forms, and each produces characteristic activities. We speak of kinetic energy, meaning the energy of a moving body, and heat energy, meaning the energy of a hot body. Other forms in which energy manifests itself are electrical-, chemical-, radiant-, and atomic energy. Each form of energy is important for some of Earth's activities (Table 3.1). We must therefore ask where all the energy comes from and, if we are to understand Earth's history, whether energy has always come from the same sources. To help answer these questions, let us first examine the four principal kinds of energy that reach the Earth: kinetic-, heat-, atomic-, and radiant energy.

Table 3.1

Activities Produced by Common Forms of Energy

Energy	Common Activity
Kinetic	Flowing water, wind, waves, landslides
Heat	Volcanoes, hot springs, rainstorms
Chemical	Decaying vegetation, forest fires, rusting, burning coal
Electrical	Lightning, aurora borealis
Radiant	Daylight, sunburn
Atomic	Heating Earth's interior

Kinds of Energy

Kinetic Energy. Every moving body has *kinetic* energy, named from the Greek word *kinetikos* (to move). The movement of Earth around the Sun and that of Earth spinning on its axis mean that Earth possesses kinetic energy. A ball moving in a tennis game, a boulder rolling downhill, a stream flowing down a valley, or tiny rock particles in a dust storm, all have kinetic energy.

All of Earth's processes involve kinetic energy. Besides running water and rolling boulders, winds, waves, icebergs drifting in the ocean, and moving glaciers are other common examples. Because every moving particle of sediment possesses kinetic energy, it is easy to see that kinetic energy is vitally important in the making of sedimentary rock.

Heat Energy. We can think of heat as a special form of kinetic energy. It is the energy possessed by the motion of atoms (App. A). All atoms move constantly. The faster they move, the more heat energy they have, and the hotter a body feels. The atoms in a crystal move within confined spaces—in a sense they rattle around their assigned places in the crystal structure—but if they move fast enough (become hot enough) they break out of their fixed positions and the crystal is said to have melted. With faster movement, which means more heat, atoms become completely free-moving bodies, and the liquid is then said to have vaporized.

Hotness, or degree of heat, is a cumbersome term; so we use the word *temperature* instead. Temperature is measured by arbitrary scales. A common one is the *Celsius scale,* in which we select as 100°C the temperature (or motions of atoms) just sufficient to boil water at sea level.

Heat energy is transmitted in two principal ways. The first is by *conduction,* which occurs when atoms pass on some of their motions to adjacent atoms.

Conduction is the process by which heat is transmitted through the wall of a cup of hot coffee; it is also the way by which most of Earth's internal heat energy reaches the surface. Conduction, therefore, is the means by which heat is transmitted through solids; it is a slow process because the transfer occurs atom by atom. The second way by which heat can be transmitted—by *convection*—is much faster. Convection occurs in liquids and gases in which the distribution of heat is uneven. When a liquid or gas is heated, it expands; so its density (mass per unit volume) decreases. The hot, less dense material floats up, with colder, more dense material sinking to replace it, setting up a *convection cell* or *convection current.*

Atomic Energy. We are all familiar with the existence of atom bombs and hydrogen bombs. They release in a destructive way vast amounts of energy locked within atoms. But the same processes of releasing energy from atoms go on continuously in nature (fortunately for us, in a controlled manner). In the Sun, atoms of hydrogen combine to form atoms of helium. Four atoms of hydrogen combine to produce one atom of helium, and in the process give off heat energy; we refer to the process as *fusion* or *nuclear burning.* Exactly how this happens is discussed in greater detail in Appendix A.

Fusion does not occur naturally in the Earth. But the natural disintegration of certain kinds of atoms does occur, and in the process heat energy is created. As noted in Chapter 2, disintegration of atoms to form new, lighter atoms is called *radioactivity.* When an atom disintegrates by radioactivity, a small amount of heat energy is released. The details of this process, too, are discussed in Appendix A. Most atoms in the Earth are not radioactive. Three kinds—certain atoms of the elements potassium, uranium, and thorium—are radioactive. In tiny amounts they are widely distributed through the crust and mantle, and the rates at which they disintegrate are very slow. Nevertheless they generate sufficient heat to keep the interior of Earth exceedingly hot—at times hot enough even to melt rock and form magma.

Radiant Energy. All the forms of energy discussed so far involve matter. How, then, is it possible for energy to pass through empty space? It is possible because energy can be transmitted by energy waves, commonly called *electromagnetic waves.* Radiant energy, therefore, *is* electromagnetic waves: visible light, X-rays, and radio waves are well-known forms of radiant energy.

The distance between the crests of two adjacent waves is a *wavelength*. The wavelengths of electromagnetic waves (Fig. 3.1) range from less than one thousand millionth of a centimeter (cosmic rays) to more than a million centimeters (radio waves). The names in Figure 3.1 do not indicate any fundamental differences among electromagnetic waves. All such waves move with the same speed, the speed of light; the names that we apply are merely convenient ways of designating various wavelength regions. The entire range of electromagnetic waves is called the *electromagnetic spectrum*.

Radiant energy and its transmission through space by electromagnetic rays is vitally important for Earth. How else, for example, could we receive the energy the Sun generates by nuclear burning?

Conversions Among Forms of Energy. When we burn gasoline in a car, we convert chemical energy to heat energy. When the car moves, the heat energy in turn is converted to kinetic energy. Conversion of heat energy to kinetic energy in a controlled manner was first accomplished in the 18th Century, when the steam engine was invented, thereby inaugurating the industrial era in which we live.

In all natural processes, energy is continually converted from one form to another. As radiation arrives from the Sun, it strikes Earth's surface, the rays are absorbed, and the ground is heated. Air in contact with the ground is warmed by conduction, expands, and rises; it is replaced by cooler, denser air from above and wind results (Fig. 3.2). The process of lifting warm air from Earth's surface to greater altitudes is an example of convection. Thus energy that began by conversion of matter to elec-

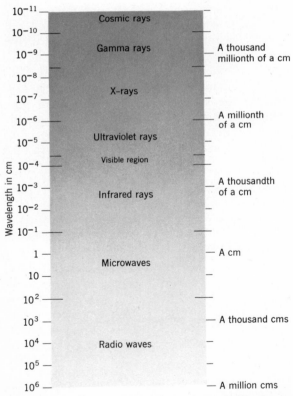

Figure 3.1 Wavelengths of radiant energy. The names applied to the different wavelengths are just convenient handles; they do not signify any fundamental differences between different parts of the spectrum. Boundaries between the differently named parts of the spectrum are gradual, not sharp.

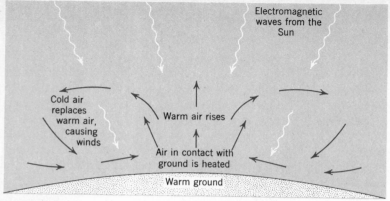

Figure 3.2 Energy in the form of electromagnetic radiation from the Sun is transformed to kinetic energy of wind. The radiation heats Earth's surface. Air in contact with the surface is warmed in turn, expands as a consequence of the heating, and rises. Cold air flows in to replace the rising hot air; winds result. The complete chain of rising warm air and inflowing cold air is called a convection cell.

tromagnetic waves in the Sun was transformed to heat energy and finally to kinetic energy in a blowing wind. The large number of Earth's internal and external activities result from the fact that many paths for such conversions are possible. For example, energy produced by natural radioactivity in the Earth causes rock to melt and form magma. Magma flows upward, converting some of the heat energy to kinetic energy. No matter what activity we examine, we find that energy drives it and that, in the process, energy is converted from one form to another.

Before people realized that forms of energy could be converted back and forth, they had already begun to use different measurement units for each form. To avoid the confusion of multiple units, we will use only the calorie, the unit of heat energy. A *calorie* is *the amount of heat energy needed to raise the temperature of one gram of water by one degree Celsius.* A calorie is so small that in one tablespoon of sugar there are approximately 50,000 calories of chemical energy. To make us feel better about the calories we consume, nutritionists use a unit of 1000 calories, which they label the Calorie. There are only 50 of these in a tablespoon of sugar! If we compare, in Table 3.2, the amounts of energy derived from different sources, it is apparent that an incomparably greater amount of energy comes from fusion and fission reactions than from burning.

Sources of Energy

Energy reaches Earth's surface from three sources. (1) Radiant energy arrives from outside the Earth, principally from the Sun; (2) kinetic energy arrives from the rotations of Moon, Earth, and Sun and appears as tides; and (3) energy reaches Earth's surface by continuous outflow of Earth's internal heat. Because the surface receives energy from three sources, we might reason that it is heating up. But this is not the case, because the average temperature of Earth's surface does not vary from year to year;

Table 3.2

Amounts of Energy Available from Different Sources. The Amount of Energy per Unit Mass from Nuclear Burning Is Vastly Greater Than the Chemical Energy Resulting from Burning

Reaction	Amount of Heat Energy
1 gram of hydrogen fuses to form helium	1.5×10^{11} calories
1 gram of U-238 decays by radioactivity	0.2×10^{11} calories
1 gram of oil burns	1.0×10^{4} calories
1 gram of coal burns	0.8×10^{4} calories
1 gram of sugar is eaten	4.2×10^{3} calories
1 gram mass moving at a velocity of 1000 cm/sec[a]	1.2×10^{-2} calories

[a]1000 cm/sec equals approximately 23 miles per hour.

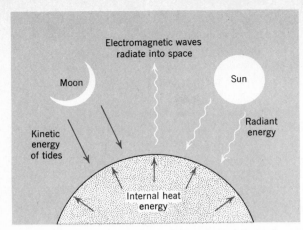

Figure 3.3 Energy reaches Earth's surface from three sources; two are external, one internal. To maintain a heat balance, long-wavelength electromagnetic waves radiate energy into space.

so some sort of energy balance must be maintained. The source of the balance is not hard to find: Earth does not merely receive radiant energy; it also emits such energy by radiating long-wave-length electromagnetic waves ("heat waves") back into space. Earth's surface is said to be in a **steady state** or

state of **dynamic equilibrium,** by which we mean *a condition in which the rate of arrival of some materials equals the rate of escape of other materials.* The rate at which energy reaches Earth's surface exactly balances the rate at which it escapes (Fig. 3.3).

Energy from the Sun. Radiant energy reaches Earth from the stars, but in amounts that are tiny by comparison with the Sun's energy. When the Sun's rays reach Earth's atmosphere, approximately 40 per cent are simply reflected back into space without change. It is this reflected radiation that the astronauts see when, standing on the Moon, they look at Earth (Fig. 3.4). The remaining 60 per cent is absorbed, partly by the atmosphere (which becomes heated in the process) and partly by the land and by the sea. The energy absorbed by the sea warms the water and causes evaporation. The resulting water vapor forms clouds and eventually rain, snow, and all other forms of precipitation. The energy absorbed by the land eventually warms the air, causes convection, and creates winds, which, blowing over the sea, create waves. Thus all the major agents of weathering and erosion that operate at Earth's

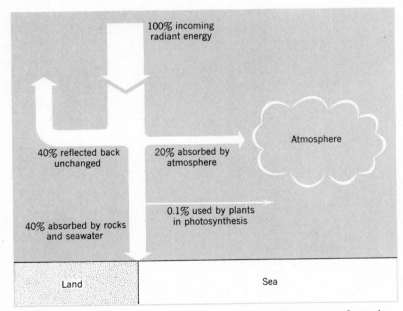

Figure 3.4 Paths followed by Earth's incoming radiant energy from the Sun. The energy used in heating the atmosphere causes winds. Most of the energy absorbed by land and sea is used up in the evaporation of water to form clouds and cause rain, snow and hail.

surface—rain, ice, streams, winds, waves, and glaciers—are activities derived from the Sun's energy.

Has the Sun always sent energy to Earth's surface? Evidence preserved in rocks—what we call the *rock record*—shows clearly that it has. Remember from Chapter 2 that sedimentary rocks extend as far back through Earth's history as we have been able to go. To have sediment we must have erosion; so we can infer that the Sun has been in something like its present state for at least as long as Earth has had the form with which we are familiar. In this respect the Sun, like our planet, reflects the Principle of Uniformity.

How much energy reaches the Earth from the Sun? The amount is enormous—far greater than the energy reaching the surface from other sources. Earth intersects a disc-shaped area of radiation coming from the Sun. The amount of energy that crosses each square centimeter of this disc, at the average distance of Earth from the Sun, is 2.88×10^3 calories every 24 hours. However, this is not the amount of energy that actually reaches each square centimeter of Earth's surface. Some rays are reflected, as we have seen; and, because Earth rotates, any given point can receive sunlight for only about 12 hours. Also, because Earth is spherical, points near the poles receive considerably fewer calories per unit of land surface than do points near the equator (Fig. 3.5). As we shall see in Chapter 12, this variation of solar energy received in various latitudes is the main driving force for the world's ocean currents.

Energy from Tides. One does not always think of tides in the context of energy. Nevertheless, tides are the mechanism by which some of the kinetic energy from the motions of Moon, Earth, and Sun reach Earth's surface. The principal effect arises from interactions between Moon around Earth; so tides are principally lunar effects.

Gravitational attraction by Moon on Earth pulls seawater toward the Moon and creates tidal bulges in the ocean (Fig. 3.6). No energy would be involved were it not for Earth's rotation about its axis and Moon's movement in orbit around Earth. As a consequence, the positions of the bulges move continuously, and at any spot on the sea we see two high tides and two low tides each day. However, as Figure 3.6 shows, the Sun also affects tides, sometimes aiding the Moon by pulling in the same direction, and sometimes opposing it by pulling at right angles. The Sun's effect is smaller than Moon's; so the two effects never cancel each other. The actual heights

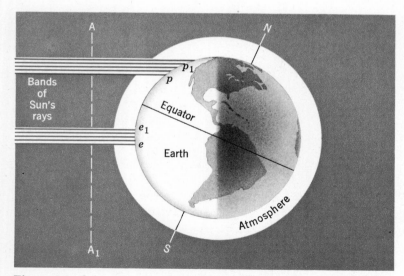

Figure 3.5 Solar radiation reaching Earth's surface varies with latitude. Radiation crossing plane AA$_1$ amounts to 2.88×10^3 calories/cm^2 every 24 hours. A square centimeter of rays crossing AA$_1$ strikes a square centimeter of Earth's surface when the rays are perpendicular to the surface (ee$_1$), but because of Earth's curvature are spread over a larger area in high latitudes (pp$_1$).

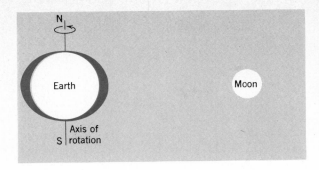

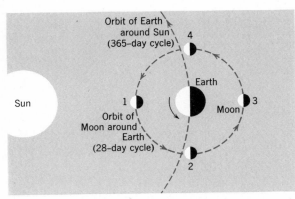

Figure 3.6 The gravitational attractions of Moon and Sun on Earth raise tidal bulges in Earth's oceans. *A.* Idealized diagram of tidal bulges relative to Earth's axis of rotation and to the position of the Moon. *B.* When Moon and Sun attract in the same direction, Moon positions 1 and 3, we observe highest tides. When Moon and Sun have opposing positions (Moon positions 2 and 4), we observe lowest tides.

of high and low tides, therefore, vary on a cycle of approximately 14 days, matching the enhancement and opposition of tides by Sun and Moon.

Tidal bulges cannot move around Earth unhindered, because continents get in the way. Water therefore piles up against the continental margins whenever a tidal bulge arrives, and then flows back to the ocean basin as the bulge passes. The piling-up effect is the reason why tides are much higher along coasts than in the open ocean. Movement of water masses in coastal tides means kinetic energy is being used, and this energy must be taken from Earth's store of kinetic energy of rotation.

If the kinetic energy of rotation is being transferred to Earth's surface to cause tides, what is happening to Earth's rate of rotation? Earth's spinning motion, and therefore its energy, apparently remain from the days of its formation; we know of

no new kinetic energy that is being added. Therefore, removing kinetic energy can have only one effect. The tides are acting as weak but steady brakes, and Earth is gradually slowing down. The rate of slowing is, fortunately, not great because the rate of energy transfer is small. Astronomers have measured the exact length of the day over the past three centuries and find that it is increasing by 0.002 seconds each century. Over hundreds of millions of years the effect of this small increase can become very large. Some billions of years into the future, Earth will stop rotating completely.

Have tides always acted on Earth—or, to say it another way, has the Moon always been there to cause the tides? The evidence comes in an interesting way. Clams and other shellfish grow a microscopically thin layer of new shell material each day. The thickness of the layer depends on the depth of water covering the shellfish—thick layers at high tide, thin at low. We have already seen that highest tides occur approximately every 14 days (Fig. 3.6)—that is, twice each lunar month. When we examine with a microscope a section of a modern clam shell and measure the thickness of the daily growth layers, we find that there are indeed variations, that thickest layers form at times of highest tides, and that a repeating pattern of 14 thick and 14 thin layers emerges (Fig. 3.7). The shell of a modern clam is therefore an accurate recorder of tides. If tides have operated in the past, the same pattern should be seen in the shells of fossil clams. This turns out to be exactly what is observed, even in fossils as old as 600 million years, with the qualification that as we look at older and older fossils, the number of thick and thin layers in a repeating pattern becomes larger. This means that Earth rotated more rapidly in the past because larger repeat numbers indicate there were more days in the lunar month that there are today (Fig. 3.8). Thus, fossil shells provide evidence that the Moon has been causing tides on Earth for an exceedingly long time, that tides have been transferring the kinetic energy of rotation to Earth's surface for an equally long time, and that Earth is indeed slowing down.

Energy from Earth's Interior. Anyone who has been down in a mine realizes that rock temperatures increase with depth. Measurements made in deep drillholes and mines show that the rate of temperature increase (the *geothermal gradient*) varies in

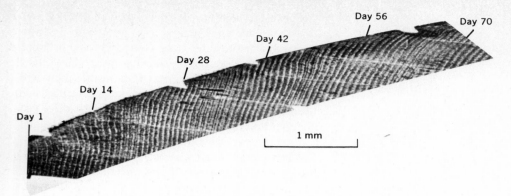

Figure 3.7 Microscope photograph of the daily growth bands in a modern clam shell. The animal lays down a thick, new layer of shell each day, the thickness varying with water depth during high tide. A 14-day repeat pattern in the growth layers corresponds to the twice-monthly highest tides, when Moon and Sun exert their tidal attractions in the same direction. (*After Pannella and MacClintock, 1968.*)

different parts of the world from 15°C to 75°C per kilometer. We cannot, however, make direct temperature measurements beyond the deepest drillholes, which are only about 10km deep; so we have to use indirect means to estimate temperatures in Earth's interior. From physical properties that vary with temperature, such as the speeds of earthquake waves, we estimate that temperatures continue to increase toward the center and eventually reach values of

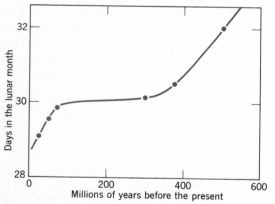

Figure 3.8 Evidence from growth layers in fossil shells indicates that Earth's rotation is slowing down and the number of days in the lunar month is decreasing. Rate of slowdown appears to have varied with time, a factor attributed to the changing positions of continents and their effects on tides. (*Pannella and others, 1968.*)

5000°C or more in the core. When we consider Earth's size, obviously a vast amount of heat energy is stored within it.

Heat flows from a hot body to a cold one; so we infer that heat must be flowing outward to the surface, by conduction from the hot interior regions of Earth. People once thought that the heat inside Earth remained from the days of Earth's formation. During the 19th Century, the famous British scientist Lord Kelvin reasoned that because Earth is apparently losing the heat energy that it had when it formed, the rate of heat loss meant Earth could be no older than 100 million years. If Earth were older, he argued, it would have to be colder. But we now know that Earth is much older, its age being closer to 4.6 billion years. When this great age was first realized, it became clear that Lord Kelvin must be wrong. Residual heat could not account for all the internal heat energy. Some energy source deep down below must, therefore, still be generating heat.

One possible explanation of the mysterious energy source seemed to lie with tides in the solid Earth. If Earth is not completely solid, but instead has some interior plastic or liquid-like properties, then tides like those in the ocean should occur within the mass of the Earth because of the Moon's gravitational pull. Small tides do occur in Earth's body, but measurements show that they are too small to provide the necessary energy. The true energy source was finally identified when chemical analyses re-

Table 3.3

Comparison of the Amount of Heat Energy Produced by Radioactivity in Three Common Kinds of Igneous Rock. Radioactive Elements Are More Concentrated in Granitic Rocks, Characteristic of Continental Crust, Than in Basalt and Peridotite, Characteristic of Oceanic Crust and Mantle

Rock Type	Rate of Heat Generation
Granite	2250×10^{-11} calories/gram/day
Basalt	328×10^{-11} calories/gram/day
Peridotite	2.5×10^{-11} calories/gram/day

vealed that natural radioactive atoms of uranium-235, uranium-238, thorium-232, and potassium-40 are widespread in common kinds of rock. As we have seen, natural radioactive decay leads to release of heat energy, and although the amounts of the natural radioactive isotopes are small, there is enough of each to keep Earth's interior hot and in a steady state, even though heat is continually flowing out at the surface. Because different kinds of rock contain differing amounts of the radioactive elements, their heat-producing abilities differ greatly. It is apparent from Table 3.3 that heat production in granite, a rock characteristically found in the continents, is much greater than in basalt, a rock characteristically found in the ocean basins.

In solid bodies heat moves by conduction. Yet Earth is not completely solid. Lava extruded on the surface brings up heat from the interior, as does a hot spring. About one-tenth of the internal heat that reaches the surface does so in lava and in the water of hot springs; the remainder arrives by conduction. Although the total amount of internal heat that reaches the surface equals only about one two-thousandth of the energy that arrives from the Sun, Earth's heat reservoir is enormous, and is sufficient to drive all of Earth's gigantic internal activities.

The average value of the heat flow that reaches Earth's surface is about 1.5×10^{-6} calories/cm²/second. To picture how small this amount of heat is, imagine that we have a gadget that can catch and use all the heat reaching a square meter of Earth's surface. We would have to gather heat for approximately four days and nights before we got enough heat energy to boil a cup of water. But the size of Earth's surface is enormous, so the total amount of heat that flows outward is also enormous. Just as the geothermal gradient varies from place to place, so does the heat flow, which is greatest near volcanoes and hot springs. Variations in surface heat flow tell us a great deal about activities down below and as we shall see in later chapters, they allow us to infer that convection may be occurring in the seemingly solid rocks of the mantle (Chap. 18).

Comparison of Earth's Energy Sources. Table 3.4 shows the amounts of energy reaching Earth's surface every 24 hours from the three different energy sources. Because solar energy is vastly larger than the energy of either tides or internal heat flow, it is clear that the temperature at Earth's surface must be controlled by the Sun. Thus, if we fill our atmosphere with dust and other pollutant particles, thereby reflecting more of the Sun's incoming radiation, we might cause a reduction of temperature. Other energy sources are too small to offset any human effects on the environment.

Solar energy does not contribute to Earth's internal activities. Just as water cannot flow uphill, heat cannot flow from a cold body to a hot one; so the geothermal gradient prevents heat from flowing from the surface down into the interior. In a real sense, therefore, Earth's surface is a surface of conflict between different energy sources. Internal activities, driven by internal heat energy, raise mountains and cause irregularities on Earth's surface. External activities, driven by solar and tidal energy, continually erode and abrade the surface irregularities. There are, however, two forces that contribute importantly to the course of events in the great energy conflict. These are gravity and magnetism, and in order to get a complete picture of Earth's energy, we must examine them too.

Table 3.4

Comparison of the Total Amount of Energy Reaching Earth's Surface Every 24 Hours from the Three Principal Energy Sources

Energy Source	Energy Each 24 Hours
Solar radiation	$37,000 \times 10^{17}$ calories
Flow of internal heat	6.6×10^{17} calories
Tides	0.6×10^{17} calories

Gravity

Gravity is a universal force of attraction; all bodies in the Universe attract each other. The force of gravitational attraction was first explained by Sir Isaac Newton in 1666, when he was only 24 years old. Newton expressed it in the following way:

$$F \text{ is proportional to } \frac{M_1 M_2}{d^2}$$

where F is the force of gravitational attraction, M_1 and M_2 are the masses of two attracting bodies, and d is the distance between M_1 and M_2.

Clearly, the larger M_1 and M_2 are, and the smaller d is, the greater F (the force of attraction) will be. Nevertheless, unless the attracting bodies are very large, gravity does not seem to be a very strong force. We cannot even sense the attraction between small objects, like knives and forks, or even somewhat larger objects like automobiles and houses. If bodies are very large, however, as Earth is, gravitational attraction becomes large too. Even the smallest objects fall to Earth's surface because the gravitational pull is so great. We take advantage of this property when we *weigh* an object. The **weight of a body** on Earth is *the force that gravity exerts on it.*

Gravity, the Sphere Maker. ***Earth's gravity*** is *the inward-acting force with which Earth tends to pull all objects toward its center* (Fig. 3.9). This force acts equally in all directions and attempts to make our planet a perfect sphere. It would succeed were it not for Earth's rotation.

Rotations give rise to a force that opposes gravity. This is centrifugal force. As children, we have whirled a weight on a string around our heads and have felt the strong pull exerted by the weight. Perhaps we also noticed that a weight on a long string exerted a stronger pull than one on a short string. The pull is a centrifugal force. The faster we rotated the string, and the farther the weight was from the axis of rotation (meaning the longer the string), the stronger was the pull. Earth rotates; so every particle on Earth is rotating, and every particle is subject to a centrifugal force. The force acts directly outward from the axis of rotation. The distance from Earth's surface to the axis of rotation is similar to the length of the string. It varies because Earth is spherical, being maximum at the Equator

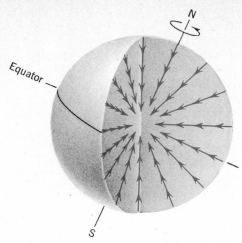

Figure 3.9 Cut-away view of Earth, demonstrating how the force of gravity pulls constantly at every object toward the center of mass. The direction of pull is radial and the result tends to create a spherical form.

and zero at the poles. Earth's actual shape is a result of interactions between the two opposing forces, centrifugal pulling out and gravitational pulling in (Fig. 3.10), and is not a perfect sphere. It is somewhat flattened. Although the pull of gravity greatly

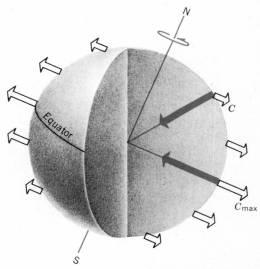

Figure 3.10 The force of gravity, shown in blue arrows, acts along a radius. The centrifugal force caused by Earth's rotation (white arrows) acts in a direction perpendicular to the axis of rotation. The centrifugal force reaches a maximum at the Equator and zero at the poles. As a result of the opposing forces, Earth bulges out at the Equator but is flattened at the poles.

exceeds the centrifugal pull, Earth's radius is 21km less at the poles than at the Equator. This departure from a perfect sphere has an interesting effect on weight: a man who weighs 200 pounds at the Equator, weighs 201 pounds at the North Pole. But we should not get the wrong idea about Earth's departure from a perfect sphere. If we could shrink Earth to the size of a basketball and keep its exact shape, it would seem to be a perfect and highly polished sphere.

Gravity, the Leveler. Earth's gravity shapes the surface of the sea to a nearly spherical form; it makes water flow down slopes; it pulls persistently on rock material on slopes and does enormous work in moving materials to lower places. But boulders sit on hillslopes for long ages; so even though gravity is pulling them down, at the same time some force is holding them up. This holding force is gravity too. On a horizontal surface, gravity holds objects in place by pulling on them in a direction perpendicular to the surface. On slopes, gravity can be resolved into two component forces. The *perpendicular component of gravity* (g_p in Fig. 3.11) acts at right angles to the slope and holds objects in place. The *tangential component of gravity* (g_t in Fig. 3.11) acts along and down the slope. When g_t exceeds g_p, objects move downhill, and we say the **angle of repose** has been exceeded. The angle of repose is *the steepest angle, measured from the horizontal, at which material remains stable.*

When the pull of gravity causes a body to move, the body acquires kinetic energy. Where does the energy come from? Actually it is there all the time, waiting to be released, and we speak of it as **potential energy,** meaning *stored energy waiting to be used.* A rock on a hilltop, for example, reached its position because it was lifted against the pull of gravity when the hill was formed by one of Earth's internal activities. Similarly, raindrops reach their positions in the clouds by energy from the Sun. When they fall, their potential energy appears as kinetic energy.

Variations in Earth's Gravity. We have seen that the pull of gravity differs between Equator and poles because Earth is slightly flattened. Using a **gravimeter** (or **gravity meter**), *a sensitive device for measuring the force of gravity at any locality,* we can easily check for any other variations on Earth's surface. Local variations turn out to be very common, and although the variations are small compared to the overall pull of gravity, they can be measured very accurately.

Local gravity variations are caused by three principal effects: First, altitude above sea level or depth below it, because altitude controls our distance from Earth's center. Second, topographic irregularities. A large mountain mass, for example, tends to attract the gravimeter, while a large valley is a "negative" mass and has the opposite effect (Fig. 3.12). Third, and most important, bodies of rock with different densities may occur beneath the surface, and the gravimeter will be more or less attracted by them depending on whether they are more or less dense than surrounding rocks.

Local variations in gravity reveal information on the sizes and depths of intrusive bodies, the depths of sedimentary basins, and the depth and shape of bedrock below a thick layer of regolith. They also reveal dramatically such features as the locations and depths of trenches in the deep ocean and many otherwise hidden features in the crust, such as the buried cores of mountain ranges.

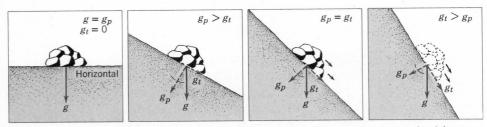

Figure 3.11 Effects of gravity on objects lying on slopes. Gravity can be resolved into two components, one perpendicular (g_p) and one parallel (g_t) to the surface. g_p creates frictional resistance to sliding. When g_t exceeds g_p the object will move.

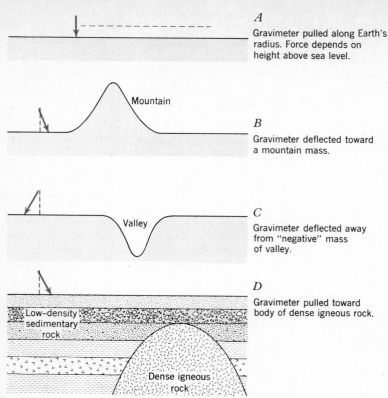

A

Gravimeter pulled along Earth's radius. Force depends on height above sea level.

Mountain

B

Gravimeter deflected toward a mountain mass.

Valley

C

Gravimeter deflected away from "negative" mass of valley.

D

Gravimeter pulled toward body of dense igneous rock.

Low–density sedimentary rock

Dense igneous rock

Figure 3.12   Local variations in the force of Earth's gravity are measured with a gravimeter (a heavy weight suspended on a sensitive spring). *A.* On a perfectly smooth surface, the gravimeter is attracted directly toward Earth's center along a radial line. The force exerted on the gravimeter decreases with increasing height above sea level. *B, C.* When the surface is not smooth, the gravimeter is affected by local masses such as mountains, or "negative" masses such as valleys. In both cases, the direction of strongest pull is deflected slightly from a radial line. *D.* A buried mass of rock more dense than its surroundings attracts the gravimeter, so that the direction of strongest pull is not radial.

Magnetism

Earth behaves as if it were a huge magnet; a compass needle points northward in response to its magnetic forces. Some minerals are naturally magnetic, and, if allowed to move freely, respond in exactly the same way as a compass needle. Both natural magnetism of the mineral magnetite, and the force of Earth's magnetism, were discovered by the Chinese more than 4000 years ago. In the Western hemisphere too, natural magnetism was known at a very early date; the very word *magnet* has ancient origins, coming from *Magnetis litho,* meaning "stone of Magnesia," a district of Thessaly where the ancient Greeks discovered magnetite. Despite the early discovery of magnetic minerals and of Earth's magnetism, the cause of magnetism, and particularly the origin of Earth's magnetism, remained a mystery until the present century.

When a magnetized needle points north, it responds to Earth's **magnetic field,** or *the magnetic lines of force surrounding the Earth.* If the needle is suspended so that it can swing freely in all directions, it aligns itself parallel to the magnetic field. We find that the needle does not point exactly toward the North Pole, and that it is tilted at an angle to Earth's surface. We call *the clockwise angle from true north assumed by a magnetic needle* the **magnetic declination,** while *the angle with the horizontal assumed by a magnetic needle* is called the **magnetic inclina-**

tion (or *dip*). Both declination and inclination vary from place to place on Earth's surface, indicating that the directions of the magnetic lines of force must also vary from place to place. When the lines of force are plotted (Fig. 3.13), we find that they seem to plunge into the Earth, being caused by a very strong magnetic dipole (meaning that the field has a north-and-south direction) buried deep in the interior. Therefore, the magnetic field is apparently caused by one of Earth's internal activities.

Source of Earth's Magnetism. The exact cause of Earth's magnetic field has not been fully established. The most probable cause is one that has been recognized only in recent years. Earth's outer core is molten (Fig. 1.4) and it is believed that fluid motions in the liquid are caused by Earth's rotation. The flowing liquid acts like a dynamo and generates a magnetic field. A magnetic field arising from the core accounts nicely for its shape. There is no reason, furthermore, why fluid motions should create a magnetic axis coinciding exactly with the Earth's axis of rotation; so the observed lack of perfect coincidence is believed to be evidence in support of the dynamo theory. Nor is there any reason why the fluid motions should be constant, and therefore why the magnetic field we measure at the surface should be constant. What sort of variations, then, do we observe?

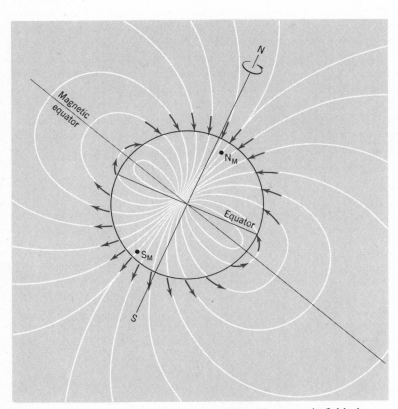

Figure 3.13 The lines of force (white) of Earth's magnetic field. A free-swinging magnetic needle would point along the nearest line of force, with the north-seeking pole in the direction of the arrows.

The axis of the magnetic field does not coincide exactly with the axis of Earth's rotation. Where the axis of the magnetic field intersects Earth's surface, a magnetic needle stands vertical and we define the points as the north and south magnetic poles. The north magnetic pole (N_M) lies in the arctic islands of Canada; the south magnetic pole (S_M) lies in Antarctica, south of Tasmania.

Changes in the Magnetic Field. Over the last 400 years scientists have made careful measurements of the declination and inclination of the magnetic field. Somewhat less precise data are available for an even longer period at certain places in Europe, but their precision is adequate to indicate how the magnetic field has varied over the past 1000 years. The variations are dramatic; Earth's magnetic field is changing constantly. Figure 3.14 shows how the declination and inclination near London have changed during historic time.

If Earth's magnetic field were the result of a strong bar magnet near Earth's center, the variations in the lines of force shown in Figure 3.14 would indicate that the bar magnet was wobbling. But solids in the Earth cannot move quickly enough to produce the rapid historic variations of the magnetic field. Only in a liquid could such rapid wobbles occur. The historic variations therefore provide additional strong evidence in favor of the dynamo theory of magnetism.

How has the magnetic field acted in the distant past? Before we state the evidence from the rock record on this point, let us briefly examine the ways in which the magnetic field affects rocks and minerals. Although the effects are weak, they create a vital record.

Effect of the Magnetic Field on Rocks. A few minerals are natural magnets, so the rocks in which they occur can become magnetized. Iron is an essential element in all magnetic minerals, but not all iron minerals are magnetic; the important magnetic ones are *magnetite* (Fe_3O_4), *hematite* (Fe_2O_3), *ilmenite* ($FeTiO_3$), and *pyrrhotite* (FeS). If all the tiny mineral magnets in a rock are randomly fixed in space, their effects cancel each other out, so that the rock as a whole is not magnetic. However, if all or most of the tiny magnets are parallel, their effects reinforce each other, and this makes the rock magnetic. How, then, do mineral magnets acquire parallel orientation?

The manner of magnetic orientation depends on the kind of rock. Each magnetic mineral has a ***Curie point,*** a *temperature above which all magnetism is destroyed.* As an igneous rock cools below the Curie point, any mineral magnets that are present acquire the magnetic orientation of Earth's field. The igneous rock thereby becomes weakly magnetic. When a sediment forms, an entirely different process occurs. Sediment grains that are magnetic do the same as a compass needle—they tend to orient themselves parallel to Earth's field as they settle through water. The oriented mineral magnets eventually make the sedimentary rock a weak magnet.

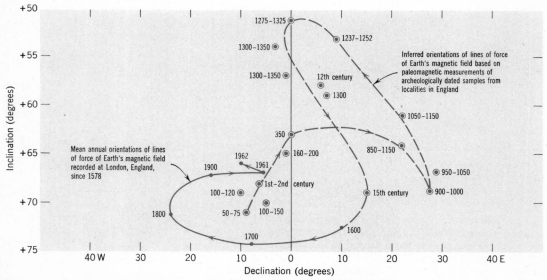

Figure 3.14 Variations in the declination and inclination of Earth's magnetic field in the vicinity of London. Direct measurements made in London are shown on the solid blue curve, inferred positions from historical and archeological data are shown on the dashed blue curve. (*After Keith Runcorn, 1964.*)

Although magnetic strength in rocks is very weak, it can be measured accurately. Also, rock magnetism is stable—it normally does not tend to change with time. If, therefore, we collect an igneous or sedimentary rock sample and determine the direction of its magnetism, we can identify the declination and inclination of Earth's magnetic field at the time the rock formed, provided the rock has not changed its position.

Magnetic Variations in the Rock Record. Careful measurements of ancient igneous and sedimentary rock prove that a magnetic field has existed on Earth for as long as rock has been forming. Magnetism in rock is called *paleomagnetism*. A very large number of paleomagnetic measurements have now been made. They reveal two extraordinary effects.

The first effect is that the positions of the magnetic poles seem to have varied much more than is suggested by Figure 3.14; in fact, they appear to have wandered all over the Earth. Not only do the indicated positions of the poles wander, but when we examine the paleomagnetic measurements from different continents, we find that the magnetic poles seem to have been in different places at the same time (Fig. 3.15). Clearly this is not possible. How, then, can we explain the data in Figure 3.15? Although the pole positions have varied in the past, they probably have not varied much more than as shown in Figure 3.14. If the poles have not varied greatly, then we must be suspicious about the measurements. Have the rocks been moving, and are they indicating false pole positions? This indeed seems to be the case. We noted in Chapter 1 that continents move when plates of lithosphere move. Ancient rocks may not, therefore, be in the same positions, relative to the magnetic poles, that they occupied when they became magnetized. The paleomagnetic measurements shown in Figure 3.15 can be viewed as records of continents moving relative to magnetic poles rather than a record of magnetic poles moving relative to fixed continents.

The second effect is even more remarkable than the first. We find many instances in the rock record when the positions of the magnetic poles were reversed—that is, when the north magnetic pole was in the position of the present south magnetic pole. Scientists have been unable to explain how this happened, or even to predict the sorts of fluid motions in the core that might cause it. We cannot even

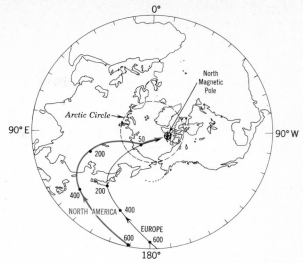

Figure 3.15 Curves tracing the inferred path followed by the north magnetic pole for the past 600 million years. Numbers are millions of years before the present. The curve determined from paleomagnetic measurements in North America (blue curve) differs from that determined by measurements made in Europe (black curve). A wide-ranging move is unlikely for the pole; a more widely accepted explanation is that the continents have moved. Divergence of the North American and European curves indicates that the two continents have at times moved independently. (*After Northrop and Meyerhoff, 1963.*)

predict when another pole reversal may occur, but reversal has evidently happened many times in the past. During the past 4 million years, for example, there have been 9 periods when the magnetic field was reversed (Fig. 3.16).

When the magnetic poles reverse themselves, they might do so in either of two ways. First, the poles can move and the whole magnetic field can move with them. Second, the field can die down to nothing, then start up again with the poles reversed. It ap-

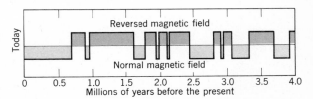

Figure 3.16 During the past 4 million years, Earth's magnetic field has experienced 9 periods when the polarity was exactly the reverse of the present polarity. (*Modified after Cox, 1969.*)

pears to be the second way that commonly occurs. Perhaps fluid motions in the core cease for a while; perhaps different motions somehow cancel each other out. For whatever reason, the magnetic field apparently just weakens and dies down, but maintains its north-south orientation as it does so. When it starts up again it has the same north-south orientation but the poles are reversed. Even though we don't understand this strange phenomenon, we can gather from it evidence that is useful in deciphering some of Earth's history. How this is done is discussed in Chapter 18.

The Magnetic Shield Around Planet Earth. Although we don't think of the magnetic lines of force as a protective shield, that is what they are, for they are essential to our well-being. The Sun constantly gives off streams of ionized particles, mostly protons. The stream is so intense that it has acquired the name *solar wind*, and at times it assumes proportions that could be lethal to anyone exposed to its particles. The shield that protects us from this barrage is Earth's magnetism.

The magnetic lines of force that surround Planet Earth extend far out into space. Meanwhile the ionized particles that stream from the Sun create magnetic fields of their own. These distort Earth's field, compressing it on the side facing the Sun and causing it to stream out in the opposite direction. When the particles crash into the magnetic shield at high velocities, they create a shock wave in space (Fig. 3.17), but they do not breach the shield. Most of the ionized particles are simply deflected *around* Earth. A few particles penetrate part of the shield and are then trapped in two radiation belts, the *Van Allen belts*. Eventually particles in the radiation belts leak out and continue their passages through space.

What happens to the magnetic shield when Earth's field reverses itself? During the short period of perhaps a few thousand years, while the field dies down and starts again in the reversed direction, the magnetic shield is apparently missing. Particles from the Sun would not be deflected, they would smash into the Earth. People have suggested that the stream of solar particles that strike Earth during reversals of the field might account for why some animals, particularly microscopic ones, have become extinct. They have suggested also that genetic alterations might occur in surviving populations during the period. We still seek answers to these suggestions, and perhaps we might some day find that Earth's magnetic field affects us all.

Synopsis

Earth's surface is a great field of conflict where activities driven by the energy of internal heat are in continual combat with external activities driven by solar and tidal energy. Continually involved in the energy battle are the two forces, gravity and magnetism. Gravity gives Earth its nearly round shape and is continually pulling material down to the lowest possible level. Gravity and Earth's exter-

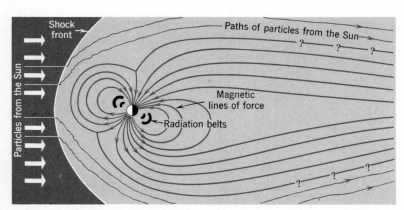

Figure 3.17 Section through Earth and its magnetic field. The stream of ionized particles from the Sun distorts the magnetic field and creates a shock wave. Particles that penetrate the magnetic field become trapped in the radiation belts.

nal activities would soon turn Earth into a smooth, leveled planet if internal activities did not combat them by continually thrusting new mountains up. Magnetism is a force, although it doesn't move material around in the way that gravity does. Magnetism provides a protective shield against the streams of dangerous ionized particles given off by the Sun.

It also leaves a weak but vital record by which we can tell where on Earth a rock was formed.

The energy battle, gravity, and magnetism have been molding Earth for a long time. Before we discuss Earth's activities, we will briefly examine, in Chapters 4 and 5, the products of the energy battle: minerals and rocks.

Summary

Energy

1. Energy is the capacity to produce activity. It appears in many forms: kinetic-, heat-, chemical-, electrical-, radiant-, and atomic energy.

2. Energy can be converted from one form to another. All of Earth's activities involve the conversion of energy.

3. Energy reaches Earth's surface from three sources. Radiant energy arrives from the Sun. Tidal energy comes from the kinetic energy of motions of Earth and Moon. Natural radioactive decay produces internal heat energy, which flows to the surface.

4. Sun, tides, and internal heat have been providing energy to Earth for as long as the rock record extends back in time.

Gravity

5. Gravity is a universal force by which bodies attract each other. When a body is as large as Earth, the force of attraction is strong.

6. All particles on Earth are pulled toward the center along a radial path.

7. Earth's shape is a slightly flattened sphere; it arises from opposition between the force of gravity and the centrifugal force of Earth's rotation.

Magnetism

8. Earth's magnetic field is caused by fluid motions in Earth's outer core.

9. The position of the magnetic field varies with time.

10. The magnetic field has a north and a south pole; at times in the past, the field has been alternately reversed and normal (as it is today).

11. The magnetic field acts as a giant shield that protects us from streams of ionized particles given off by the Sun.

Selected References

Clark, S. P., Jr., 1971, Structure of the Earth: Englewood Cliffs, New Jersey, Prentice-Hall.

Cox, A., Dalrymple, G. G., and Doell, D. R., 1967, Reversals of the earth's magnetic field: Scientific American, v. 216, no. 2, p. 44–54.

Cox, A., and Doell, D. R., 1960, Review of paleomagnetism: Geol. Soc. America Bull., v. 71, p. 645–768.

Garland, G. D., 1965, The Earth's shape and gravity: New York, Pergamon Press.

Heiskanen, W. A., and Vening Meinesz, F. A., 1958, The Earth and its gravity field: New York, McGraw-Hill.

Irving, E., 1964, Paleomagnetism and its application to geological and geophysical problems: New York, John Wiley.

Lee, W. H. K. (ed.), 1965, Terrestrial heat flow: Geophysical monograph No. 8, Washington, D. C., American Geophysical Union.

Scientific American, 1971, Energy and power, v. 224, no. 3, p. 36–200 (special issue consisting of 11 articles devoted to energy).

Strahler, A. N., 1969, Physical geography: New York, John Wiley, chaps. 6, 7, 8.

Chapter 4

Earth's Basic Constituents

The Building Blocks

Walk outside and pick up a rock. If you can't find a rock, pick up some sand or gravel, or even a handful of soil. You will be holding a handful of minerals, the building blocks of the Earth. Wherever you look you see minerals, and whatever you do you use products made from minerals. Most minerals we see are common and have little value. Some, like diamonds and rubies, are extremely valuable because they are rare and greatly prized for their beauty. Other minerals supply the raw materials for industry and national wealth. Empires have been won and lost in the search for useful minerals, and powerful countries have collapsed when their deposits of valuable minerals became exhausted. The Romans conquered most of Eurpoe and the Near East in their search for minerals containing copper, gold, tin, iron, silver, and lead. They built one of the most remarkable empires the world has ever seen, largely by using the mineral wealth they found. When the mineral deposits became exhausted, or were captured by local tribes, Rome was deprived of its great sources of wealth. Partly for this reason its empire slowly died.

The word *mineral* has many meanings and uses. We can read advertisements for plant foods that provide "minerals" for plant growth, and for vitamins that provide "minerals" for the growth of our bodies. We will use the word mineral in a more restricted way, and in order to do so we will give it an exact definition. Before attempting a definition, however, we must examine the two most important characteristics of minerals: composition and structure. And because all minerals are made up of one or more kinds of atoms, it is helpful to discuss atoms as well. Appendix A contains a more detailed discussion of atoms; here we mention only the essential points needed for our present discussion.

Atoms and Elements. If we ask a chemist to analyze a rock or any other piece of matter, including liquids and gases as well as solids, he will report the kinds and amounts of the **chemical elements** present, because the chemical elements are *the most fundamental substances into which matter can be separated by chemical means.* At present there are 103 of them. Each is separately named and identified by a symbol, such as H for hydrogen, Ag for silver, and E for einsteinium. The known elements and their symbols are listed in Appendix A.

If we ask what an element is made of, the chemist will reply that each element he analyzes consists of a large number, or mass, of identical particles called atoms. An **atom** is *the smallest individual particle that retains all the properties of a given chemical element.* We speak of an atom of hydrogen or an atom of lead, but we cannot see one because it is too small. When we handle a pure chemical element we are seeing instead an aggregation of a vast number of identical atoms. A cube of pure silver 1cm on edge contains 58×10^{21} atoms, a number so large that its significance is almost impossible to grasp. A faint idea of its magnitude may be conceived by imagining that we had somehow spread the cubic centimeter of silver thinly and evenly over the face of the entire Earth. Each square centimeter of Earth's surface would then be covered by approximately 10,000 atoms.

Rocks, minerals, and all other forms of matter on Earth are composed of atoms, but atoms in turn are built up from still smaller sub-atomic particles. The principal sub-atomic particles are *protons* (which have positive electrical charges), *neutrons* (which are electrically neutral), and *electrons* (which have negative electrical charges that balance exactly the positive charges of protons). Protons and neutrons are dense but very tiny particles and they join together to form the core or nucleus of an atom. Protons give a nucleus a positive charge, and we call *the number of protons in the nucleus of an atom* the **atomic number.** Electrons are even tinier particles; they move, like a distant and diffuse cloud, in orbits around the nucleus.

Elements are built systematically, beginning with hydrogen, which has one proton and one electron, then helium, with two protons and two electrons, and so on. A new element is formed each time another proton is added to the nucleus. As a consequence, another electron is added to balance the electrical charge.

All atoms having the same atomic number are atoms of the same element and therefore have the same chemical properties. The number of neutrons that join with the protons in a nucleus can vary, within small limits, without affecting the chemical properties. Any chemical element may therefore have several **isotopes,** *atoms having the same atomic number but differing numbers of neutrons.* For some elements as many as ten isotopes have been discovered. But not all combinations of protons and neutrons are completely stable, so that some isotopes break up spontaneously. In the process energy is emitted and new isotopes and new elements are formed. *The decay process by which an unstable atomic nucleus spontaneously disintegrates is* **radioactivity.** As we learned in Chapter 2, the rate at which any radioactive isotope decays is constant, and can be used as a sort of clock.

Compounds and Ions. A few minerals, such as gold and platinum, contain only atoms of one element. Most minerals are groupings of atoms of several elements. *A combination of atoms of different elements* is called a **compound.** The reason atoms combine, or *bond* together, is discussed more fully in Appendix A; combination depends on atoms transferring and sharing orbiting electrons. *An atom that has excess positive or negative charges caused by electron transfers* is called an **ion.** When the charge is positive (meaning that the atom gives away electrons) the ion is called a *cation;* when negative, an *anion.* The convenient way to indicate ionic charges is to record them as superscripts. For example, Ca^{+2} is a cation, while S^{-2} is an anion. Compounds contain one or more elements that are cations and one or more elements that are anions; for a compound to be stable, the sum of the positive charges on the cations and the negative charges on the anion must equal zero.

Some atoms form such strong bonds that they seem to act as a single atom. A strongly bonded pair is said to form a complex ion. Complex ions act in the same way as single ions, forming compounds by bonding with other elements. For example, carbon and oxygen combine to form the very stable carbonate anion $(CO_3)^{-2}$. Other examples of important complex anions are the sulfate $(SO_4)^{-2}$, nitrate $(NO_3)^{-1}$ and silicate $(SiO_4)^{-4}$ groups.

Two broad classes of compounds are recognized. *Organic compounds* are made from carbon and hydrogen, with or without other elements such as nitrogen and oxygen. Organic compounds can form by direct combination of carbon and hydrogen, but most come directly or indirectly from the activities of living organisms. Mixtures of organic compounds are called organic matter. All other matter is said to be *inorganic* and its compounds are *inorganic compounds*. All minerals are inorganic. A few, such as gold and silver, are chemical elements, but all the rest are inorganic compounds. We have therefore defined one property of a mineral. But one property alone does not define a mineral. We must consider other properties as well.

Minerals

Composition. Approximately 2200 minerals are known, all of which have been found in the crust, because that is the only part of Planet Earth that is accessible to us. In view of the large number of known chemical elements, the number of inorganic compounds it is theoretically possible to form is enormous. But the number of inorganic compounds that occur naturally is limited, because only a few chemical elements comprise the great bulk of the crust (Table 4.1).

Looking at Table 4.1, we see that two elements, oxygen and silicon, make up more than 70 per cent of the crust. Oxygen forms a simple anion, O^{-2}, and

Table 4.1
The Most Abundant Elements in the Continental Crust (After K. K. Turekian, 1969)

Element	Weight %
Oxygen (O)	45.2
Silicon (Si)	27.2
Aluminum (Al)	8.0
Iron (Fe)	5.8
Calcium (Ca)	5.1
Magnesium (Mg)	2.8
Sodium (Na)	2.3
Potassium (K)	1.7
Titanium (Ti)	0.9
All other elements	1.0
Total	100.0

silicon and oxygen together form the complex ion, $(SiO_4)^{-4}$, commonly called the *silicate anion*. In view of the abundances of silicon and oxygen, it is not surprising that silicate minerals are the most common naturally occurring inorganic compounds, and that oxides are the next most abundant. Other common natural compounds, although much less common than silicates and oxides, are sulfides, chlorides, carbonates, sulfates, and phosphates.

Structure. Minerals are solids. Whereas atoms in gases and liquids are randomly jumbled, in most solids the atoms are organized in regular geometric patterns, like eggs in a carton. *The geometric pattern that atoms assume in a mineral* is called **crystal structure.** The crystal structure of a mineral is a unique property, and all specimens of a given mineral have identical structures.

The packing of atoms in the mineral *galena*, PbS, the most common lead mineral, is shown in Figure 4.1. Notice that the sulfur atoms are larger than the lead atoms. Now anions tend to be large; cations tend to be small. The crystal structures of minerals, therefore, are largely determined by the packing arrangements of anions. The radii of some common ions are shown in Figure 4.2 and are given in *Angstroms* (abbreviated *A*), a unit of length used for atomic measurements.

It is apparent from Figure 4.2 that some ions have the same electrical charge and are nearly alike in size; for example Fe^{+2} and Mg^{+2} have radii of 0.83A and 0.78A respectively. Because of their similarity in size and charge, ions of Fe^{+2} are often found substituting for ions of Mg^{+2} in magnesium-bearing minerals. The *structures* of the magnesium minerals are not changed as a result of the substitution, but of course the *composition* of the mineral is affected. *The substitution of one atom for another in a random fashion throughout a crystal structure* is known as the principle of **solid solution.** There is a special way of depicting solid solutions in chemical formulas. When Fe substitutes for Mg in the mineral olivine Mg_2SiO_4, for example, we simply write the formula $(Mg,Fe)_2SiO_4$, which indicates that the Fe substitutes for Mg but not for any other atoms in the structure. Variations in mineral composition caused by solid solution are not usually large, but, as we shall see, they are important in the formation of common minerals and rocks.

Each mineral has a unique crystal structure, but

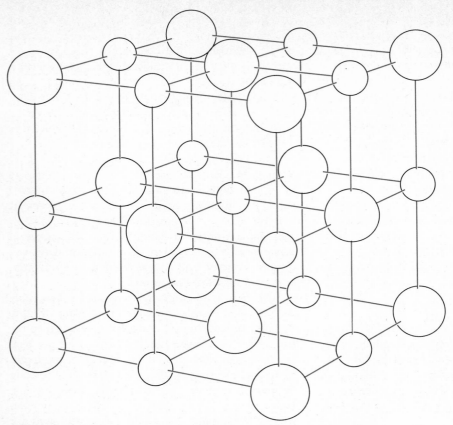

Figure 4.1 Arrangement of atoms in galena, PbS, the common lead mineral. Pb is a cation with a charge of +2, S is an anion, charge −2. Therefore, for every Pb atom in the structure, there must be one S atom. The atoms are shown pulled apart along the blue lines, so that we can see how they fit together.

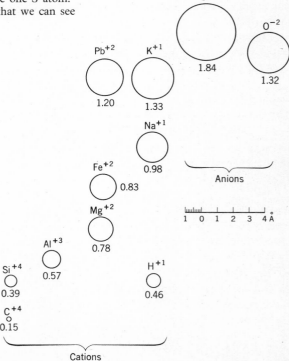

Figure 4.2 Radii of ions of 11 common elements range from C^{+4} at lower left to O^{-2} at upper right. Ions are arranged in vertical groups based on charge, from +4 at left to −2 at right. Ions in each of the pairs Si^{+4} and Al^{+3}, Mg^{+2} and Fe^{+2}, and Na^{+1} and Ca^{+2} are about the same size and commonly substitute for each other in crystal lattices. Radii expressed in Angstroms (A), where one Angstrom equals 10^{-8}cm. (*Data from Rankama and Sahama, 1950.*)

Table 4.2

Minerals with Identical Compositions but Different Crystal Structures Are Called Polymorphs. Some Well-known Polymorphs Are Listed Below, with the Most Common Variety Listed First

Compound	Mineral Name
C	Graphite
	Diamond
$CaCO_3$	Calcite
	Aragonite
FeS_2	Pyrite
	Marcasite
SiO_2	Quartz
	Cristobalite
	Tridymite
	Coesite
	Stishovite

some compounds are known to form two or more different minerals. The compound $CaCO_3$, for example, forms two different minerals. One is *calcite,* which is the mineral of which marble is composed; the other is *aragonite,* which is most commonly found in the shells of clams, oysters, snails, and other aquatic life. Calcite and aragonite have identical *compositions,* but entirely different *crystal structures. A compound that occurs in more than one crystal form is called a **polymorph.*** Some common polymorphs are listed in Table 4.2.

Definition. Minerals are inorganic compounds, and each mineral has a unique crystal structure. We can now give an exact definition: ***minerals*** are *all naturally occurring, crystalline, inorganic compounds.* The definition does not include all naturally occurring solid compounds. Some substances lack a systematic arrangement of atoms and do not have fixed compositions; they are called *mineraloids* and because they do not have crystal structures, they are said to be *amorphous* ("without form").

Properties. The properties of minerals are determined by their compositions and their crystal structures. Once we know which properties are characteristic of which minerals, we can use the properties to identify the minerals. It is not necessary, therefore, to analyze a mineral chemically and determine its crystal structure in order to identify most com-

mon ones. The characteristics most often used in identifying minerals are the obvious physical properties, such as color, shape of crystal, and hardness, plus some less obvious properties, such as cleavage and specific gravity. Each property is discussed below and, where that is helpful, further detailed in Appendix B, which also contains a table of the common minerals together with the most characteristic physical properties used in their identification.

Crystal Form and Habit. When a mineral grows freely, without obstruction from adjacent minerals, it forms a characteristic *geometric solid that is bounded by symmetrically arranged plane surfaces.* The characteristic solid is called the **crystal form** of a mineral. The plane faces of a crystal are an external expression of the strict, internal geometric arrangement of the constituent atoms. Each plane surface corresponds to a plane of atoms in the crystal structure. Unfortunately, crystals do not grow commonly in open, unobstructed spaces. Well-formed crystals are therefore rare. When nice crystals are found, however, an examination of the crystal form tells much about the crystal structure and immediately aids us in identification.

The sizes of individual crystal faces differ. Under some circumstances a mineral may grow a long, thin crystal; under others, a short, fat one. Superficially the two crystals may look very different; however, the unique characteristic of crystals is not the relative sizes of the individual crystal faces, but the angles between the faces. The angle between any two adjacent crystal faces in a mineral is a constant and is the same for all specimens of the mineral. Two crystals of quartz (SiO_2) are shown in Figure 4.3. One is flattened, the other elongate, but it is clear that the same sets of crystal faces occur on both minerals. It is also clear, however, that on the two crystals the sets of faces are parallel and therefore that the angle between any two equivalent faces must be the same on each crystal.

Every mineral has a unique crystal form. Some minerals also form crystals with distinctive shapes. For example, the mineral pyrite (FeS_2) is commonly found as intergrown cubes (Fig. 4.4) with markedly striated faces, while the mineral stibnite (Sb_2S_3) almost invariably forms long, needle-like crystals (Fig. 4.4).

As noted earlier, most minerals do not grow freely into open spaces and therefore do not develop well-

Figure 4.3 Two quartz crystals with the same crystal forms. Although the size of the individual faces differ markedly between the two crystals, it is clear that each face on one crystal is parallel to an equivalent face on the other crystal. The angles between adjacent faces are identical for all crystals of the same mineral. (*Yale Peabody Museum.*)

shaped crystals. Instead, the growing minerals usually encounter other minerals and obstructions that prevent the development of crystal faces. Usually, then, we cannot use crystal form to identify minerals, but nevertheless we can sometimes make use of

distinctive growth habits to aid in identification. For example, Figure 4.5 shows asbestos, a variety of the mineral serpentine that characteristically grows as fine, elongate threads. Figure 4.5 also shows psilomelane, a common manganese oxide that forms smooth, rounded surfaces, usually referred to as being *botryoidal*.

Cleavage. If we break a mineral specimen with a hammer, or drop the specimen on the floor so that it shatters, the broken fragments are seen to be bounded by smooth, plane faces, so that the fragments resemble small crystals. A closer look shows that all fragments break along similar planes. *The tendency of a mineral to break in preferred directions along plane surfaces* is called **cleavage.** The plane surfaces along which cleavage occurs are governed by the crystal structure (Fig. 4.6). They are planes along which the bonds between atoms are relatively weak. Because the cleavage planes are direct expressions of the crystal structure, they are valuable guides for the identification of minerals.

Many minerals have distinctive cleavage planes. One of the most distinctive is found in the mineral *muscovite* (Fig. 4.7). Clay minerals also have distinctive cleavage (Fig. 4.7) and it is this easy cleavage direction that makes them feel smooth and slippery when rubbed between the fingers. Another mineral

A   *B*

Figure 4.4 Crystals commonly have distinctive habits. *A.* Pyrite (FeS_2) is often found as intergrown cubic crystals with striated faces. *B.* Stibnite (Sb_2S_3) almost invariably occurs as long, thin needle-like crystals. The white, tabular-shaped crystals intergrown with the needles are calcite ($CaCO_3$). (*A, B. M. Shaub; B, Yale Peabody Museum.*)

A

B

Figure 4.5 Some minerals have distinctive growth habits even though they do not develop well-formed crystal faces. *A.* Asbestos, a variety of the mineral serpentine, grows as fine, cotton-like threads that can be separated and woven into fireproof fabric. *B.* Psilomelane, a manganese oxide, commonly forms botryoidal surfaces. This form is also common in various iron oxide minerals. (*A, B. M. Shaub.*)

with a highly distinctive cleavage is calcite, which breaks into perfect rhombs (Fig. 4.8).

Not all minerals have distinctive cleavages, and a few minerals lack cleavage planes altogether and thus are distinctive in the opposite sense. Quartz (Fig. 4.9), a common mineral that lacks cleavage

planes, breaks along irregular fracture surfaces and can be easily distinguished from other similar-looking minerals by the fractures.

Color. As we pick up a mineral in order to identify it, we first of all notice its color. Unfortunately,

A

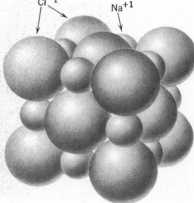

B

Figure 4.6 *A.* The mineral halite (NaCl) has a set of well-defined cleavage planes causing it to break into fragments bounded by perpendicular faces. *B.* A representation of the crystal structure, drawn in the same orientation as the cleavage fragments in *A*, shows that the plane of breakage is parallel to plane in the crystal in which sodium and chlorine atoms occur in equal numbers. (*A, B. M. Shaub.*)

0   2 cm

A

0   0.1 μ

B

Figure 4.7 *A.* Perfect cleavage of the mica mineral, muscovite shown by very thin, plane flakes into which this six-sided crystal has been split. The cleavage flakes suggest leaves of a book, a resemblance embodied in the name "books of mica" for crystals elongated in a direction perpendicular to the cleavage flakes. (*Ward's Natural Science Establishment.*) *B.* Electron micrographs (pictures made with an electron microscope) of the clay mineral *dickite* (hydrous aluminum silicate). Crystal enlarged 37,500 times, to show prominent horizontal cleavage planes. (*T. F. Bates.*)

color is the one obvious property of minerals that is not always reliable for identification. Color in minerals is determined by their composition, and, as we have seen, solid solution causes the compositions of minerals to vary within small ranges. Some elements can create strong color effects even when they are present in very small amounts. For example (Fig. 4.10) the mineral corundum (Al_2O_3) is commonly white or grayish in color, but when small amounts of Cr have replaced Al by solid solution, this mineral is blood red, forming *ruby*, a prized gem variety of corundum. Similarly, when small amounts

Figure 4.8 Perfect rhombs of *calcite* (calcium carbonate) formed by cleavage planes in three directions. (*Ward's Natural Science Establishment.*)

Figure 4.9 Irregular fracture of quartz. Curved fracture surfaces at end of quartz crystal from Arkansas. (*B. M. Shaub.*)

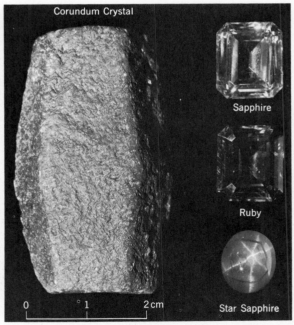

Figure 4.10 Corundum (Al_2O_3) commonly occurs as barrel-shaped, six-sided crystals with a variety of colors. Transparent crystals of corundum are the familiar gemstones ruby (red) and sapphire (blue). Sometimes tiny inclusions of other minerals are trapped within a corundum crystal. In star sapphires the inclusions are arranged parallel to the six crystal faces and disperse light so as to form a striking, six-pointed star. (*Yale Peabody Museum.*)

Table 4.3
Scale of Hardness

Relative Number in the Scale	Mineral	Hardness of Reference Objects
10.	Diamond	
9.	Corundum	
8.	Topaz	
7.	Quartz	
6.	Orthoclase	Pocket knife; glass
5.	Apatite	
4.	Fluorite	
3.	Calcite	Copper penny
2.	Gypsum	Fingernail
1.	Talc	

(Decreasing ↓)

of Fe and Ti are present, the corundum is deep blue and another prized gemstone, *sapphire,* is the result.

Hardness. *Relative resistance of a mineral to scratching* is **hardness,** another distinctive property of minerals. Hardness, like crystal form and cleavage, is governed by crystal structure and by the strength of the bonds between atoms. The stronger the bonds, the harder the mineral. Degree of hardness can be decided in a relative fashion by determining the ease or difficulty with which one mineral will scratch another. Talc, the basic ingredient of most body ("talcum") powders, is the softest mineral known, and diamond the hardest. A relative hardness scale between talc (number 1) and diamond (number 10) is divided into ten steps, each marked by one of ten common minerals (Table 4.3). The steps do not represent equal intervals of hardness, but the important feature of the hardness scale is that any mineral on the scale will scratch all minerals below it. Minerals on the same step of the scale are just capable of scratching each other. For convenience we often test relative hardness by using a common object such as a penny, or a penknife, as the scratching instrument.

Density. The final obvious physical property of a mineral is its density, which in practical terms means how heavy it feels. We know that equal-sized baskets of feathers and of rocks have different weights: feathers are light, rocks are heavy. The property that causes this difference is **density,** or *the average weight per unit volume.* The units of density are numbers of grams per cubic centimeter. Minerals

with a high density, such as gold, have their atoms closely packed. Minerals with low density, such as ice, have loosely packed atoms.

Minerals are divided into a heaviness or density scale. Gold has the highest density of all minerals, 19.3gms per cubic centimeter, but many others such as galena (7.5), magnetite (5.2), and hematite (5.3) feel heavy by comparison with most silicate minerals, which have densities between 2.5 and 3.0.

The Rock-Forming Minerals. A few minerals—fewer than 20 kinds—are so common that they account for more than 95 per cent of the crust. These are the *rock-forming minerals,* so called because all rocks contain one or more of them. We have already seen that silicates and oxides are the most common minerals of the crust, but that silicates are much more abundant than oxides. Most rock-forming minerals, therefore, are silicates.

The Silicates. Silicon and oxygen form the complex anion $(SiO_4)^{-4}$. The four oxygen atoms in the complex anion are tightly bound to the single silicon atom. Oxygen is a large ion (Fig. 4.2), while silicon is a small ion. The oxygens pack into the smallest packing space possible for four large spheres. As can be seen in Figure 4.11, the four oxygens sit at the corners of a tetrahedron—a pyramid—and the small silicon sits at the center of the tetrahedron. The shape of the complex silicate anion is therefore a

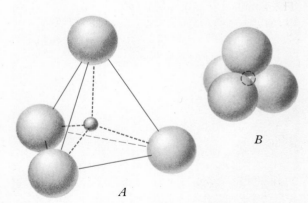

Figure 4.11 Silica tetrahedron. *A.* Expanded view showing large oxygen ions at the four corners, equidistant from a small silicon ion. Dotted lines show bonds between silicon and oxygen ions: solid lines outline the tetrahedron. *B.* Tetrahedron with oxygen ions touching each other in natural positions. Silicon ion (dashed circle) occupies central space.

tetrahedron and the structures and properties of silicate minerals are determined by the manner in which silica tetrahedra pack together.

Silica tetrahedra combine to form compounds in two ways. First, the oxygens of the tetrahedra can form bonds with unlike atoms, so that the tetrahedra are simply acting like ordinary anions, isolated from each other in the structure and surrounded by cations. An example of this is found in olivine, in which two Mg^{+2} atoms satisfy the charges of each isolated $(SiO_4)^{-4}$ tetrahedron, yielding the formula Mg_2SiO_4. The second, and entirely different way, is for two adjacent tetrahedra to share an oxygen. As a consequence, two or more tetrahedra are bound into larger anion units in the same way that beads are joined to form a necklace. *The process of linking silica tetrahedra into larger groups* is called a **polymerization.** The simplest case of polymerization is that of two tetrahedra sharing a single oxygen atom. As Figure 4.12 shows, a large anion $(Si_2O_7)^{-6}$ results. There is one rare mineral that contains the $(Si_2O_7)^{-6}$ anion, but unfortunately this simplest case of polymerization does not occur in any common minerals.

All the common silicate minerals have more complicated polymerizations. When a tetrahedron shares more than one oxygen with adjacent tetrahedra, large groupings, and even endless chains, sheets, and networks of tetrahedra can be formed. The common polymerizations, together with the rock-forming minerals containing them, are shown in Figure 4.13. The bonds between silicon and oxygen atoms are exceedingly strong. The physical properties of sili-

cate minerals therefore strongly reflect the way tetrahedra are packed in the crystal structure. Cleavage, for example, always occurs in a direction that avoids disruption of a tetrahedron, because that would involve breaking a silicon/oxygen bond. Quartz lacks cleavage planes because its entire crystal structure is formed by a three-dimensional polymerization of silica tetrahedra. All directions of breakage are therefore equally difficult.

Now that we have discussed the principles that determine the structure and properties of silicate minerals, we are in a position to discuss the common varieties of silicate minerals. We shall treat them in order of increasing complexity of polymerization.

Olivine. Two important minerals have crystal structures containing isolated silica tetrahedra. The first is a glassy-looking mineral, usually pale green in color, called *olivine.* Actually olivine is a family of minerals because Fe and Mg substitute for each other in solid solution, giving a general formula $(Mg,Fe)_2SiO_4$. Olivine sometimes occurs in such flawless and beautiful crystals that it can be used as a gem, the gem *peridot.* Most commonly, however, it occurs in irregularly shaped grains in igneous rock that was formed by cooling of magma from the mantle.

Garnet. The second important mineral with isolated silica tetrahedra is *garnet.* As with olivine, garnet is actually the name of a family of minerals. The garnets have the complex formula $A_3B_2(SiO_4)_3$ where A can be the cations Mg^{+2}, Fe^{+2}, Ca^{+2}, and Mn^{+2}, or any solid solution mixture of them, while B can be either of the triply charged cations Al^{+3} or Fe^{+3}. Garnets are more widespread than olivines, being characteristic of both mantle and crustal rocks. One of the most characteristic features of garnets is their tendency to form beautiful crystals (Fig. 4.14). The iron-rich garnet, almandine, is deep red and is well known as a gemstone. Another useful property of the garnets is their hardness, which makes them useful as abrasives for grinding and polishing.

Pyroxene and Amphibole. The *pyroxenes* and *amphiboles* are two families that have continuous chains of silica tetrahedra. They differ in that pyroxenes are built from a polymerized chain of single tetrahedra, each of which shares two oxygens, while the amphiboles are built from double chains of tetra-

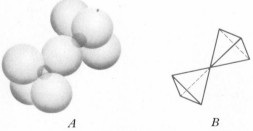

A   *B*

Figure 4.12 Two silica tetrahedra can share an oxygen, thereby satisfying some of the unbalanced electrical charges, and, in the process, forming a larger and more complex anion group. *A.* The arrangement of oxygen and silicon atoms in double tetrahedra, giving the anion $(Si_2O_7)^{-6}$. *B.* A geometrical representation of the silica tetrahedra, drawn so that an oxygen atom would sit at each apex and a silicon atom at the center of each tetrahedron.

	Arrangement of silica tetrahedra	Formula of the complex anions	Typical mineral
Isolated tetrahedra		SiO_4	The olivine family
Isolated polymerized groups		$(Si_2O_7)^{-6}$	Lawsonite
	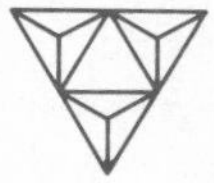	$(Si_3O_9)^{-6}$	Wollastonite
	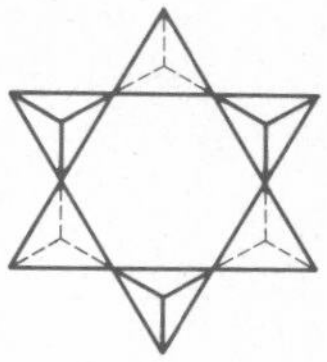	$(Si_6O_{18})^{-12}$	Beryl
Continuous chains	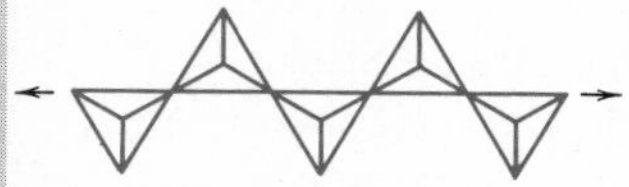	$(SiO_3)_n^{-2}$	The pyroxene family
	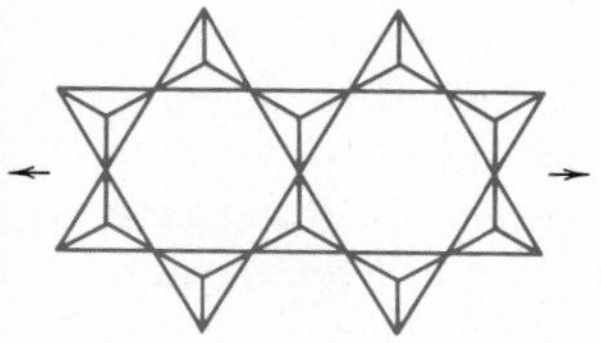	$(Si_4O_{11})_n^{-6}$	The amphibole family
Continuous sheets	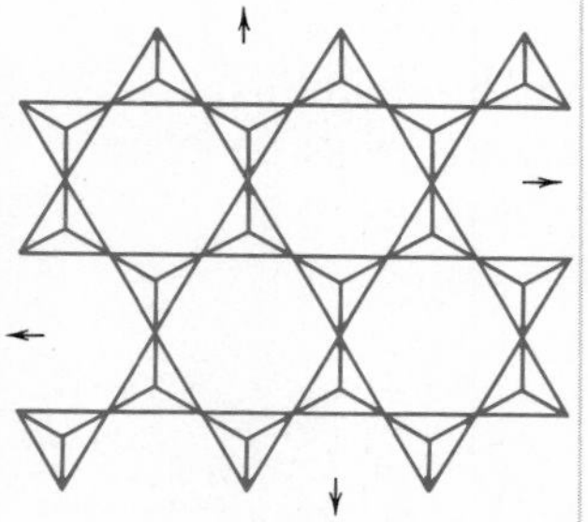	$(Si_4O_{10})_n^{-4}$	The mica family
Three-dimensional networks	Too complex to be shown by a simple two-dimensional drawing	(SiO_2)	Quartz

Figure 4.13 The way in which silica tetrahedra polymerize by sharing oxygens determines the structures and compositions of the rock-forming silicate minerals. Polymerizations other than those shown below are theoretically possible, but have not yet been found in minerals.

Figure 4.14 Crystals of garnet have a characteristic shape, bounded by four-sided crystal faces. This striking group of crystals was found near Russell, Massachusetts. (*Yale Peabody Museum.*)

hedra, equivalent to two pyroxene chains in which half the tetrahedra share two oxygens and the other half share three oxygens. These relations can be clearly seen in Figure 4.13.

Minerals in the pyroxene and amphibole families look somewhat alike because their building blocks are similar. But the arrangement and shapes of the single and double chains of tetrahedra lead to a very distinctive difference of property. Both have pronounced cleavages produced by breakages parallel to the silica chains. Because of the shapes of chains, however, the angle between the cleavage surfaces is almost 90° in pyroxenes but only 56° in amphiboles (Fig. 4.15).

The general formula for pyroxenes is A.B. $(SiO_3)_2$, where A and B can be any of the cations Mg^{+2}, Fe^{+2}, Ca^{+2}, and Mn^{+2}. The pyroxenes are most abundantly found in rocks of the mantle, but are common also in many rocks of the crust. The most common pyroxene mineral is a shiny black species called *augite.*

Amphiboles are perhaps the most complicated family of all silicate minerals, with the general formula $A_2B_5(Si_4O_{11})_2(OH)_2$, where A can be either Ca^{+2} or Mg^{+2}, and B can be Mg^{+2} or Fe^{+2}. The amphiboles are characteristically found in crustal rocks but are also known from some mantle rocks. The most abundant mineral species is *hornblende,* a dark green to black mineral that looks very like augite.

Mica and Clay. Members of the *mica* and *clay* families have one marked property in common: because their basic building unit is a polymerized sheet of silica tetrahedra, they have one perfect cleavage parallel to the sheet, as can be seen in Figure 4.7.

The electrical charges in the sheets are balanced by Al^{+3} cations in clays, leading to the formula $Al_4Si_4O_{10}(OH)_8$ for the clay mineral *kaolinite.* With the micas, however, a new principle must be mentioned to explain their compositions. Al^{+3} cations are only a little larger than Si^{+4} cations, and under some conditions solid solution by Al^{+3} ions can replace the Si^{+4} ions in silica tetrahedra without affecting the polymerization. Because Al^{+3} has a smaller charge than Si^{+4}, a tetrahedron in which an aluminum has substituted for a silicon has an extra negative charge to be satisfied. This charge cannot be satisfied by polymerization; so extra cations must be added to satisfy the charge. Approximately one-quarter of the silicon atoms in tetrahedra are substituted by aluminum atoms in the micas, and cations such as K^{+1}, Mg^{+2}, and even some extra Al^{+3} must be added outside of the tetrahedra to balance the charges. The mica mineral *muscovite,* for example, has the formula $KAl_2(Si_3Al)O_{10}(OH)_2$.

The most common micas are muscovite, a clear and commonly colorless variety that derives its name from Muscovy, an old name for Russia, where it was widely used as a substitute for glass, and *biotite,* a dark variety rich in iron and magnesium. Micas and clays are most typically found in crystal rock.

Quartz. The only common mineral composed exclusively of silicon and oxygen is *quartz.* It is the classic example of a crystal structure that has all its charges satisfied by polymerization of the tetrahedra into a three-dimensional network.

Quartz characteristically forms beautiful hexagonal crystals (Fig. 4.16), but, as we noted earlier, it lacks cleavage. Quartz is also found with many beautiful colors, and it is one of the most widely used gem and ornamentation minerals. Common names for some gemstone varieties of quartz are rock crystal (colorless), citrine (yellow), amethyst (violet), and agate (banded structure, a variety of colors). Quartz is a particularly abundant mineral in rocks of the continental crust. Indeed, it is so abundant that certain sedimentary rocks are composed entirely of quartz.

Feldspar. The name given to the *feldspar* family is derived from two German words, *feld* (field), and

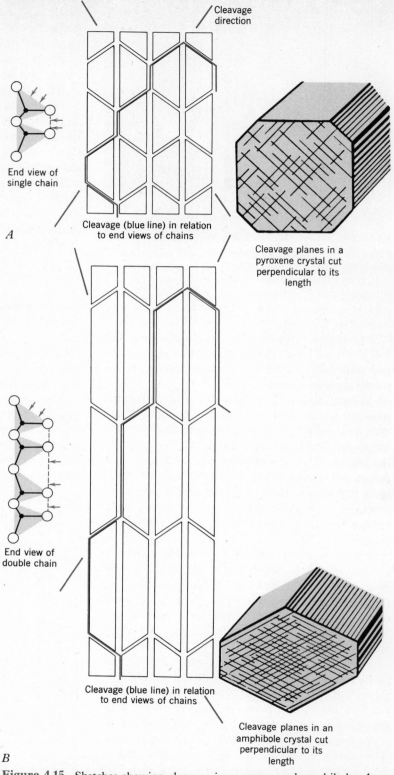

Cleavage
direction

End view of
single chain

Cleavage (blue line) in relation
to end views of chains

A

Cleavage planes in a
pyroxene crystal cut
perpendicular to its
length

End view of
double chain

Cleavage (blue line) in relation
to end views of chains

Cleavage planes in an
amphibole crystal cut
perpendicular to its
length

B

Figure 4.15 Sketches showing cleavage in pyroxene and amphibole. *A. Pyroxene. B. Amphibole. (Cleavage sketches: Fyfe, 1964.)*

Figure 4.16 Small hexagonal quartz crystals showing three of the six prismatic sides. The characteristic crystal shape of quartz is one of its most diagnostic characteristics. (*B. M. Shaub.*)

spar (mineral). Early German miners were so struck by the abundance of feldspar that they chose a name to indicate that their fields seemed to be always growing new crops of feldspar crystals. Of course they were mistaken about fields growing new crystals, but the German miners were not mistaken about the abundance of the feldspars: they are the most common group of minerals in the crust. The feldspars account for 60 per cent of all minerals in the continental crust, and together with quartz comprise about 75 per cent of the volume of the continental crust. Unlike quartz, which is rare in rocks in the mantle, feldspars are abundant in mantle rocks.

Like quartz, the feldspars have structures formed by complete polymerization of all oxygen atoms in the tetrahedra. Unlike quartz, however, some of the tetrahedra contain Al^{+3} substituting for Si^{+4}, so that other cations must be added to the structures to preserve the charge balances.

There are three principal feldspar minerals: *orthoclase* ($KAlSi_3O_8$), *albite* ($NaAlSi_3O_8$), and *anorthite* ($CaAl_2Si_2O_8$). Although K^{+1} and Na^{+1} substitute for each other to some extent, there are limits to the process of solid solution. The most important substitution is Ca^{+2} for Na^{+1} because, as can be seen in Figure 4.2, the two ions are much closer in size than either is to the size of K^{+1}. However, Ca^{+2} and Na^{+1} have different charges, so the actual substitution involves two ions: ($Na^{+1} + Si^{+4}$) for ($Ca^{+2} + Al^{+3}$). The substitution is so effective that a continuous group of feldspar minerals, the *plagioclase* group, has compositions ranging from albite to anorthite.

Other Minerals. Although silicates are the most abundant minerals on Earth, a number of others—principally oxides, sulfides, carbonates, phosphates and sulfates—are common.

Some common oxide minerals are the compounds of iron, *magnetite* (Fe_3O_4) and *hematite* (Fe_2O_3); the oxide of aluminum, *corundum* (Al_2O_3); that of titanium, *rutile* (TiO_2); and of course the solid form of H_2O, *ice*. Oxides are important ore minerals and the principal source of tin, iron, chromium, manganese, uranium, niobium, and tantalum.

The most common sulfide minerals are *pyrite* (FeS_2), *pyrrhotite* (FeS), *galena* (PbS), *sphalerite* (ZnS), and *chalcopyrite* ($CuFeS_2$). The sulfide minerals are exceedingly important as ore minerals, being the principal source of copper, lead, zinc, nickel, cobalt, mercury, molybdenum, silver, and many other elements.

The complex carbonate anion $(CO_3)^{-2}$ forms three important and common minerals: *calcite, aragonite,* and *dolomite.* We have already seen that calcite and aragonite have the same composition, $CaCO_3$, and are polymorphs. Calcite is much more common than aragonite and can always be distinguished by its characteristic cleavage (see Fig. 4.8). Aragonite, however, is widespread because it is commonly found in shells, coral reefs, and stalactites in caves. Dolomite has the formula $CaMg(CO_3)_2$.

One very important phosphate mineral contains the complex anion $(PO_4)^{-3}$. The mineral is *apatite,* $Ca_5(PO_4)_3OH$, and it is the substance that our bones and teeth are made from. It is also a common mineral in many varieties of rocks and is the main source of phosphorus used for making plant fertilizers.

Sulfate minerals contain the complex anion $(SO_4)^{-2}$. Although many sulfates are known, only two are common, and both are calcium sulfate minerals: *anhydrite,* $CaSO_4$; and *gypsum,* $CaSO_4 \cdot 2H_2O$. Both form when sea water evaporates; they are the raw material used for making plaster of all kinds. The famous Plaster of Paris got its name from a quarry near Paris where a very desirable pure-white form of gypsum was mined in former centuries.

63

Summary of Minerals. We now know that minerals are naturally occurring elements, or inorganic compounds, and that each has distinctive properties arising from its crystal structure and its composition. All rocks are made from minerals, but we now know that only a few minerals and mineral families form more than 95 per cent of the mass of the crust. The common rock-forming minerals are listed in Table 4.4 which can be conveniently referred to as we proceed in later chapters to discuss rocks and how they are put together to form the Earth. For convenience, the silicate minerals in Table 4.4 are divided into two categories: the *iron-and-magnesium-rich minerals* (those that contain iron and magnesium as their principal cations) and the *iron-and-magnesium-poor minerals.*

Table 4.4

The Common Rock-Forming Minerals

Silicates	Oxides	Sulfides	Carbonates	Sulfates	Phosphates
The iron- and magnesium-rich minerals	Hematite	Pyrite	Calcite	Anhydrite	Apatite
Olivines	Magnetite	Marcasite	Aragonite	Gypsum	
Pyroxenes	Corundum	Sphalerite	Dolomite		
Augite	Rutile	Galena			
Amphiboles	Ice	Chalcopyrite			
Hornblende					
Garnets					
The iron- and magnesium-poor minerals					
Quartz					
Feldspars					
Orthoclase					
Plagioclase					
Micas					
Muscovite					
Biotite					
Clays					
Kaolinite					

Summary

1. The Universe is composed of matter, and matter is composed of protons, neutrons and electrons combined to form atoms.

2. Atoms form 103 types of elements, and 88 of these occur naturally on Earth.

3. Each element has two or more isotopes. Isotopes are varieties of atoms having the same chemical properties but differing in their masses because of differing numbers of neutrons in their nucleii.

4. Some isotopes are not stable and disintegrate spontaneously by radioactivity. During radioactive decay, energy is emitted and new isotopes and elements are formed.

5. The atoms of some elements occur by themselves, but most elements combine to form compounds.

6. The transfer or sharing of electrons between atoms causes the atoms to have unbalanced electrical charges that bond atoms together. An atom that has gained or lost an electron, thereby acquiring an electrical charge, is called an *ion.*

7. Minerals are naturally occurring, crystalline, inorganic compounds. The atoms in each mineral are arranged in a definite geometric array, called a *crystal structure,* that is unique to the mineral.

8. Some compounds form more than one crystal structure, and the resulting different minerals with the same compositions are called *polymorphs*.

9. Minerals are identified by their physical properties, such as crystal form, habit, cleavage, hardness, streak, and specific gravity. The physical properties of minerals are controlled by their crystal structures and compositions.

10. Approximately 2200 minerals are known, but of these about 20 of them make up more than 95 per cent of the Earth's crust and are called the *rock-forming minerals*.

11. Silicate minerals are the most common rock-forming minerals, followed by oxides, sulfides, carbonates, sulfates, and phosphates.

Selected References

Matter: Atoms, Elements and Compounds

Fyfe, W. S., 1964, Geochemistry of solids, an introduction: New York, McGraw-Hill.
Lapp, R. E., 1963, Matter, in *Life* Science Library: New York, Time Inc.
Pauling, Linus, 1953, General chemistry: San Francisco, W. H. Freeman.
Turekian, K. K., 1972, Chemistry of the Earth: New York, Holt, Rinehart and Winston.

Minerals

Ernst, W. G., 1969, Earth materials: New York, Prentice-Hall.
Hurlbut, C. S., 1969, Minerals and man: New York, Random House.
Hurlbut, C. S., 1971, Dana's manual of mineralogy, 18th edition: New York, John Wiley.
Mason, Brian, and Berry, L. G., 1968, Elements of mineralogy: San Francisco, W. H. Freeman.

Chapter 5

Rocks

Minerals are like words: informative by themselves, but much more informative when gathered together—words into a sentence, minerals into a rock. In order to read history from rocks, we first learn "rock language," and this means that we must know a few common rocks, because it is in the common kinds of rock that Earth's history is recorded. In this chapter, therefore, we shall see how minerals combine to form rock, how certain features in each kind of rock tell us the way in which the rock was put together, and how we use the information to classify common rocks.

The Rock Families

In appearance rocks are varied. Here we see a platy rock showing abundant cleavage surfaces of mica, there a rock with coarse grain and no pronounced layering, and in other places fine-grained varieties with distinct layering, or perhaps rocks that look like cemented beach sand. Yet despite their obvious diversity, we can group all rocks into three families. The first family consists of **igneous rock,** named from the Latin word *ignis,* meaning fire. It is *rock formed by the cooling and solidification of magma; an interlocking aggregate of silicate minerals.* The second family consists of **sedimentary rock,** named from the Latin word *sedimentum,* meaning *settling,* and is defined as *rock formed from sediment by cementation or by other processes acting at ordinary temperatures at or near Earth's surface.* The third and final family is made up of **metamorphic rock,** named from the Greek words *meta,* meaning change, and *morphe,* meaning form, hence: change of form. Metamorphic rock is *rock formed within Earth's crust by transformation, in the solid state, of pre-existing rocks as a result of high temperature, high pressure, or both.*

Within each family, the various rocks share a common origin—by solidification of magma, by cementation of sediment, or by transformation of pre-existing rock. Yet the rocks within each family do not necessarily look alike. One igneous rock contains only big mineral grains 20cm or 30cm in diameter; while another may consist entirely of grains as small as the head of a pin. One sedimentary rock has pronounced layered structure, while another is almost devoid of layering. The more closely we look at the kinds of rock within a family, the more differences we see. These differences are the clues we use to decipher how and where a rock formed. For example, large mineral grains in an igneous rock suggest slow cooling, while very small grains suggest rapid cooling. Slow cooling suggests, in turn, that an igneous rock formed deep underground, where the rock above it acts as a blanket that prevents rapid chilling. In contrast, rapid cooling suggests that the rock formed from lava poured out on the surface and cooled quickly.

From a simple and straightforward clue, like the diameters of mineral grains, then, we can learn much about place of origin. But there are many other clues, from each of which we can learn more. Next, therefore, we talk about how to study rocks and find the clues.

How To Study Rocks

Studying rock is rather like buying a new car. First, the buyer views the car from a distance, assessing its external features; then he moves closer and examines the small details; finally he puts all his observations together and reaches a decision. With rock, we follow the same sequence: distant viewing, close-up examination, and a decision.

Exposures. **Bedrock,** *the continuous body of solid rock that underlies the regolith,* projects through its overlying cover to form an **exposure,** *a place where solid rock is exposed at Earth's surface.* Exposures (sometimes also called *outcrops*), both natural and man-made (such as road cuts, quarries, and mines) are places where we find and study rock. Unfortunately, less than 10 per cent of the surfaces of continents consist of bedrock exposure; so the bedrock

record available at Earth's surface is fragmentary. Like a detective with only a few clues available to him, we have to extract as much information as possible from every exposure.

Exposures range in diameter from a few meters to a whole mountainside. As we approach an exposure, we see large-scale features long before we are near enough to distinguish the individual minerals. We may notice that the exposure consists of **massive rock,** meaning that it is *rock that is fairly uniform in appearance and lacks any breakage surfaces* (Fig. 5.1). Many igneous rocks are massive. Or we may notice that rock in the exposure is distinctly layered (Fig. 5.2) because it consists of *sedimentary strata* (Chap. 2). Even before reaching an exposure, therefore, we might already have formed an opinion about which family our rock belongs to. If we see, cutting across the sedimentary layers, surfaces that cause the rock to break into innumerable thin slabs and platy fragments (Fig. 5.3), we infer that metamorphism has occurred. For even though sedimentary layers may still be obvious, the pronounced breaking direction indicates that new micaceous minerals have grown within the rock, in response to increased temperature and pressure down within Earth's crust. *The property by which a rock breaks into plate-like fragments along flat planes* is called **rock cleavage.** It should not be confused with the mineral cleavage discussed in Chapter 4. The exposure in Figure 5.3, then, provides two clues. First, the rock was originally a sediment; second it was metamorphosed.

Figure 5.1 Exposure of massive igneous rock, uniform in appearance and lacking obvious layering or small-scale breaking surfaces. Guilford, Conn. (*Yale Peabody Museum.*)

Figure 5.2 Exposure of two units of sedimentary rock, showing pronounced sedimentary layering. The layering is less well marked in the lower unit, which consists mainly of grains of quartz, than in the upper unit, which consists mainly of grains of calcite. Between the two units is a **contact,** *a surface along which two rock units meet.* Contacts occur between rocks of different families or between two rocks of the same family. A contact is a record of a time in the past when conditions changed, stopping the formation of one kind of rock and beginning the formation of another. Yorkshire, England. (*Crown Copyright.*)

Joints. Nearly every rock exposure displays one or more fracture surfaces, and we must take care not to confuse them with rock cleavage. As rock that has been deeply buried is slowly uncovered by erosion, it is relieved of the confining pressure exerted by overlying material, and so expands, and in the process develops small fractures. Such fractures are

Figure 5.3 Exposure of metamorphic rock in northern Vermont, showing well-defined cleavage, a plane along which easy breakage occurs because of the growth of new micaceous minerals. Original sedimentary layering, somewhat folded, is still visible. The cleavage cuts across the layering from upper left to lower right. (*B. J. Skinner.*)

joints, defined as *fractures on which movement has not occurred in a direction parallel to the plane of the fracture.* Rarely do joints occur singly. Most commonly they form *a widespread group of parallel joints,* called a **joint set** (Fig. 5.4). Joints do not let us infer much about the origins of rocks—they are found in all three families—but as we shall see in later chapters, they are extremely important in the control of weathering of rock, because they are passageways by which rainwater can enter it.

One special class of joints is restricted to certain igneous rocks, and in this one case they do afford a clue to origin. When a body of igneous rock cools, it contracts and sometimes fractures into small pieces, in the same way that a very hot glass bottle, plunged into cold water, contracts and shatters. Cooling joints are found in igneous rock that cooled rapidly. When we see cooling joints, we infer that the rock cooled at or close to Earth's surface. Just as the hot bottle, if allowed to cool slowly, does not fracture, so a deeply buried body of igneous rock does not develop cooling joints. Unlike shattered glass, cooling fractures in igneous rock form regular patterns. For *joints that split igneous rocks into long prisms or columns,* we use the special term **columnar joints** (Fig. 5.5).

Hand Specimens. After we have noted the obvious features of an exposure, we next look closely at a piece of the exposed rock. To test and examine a rock, we must take a hammer and break out a **hand specimen,** *a piece of rock that can be conveniently held in the hand for close study.* In a hand specimen we can see two kinds of small-scale features that provide many clues.

The first feature is **texture.** This means *the sizes and shapes of the individual particles in a rock, and the mutual relationship between them.* For example, the mineral grains may be flat and parallel to each other, giving the rock a pronounced platy or flaky texture, like a pack of playing cards. In addition, the various minerals may be unevenly distributed and concentrated into specific layers. The rock texture is then both layered and platy. A texture that is both layered and platy is characteristic of metamorphic rock that was once a layered sedimentary rock.

Texture indicates much about the history of a rock. As we have noted, texture can tell us whether an igneous rock has cooled rapidly or slowly; but it can also tell us whether the grains in a sedimentary

Figure 5.4 This exposure of sedimentary rock in Alberta, Canada contains two joint sets. The sedimentary layers, now tilted at a high angle, are broken by two sets of joints approximately perpendicular to each other. The joints in each set run at right angles to the sedimentary layering and tend to break the exposure into roughly cube-shaped fragments. *A combination of two or more intersecting sets of joints* is a **joint system.** (*B. J. Skinner.*)

Figure 5.5 Cooling igneous rock contracts, and in many cases develops shrinkage fractures. In some fine-grained igneous rock the cooling fractures occur in a system of joints that divide the rock into long, thin columns. Such joints are *columnar joints*. The elongate columns in Devil's Post Pile National Monument, California, are 10cm to 20cm across and several meters long. (*H. L. Mackay, from Design Photographers International, Inc.*)

rock settled slowly down through still lake water, or were tumbled about by a rushing stream or by the surf along a coast. We shall discuss the most common textures later in this chapter, as we see how rocks are classified.

The second feature we see in a hand specimen is the minerals the rock contains. A few kinds of rock contain only one mineral, but most contain two or more of the common rock-forming minerals. The minerals, and their percentages, immediately tell us the composition of the rock. *The varieties and abundances of the minerals present in rocks,* commonly called **mineral assemblages,** are important pieces of information in our reading of the rock record. They tell us, for example, whether an igneous rock was formed in the mantle or in the crust, because the compositions of rocks from crust and mantle are quite different. Mineral assemblages can also aid us in distinguishing between different rock families. For example, calcite is a common mineral in sedimentary rock, but is rarely observed in igneous rock.

Sometimes the features in a hand specimen must be magnified so that they can be seen clearly. For this, a low-power magnifying glass can be a great help. Another convenient way to help with the examination of rock texture is to polish the surface of a hand specimen. Yet another technique—one we use

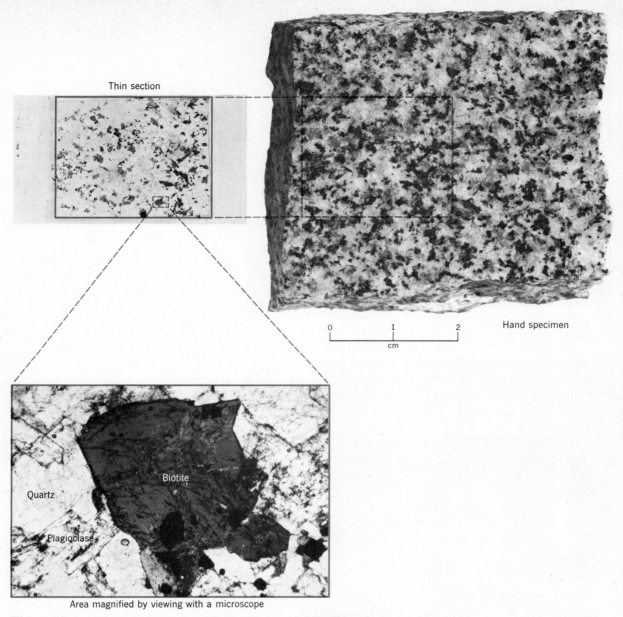

Thin section

Hand specimen

0   1   2
        cm

Quartz

Biotite

Plagioclase

Area magnified by viewing with a microscope

Figure 5.6 In the study of rock, polished surfaces and thin slices of rock reveal texture and distribution of minerals to great advantage. The specimen here is igneous rock containing quartz, plagioclase, hornblende, and biotite. (*Yale Peabody Museum and B. J. Skinner.*)

for illustration many times in this book—is to make a *thin section*. A thin section is prepared by first grinding a smooth, flat surface on a piece of rock. The flat surface is then glued to a glass slide, and is ground down to a slice so thin that light passes through it easily. The appearance of the same rock on a polished surface and in a thin section is shown in Figure 5.6. Although we don't expect many people will have a chance to make or study thin sections, the photographs in Figure 5.6 should prove that the technique only makes more obvious the features that can already be seen in a hand specimen.

71

Igneous Rock

Now that we know how to look for clues in a rock, we can ask the meaning of the evidence we gather. Discussing rocks, family by family, we start with the igneous rocks because they are the most abundant family. Seventy-five per cent of all exposures contain sedimentary rock, but if we drill through the sedimentary rock, we always find igneous—or metamorphic—rock down below. The great bulk of rock in the crust—at least 95 per cent—is igneous (Fig. 5.7).

The adjective *igneous,* as we have seen, pertains to fire. We use it for all the activities associated with the formation, movement, and cooling of magma in the Earth. Igneous events have apparently played important roles in shaping our ancestors' beliefs, including mythologies and religions. Even Stone-Age people were aware of igneous activity, because they saw lava pouring out of volcanoes. But they did not understand what caused these awesome eruptions, and they visualized volcanic "fires." The very word **volcano,** *the vent from which molten igneous matter, solid rock debris, and gases are erupted,* comes from *Vulcan,* the Roman god of fire. Let us, then, consider first the most obvious igneous rocks,

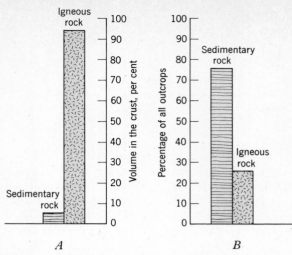

Figure 5.7 Relative amounts of sedimentary and igneous rock in Earth's crust. Metamorphic rocks are included in the sedimentary or igneous category, depending on their origin. *A.* The great bulk of the crust consists of igneous rock (95 per cent), while sedimentary rock (5 per cent) forms a thin covering at and near the surface. *B.* The extent of sedimentary rock at the surface is much larger than that of igneous rock; so 75 per cent of all outcrops are sedimentary and only 25 per cent are igneous. (*After Clarke and Washington, 1924.*)

Figure 5.8 Volcanic ash, blasted out of a volcano on Heimaey Island, Iceland. The ash was distributed downwind from the vent, covering houses and roads in a small fishing village to a depth of nearly 5m. (*Wide World photo.*)

those formed at the Earth's surface where igneous matter from volcanoes cools and hardens.

Volcanic- or Extrusive Igneous Rock. Igneous rock formed by volcanic eruption is called *extrusive igneous rock,* because it is *rock formed by the cooling of magma poured out onto Earth's surface.* Early dwellers in the Mediterranean region witnessed the repeated eruption of magma from several volcanic vents, and were well aware that the "fiery" liquid cools and hardens to form the rocks that surround most volcanoes. Also they realized that the dust and oddly shaped rock fragments spread over the countryside during an explosive eruption need only be compacted and cemented so as to become identical with much of the bedrock seen in Mediterranean lands. Although those early dwellers had no organized science, some individuals had great curiosity. Pliny the Elder, a Roman who wrote a famous book on natural history (including the origins of volcanoes), lost his life while trying to observe at close quarters the destructive eruption of the famous volcano Vesuvius, in A.D. 79.

Not all magma comes out of a volcano as a smooth-flowing liquid. Often gases escape from a volcanic vent so violently that they splatter magma into small, hot fragments, and rip pieces of solid rock off the walls of the vent. Fragments from both sources form a pile or blanket around the vent. Dramatic photographs of erupting volcanoes show clouds of fine rock particles ("ash") thrown out by gases. Figure 5.8 shows the "ash" that was thrown out by an erupting volcano on Heimaey Island, Iceland, in April 1973. The "ash" was carried many kilometers away from the vent by the wind, covering and destroying roads and buildings. Volcanic ash has been found to travel many hundreds of kilometers from its source (Fig. 5.9).

When applied to volcanoes, the word *ash* is misleading, because ash means, strictly, the solids that are left after something inflammable, such as wood, has burned. But the fine particles thrown out by volcanoes look so like true ash, that it has become a convenient custom to use the word for them too. Volcanic particles, regardless of size, become packed together to form rock. *An extrusive igneous rock formed by the agglomeration of small volcanic particles ("ash") is* **tuff** (Fig. 5.10). Agglomeration means the gathering, or clustering together, of particles. It occurs in two ways. First, when ash is ejected from

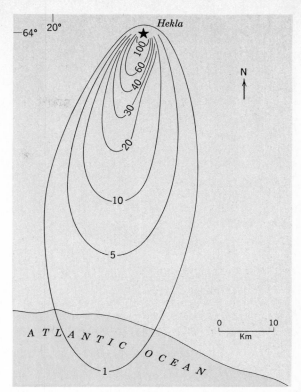

Figure 5.9 Map showing thickness of volcanic ash deposited downwind from Hekla Volcano, Iceland, during eruption of March 29, 1947. Lines are contours showing thickness of ash in centimeters. Most of the ash was deposited within an hour after the eruption began. (*After Thorarinsson, 1954.*)

a volcano and settles on the ground, it may be so hot that its particles simply weld or fuse together. In many parts of western United States, welded tuff is common. The second way in which agglomeration occurs is when ash is not hot enough to weld; in this case the particles become tuff by later cementation. How this happens will be explained in the course of our examination of how the particles in sedimentary rock are held together, because tuff is in fact a sedimentary rock, all of whose particles are volcanic.

The sudden eruption, from Vesuvius, of hot volcanic particles, and their subsequent settling and agglomeration into tuff, was once responsible for a remarkable human tragedy. During the eruption in which Pliny the Elder died, the wealthy Roman resort town of Pompeii, at the southern base of the volcano, was buried beneath a "rain" of ash. The ash became converted to tuff, and for nearly 1800

Figure 5.10 Hand specimen of tuff, an extrusive igneous rock formed by the packing of small rock- and mineral particles ejected violently from a volcano. In this specimen, fragments of igneous rock several centimeters across are trapped in a matrix of similar particles that range down to the size of dust grains. Clark County, Nevada. (*Yale Peabody Museum.*)

years Pompeii lay buried beneath it, until archeologists excavated the town and revealed a vivid picture of Roman culture and of some of the everyday activities of a Roman citizen.

Although the eruption of ash is dramatic, the volume of rock formed from ash and larger volcanic particles is small compared to that formed by the liquid magma that flows from volcanoes. *Magma that reaches Earth's surface through a volcanic vent, and flows out as hot streams or sheets,* is **lava** (Fig. 5.11). We have already noted that lava tends to cool rapidly by comparison with the magma that cools slowly, deep underground.

Some lava cools and solidifies too quickly for its atoms to organize themselves into minerals; so, instead, the lava forms natural glass, the substance we call *obsidian*. Most lavas, however, cool slowly enough for crystals to form, taking months or even a few years for complete cooling. In such cases the resulting rock consists largely or entirely of small, crystalline mineral grains.

What are the clues that tell us a rock is volcanic; and, if it is, how do we decide whether it is tuff or frozen lava? Although they differ among themselves, volcanic rocks have unmistakable characteristics that identify them. Nearly all the particles that constitute

Figure 5.11 Streams of fluid lava that flow like streams of water. Mauna Loa Volcano, Island of Hawaii. (*Sawders from Cushing.*)

tuff are bits of igneous rock, although a few may be fragments of other rocks torn from the walls of the volcanic vents. The fragments, both small and large, are rough-edged and irregular in shape, two features that indicate that the explosions were violent. On the other hand, frozen lava shows features that record the way in which the lava flowed (Fig. 5.12). Sometimes, too, frozen lava contains features that formed when gases bubbled out of the former liquid. When a bottle of soda water is opened, the water effervesces, because the carbon dioxide gas, formerly held in solution under pressure, forms bubbles and escapes. The same thing happens when magma rises to Earth's surface from deep within the crust. Because magma is sticky and viscous, gas bubbles cannot escape from it nearly as quickly as they can from soda water. Therefore, much of the lava cools before all the bubbles have escaped, leaving the upper parts of most former lava flows marked by "frozen" bubbles (Fig. 5.13).

How common is volcanic rock? Anyone who has seen an active volcano such as Mount Etna in Sicily, Mauna Loa and Kilauea in Hawaii, and the volcanoes in Iceland and Central America, knows that these great mountain masses consist of volcanic rock,

Figure 5.13 Frozen bubbles (*vesicles*) in lava are *small openings made by escaping gas originally held in solution under high pressure while the magma was underground.* The presence of vesicles in a rock is clear evidence that the rock is lava. Some lavas are so vesicular that they are frozen, rocky froth, (called *pumice*). Mauna Loa Volcano, Hawaii. (*Yale Peabody Museum.*)

and that they must have been built up by numerous eruptions of volcanic fragments and lava. Furthermore, once we have recognized the features of active volcanoes, we can easily recognize old volcanic rock, even though the volcanoes that formed it have long since died. Almost everywhere we look, we find at least some evidence of former volcanic activity. In the Columbia Plateau in Washington, in much of Ontario and Quebec, along the Blue Ridge in Virginia, in central Connecticut, southern Colorado, central Nevada, Texas, Oklahoma—in fact, in every one of the United States and in every province of Canada, we find volcanic rock. Even on the Moon, astronauts have found many former lava flows.

Intrusive Igneous Rock. With so much magma poured out onto Earth's surface, how much magma fails to reach the surface and so solidifies down below? A single volcano such as Mount Etna has thrown out enough lava to build a cone-like mound whose volume is many cubic kilometers. Such persistent outpouring indicates that beneath the volcanic mound lies a large reservoir of magma. When a volcano becomes extinct, we infer that all the magma remaining in the reservoir has cooled and solidified.

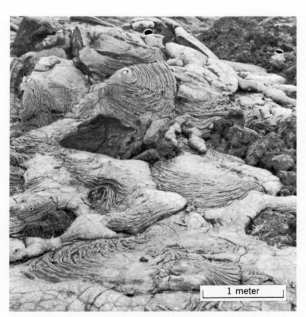

Figure 5.12 Frozen lava from a 19th-Century eruption of Mauna Loa Volcano, Hawaii. The rounded shapes developed when cooled, crusted-over lava cracked open, allowing still-liquid lava to ooze out. (*R. S. Fiske, U.S. Geological Survey.*)

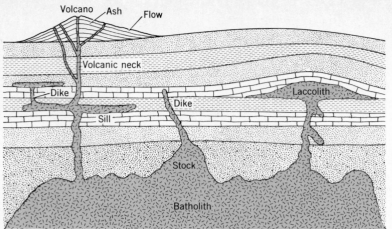

Figure 5.14 (*A*) *Intrusive igneous bodies, regardless of shape, size, or composition, are* **plutons.** Many plutons were once connected with volcanoes, and indeed there is a close relationship between intrusive and extrusive igneous rocks. The common plutonic forms are: **dike,** *a sheet of intrusive igneous rock cutting across the layering of pre-existing rock;* **sill,** *a sheet that is parallel to the layering;* **laccolith,** *a lenticular intrusive igneous body above which the layers of invaded bedrock have been bent upward to form a dome;* **batholith,** *a very large intrusive igneous body that cuts across the layering of the intruded rocks and has an irregular shape; and* **stock,** *a small body with the same characteristics as a batholith.*

A

(*B*) Shiprock, New Mexico, the eroded remains of a volcanic neck 400m high. The three prominent ridges radiating outward are made by dikes of intrusive igneous rock. *B* (*J. S. Shelton.*)

(*C*) Finger Mountain, Victoria Land, Antarctica, seen from the air, reveals a massive sill (dark-colored, horizontal rock) 200m thick, and a dike of the same thickness that cuts across the well-marked horizontal sedimentary layering. (*Warren C Hamilton.*)

Although we cannot slice down through a volcanic cone to see what lies beneath, erosion has done the slicing for us, and has laid bare many former reservoirs of magma. We can see that they are full of frozen magma. Despite the enormous volumes of extrusive igneous rock, far more magma has cooled below the surface, where it has formed intrusive igneous rock. *Any igneous rock formed by cooling and solidification of magma below Earth's surface is an **intrusive igneous rock.***

Apparently, then, intrusive and extrusive igneous rock are closely related in composition; some of them even have similar textures. But intrusive and extrusive igneous rocks are commonly found in masses that have very different shapes. Extrusive rocks form great blankets, consisting of both fallen ash and former streams of flowing lava. Intrusive rock forms bodies with many complex shapes, because magma can flow and squeeze into innumerable openings in a rock. Although the shapes and sizes of intrusive masses are varied, in many cases the shapes were controlled by the viscosity—the stiffness—of the magma. The shapes of intrusive masses, then, are closely related to the kind of igneous rock that was formed (Fig. 5.14).

Textures. The irregular shapes of the mineral grains in igneous rock rarely show the perfect crystal form we see in the mineral specimens exhibited in museums (Fig. 5.15). This is because during their growth, the mineral particles crowd against each other and prevent the formation of smooth crystal faces.

The average size of mineral grains varies widely from one kind of igneous rock to another, but in all kinds of rock, the main control of grain size is the rate at which the magma cooled. Figure 5.16 illustrates that control.

In most igneous rocks, all the grains are about the same size, but there is a special texture class called **porphyry,** *an igneous rock consisting of coarse mineral grains scattered through a mixture of fine*

0   5 cm

Biotite   Quartz   Hornblende   Feldspar

Figure 5.15 Hand specimen of granite, an intrusive igneous rock consisting of biotite, quartz, feldspar, and hornblende. The shapes of the mineral grains are irregular, unlike the perfect forms of the specimens shown below. The texture formed by the close intergrowth of irregularly shaped grains of silicate minerals is typical of many igneous rocks.

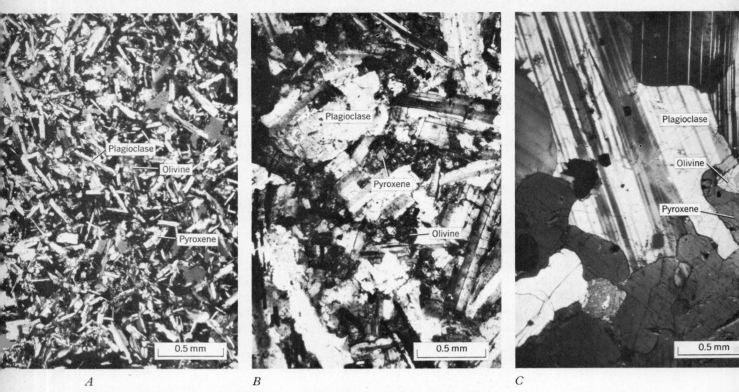

A   B   C

Figure 5.16   The very different grain sizes that are visible in these thin sections of (*A*) *basalt*, (*B*) *diabase*, and (*C*) *gabbro* reflect the rate at which each magma cooled and thereby the control that depth plays on rate of cooling. The three rocks shown here have the same composition (plagioclase, pyroxene, and olivine). Gabbro is a deep-seated intrusive igneous rock, diabase a shallow intrusive rock, and basalt a frozen lava. (*B. J. Skinner.*)

mineral grains (Fig. 5.17). The texture of a porphyry is a revealing clue. Porphyry forms when magma in a deep reservoir has partly crystallized to form a few large crystals, and is then rapidly moved upward. In its new cooling place it cools more rapidly, and all new crystals are tiny. *The isolated large crystals in porphyry* are called **phenocrysts.** Because phenocrysts grow within a fluid, and encounter no interference from crystals growing adjacent to them, most of them have perfect crystal forms.

Kinds of Igneous Rock.   The texture of igneous rock tells us how its parent magma cooled. But texture tells us little about the mineral assemblages, and therefore little about the magma itself. As we shall see in Chapter 16, the composition of magma can provide important clues about many of Earth's internal activities. To decipher them, we must work with igneous rock—frozen magma—and classify the mineral assemblages it contains.

When we examine a hand specimen, the first thing we note is its texture. We should be able to decide whether we are seeing an intrusive- or an extrusive rock. Next we determine the minerals present, and roughly estimate their amounts. From this we get the information necessary to give a specific name to the rock, and thereby to say what kind of magma it represents. We determine the kinds of igneous rock by the use of Figure 5.18. In that figure, the classification is less complicated than it looks, and is so helpful that it is worth examining closely.

All the common igneous rocks are mixtures of one or more of six mineral families: quartz, feldspar, mica, amphibole, pyroxene, and olivine. When the percentages of each mineral in a hand specimen have been estimated, we find the correct place in Figure 5.18, and read the corresponding rock name, selecting the fine-grained or coarse-grained variety, whichever is appropriate.

The common igneous rocks, with suggestions as

Figure 5.17 Phenocrysts of plagioclase trapped in a matrix of sub-microscopic minerals reveal typical *porphyritic texture.* The porphyry seen in this thin section has the composition of basalt. It formed when the partly crystallized magma was extruded and suddenly chilled, preventing further growth of large crystals. (*B. J. Skinner.*)

to how to identify them, are discussed more fully in Appendix C. Here we mention only the most common varieties.

First, consider granites and granodiorites, coarse-grained intrusive igneous rocks found only in the continental crust. These contain abundant feldspar and quartz, and are usually light-colored. Granites and granodiorites form huge batholiths in mountainous areas, and their origins seem to be associated with the origins of mountains. In Chapters 17 and 18 we shall see how this happens. Although bodies of granite and granodiorite are common, masses of extrusive igneous rock with the same composition—rhyolite—are less common. This fact provides an important clue that we shall use in Chapter 16.

Second, consider basalt, the fine-grained extrusive igneous rock that is characteristic of oceanic crust, but is also found on the continents. The abundance of basalt on the floor of the ocean suggests its formation might have something to do with the origin of ocean basins. In Chapter 18 we shall see how this might happen. Basalt is dark colored because it contains dark-colored minerals such as pyroxenes. In complete contrast to the magma that forms granites, a magma that seems rarely to reach the surface, basaltic magma almost always gets up to Earth's surface. Basalt is therefore common, whereas deeply buried intrusive igneous rock with the same composition, gabbro, is rare. This fact too provides an important clue that we will use in Chapter 16.

Origins. Why do igneous rocks have the wide ranges of composition shown in Figure 5.18? An answer is essential to an understanding of how the lithosphere, with its great volume of igneous rocks, formed. The answer involves two processes: how magma crystallizes to form rock, and how rock melts to form magma.

How Magma Crystallizes. Magma of a given composition can crystallize into several different kinds of igneous rock. This is true because magma is a complicated liquid. It does not crystallize, like water freezing to form ice, into a single compound. Freezing magma forms several different minerals, and, again unlike water, the minerals crystallize at different temperatures. As the temperature slowly falls and a magma freezes, first one mineral crystallizes, then another. Therefore, a freezing magma soon consists of a mixture of already-crystallized minerals and still unfrozen liquid. The combination is like a partly frozen bottle of cider. When cider cools, crystals of ice form from the water it contains. All the other ingredients—sugar, alcohol and flavorings—become concentrated in the remaining liquid. Similarly, in cooling magma, the first minerals that crystallize have different compositions from the remaining liquid. Because different minerals begin to crystallize at different temperatures, the composition of the remaining liquid changes continually as the temperature changes. If, at any time during the crystallization process, the remaining liquid becomes separated from the crystals, the liquid can continue to cool as a magma with a brand-new composition.

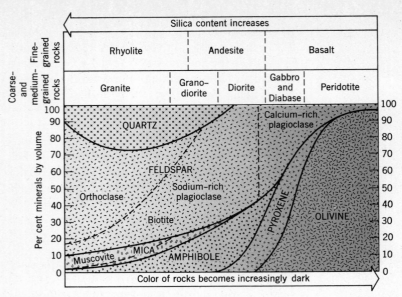

Figure 5.18 Diagram showing the textures and proportions of the principal minerals in the common igneous rocks. Boundaries between kinds of adjacent rock in the table are not abrupt but gradational, as suggested by broken lines. In granites, for example, note the wide range in the proportions of minerals present: granites with nearly 75 per cent potassium feldspar belong at the left side of the diagram; others with only 20 per cent are near the boundary with granodiorite. To see the general range in composition for any granular rock, project the broken-line boundaries vertically downward; then estimate the percentage of a given mineral component by means of the figures at the right and left edges of the diagram. Only three kinds of fine-grained rock are included. Without considerable magnification it is not possible to estimate proportions of minerals in these rocks. (*Modified from R. V. Dietrich, Virginia Minerals and Rocks, Virginia Polytech. Inst.*)

The great importance of separation in magmas was first recognized by the American scientist N. L. Bowen. In 1922, Bowen showed that a wide range of igneous rocks could be developed from a single magma, depending on whether or not the early-formed crystals remained in, or were separated from, the remaining liquid. We call *the compositional changes that occur in magmas by the separation of early-formed minerals from residual liquids,* **magmatic differentiation by crystallization.** Differentiation occurs most commonly when crystals simply sink to the bottom of a pool of magma, leaving a crystal-free liquid above. In the process, two kinds of igneous rock are formed. Crystals that sink and accumulate form one kind; the remaining liquid eventually crystallizes and forms the other (Fig. 5.19).

Bowen discovered also that as the magma continues to cool, the early-formed minerals undergo continuous changes in composition. If crystals remain suspended in liquid, they react continuously with the liquid as it cools and changes its composition. The reactions are very complicated, but Bowen grouped them into two series. In the first, as crystallization proceeds, some minerals such as plagioclase feldspars continually change their compositions but not their crystal structures. These reactions he called a *continuous reaction series.* In the second series, some early-formed minerals later react with the cooling liquid to form entirely new minerals. For example, olivine reacts with the cooling residual liquid to form pyroxene. Bowen called such reactions, that lead to the transformation of early-formed crystals into new minerals, a *discontinuous reaction series.*

Figure 5.20 is an example selected by Bowen to

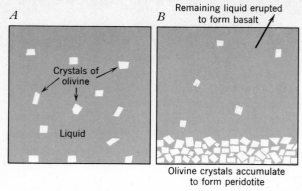

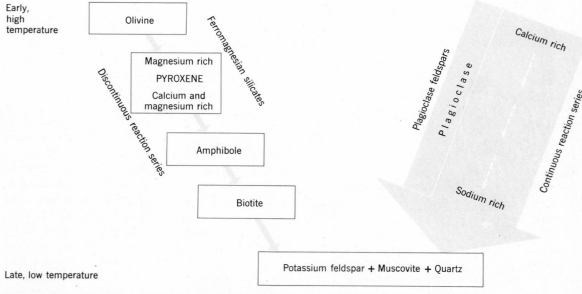

Crystals of olivine

Liquid

Remaining liquid erupted to form basalt

Olivine crystals accumulate to form peridotite

Figure 5.19 Cooling of magma to form two different kinds of igneous rock. *A.* Magma in an underground chamber. Magma cools and begins to crystallize. In this example, the first crystals that form are olivine. Heavier than the remaining liquid, the olivine crystals sink. *B.* Olivine crystals accumulate to form the igneous rock peridotite. The remaining liquid, if erupted on the Earth's surface, crystallizes to form basalt, different in composition from peridotite (Fig. 5.18).

demonstrate a continuous and a discontinuous reaction series operating side by side in the same magma. In a magma with the composition of basalt, the earliest crystals to form are olivine and anorthite, the calcium-rich plagioclase feldspar. As more and more crystals appear, the plagioclase grains continuously change their composition, becoming richer in sodium. Thus, the feldspar always retains the feldspar crystal structure, as the composition changes continually by solid solution (Chap. 4). The feldspars in a cooling basaltic magma form part of a continuous reaction series. Olivine, however, is part of a discontinuous reaction series. Olivine forms, and provided it remains in contact with the residual liquid, it soon reacts to form pyroxene. The pyroxene in turn reacts to form amphibole, and the amphibole finally reacts to form biotite. The final small fraction of residual liquid does not react with or crystallize into any of the earlier minerals, but instead forms an aggregate of orthoclase, the potassium feldspar, quartz, and muscovite. If we refer again to Figure 5.18, we can see that a rock consisting of orthoclase, quartz, and muscovite is granite. The compositions of igneous rock formed by cooling processes of basaltic magma can therefore range

Early, high temperature

Olivine

Ferromagnesian silicates

Discontinuous reaction series

Magnesium rich
PYROXENE
Calcium and magnesium rich

Amphibole

Biotite

Late, low temperature

Potassium feldspar + Muscovite + Quartz

Plagioclase feldspars

Plagioclase

Calcium rich

Sodium rich

Continuous reaction series

Figure 5.20 The earliest minerals that crystallize from a cooling magma of basaltic composition are olivine and plagioclase. As crystallization proceeds, olivine reacts with the remaining liquid to form a new mineral, pyroxene. Pyroxene reacts to form amphibole and amphibole reacts to form biotite. The early plagioclase also reacts with the remaining liquid, but, instead of forming a new mineral, continuously changes its composition.

widely; the compositions depend, naturally, on the moment within the cooling process when the remaining liquid becomes separated from the already-formed crystals. Bowen showed that if one starts with a basaltic magma, the process of magmatic differentiation could even develop granite. This is not the only way to form granite, however, and as we shall see in Chapter 16, most granites apparently form in an entirely different manner.

How Rock Melts. Now we face the second part of the question: why do igneous rocks have a wide range of composition? If a whole rock melts, the resulting magma will have the same composition as its parent. But the melting of rock is a process that is just the reverse of the process of crystallization of magma. Once a rock is heated sufficiently, first one mineral melts, and then another, so that at any instant the aggregate of unmelted crystals has a composition different from the already-formed liquid. Suppose now that the liquid from a partially melted aggregate is squeezed out of the remaining pile of unmelted crystals. The liquid is a magma, with a composition that differs from that of its parent rock. *The process of forming magmas with differing compositions by the incomplete melting of rocks* is known as ***magmatic differentiation by partial melting*** (Fig. 5.21). It is not difficult to see how the composition of a magma that develops by partial melting

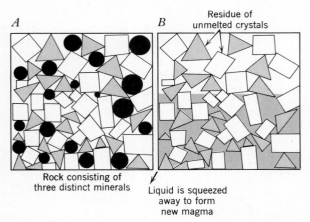

A   *B*   Residue of unmelted crystals

Rock consisting of three distinct minerals

Liquid is squeezed away to form new magma

Figure 5.21 Creation of magma by partial melting of rock. *A.* Rock consisting of three distinct minerals. *B.* As rock is heated, one mineral melts and dissolves small portions of the others. The composition of the newly formed liquid, therefore, differs from the composition of the remaining minerals. When the liquid is squeezed out, it forms magma of one composition and leaves a residue of unmelted crystals having a different composition.

depends on both the composition of the parent rock and the percentage of it that melts. Basalt forms, apparently, by partial melting of rock in the mantle; some people estimate that as much as 30 per cent of the parent rock in the mantle may melt at one time or another. Granite, too, appears to form by partial melting, but by melting of rock of the continental crust rather than of the mantle. Then, by further cooling of basalt and granite, many other kinds of igneous rock are probably created by magmatic differentiation by crystallization.

Synopsis. Although igneous rocks may seem to be a confusingly large and difficult family of rocks to deal with, in fact they are not. In later chapters we shall see how the few magma compositions and the rocks discussed here, plus the basic properties that govern the formation of igneous rocks, provide clues to a great deal of Earth's history. Let us state two important points about igneous rocks. First, rock formed from magma of granitic composition is confined largely to the continental crust, while rock formed from magma of basaltic composition occurs most commonly in oceanic crust. This distribution must be a clue to some of Earth's fundamental internal processes. The second point is that the many kinds of igneous rock are the products of internal processes. They are being formed continuously and brought to the surface, where they become the principal source of new sediment. Therefore, igneous rock must be an important link through which internal and external processes interact.

Sedimentary Rock

Like a perpetually restless housekeeper, nature is ceaselessly sweeping regolith off bedrock, carrying the sweepings away, and depositing them as sediment in river valleys, lakes, seas, and innumerable other basins. We can see sediment being transported by every trickle of water after a rainfall and by every wind that carries dust. The mud on a lake bottom, the sand on the beach, even the dust on a window sill is sediment. Because erosion of bedrock and deposition of the products of erosion are continuous, we find deposited sediment almost everywhere.

When a thick pile of sediment accumulates, the particles near the base of the pile become compacted. Eventually they become cemented together to form a solid aggregate rock. Most commonly the cementation occurs when films of new mineral substance are deposited from water trapped in the spaces between particles of sediment. By compaction and cementing, then, sediment is transformed into sedimentary rock.

Sedimentary rock, as we have seen, is the rock most commonly seen at Earth's surface. Sedimentary rock covers more than 75 per cent of Earth's lands, where it forms a thin blanket over the igneous rocks below. Its most obvious feature is sedimentary layering, strikingly exposed on mountainsides, in the walls of canyons, or in artificial cuts along many highways (Fig. 5.2). Such layers drew the attention of observant people even in ancient times. The most thoughtful observers realized that many sedimentary layers are composed of fragments of other rocks, spread out as loose sediment and eventually cemented to form new rock. Writings by Greek philosophers long before the time of Christ show that the meaning of sedimentary layers was understood even at that early time. Later, in 15th-Century Italy, Leonardo da Vinci wrote an explicit statement of the connection between erosion, sediment, and sedimentary rock. In his notebooks he recorded the close similarity between sedimentary rock high in the mountains of northern Italy and the sand and mud he saw along the seashore.

When we study sedimentary rock, we see the same kinds of things we see in igneous rock: its constituent minerals and its texture. But because sedimentary rock does not resemble igneous rock, the questions we ask about it differ from those we ask about igneous rock. Some of the questions are: where did the sediment come from? How was it transported? What led to its deposition? The terms we use in describing and classifying the kinds of sedimentary rock reflect these questions.

Figure 5.22 *Conglomerate, a sedimentary rock that contains numerous rounded pebbles or larger particles,* is coarse-grained example of many sedimentary rocks. Each fragment is itself a piece of some older rock or mineral, more or less rounded during transport. Canyon Range, Utah. (*R. L. Armstrong.*)

Sediment. Looking closely at a sediment, we see that its pebbles or sand grains are simply bits of rocks and minerals (Fig. 5.22). With a magnifying glass, we see that the finer sedimentary particles, too, are derived from broken-up rock, but that generally the particles have undergone chemical changes; for instance, feldspars have been partly altered to clay. All sediment of this kind is known as ***clastic sediment,*** from the Greek work *klastos* (broken), meaning *the accumulated particles of broken rock and of skeletal remains of dead organisms.* There is, of course, a continuous gradation of particle size, from the largest boulder down to submicroscopic clay particles. This range of particle size is embodied in the classification in Table 5.1.

If a sedimentary rock is made up of mineral parti-

Table 5.1

Definition of Detrital Particles, Together with the Sediments and Sedimentary Rocks Formed from Them (After C. K. Wentworth, Jour. Geol. Vol. 30 p. 377, 1922)

Name of Particle	Range Limits of Diameter		Name of Loose Sediment	Name of Consolidated Rock
	mm.	inches (approx.)		
Boulder	More than 256	More than 10	Gravel	Conglomerate
Cobble	64 to 256	2.5 to 10	Gravel	and
Pebble	2 to 64	0.09 to 2.5	Gravel	sedimentary breccia
Sand	1/16 to 2	0.0025 to 0.09	Sand	Sandstone
Silt	1/256 to 1/16	0.00015 to 0.0025	Silt	Siltstone
Clay[a]	Less than 1/256	Less than 0.00015	Clay	Claystone, mudstone, and shale

[a] Clay, used in the context of this table, refers to a particle size. The term should not be confused with clay minerals, which are definite mineral species. Many geologists prefer to use the term clay-sized particle to avoid confusion.

cles derived from the erosion of igneous rock, how can we tell it is sedimentary and not igneous? Besides obvious clues such as sedimentary layering, there are clues in texture as well. A typical clastic sedimentary rock contrasts strongly with igneous rock in the shapes and arrangements of the grains (Fig. 5.23). In igneous rock the grains are irregular and are interlocked. In sedimentary rock the particles are commonly rounded and show signs of the abrasion they received during transport. Clastic sedimentary rock also reveals cement that holds the particles together, whereas the grains in igneous rocks are held together by interlocking crystals. Another important feature of many sedimentary rocks is the presence of fossils (Chap. 2). Life surely cannot tolerate the high temperatures under which igneous rocks form; so the presence of fossils is an excellent clue to sedimentary origin. Similarly, the presence of features such as ancient ripple marks, marks of erosion by fast-flowing water, and mud cracks in the floors of ancient lakes tell us a rock was once a sediment. We shall discuss these features in Chapter 14.

Certain kinds of rock contain fossils and other evidence of sedimentary origin, yet seem to be free of clastic sediment. The origin of such rock might be a puzzle were it not possible to find places on Earth, like the Bahama Banks and the Persian Gulf, where similar rock is forming today. The rock is indeed sedimentary and the material deposited has indeed been transported. But the sediment is not clastic because its components were dissolved, transported in solution, and precipitated chemically instead of mechanically. *Sediment formed by precipitation of minerals from solution in waters of the Earth is* **chemical sediment.** It forms in two principal ways. One of these consists of biochemical reactions within the water. Biochemical reactions result from the activities of plants and animals living in the water; for example, tiny plants living in seawater can influence the amount of carbon dioxide that is dissolved in the water and so cause calcium carbonate to precipitate.

The other way in which chemical sediment forms consists of inorganic reactions within the water. For instance, when the water of a hot spring cools, it may precipitate opal or calcite. Another common example is simple evaporation of seawater or lake water. After evaporation of the water, the salts that were formerly in solution remain as a residue of chemical sediment. All the table salt we eat comes from sedimentary rock formed in this way.

Transport and Deposition of Sediment. Sediment is transported in many ways. It may simply slide down a hillside, or may be carried by wind, by a glacier, or by running water. In each case, when transport ceases, the sediment is deposited in a fashion characteristic of the transporting agency. When sediment transported by sliding or rolling downhill stops moving, the result is a mixture of particles of all sizes. Much of the sediment carried by a glacier

A   *B*

Figure 5.23 *A.* Thin section of sandstone, showing rounded sand grains that are cemented by calcite. *B.* Broken fragments of sea shells stuck together with calcite cement make a clastic sedimentary rock called *coquina.* St. Augustine, Florida. (*Yale Peabody Museum.*)

is deposited when and where the transporting ice melts. Such sediment also is a mixture of sedimentary particles of all sizes.

In the transport of sedimentary particles by wind or water, deposition happens when the flowing water or moving air slows down to a speed at which particles can no longer be moved. In a general way, therefore, the size of the grains in clastic sediment indicates the speed of the transporting medium. Coarse-grained sediment indicates deposition in fast-moving wind or water; fine-grained sediment indicates deposition by slow-moving wind or water.

The size of particles in sedimentary rock, and the way they are packed together, lets us draw inferences about the environment in which the original sediment was deposited. The existence of ancient oceans, coasts, lakes, streams, swamps, and all the other places where sediment accumulates today, can be demonstrated from clues in sedimentary rock. Clues to former climates, likewise, are found in the fossils within a sedimentary layer. Some animals and plants are restricted to warm, moist climates, others to hot, dry climates, and yet others to very cold climates. By using modern plants and animals as guides to the preferences of their fossil ancestors, we can draw inferences about the climate in which the latter lived.

Sedimentary rock is classified on the basis of features such as size of clastic particles, shape of particles, and composition of chemical sediment. The names given to the varieties of sedimentary rock are listed in Table 5.1. Each name is associated with particles of a given size. Common rocks formed by chemical precipitation are limited to limestone and dolostone, which contain the minerals calcite and dolomite, respectively. The names and identifying features of all common sedimentary rocks are listed in Appendix C.

Metamorphic Rock

As we have now seen, igneous rock is the principal product of internal, sedimentary rock of external, processes. In the rock cycle (Chap. 2) sedimentary rock becomes so deeply buried and so greatly heated that it melts and becomes magma. But long before the rock begins to melt, new minerals start to grow in it, replacing some of the original minerals of the sedimentary rock. At the same time the new-mineral growth imprints new textures on the sedimentary rock. *The changes in mineral assemblage and rock texture, or both, that take place in the solid state within Earth's crust as a result of high temperatures and high pressures* are termed **metamorphism.**

Metamorphism affects igneous rock also, but in a different way, because such rock responds differently to increased temperature and pressure. The explanation of the difference concerns water content. Sedimentary rock has, between its grains, innumerable open spaces that are saturated with water. When metamorphism takes hold, this intergranular "juice" serves to speed up chemical reactions, in much the same way that water in a stew pot helps cook a tough piece of meat. In contrast, igneous rock has very little open space and contains little water. Before new minerals can grow within dry rock, the rock may have to be heated to much higher temperatures, and squeezed under higher pressures, than would be needed for mudstone or limestone.

Metamorphic rocks allow us to infer a great deal about the histories of mountain ranges. They contain clues as to the strength of the forces that deform rocks deep within the crust, about the temperatures that change sedimentary rocks, and about the complicated events that produce the huge volumes of altered and crumpled rock seen in mountain ranges.

Composition. In terms of the chemical elements present, the compositions of metamorphic rocks are essentially the same as those of igneous and sedimentary rock. Small differences exist because metamorphic heating eventually drives off the watery juices, which carry with them, in solution, small amounts of sodium, calcium, and other elements. But the chemical changes are small, and the principal differences between metamorphic rock on the one hand, and sedimentary and igneous rock on the other, are in mineral assemblages and textures.

Mineral Assemblages. As sediment is deposited layer on layer, the accumulating pile of rock may eventually reach thicknesses greater than 20,000m. Under the influence of rising temperature and pressure, new minerals grow, and water is driven off. A new, water-scarce environment comes into existence, and in it, water-free minerals begin to form. Among these are garnet, kyanite, and sillimanite. Finally, melting takes place, converting the metamorphosed rock into magma.

Metamorphism, therefore, creates progressive changes in mineral assemblages, and the changes depend on initial composition of the rock, temperature, and pressure. Each mineral assemblage is characteristic of a specific range of temperature and pressure. Therefore, by studying mineral assemblages, we can draw many inferences about ancient rock temperatures, and sometimes about how deeply a rock body has been buried. But changes in temperature and pressure may occur otherwise than by burial, and metamorphic rock provides clues about this too. For instance, rock adjacent to an intrusive igneous rock becomes heated, and adjacent to the contact, a rim of metamorphic rock develops. Changes in mineral assemblages are not the only obvious changes that take place in metamorphic rocks. Textures, too, change dramatically, and in some cases are even more evident than mineral changes. We have already seen, in Figure 5.3, how important rock cleavage can be for identifying a metamorphic rock.

Texture. The earliest minerals that grow within a metamorphic rock are silicates with layer structures: the micas. These minerals are distinctly platy because they grew in positions that minimize pressure; that is, with their flat sides perpendicular to the direction of maximum pressure. A similar thing happens when we press down on a heap of jumbled playing cards in a box. The more we press, the more the cards tend to line up parallel to the bottom of the box. So too, with metamorphic rock; the more it is squeezed, the more the platy minerals line up.

In the earliest stage of mineral growth, pressure is caused by the weight of the overlying rock; so the mineral plates grow parallel to the sedimentary layering. But with deeper burial, the sedimentary layers may become folded and contorted and the growing platy minerals no longer develop parallel to the layering. As Figure 5.3 shows, rock cleavage develops parallel to the new mineral plates, often at an angle to the sedimentary layering. Eventually each sedimentary layer develops its own new mineral assemblage. Because the sedimentary layers differ slightly in composition, the new mineral assemblages differ also, and create *foliation* in the rock. Foliation is *a parallel or nearly parallel structure in metamorphic rock, caused by compositional layering* (Fig. 5.24).

Kinds of Metamorphic Rock. The most obvious changes in metamorphic rock are textural. Although

Figure 5.24 Strongly foliated metamorphic rock. The light-colored bands are principally quartz, feldspar, and epidote, the dark bands quartz, hornblende, and biotite. The direction of foliation marks a direction of easy breakage. Because the layers are curved and contorted, flakes of rock that break along foliation surfaces also tend to be curved. Cornwall, England. (*Crown Copyright.*)

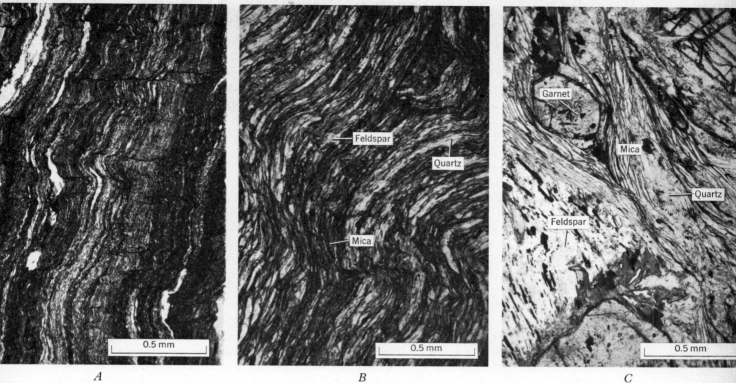

Figure 5.25   The grain sizes in these thin sections of (*A*) *slate,* (*B*) *phyllite,* and (*C*) *mica schist* show how continued mineral growth occurs during metamorphism. The three rocks have the same composition. Mineral grains in the slate are barely visible. Grains in the phyllite are large enough to be seen easily, while those in the schist are large and obvious. (*B. J. Skinner.*)

compositions vary as widely as the entire compositional range of sedimentary and igneous rock, textures of metamorphic rocks can be grouped into a few distinctive classes. Therefore we use textures as the basis of describing metamorphic rocks, although where a particular mineral is very obvious, we modify the rock name by using the mineral name as a prefix.

The four commonest kinds of metamorphic rock are: *slate, phyllite, schist,* and *gneiss.* Complete descriptions of these and other metamorphic rocks, with suggestions as to how to identify each kind, are given in Appendix C. The sequence: *slate, phyllite, schist, gneiss* illustrates an important point about metamorphic rocks: it provides clues about changes that happen systematically (Fig. 5.25). For example, when mudstone is metamorphosed, tiny new grains of mica grow. The rock still looks rather like mudstone because it is fine grained, but the new mica grains give it a pronounced rock cleavage. The new rock is *slate.* Continuing metamorphism makes

the mica grains grow larger, so that they can be seen with a magnifying glass; the rock is now *phyllite.* Eventually, the mica grains reach a size visible to the unaided eye, and new minerals such as garnet have formed; the rock has become *schist.* But as new minerals, such as garnet, grow, we observe that original differences in composition between layers of the mudstone are exerting a control. Some layers in the metamorphic rock are becoming rich in grains of feldspar, quartz, and other minerals. The rock becomes markedly layered, with alternating layers of coarse mica grains and coarse grains of quartz and feldspar; it is now *gneiss.* As we shall see in Chapter 17, these sequential changes are helpful clues to understanding how mountains are built.

What Holds Rock Together?

We need not have had much experience with rocks in order to realize that some kinds of rock hold together with great tenacity, whereas other kinds are easily broken apart. The most tenacious rocks are igneous and metamorphic, because these possess intricately interlocked mineral grains. The growing minerals crowd against each other, filling all spaces and forming an intricate three-dimensional jigsaw puzzle. Similar interlocking of grains holds together steel, ceramics, and bricks.

The forces that hold together the grains of sedimentary rock are less obvious. Yet even the particles in loose sediment have a tendency to hold together. Why? Cohesion among particles is the result of two factors, electrostatic force and capillary tension. In gravel and coarse sand, the particles and the pore spaces between them are so large that neither force can act effectively. Hence such materials lack cohesion. This is why dry, coarse sand falls apart when it is dug. But as grain size decreases, the size of pore space decreases, surface area increases conspicuously, and cohesion can reach large values. This is why silt and clay, when dug, hold together to form chunks or clods.

Sediment is transformed into sedimentary rock in four ways. (1) By pressure from overlying sediment or by vibrations of the ground arising from earthquakes, the irregular-shaped grains in a sediment can be packed into a tight, coherent mass. The interlocking of grains that results from this kind of packing is not strong, as in igneous rock, but under some circumstances it can hold sediment together. (2) Water that circulates slowly through the open spaces between grains deposits new minerals such as calcite, quartz, and iron oxide, which cement the grains together. (3) The weight of overlying deposits can squeeze water out of deeply buried sediment, com-

pacting the sediment and reducing pore space. Compaction forces small grains close together and makes more effective the electrostatic- and capillary forces between the grains. (4) As sediment becomes deeply buried, its mineral grains begin to be recrystallized. Recrystallization resembles very low-grade metamorphism and has a somewhat similar effect on the texture of the rock. The newly growing minerals interlock and form strong aggregates, like those in igneous rock.

Conclusion

The three rock families contain keys for understanding the internal and external activities of Planet Earth. Igneous rock results from internal processes, sedimentary rock from external. Metamorphic rock records intermediate steps by which both sedimentary and igneous rock are changed and influenced by internal processes. Our study of Earth's internal and external activities reveals that new rock is being made continuously. Extrusive igneous rock is forming at all active volcanoes, sedimentary rock along the margins of ocean basins, and metamorphic rock in the cores of high mountains (Himalaya) and adjacent to recently intruded masses of igneous rock (beneath the Imperial Valley in southern California).

In subsequent chapters we shall examine the many paths and discuss the many processes involved in the making of rock. First we discuss, one by one, Earth's external activities and the processes by which sediment is created and transported. Then we deal with the internal processes by which metamorphic rocks are made and by which magma is generated.

Summary

1. The three rock families are igneous, sedimentary, and metamorphic. Igneous rocks form by Earth's internal processes, sedimentary rocks by external processes. Metamorphic rocks form when sedimentary or igneous rocks are buried and heated so that new minerals grow.

2. Beneath the regolith is a continuous solid body of bedrock. Where bedrock forms exposures that project through the regolith, we can study the kinds and structures of rocks.

3. Exposures are commonly broken by one or more surfaces of easy breakage. Rock cleavage is breakage caused by the growth of new minerals in metamorphic rock; joints are breakages formed in rock by release of pressure, as the load of overlying rock is removed by erosion.

4. Mineral assemblages indicate the composition of a rock and in some cases tell us how the rock formed. Mineral assemblages are very useful for classifying igneous rocks.

5. Texture—the size, shapes, and mutual relationships of particles in rock—results from rock-forming processes. Crystalline textures predominate in igneous and metamorphic rock; clastic texture is common only in sedimentary rock.

6. Igneous rock may be intrusive (meaning it formed within the crust) or extrusive (meaning it formed on the surface). The texture and grain size of igneous rock indicate how and where the rock cooled.

7. Igneous rock rich in quartz and feldspar, such as granite, granodiorite, and rhyolite, is characteristically found in continental crust. Basalt, rich in pyroxene and olivine derived from the mantle, is common beneath the ocean basins.

8. The wide range of composition of igneous rock results from magmatic differentiation by partial melting and crystallization.

9. Sediment is transported by wind, streams, seawater, glaciers, and landslides. Then it is deposited, compacted, and cemented to form sedimentary rock.

10. Clastic sediment consists of broken, fragmental debris derived from weathering. Chemical sediment forms where materials carried in solution are precipitated.

11. Mineral assemblages and textures in metamorphic rocks change continuously because of rising temperatures and pressures. Metamorphic temperatures may eventually become so high that rock begins to melt.

12. Metamorphic rocks are named for their most obvious feature, texture, and the name of the most prominent mineral present may be added as a prefix. We speak of *mica schist* and *garnet gneiss.*

13. The effects of friction and gravity hold coarse particles of regolith together; the effects of cohesion hold fine particles together.

14. By various processes of compaction and cementation, sediments become sedimentary rocks. Common cementing agents are quartz, calcite, and limonite.

Selected References

General

Bayly, Brian, 1968, Introduction to petrology: Englewood Cliffs, N.J., Prentice-Hall.

Ernst, W. G., 1969, Earth materials: Englewood Cliffs, N.J., Prentice-Hall.

Loomis, F. B., 1948, Field book of common rocks and minerals, rev. ed.: New York, G. P. Putnam's Sons.

Pearl, R. M., 1965, How to know the minerals and rocks: New York, New American Library.

Pirsson, L. V., and Knopf, Adolph, 1947, Rocks and rock minerals, 3rd ed.: New York, John Wiley.

Igneous Rock

Daly, R. A., 1933, Igneous rock and the depths of the Earth: New York, McGraw-Hill (reprinted in 1968).

Hess, H. H., and Poldervaart, A. (ed.), 1967, Basalts, v. I, II: New York, Wiley-Interscience.

MacDonald, G. A., 1972, Volcanoes: Englewood Cliffs, N.J., Prentice-Hall.

Sedimentary Rock

Degens, E. T., 1965, Geochemistry of sediments: a brief summary: Englewood Cliffs, N.J., Prentice-Hall.

Garrels, R. M., and Mackenzie, F. T., 1971, Evolution of sedimentary rocks: New York, W. W. Norton.

Krumbein, U. C., and Sloss, L. L., 1963, Stratigraphy and sedimentation: San Francisco, W. H. Freeman.

Laporte, L. F., 1968, Ancient environments: Englewood Cliffs, N.J., Prentice-Hall.

Pettijohn, F. J., 1957, Sedimentary rocks: New York, Harper & Row.

Metamorphic Rock

Harker, Alfred, 1932, Metamorphism: London, Methuen.

Turner, F. J., 1968, Metamorphic petrology: New York, McGraw-Hill.

Winkler, H. G. F., 1967, Petrogenesis of metamorphic rocks: Berlin, Springer-Verlag.

The Iguassú River creates a falls more than 4 km wide, over a layer of resistant rock just before it joins the Paraná River. The Brazil-Argentina boundary runs down center of stream. (*George Holton/Photo Researchers.*)

Part II External Processes on Land

Chapter 6

Weathering and Soils

Up to this point our view has been a general one, embracing Earth's solid surface and its envelopes, the minerals and rocks of the crust, the energy that drives internal and external activities, the movement of matter through the rock cycle, and finally the vital factor of geologic time, which makes it possible for very slow processes to accomplish results that are enormous. Now we begin to look at each of the external processes by which rock is broken up and moved in the form of sediment, which is then deposited as strata. We start logically with weathering, the process by which rock is broken down and prepared for transport.

Environment of Weathering

Since very early times people have tried to select for their buildings, tombstones, and other structures stone that would be durable. Their success has been mixed. The durability of rock varies with climate, composition, and degree of exposure to weather. If gravestones (Fig. 6.1) and buildings (Fig. 6.2) made of firm rock have begun to crumble within a few centuries, what would have happened to rock exposed to the atmosphere through thousands or even millions of years? Fast or slow, chemical and mechanical alteration occurs everywhere at the interface between lithosphere and atmosphere.

The interface, as we noted in Chap. 1, is not sharply drawn. It is a zone rather than a surface, and extends down into the ground to whatever depth air and water penetrate. In this critical zone both hydrosphere and biosphere are involved also; in fact all the "spheres" are continually interacting. Within

1806   1779   1769

Figure 6.1 Three dated gravestones in a Connecticut cemetery show the strong influence of rock composition on rate of chemical weathering. (*R. F. Flint.*) *Left:* Marble, consisting of very soluble calcite, is greatly corroded. Urn (in medallion at top) is almost obliterated; inscription is hard to read. Entire surface is rough. *Center:* Medium-grained quartz sandstone containing feldspar and micas, much less soluble than marble, is slightly roughened overall and has flaked off in a patch just above center. *Right:* Fine-grained sandstone consisting almost wholly of quartz, very insoluble, is almost unaltered. Incised lettering is sharp and clear.

Figure 6.2 Seventeenth-century wall of limestone blocks, Trinity College, Oxford, photographed in 1961. Detail showing two or three shells peeling off the surface, most effectively at corners. In some places the stone has lost as much as 1cm of thickness during 300 years of weathering. (*R. F. Flint.*)

the interface, the rock constitutes a framework, full of joints, cracks, and other openings, some of which are tiny but all of which make the rock vulnerable. This framework is continually being attacked, chemically and physically, by solutions. The result is conspicuous alteration of the rock.

We see the effects of such alteration exposed rather commonly in cuts along highways and in other large excavations. Figure 6.3 shows fresh, unaltered bedrock (1) imperceptibly grading upward through rock (2) that has been altered but that still retains its organized appearance, into loose, unorganized, earthy regolith (3) in which the texture of the original rock is no longer present. Evidently alteration of the fresh rock is progressing from the surface downward. When exposed to the atmosphere no rock, whether bedrock or the stone in a man-made structure, escapes the effects of **weathering,** *the chemical alteration and mechanical breakdown of rock materials during exposure to air, moisture, and organic matter.*

The regolith we see in Fig. 6.3 was formed *in place* (on the spot) by conversion of bedrock, and

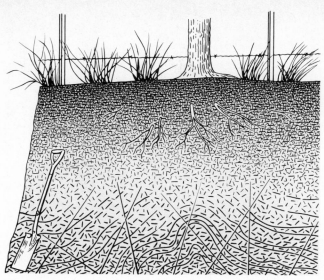

③ Loose, earthy regolith; texture and structures disappear as rock particles are slowly churned by roots, worms, and other agents.

② Bedrock weakened by chemical alteration.

① Fresh, unaltered granite gneiss with crystalline texture, wavy foliation, and joints.

Figure 6.3 Gradation upward from fresh bedrock (granite gneiss) to earthy regolith.

so we say it is *residual*. In many places, however, regolith is so different from the bedrock below that it cannot have resulted from chemical alteration. Instead, former regolith has been removed, and sediment transported from elsewhere has been deposited in its place. Both removal of the former residual material and deposition of the sediment could have been performed by a single agency such as a river, surf along a coast, or a glacier, or by two or more agencies working together.

Processes of Weathering

If we could look closely at the bedrock in Fig. 6.3, we would see that near the bottom of the exposure (1) the cleavage surfaces of the grains of feldspar flash brightly between the grains of quartz. Higher up (2) these surfaces are lusterless and stained. Near the top (3) the grains of quartz, although still distinguishable, are separated by soft, earthy material that in no way resembles the former feldspar, which has rotted away. Evidently the changes that have occurred are mainly chemical. However, some regolith consists of fragments identical with the adjacent bedrock. In them the minerals are quite fresh or have been only slightly altered. This relationship is commonly seen in the aprons of loose sliderock that mantle the lower parts of bedrock cliffs from which the sliderock must have been derived (Fig. 7.18). Because the coarse fragments of sliderock show little

or no chemical change as compared with the bedrock, we conclude that bedrock can be broken down mechanically as well as decayed chemically. So we speak of *mechanical weathering* as distinct from *chemical weathering*, even though the two processes work hand in hand, and even though their effects are inseparably blended. **Disintegration,** *the mechanical breakup of rocks,* exposes additional fresh surfaces to air and water (Fig. 6.4). So disintegration aids **decomposition,** *the chemical alteration of rock materials.*

At this point we should note two significant relationships. (1) The effectiveness of chemical reactions increases with increased surface area available for reaction. (2) Increased surface area results simply from subdivision of large blocks of rock into smaller blocks. Take, for example, a cubic block 1cm on a side. Its volume is 1cm^3, and its surface area is 6cm^2, or 600mm^2. If this block is cut along planes that bisect each of its edges, the result is 8 cubes each 5mm on a side (Fig. 6.4). The area of each side is 5×5mm, or 25mm^2. The number of blocks has been multiplied by 8, the total number of sides is now 8×6, or 48, and the aggregate surface area (the sum of the areas of all the sides) has become 25mm$^2 \times 48$, or 1200mm^2 (12cm^2). In other words, by merely subdividing the cube, while adding nothing to its volume, we have doubled the surface area available for reaction.

However, if we continue to subdivide the smaller

93

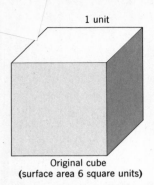

1 unit

Original cube
(surface area 6 square units)

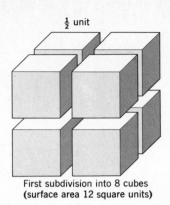

$\frac{1}{2}$ unit

First subdivision into 8 cubes
(surface area 12 square units)

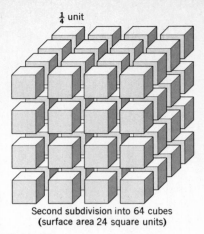

$\frac{1}{4}$ unit

Second subdivision into 64 cubes
(surface area 24 square units)

Figure 6.4 Subdivision of a cube into smaller cubes. Each time a cube is subdivided by slicing it through the center of each of its edges, the aggregate surface area doubles. This greatly increases the speed of chemical reaction.

cubes again and again, each time doubling the surface area, the result is startling. One cubic centimeter of rock subdivided into particles the size of the smallest clay minerals results in an aggregate surface area of nearly one acre. This is one of the changes we see happening from (1) to (3) in Fig. 6.3.

Chemical Weathering: Decomposition

Effects on Potassium Feldspar. The active agents of decomposition consist of chemically active water solutions and water vapor. As it falls through the atmosphere, rainwater dissolves small quantities of carbon dioxide, so that when it reaches the ground it is a weak carbonic acid. As the water percolates down through the soil, the strength of the acid solution is increased many times by addition of more carbon dioxide created during the bacterial decay of vegetation.

The carbonic acid ionizes to form hydrogen ions and bicarbonate ions:

$$H_2O \ + \ CO_2 \ \rightarrow \ H_2CO_3 \ \rightarrow$$
(Water)   (Carbon   (Carbonic
dioxide)   acid)

$$H^{+1} \ + \ (HCO_3)^-$$
(Hydrogen   (Bicarbonate
ion)   ion)

Hydrogen ions are extremely effective in decomposing minerals. Being very small (Fig. 4.2), they can

squeeze in between the atoms of a crystal and disrupt its structure. The effectiveness of H^+ ions is illustrated by the way in which potassium feldspar (orthoclase) is decomposed by hydrogen ions and water:

$$2KAlSi_3O_8 \ + \ 2H^{+1} \ + \ H_2O \ \rightarrow$$
(Potassium   (Hydrogen   (Water)
feldspar)   ions)

$$2K^{+1} \ + \ Al_2Si_2O_5(OH)_4 \ + \ 4SiO_2$$
(Potassium   (Kaolinite)   (Silica)
ions)

Here the H^{+1} ions forcibly enter the potassium feldspar and displace potassium ions, which then leave the crystal and go into solution. Water combines with the remaining aluminum silicate radical to create the clay mineral kaolinite. This combination of water with other molecules is *hydrolysis,* one of the chief processes in chemical weathering. The resulting kaolinite we call a *secondary mineral,* because it was not present in the original rock. Kaolinite is the most conspicuous of the three products of the reaction. It is a common member of the group of very insoluble minerals that constitute clay, and as clay it accumulates and forms a substantial part of the regolith. Many of the potassium ions released during the decomposition of orthoclase are eventually taken up by plants; others enter into clay minerals other than kaolinite.

The silica, less insoluble than clay minerals, in part remains in the clay-rich regolith and in part moves away in solution. Many of the potassium ions likewise escape in water solution, and some of them, together with some of the dissolved silica, eventually find their way through streams to the sea. Some, however, are held in the clay-rich regolith and are utilized by plants as food.

We speak of the matter carried away in solution as having been *leached* out of the parent rock. **Leaching** is *the continued removal, by water, of soluble matter from bedrock or regolith.*

Effects on Granite, Basalt, and Limestone. Our look at the chemical weathering of potassium feldspar has prepared us for a broader look at the weathering of two kinds of igneous rock: granite and basalt. Chemical weathering converts granite to a mixture of clay minerals, some of the original micas, most or all of the original quartz, and some iron oxides. As the feldspar decays, the quartz grains are loosened like bricks in a wall when the mortar between them crumbles.

On the other hand, when basalt is weathered, it is converted into clay without quartz, simply because little or no quartz is present in the original rock. Besides clay, the products of weathering of basalt include yellowish or brownish iron oxides collectively called *limonite*. The rust on a piece of old iron is an example of limonite.

Because pure limestone consists mainly of calcium carbonate, which is highly soluble in carbonic acid, the chemical weathering of limestone is a simple process. Carbonic acid in the ground readily dissolves the carbonate, leaving behind only the nearly insoluble impurities (chiefly clay and quartz) that are always present in very small amounts. As limestone becomes weathered, then, the residual regolith that develops from it consists mainly of clay with particles of quartz.

Rates of Chemical Weathering of Minerals. One can see in Table 6.1 (upper box) that quartz and some of the micas decompose much less rapidly than feldspars, which are transformed into clay. Indeed minerals can be arranged in order of decreasing susceptibility to chemical decomposition. This order is dependent, not on the kinds of rock in which the minerals occur, but on the stability of chemical constituents in the minerals themselves. Calcium, sodium, magnesium, and potassium are the most susceptible, followed by silica. Alumina and iron are affected least rapidly. In Fig. 6.5 we see several common minerals listed in order of increasing stability in the environments in which weathering takes place.

This **Y**-shaped list is identical with the reaction-series diagram (Fig. 5.20) representing the order in which minerals crystallize from a cooling magma. This means that the first minerals to be created from a magma show the fastest rate of decay when attacked by weathering, whereas the last mineral to crystallize from the magma shows the slowest rate of decay. Obviously, then, minerals such as olivine and calcic plagioclase, which were most stable in the high-temperature, high-pressure environment early in the life of the magma, are least stable under the low-pressure, low-temperature environment of weathering. Hence they decompose most readily. In contrast quartz, stable under the much lower temperatures and pressures that prevailed as the magma finally solidified, is more "at home"—that is, stable—in the environment of weathering.

Just why do the first rock-making silicate minerals to crystallize in a magma become unstable first when they are weathered? The explanation lies in their crystal chemistry. They contain the cations $^2Mg^+$ and $^2Fe^+$, built into their structures in various ways. For example, olivine contains these two cations between successive silica tetrahedra; pyroxene contains them between adjacent single chains of linked silica tetrahedra; hornblende contains them between adjacent double chains of linked silica tetrahedra. These cations are weak spots in the crystal structures, because they are easily removed by weathering. $^2Fe^+$ is removed by oxidation, $^2Mg^+$ by dissolution. Once these cations are removed, the crystal structure collapses. In other words, the mineral is decomposed.

The stubbornness with which quartz yields to decomposition is a part of the reason why quartz is the commonest mineral in sedimentary rocks. Loose grains of quartz in the clayey regolith can be picked up by running water, carried far away, spread out in layers, and eventually cemented to form sandstone. If the sandstone is later exposed to erosion and broken up, the quartz grains will slowly grow smaller with abrasion by water and wind but will persist without chemical change for long ages. Much of the sand now on beaches or in desert dunes consists of quartz grains that were initially set free by the weathering of ancient granitic rocks.

Durability of Minerals. Not only quartz but other minerals as well are stable in the environment at the

Table 6.1

Chemical Weathering of Two Great Groups of Igneous Rocks, Represented by Granite and Basalt

GRANITE

Primary Constituents			Weathering Products		
Minerals	Cations	Colloids	Secondary minerals that form from colloids and ions	Primary minerals that persist	Soluble cations removed in solution
FELDSPARS	K^{+1} Na^{+1}	Silica, alumina	Clay minerals		Na^{+1} K^{+1}
QUARTZ				Quartz	
MICAS	K^{+1} Fe^{+2} Mg^{+2}	Silica, alumina	Clay minerals	Some mica	
FERRO-MAGNESIAN MINERALS	Mg^{+2} Fe^{+2}	Silica, alumina	Clay minerals		Mg^{+2}
		Iron oxides	Hematite, "limonite"		

Biotite — Biotite — Orthoclase — Plagioclase — Hornblende

1 mm

BASALT

Minerals	Cations	Colloids	Secondary minerals that form from colloids and ions	Primary minerals that persist	Soluble cations removed in solution
FELDSPARS	Ca^{+2} Na^{+1}	Silica, alumina	Clay minerals		Na^{+1} Ca^{+2}
FERRO-MAGNESIAN MINERALS	Mg^{+2} Fe^{+2}	Silica, alumina	Clay minerals		Mg^{+2}
MAGNETITE	Fe^{+2}	Iron oxides	Hematite, "limonite"		

Plagioclase — Pyroxene — Pyroxene

1 mm

Primary minerals are shown in capital letters; minerals constituting the product of complete weathering are in shaded boxes. Photographs (left) are thin sections of granite and basalt, with key minerals identified. Uncommon primary minerals (such as rutile) are not included, because they show up in the weathered product in very small quantities. (Upper photo G. M. Friedman; lower photo B. M. Shaub.)

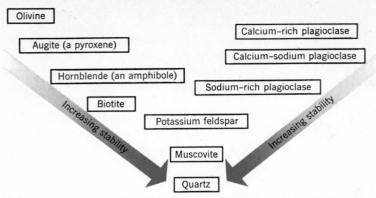

Figure 6.5 Mineral-stability series in weathering. Minerals common in igneous rocks are arranged in order of decreasing rate of chemical weathering—in other words, in order of increasing stability in the environment where weathering occurs. (*Goldich, 1938.*)

Earth's surface, and so resist destruction by weathering. Like quartz, such minerals as gold, platinum, and diamond persist in weathered regolith and so are eroded and become sediment. Because these minerals are unusually heavy, they drop out and become concentrated on the beds of streams, forming the *placers* (Fig. 20.27) from which they are mined. The widespread occurrence of these minerals in sedimentary strata of all ages indicates that weathering, as part of the rock cycle, has been going on through much if not all of Earth's history.

Mechanical Weathering: Disintegration

In many places regolith consists wholly of rock material that is identical in every way with the local bedrock. Hence, no chemical changes like those just described can have occurred; the weathering processes that formed such regolith must have been mechanical rather than chemical. Mechanical breakdown occurs commonly. It is brought about by freezing and thawing, by activities of plants and animals, and by the heat of fires and probably of the Sun as well.

Frost Wedging and Frost Heaving. In cold and cold-temperate climates, water freezes in both bedrock and regolith. Thermometers installed in such places show that in winter, at least, the water alternately freezes and thaws, in many places at least once daily. Laboratory experiments show that when water

crystallizes into ice, a volume increase of 9 percent occurs. Consequently, when ice forms in an opening within rock, rock material is forcibly pushed up or pried apart—not only tiny particles but also blocks small and large, some of them weighing many tons. This mechanism is aptly called *frost wedging*. In Fig. 6.6 we can see that movement of the largest block of rock has been sideways, toward the left. This is the direction of easiest relief of the pressure exerted by water freezing in a former crack (now a wide gap)—the direction in which the least energy would be needed to move the block. This is why frost wedging is particularly effective at the faces of cliffs. But even with no cliffs present, it breaks up bare bedrock, littering the ground with broken pieces that in some places conceal the bedrock beneath (Fig. 6.7).

Because most cracks and cavities in rock are open to the atmosphere, it might be supposed that freezing water could easily expand outward along the cracks. But water in the upper and outer parts of the openings tends to freeze first, creating confined spaces in which further freezing can cause the rock to burst. As nearly all rock is cut by cracks of various kinds, frost wedging is believed to be an effective rock-breaking process. Probably the wedging apart of tiny mineral grains loosens and, in the long run, moves more rock material than the more spectacular pushing of large blocks.

Freezing water affects not only bedrock but also fine-grained regolith, in which it causes *frost heaving*. Early in winter, the water in such regolith freezes

just below the surface as the ground loses heat to the atmosphere. By additions from rainwater and melting snow the ice thickens. During freezing, expansion heaves up the material above the frozen zone and creates bulges in the surface. The bulges, "frost boils," are conspicuous in roads at the end of winter.

Mechanical Work of Plants and Animals. As we have seen, chemical weathering concentrates in cracks and other openings in rock. It is in such places, then, that K^{+1} and other fertilizer ions go into solution. Because of the fertilizer, plants tend to concentrate in the same places and extend their roots into cracks. As they grow, tree roots wedge apart the adjoining blocks. Shrubs and smaller plants also send their rootlets into tiny openings, which they enlarge. Although it would be hard to measure, the total amount of rock breaking done by plants must be enormous. Much of it is obscured by chemical decay, which takes advantage of the new openings as soon as they are created.

Burrowing animals large and small (ants, for example) bring quantities of partly decayed rock particles to the surface to be exposed more fully to chemical action. More than a hundred years ago, Charles Darwin made close observations in his English garden, and calculated that every year earthworms brought particles to the surface at a rate of more than ten tons per acre. After a study in the basin

Figure 6.6 Mechanical weathering (chiefly frost wedging along joints) is disintegrating diabase bedrock in East Greenland, forming large, angular blocks. Curved surface (skyline of foreground) was created by glacial erosion during latest glacial age (compare glacially abraded rock, Fig. 11.9). The carbon-14 dates of samples collected nearby show that glaciers had melted back and that this locality was exposed to weathering about 8000 years ago. Thus the disintegration shown is the work of about 8000 years of weathering. Mesters Vig, Greenland. (*A. L. Washburn.*)

Figure 6.7 Result of frost wedging. This hillslope in East Greenland is littered with slabs of the underlying sandstone bedrock (not visible). They show no sign of having been weathered chemically. Hammer, center, gives scale. (*A. L. Washburn.*)

of the Amazon River, the geologist J. C. Branner wrote that the soil there "looks as if it had been literally turned inside out by the burrowing of ants and termites" (Fig. 6.8). The amount of rock material moved by burrowing organisms, cumulatively through hundreds of millions of years, must be huge. It illustrates again the cumulative effect of small forces acting through a very long time.

Effects of Heat. The heat of forest and brush fires breaks large flakes from exposed surfaces of bedrock. Because rock is a poor conductor of heat, fire heats only a thin outer shell, which expands and breaks away. Fires set by lightning must have been common during long ages before man began to disturb nature's economy, and fire has probably been an important factor in the mechanical breaking of rock.

Figure 6.8 A colony of termites in northern Australia built this pinnacle-like structure by bringing up tiny particles, one by one, from below ground and cementing them with a liquid they secrete. In some areas dozens of such pinnacles are visible in a single view. Detail photo shows two termites building one compartment of a similar mound. (*B. J. Skinner.*)

There is also reason to believe that in hot regions, daily changes in the amount of heat received from the Sun may be great enough to disrupt rock, although little is yet known about this activity.

Exfoliation and Spheroidal Weathering. A process that occurs commonly in rock that is cut by joints is *exfoliation, the separation, during weathering, of successive shells from rock*, like the "skins" of an onion (Figs. 6.9, 6.10). In some cases only a single shell is present, in others ten or more. The outermost shells tend to be flattish, whereas the innermost ones are spheroidal as corners become more and more rounded. Evidently exfoliation can take place not only at the surface but below ground as well, because we see it exposed in newly made cuts along roads

(Fig. 6.9). The process is not restricted to a particular type of climate, although it is seen most commonly in dry climates. The spheroids created by exfoliation usually form a distinct pattern of nearly straight rows running in two or more directions. This is because the positions of the spheroids are controlled by joints (visible in Fig. 6.9) that existed in the rock long before weathering began. The joints were avenues of slow movement of water solutions that attacked the rock, causing chemical weathering.

Although the outer surface of exposed rock dries rapidly after a wetting, moisture that penetrates between mineral grains and into crevices remains long enough to cause some decomposition of feldspars and create clay. The process is accompanied by increased volume of the weathered rock. Probably the

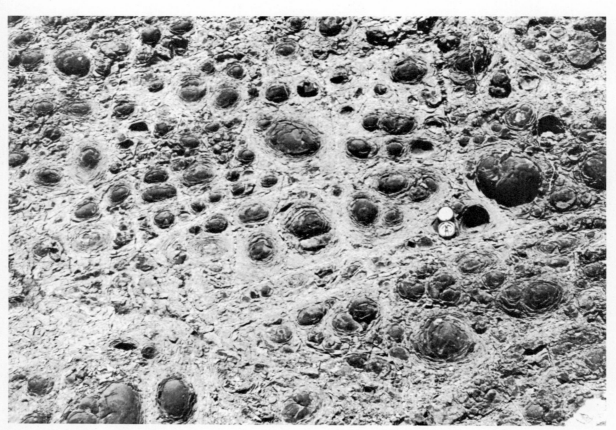

Figure 6.9 Basalt, exposed in a road cut in western Argentina, is cut by three sets of joints, one nearly vertical, a second inclined gently from right to left, and a third (not visible) paralleling the plane of the photograph. Solutions penetrating inward from the joints have converted nearly cubic blocks of rock to spheroids, each with several successive shells. Diameter of mirror on the compass (*right center*) is 7.5cm. (*R. F. Flint.*)

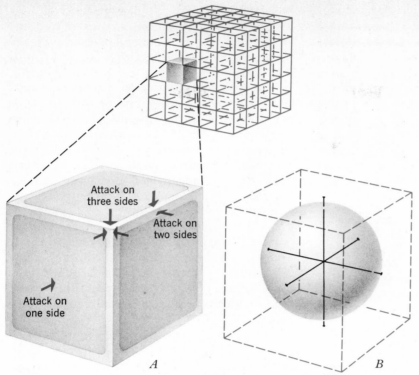

Figure 6.10 Geometry of spheroidal weathering. *A*. Solutions that occupy joints separating nearly cubic blocks of rock attack corners, edges, and sides at rates that decline in that order, because the numbers of corresponding surfaces are 3, 2, and 1. Corners become rounded; eventually the blocks are reduced to spheres (*B*). Energy of attack has now become distributed uniformly over the whole surface, so that no further change of form can occur.

increase sets up small forces that cause shells to separate from the main body of the rock. Thus we have a *mechanical* effect that is a result of chemical weathering.

Factors That
Influence Weathering

Kind of Rock. As Fig. 6.1 shows, the minerals of which a rock is composed influence its decomposition. Quartz is so resistant to decomposition that rocks (such as granite) that are rich in quartz stoutly resist chemical weathering. For example, in the eastern United States, from Maine to Georgia, granite hills and mountains are distinctly higher than surrounding areas underlain by weaker rocks. This relationship between mineral composition and the

details of form of land surfaces is the result mainly of differences in rate of weathering.

However, rate of weathering of a rock is influenced not only by minerals but also by structure. If a rock, even though it consists entirely of quartz (sandstone, quartzite), contains closely spaced joints or other partings, it can disintegrate very rapidly, especially when attacked by frost wedging.

Slope. When a mineral grain is loosened by decomposition on a steep slope, it is washed down the hill by the next rain. With the solid products of weathering moving quickly away, fresh bedrock is continually exposed to attack, so that weathered rock extends only to slight depth below the surface. In contrast, on gentle slopes weathering products are not readily washed away; so they accumulate to depths that in places reach 50m or more.

Climate. Because abundant moisture and heat promote chemical reactions, weathering generally goes deeper in moist and warm than in dry and cold climates. In high mountains and in cold high latitudes, frost-wedged rock fragments litter large areas of the surface, partly obscuring any effects of decomposition. Rocks such as limestone and marble (Fig. 6.2), which consist almost entirely of calcite, decompose quickly in moist climates because calcite is very soluble in carbonic acid. In dry climates, however, such rocks are very resistant, because with little rainfall, rock comes into contact with carbonic acid only rarely.

Time. Study of decomposition of stones in ancient buildings and monuments readily shows that hundreds or even thousands of years are required for hard rock to decompose to depths of only a few millimeters. Granite and other kinds of bedrock in the Sierra Nevada, New England, northern Europe, and elsewhere still exhibit polish and fine grooves made by glaciers during the latest glacial age, some 10,000 to 25,000 years ago (Fig. 11.8). We can reason that in such rocks and in such cool-temperate climates, it could take many tens of thousands of years, at the very least, to create weathered regolith like that shown in Fig. 6.3.

But in some regions weathering extends far deeper and is far older. Mining operations have exposed bedrock that has been thoroughly weathered down to depths of 100m continuously through many tens of millions of years.

The most promising way to measure the time involved in developing a zone of weathering is by radiometric dating. At present we have barely made a start at measuring rates of weathering.

Soils

One of the significant aspects of weathering is that it creates soil, "the great bridge between the inanimate and the living." Soil is the principal natural resource of any country, and the special field of study, *soil science,* is concerned with its origin, use, and protection.

The term *soil* is used in more than one sense. To an engineer it is a synonym for regolith, the accumulation of all loose rock material above bedrock. Most geologists, however, adopt the definition used in soil science and agriculture: *soil is that part of the regolith which can support rooted plants.*

Soil Profiles. Three recognizable layers, or *horizons,* are formed in soil. *The succession of distinctive horizons in a soil, from the surface down to the unchanged parent material beneath it,* is a **soil profile** (Fig. 6.11).

The uppermost horizon, or *A* horizon, is commonly grayish or blackish because of the addition to it of *humus, the decomposed residue of plant and animal tissues.* The *A* horizon has lost part of its original substance through mechanical removal (by soaking downward through the ground) of clay particles and, more importantly, through the chemical removal of soluble minerals. Such removal, as we noted earlier, is *leaching.*

The *B* horizon, although poor in organic content, is a site of accumulation, for it has gained part of what has been leached from the *A.*

Although it is part of the soil profile, the *C* horizon is not part of the soil itself. It consists of parent material, usually bedrock, and has no distinct lower limit.

The parent material from which a soil is developed can be bedrock or regolith of any kind. In the "bottom lands" of large valleys the soil is exceptionally rich and thick because the parent material itself consists of topsoil washed down into the valleys

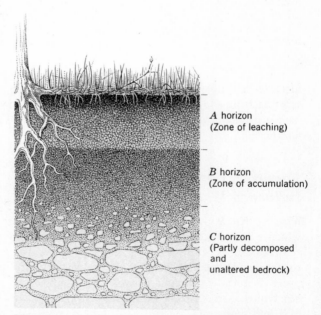

A horizon
(Zone of leaching)

B horizon
(Zone of accumulation)

C horizon
(Partly decomposed
and
unaltered bedrock)

Figure 6.11 The *A, B,* and *C* horizons of a typical soil profile. Each horizon grades down into the one beneath it.

from nearby uplands. In much of the northern United States and Canada, the parent material of soils is regolith consisting of glacial deposits (Chap. 11) created as recently as 10,000 years ago by former ice sheets. Because of their rather short history, the soils of glaciated New England and the Great Lakes region are only partly developed. Soil scientists call them *immature*. The soils farther south, undisturbed by glaciation, are generally mature. A ***mature soil*** is *a soil that has a fully developed profile*.

Climatic Soil Groups. Parent materials differ widely and strongly influence the character of soils, especially during the earlier part of soil development. But through a long period of time the influence of climate is even stronger than that of bedrock in determining the character of soil. Under similar climatic conditions the profiles of mature soils developed on widely different kinds of rock become surprisingly alike.

In a general way the soils in the United States are divided into two groups on the basis of degree of leaching by rainwater. The most effective leaching is found in areas with average annual rainfall of more than 25 inches. An irregular line (Fig. 6.12) extending northward from central Texas to western Minnesota divides the more humid eastern half of the country from the drier western half. East of this line rainwater has leached from *A* horizons of mature soils large fractions of the calcium and magnesium carbonates that formed part of the parent material, and has deposited much clay and iron in the *B* horizons. West of the line, where rainfall is less abundant, carbonates accumulate in the upper part of the soil, making it strongly alkaline in contrast to the acidic soils of humid regions. An important part of the accumulation results from evaporation of water that rises through the ground, bringing dissolved salts from below. In large areas of Texas and adjoining states carbonates have, in this way, built up a solid, almost impervious layer. Such *a whitish accumulation of calcium carbonate developed in a soil profile* is generally known in western North America as ***caliche*** (kă lē′ chē).

A soil with calcium-rich upper horizons is a ***pedocal*** (Greek *pedon,* "soil," and the first syllable of *calcium*). *A soil in which much clay and iron have been added to the B horizon* is a ***pedalfer*** (symbols for aluminum and iron—Al and Fe—added to the Greek root).

Rate of Soil Formation. Although the making of soil constitutes a part of the complex process of weathering, soil formation and weathering are not synonymous. The time factor in weathering chiefly concerns the decomposition of bedrock, which involves great lengths of time and is essentially a geologic process. The length of time required to form an orderly series of soil horizons in regolith involves far shorter periods. Soil scientists believe

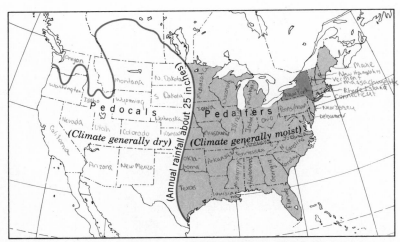

Figure 6.12 General division of soils in the United States into two major classes, separated by a line along which annual rainfall is about 25 inches.

that on loose sandy material (not bedrock) having a forest cover, a soil might develop within 100 to 200 years.

Role of Weathering in the Rock Cycle

The group of processes that together constitute weathering are part and parcel of erosion and can not be separated from it. This fact is emphasized in the definition of erosion given in Chap. 1. Indeed weathering occupies a key place in the rock cycle.

The place where weathering occurs is like the post office of origin in a worldwide sorting system, where outgoing pieces of mail get their first sorting and are sent off in various directions, only to be sorted again and again as they approach their destinations. In the sorting system begun by weathering, particles are detached or extracted from rocks and are sent on their way down the nearest slope, into a stream, and so onward into a basin, where they are deposited. At every step along the way the particles are sorted according to their size, their weight, their shape, their durability, and other factors. Some are deposited at various points en route; others make it to a basin of some kind—the end of the particular system through which they are moving.

The products of chemical weathering are more stable than are the materials of rocks that have not been weathered. We could state this differently by saying that weathering results in an increased adjustment of rock materials to environments at Earth's surface. This is because more and more of the fresh unweathered rock, including minerals possessing various degrees of stability, gives place to chemically weathered products that are quite stable with respect to the surface environment. Of these products the most abundant are the clay minerals. Broadly speaking, then, chemical weathering is a huge clay-making process. It is also a process in which quartz is unlocked from granites and other quartz-bearing rocks; the quartz, being comparatively stable, persists in an unchanged state. The two most conspicuous products of weathering, new-made clay and old (but now free) quartz, go on into the next phase of the rock cycle, are transported and deposited, and eventually are transformed into claystone and sandstone.

Meanwhile, other products of weathering are the cations Na^{+1}, K^{+1}, Ca^{+2}, Mg^{+2}. Most of the K^{+1} goes immediately into plants and certain clay minerals. The other cations move slowly through the ground in solution (Chap. 9) and emerge at all points along the line into streams and so eventually reach the sea. There Ca^{+2} and Mg^{+2} are deposited, mostly as carbonates, eventually to form the limestone and dolostone that with claystone and sandstone constitute the common sedimentary rocks. The Na^{+1} mostly remains in solution.

In the final analysis, therefore, weathering is one of two great links between rocks and the sediments derived from them. The other link consists of the processes that transport weathered detritus from the places of weathering to the places of deposition. Having examined weathering, we shall consider the transporting processes in the chapters that follow.

Summary

Environment

1. The ground surface and a shallow zone beneath it represent an environment of low temperature and pressure, in which water and organic matter are present.

Processes

2. Chemical weathering and mechanical weathering are very different but generally work in close cooperation.

3. Subdivision of large blocks into smaller particles increases surface area and thereby accelerates weathering.

Chemical Weathering

4. Carbonic acid is the prime agent of chemical weathering; heat and moisture speed chemical reactions.

5. Chemical weathering converts feldspars into clay. Grains of quartz, however, escape chemical decomposition, survive, and are carried away to be deposited as sand.

Mechanical Weathering

6. The most obvious process of mechanical weathering is frost wedging, in which mineral composition is not altered.

7. Decomposition, however, is one of the causes of mechanical weathering.

Soils

8. A soil profile consists of horizons A, B, and C.

9. In moist climates, profiles of mature soils are similar even though they are developed in bedrock of different kinds.

10. Pedalfers form in areas of high rainfall, pedocals in areas of low rainfall.

Effect of Biosphere

11. Animals and plants play a significant part in many processes of weathering.

The Rock Cycle

12. Weathering transforms bedrock into residual regolith, the chief source of sediments, which eventually become sedimentary rocks.

13. Weathering is the source of sodium in the sea and of calcium in limestone.

Selected References

Bradley, W. C., 1963, Large-scale exfoliation in massive sandstones of the Colorado Plateau: Geol. Soc. America Bull., v. 74, p. 519–528.
Carroll, Dorothy, 1970, Rock weathering: New York and London, Plenum Press.
Cruickshank, J. G., 1972, Soil geography: New York, Halsted Press Div., John Wiley.
Goldich, S. S., 1938, A study in rock weathering: Jour. Geology, v. 46, p. 17–58.
Hunt, C. B., 1972, Geology of soils. Their evolution, classification, and uses: San Francisco, W. H. Freeman.
Jenny, Hans, 1950, Origin of soils, *in* Trask, P. D., Applied sedimentation, a symposium: New York, John Wiley, p. 41–61.
Keller, W. D., 1957, The principles of chemical weathering: Columbia, Mo., Lucas Bros.
Lyon, T. L., Buckman, H. O., and Brady, N. C., 1960, The nature and properties of soils, 6th ed.: New York, Macmillan.
Ollier, C. D., 1969, Weathering: New York, American Elsevier Publishing Co.
Reiche, Parry, 1950, A survey of weathering processes and products, rev. ed.: Albuquerque, Univ. of New Mexico Publications in Geology.
U.S. Department of Agriculture, 1938, Soils and men: Yearbook for 1938: Washington, U.S. Gov. Printing Office.
U.S. Department of Agriculture, 1957, Soil: Yearbook for 1957: Washington, U.S. Gov. Printing Office.

Chapter 7   Mass-wasting

Downslope Movement

Universality of Mass-Wasting. Once a particle is loosened by weathering, it ceases to be bedrock and becomes regolith. Because it is loose it becomes a prey to forces that move it downslope under the influence of gravity (Fig. 7.1). Sooner or later the particle reaches a stream or some other carrier, which transports it farther. Finally it ends up in the sea.

The beginning of the long journey, the trip down the nearest slope, can be very slow or very fast, but in either case it is controlled primarily by gravity. The movement is of several kinds, known collectively as **mass-wasting,** *the movement of regolith downslope by gravity without the aid of a stream, a glacier, or wind.* Although our definition excludes flowing water as a medium of transport, water nevertheless plays an important part in mass-wasting. Saturation of regolith with water makes movement easier. This is the main reason why some mass-wasting activities are especially common after long rains.

Mass-wasting affects whole bodies of regolith. Wide blankets and narrow streams of rock particles, large or small, move down hillsides, cliffs, and other slopes. But at times single particles, small boulders for instance, are detached from a cliff and fall, roll, or bounce downslope. This, too, is mass-wasting, because such movement depends on the force of gravity directly, rather than indirectly through a carrying medium.

Mass-wasting is not confined to the lands. Regolith, in the form of transported sediments,

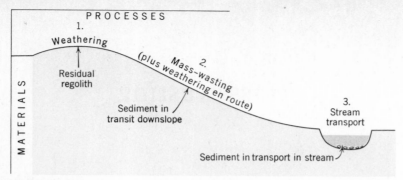

Figure 7.1 Relation of mass-wasting (2) to weathering (1) and to stream transport (3).

covers vast areas of the sea floor. Such material is seen moving down submarine slopes. Because such movements resemble mass-wasting, we think of mass movement as a universal process, submarine as well as subaerial, active wherever slopes exist. On any slope mass-wasting depends on the interplay of the components of gravity g_p and g_t (Fig. 3.11). When g_t exceeds g_p reinforced by cohesion and friction among the particles of rock, movement occurs. Even where movement is far too slow to be seen without making very careful measurements, one can identify, in the regolith, particles derived from kinds of bedrock exposed only in areas that lie farther up the slope; so we know that movement has occurred.

Mass-wasting is nearly universal because the Earth's land areas (and sea floors too) consist almost entirely of slopes, mostly arranged in systematic groups (Fig. 8.23). Natural flat surfaces that are truly horizontal are very rare. Furthermore most land slopes are gentle; probably the majority lie at less than about 5° from the horizontal (Fig. 7.2). Although steep slopes attract attention because they are conspicuous, in terms of the amount of land surface they represent, they are rare.

The best way to visualize mass-wasting as a whole is to think of it as a natural system. Its input consists of the solid products of weathering contributed all the way down each slope; its output consists of sediment discharged, at the bases of the slopes, into carriers such as streams. As soon as regolith begins to move downslope it becomes sediment by definition. A steady state in the system is represented by a balance between input and output, and also by alteration of the slope itself to an angle that will just permit the quantity of regolith moving down it to maintain that balance.

Practical Importance of Mass-Wasting

The broad complex of activities we call mass-wasting is subdivided into a varied group of separate processes. Although they differ in other respects, these processes share one characteristic: they all take place on slopes. Many of them were recognized even before the beginning of the present century, but study of them was accelerated by the soil-conservation movement in North America, which gained impetus in the early 1930s. Scientists began to give close attention to the soil of farmlands and pastures and, hence, to the regolith as a whole. Their first objective was to help reduce rates of soil erosion by

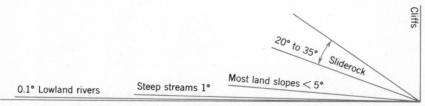

Figure 7.2 Representative angles of natural slopes. Most natural slopes are gentle; steep slopes are rare.

running water and wind. Such rates had been increasing alarmingly, partly through lack of understanding of the various factors that determine the stability of soil on slopes.

An additional reason for research on mass-wasting was the recognition that such research could help solve problems in engineering. Some important engineering questions are: How stable are the foundations of proposed buildings, dams, and other structures built on or at the bases of slopes of various angles and in various materials? If they are unstable, what kind of engineering treatment will create stability, or should the proposed structure be built elsewhere?

Classification of Processes

Studies made to help answer such questions led to attempts to define and classify the processes of mass-wasting. Some of the more obvious processes are illustrated in Table 7.1, in which the sketches are almost self-explanatory. It would be satisfying to be able to classify the various processes entirely according to the kind of motion each displays. But this is not possible because some processes involve two or more distinct kinds of motion. Again, it would help if we could classify the processes according to their velocities. But we cannot, because the velocity of a single process at a single locality can vary with time. Even the gaps between the bars in Fig. 7.3 may not be correct, because the lengths of some of the bars are based on very few data. Let us take up the more rapid processes first, and then the slower ones.

Falling and Sliding. *Rockfall* and *debris fall, rockslide* and *debris slide,* and *slump* are generally small-scale processes, seen on cliffs and steep slopes where rock particles both small and large fall or slide downward. Slump is particularly common in places where slopes are kept steep and clifflike by erosion at their bases, as along stream banks (Fig. 7.4) and coastal cliffs (Fig. 7.5).

Rock Avalanche. Early in 1903 a remarkable event destroyed much of the coal-mining town of Frank, Alberta, killing 70 people. With a great roar, a mass of rock having a volume of some 40 million cubic yards rushed down the face of Turtle Mountain, 3000 feet high, at a speed estimated at about 60 miles per hour. Its momentum carried parts of the moving mass across a valley 2 miles wide and 400 feet up the opposite side. Later investigation showed that the configuration of Turtle Mountain almost guaranteed an unstable situation (Fig. 7.6). Joints in the limestone strata were inclined toward the valley and were being subjected to solution and frost wedging. When the limestone becomes further weakened, the rock near the top of the mountain will be in serious danger of sudden movement, in another event similar to that of 1903.

Formerly the Frank event was referred to as a "landslide," a term we do not use here because it is ambiguous, implying that only sliding occurred. Rockslide and rockfall certainly were involved, but the dominant process is thought to have been *rock avalanche,* in which the dry, moving debris actually flows. The high velocity attained and the fact that

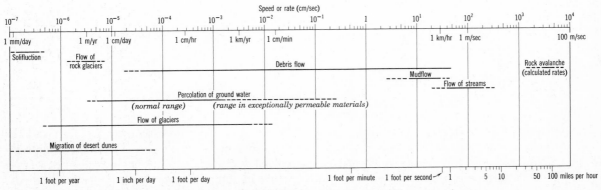

Figure 7.3 Ranges of measured rates of movement of selected processes of mass-wasting, compared with those of some other external processes.

Table 7.1
Processes Involved in Some Kinds of Mass-Wasting

Process	Definition and Characteristics	Illustration
Rockfall and debris fall	*The rapid descent of a rock mass, vertically from a cliff or by leaps down a slope.* The chief means by which taluses are maintained.	
Rockslide and debris slide	*The rapid, sliding descent of a rock mass down a slope.* Commonly forms heaps and confused, irregular masses of rubble.	
Slump	*The downward slipping of a coherent body of rock or regolith along a curved surface of rupture.* The original surface of the slumped mass, and any flat-lying planes in it, become rotated as they slide downward. The movement creates a scarp facing downslope.	
Debris flow	*The rapid downslope plastic flow of a mass of debris.* Commonly forms an apronlike or tonguelike area, with a very irregular surface. In some cases begins with slump at head, and develops concentric ridges and transverse furrows in surface of the tonguelike part.	
Variety: Mudflow	*A debris flow in which the consistency of the substance is that of mud;* generally contains a large proportion of fine particles, and a large amount of water.	

debris was carried uphill in places suggest that mixtures of rock particles and air flowed on a cushion of pure air trapped and compressed within and beneath the rushing mass. Apparently air pressure forces the moving rock particles apart, exercising a lifting force that overcomes the force g_p, and the mass may become fluidized. Where the heights involved permit great acceleration, velocities well in excess of 100 miles per hour are thought to be possible.

110   Mass-wasting

Figure 7.4 Slump, on a small scale, in regolith consisting of loose silt, clay, and sand, in the bank of a stream. Slump was started when erosion by the stream steepened the bank and made it unstable. Spartanburg, South Carolina. (*C. F. S. Sharpe, U.S. Soil Conservation Service.*)

Figure 7.5 Slump, on a large scale, that occurred in coastal cliffs at Point Firmin, California. Principal surface of slip is curved; it cuts cliff and displaces highway at two points. Other surfaces of slip, also curved, are shown by arrows. Erosion of foot of cliff by surf may have been the chief factor in removing support and causing movement. (*Spence Air Photos.*)

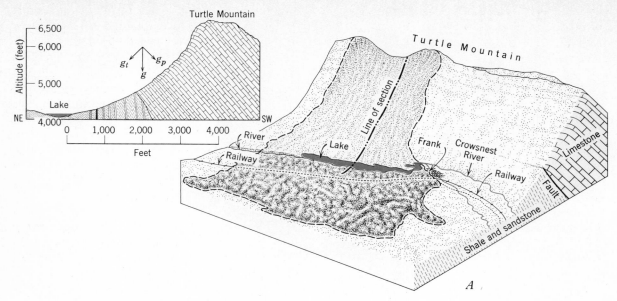

Figure 7.6 *A.* The rock avalanche at Frank, Alberta shot 40 million cubic yards of rock downward 3000 feet to the base of Turtle Mountain, dammed a river to form a lake, destroyed 7000 feet of railway line and part of the town of Frank, and spread a great apron of debris up the gentle slope beyond, reaching as high as 400 feet above the lake. Dotted line shows segment of railway that was destroyed. Section at left was made along a line running directly down the mountain slope north of the avalanche area. (*Data from Geol. Survey of Canada, modified by D. M. Kruden and J. Krahn.*)

B. Photo taken in 1972, looking from debris toward Turtle Mountain, shows large size of blocks involved in the slide. (*B. J. Skinner.*)

Vaiont Event. A colossal mass-wasting event that occurred in 1963 attained disaster proportions, because in it 2600 people lost their lives. This occurrence was in the deep Vaiont Canyon, which traverses mountainous country in the Italian Alps (Fig. 7.7). A great dam, 265m high, had been built in the canyon in 1960, impounding a deep reservoir. On October 9, at 10:41 P.M., a tremendous body of rubble 1.8km long and 1.6km wide, consisting of an estimated 240 million cubic meters of rock debris, thundered down the slopes on the south side of the canyon, setting up vibrations in the ground that were recorded throughout much of Europe. Within about 1 minute, debris had filled the reservoir and had piled up to more than 150m higher than the water surface. Air compressed by the rapidly flowing debris moved water and rock material more than 260m up the opposite (northern) side of the canyon.

Had the reservoir not been present, probably the disaster would have been far less serious. But the sudden filling with debris displaced a huge volume of water, which struck the dam in waves so large that they overtopped it by 100m. Although the dam itself withstood the impact, the water swept on as a flood. It destroyed every structure down the valley through 20km or more, all within a period of 7 minutes.

Thorough investigation showed that two conditions had contributed greatly to the movement. One was an unfavorable geologic situation; the other was the presence of the reservoir. In other words, both natural conditions and human activity were involved.

The bedrock consists of layers of limestone and claystone, weakened by much ancient deformation that has bent the layers and created many joints. Back in prehistoric times these conditions had caused earlier mass-wasting on a large scale, as inferred from debris. Then, in 1960, when the reservoir was first filled, water seeped from it into the rock of the valley sides, saturating the rock and changing the environment below ground. Moistening of rock layers previously dry caused the clay to increase in volume and become plastic, weakening the rocks. In 1960 rockslide occurred on a small scale, showing that conditions were becoming worse.

During the period between the 1960 and 1963 events, surveys showed that the ground was moving downslope at a rate of around 1cm per week. In September 1963 heavy rains lasting 2 weeks intensified the movement, which increased to about 1cm per day and finally ended in catastrophe.

The activity in the Vaiont event was complex. Probably the slow movement between 1960 and 1963 was principally rockslide accompanied by debris flow, but the 1963 catastrophe was mainly rock avalanche, as indicated by the evidence of the action of compressed air.

Debris Flow. Many if not most of the conspicuous events in mass-wasting involve debris flow (Table 7.1). In many of them movement begins with slump and continues as debris flow. Figure 7.8 shows the idealized relationship between slump and debris flow, with well-defined masses, created by slump,

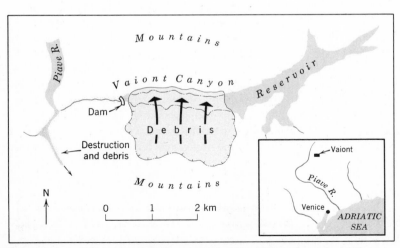

Figure 7.7 Sketch map showing setting of the Vaiont event in 1963.

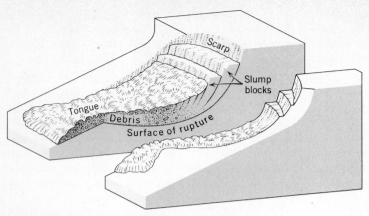

Figure 7.8 Idealized sketch showing chief characteristics of many debris flows. Compare Fig. 7.9.

breaking up into a single flowing mass that is ordinarily tonguelike in shape. Many such features are 2 to 5km in length, although some are larger and others much smaller. Figure 7.9 shows the same features: slumped blocks at head, tonguelike flowing mass downslope. Rates of flow range from one inch per day to one foot per second.

Nicolet Debris Flow. A clear example of debris flow combined with slump occurred at Nicolet, Quebec, a town built on the flat surface of a body of silt and clay. The Nicolet River had cut a cliff in this material, and late in 1955 the widening of a road along its base steepened the cliff artificially. Following heavy rains in November of that year, the mass of sediment became unstable, and on November 12, at 11:40 A.M., slump like that shown in Fig. 7.4 occurred back of the cliff. The slump blocks broke up immediately and flowed toward the river as jumbled masses. As rapidly as the flowing away of each mass weakened the support of the slump scarp (Fig. 7.6), further slump occurred, and the blocks in turn broke up and flowed away. Buildings and pavements were undermined, promptly disintegrated, and were incorporated into the flowing mass. The whole event lasted 7 minutes. A 7-minute movement that transports material through 500 feet would have a velocity of about 70 feet per minute, or more than 1 foot per second. Yet despite all this activity only three lives were lost.

The movement stopped because the long gentle slope from scarp to river, created by debris flow, had restored stability. The repeated slump destroyed a

garage, a school, and several houses but stopped just short of the cathedral, which was left intact. The final result was a bite 500 feet long, 300 feet wide, and 10 to 20 feet deep taken out of the town (Fig. 7.10). The bite had a concave scarp at its head and was floored with a lobate tongue of flowed debris more than 800 feet long, which dammed the river temporarily.

Mudflow. A fast variety of debris flow, mudflow is characterized by the presence of mostly fine particles and a water content that can amount to as much as 30 per cent. It does not originate in slump blocks. As a result of its fine grain size and large water content, flowing mud tends to follow valleys as streams do.

Mudflow material grades from mud as stiff as freshly poured concrete to a souplike mixture not much thicker than very muddy water. In fact, after heavy rains in mountain canyons, mudflow can start as a muddy stream that continues to pick up loose material until its front portion becomes a moving dam of mud and rubble, extending to each steep wall of the canyon and urged along by the pent-up water behind it. On reaching open country at the mountain front, the moving dam collapses, floodwater pours around and over it, and mud mixed with boulders is spread out as a wide, thin sheet (Fig. 7.11) with destructive effects on farms and towns.

Because of its great density, which enables it to move large, heavy objects, mudflow is destructive. Houses and barns in the paths of some mudflows have been carried from their foundations, and large blocks of rock are pushed along, rolling and sliding in the slimy mixture, some of them finally coming to rest on gentle slopes well out beyond the foot of a mountain range. The occurrence of huge, isolated boulders in such positions is not uncommon, and has led to the erroneous inference that the boulders were carried by former glaciers or by huge stream floods. Certainly some and probably many of them are the work of mudflow, the mud having been later removed by erosion.

Such removal is not difficult, because the sediment deposited by mudflow is seldom thick at its lower end—commonly only a few centimeters and rarely much more than 1m, although in a narrow valley near the upper end, thickness can be far greater.

Volcanic Mudflow. In regions of explosive volcanic activity, layers of volcanic ash are common. Noncemented fine-grained ash, mantling slopes, is peculiarly susceptible to mudflow. A sudden heavy rain can bring this loose, easily transportable material into and down valleys in huge volume, as mudflow. Where downvalley slopes are gentle the mushy material, having the consistency of fresh, fluid concrete, can flow long distances and can build up thick, flat-topped deposits of sediment (Fig. 7.12) that fill the containing valley from side to side. A valley fill

Figure 7.9 Slump and debris flow, near Oakland, California, December 9, 1950. The slopes, although steep, are exaggerated somewhat by the angle of view. The whole feature is more than 700 feet long. (*Bill Young, from San Francisco Chronicle.*)

115

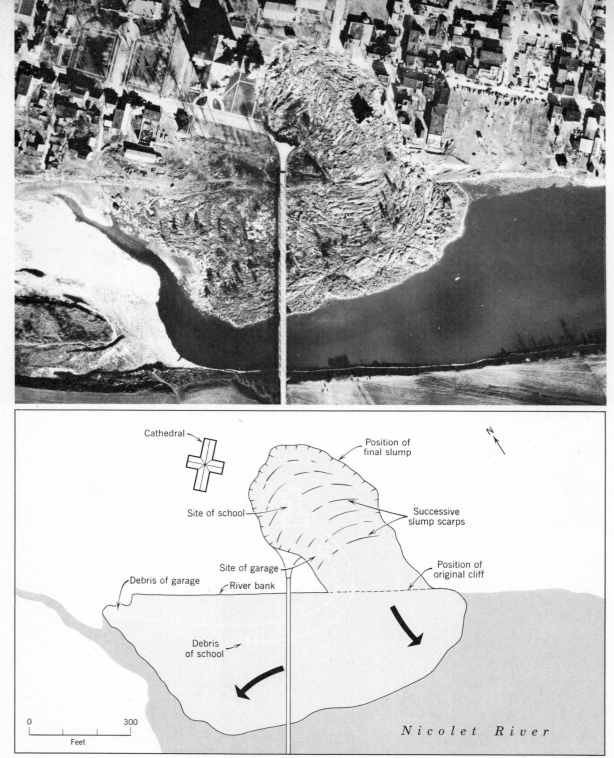

Figure 7.10 Vertical view from the air, showing result of the Nicolet, Quebec debris flow in November 1955, a few days after the event. Large arrows indicate directions of flow. The outer edge of the displaced mass is poorly defined because part of it has been washed over by the river. (*Spartan Air Services, Ltd.*)

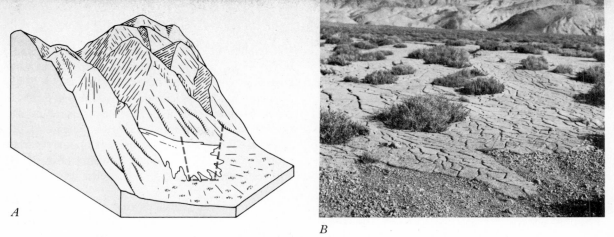

A

B

Figure 7.11 Mudflow. *A.* Typical mudflow setting, at base of mountains in an arid climate. Dashed line shows area like that in *B. B.* Thin mudflow sediment at its downslope limit, forming small lobes on a slightly uneven surface of gravel. Mud has shrunk during drying and has been split by cracks. East base of Stillwater Range, Nevada. (*Eliot Blackwelder.*)

Figure 7.12 Vertical section of sediment of volcanic mudflow, Mud Mountain, east of Tacoma, Washington. Large stones are concentrated near base, marked by small pickaxe. Mudflow sediment overlies stratified sand (dark color) deposited by a stream before the mudflow occurred. Logs found in mudflow sediment yielded a ^{14}C age of about 4800 years. (*D. R. Crandell, U.S. Geol. Survey.*)

of this kind, on the north side of Mount Rainier in western Washington, is 45 miles long and 20 to 350 feet thick; its estimated volume is 1.5 billion cubic yards. A similar though smaller volcanic mudflow buried and destroyed the Roman city of Herculaneum during the famous eruption of Vesuvius in A.D. 79 (Chap. 5).

Creep. In contrast with those described so far, another group of mass-wasting processes moves at rates that are generally imperceptible. In this group the most widespread process is *creep, the imperceptibly slow downslope movement of regolith.* Its effects are evident in the consistent leaning of old fences, poles, and gravestones and in the derangement of roads (Fig. 7.13). Cuts that expose bedrock show that steeply inclined layers of rock curve strongly downslope just below the surface of the ground.

On sloping ground a cover of close-growing grass or other vegetation forms a protective armor against the cutting of gullies by running water and so keeps the slopes smooth. One might therefore suppose that ground so protected loses nothing to erosion except for a little mineral matter dissolved and carried off underground. But closer observation shows clearly that the regolith is creeping downslope, carrying the vegetation with it. In some places boulders of granite have been softened by weathering and drawn out into long thin sheets by movement of the enclosing material (Fig. 7.14). Many factors contribute to the

effectiveness of creep. In regions with cold winters, water in the pore spaces of regolith freezes, increases in volume, and pushes up the surface. This *lifting of regolith by freezing of contained water is* **frost heaving.** On a hillside the surface of the ground is lifted essentially at right angles to the slope; but when thawing occurs, each point tends to drop vertically and so moves downhill (Fig. 7.15). The movement consists of a complex series of zigzags.

The persistent activities that combine to make particles in the regolith creep downslope are listed in Table 7.2. Through countless repetitions on every slope the effects of these activities add up to slow persistent transport of regolith downhill. On a worldwide basis probably all the mass-wasting visibly accomplished by rock avalanche, rockslide, and other rapid processes, no matter how spectacular, is but a small fraction of what is accomplished by creep in the same period of time.

Solifluction. In cold regions, including some high mountains, another imperceptible activity carries regolith downhill. In such places the ground freezes to great depth, and when the weather warms enough to thaw the upper part of the regolith, the deeper part stays frozen. Surplus water then can not percolate downward, and the thawed layer, a few inches to several feet thick, becomes saturated and will not bear the weight of a man. On slopes the water-soaked material flows like an extremely stiff liquid,

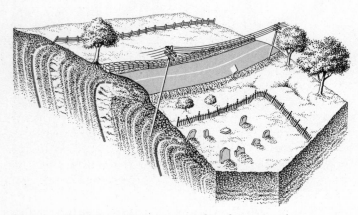

Figure 7.13 Effects of creep on surface features and on bedrock. Blue lines emphasize strata bent over by drag of creeping regolith. (*Modified from C. F. S. Sharpe, 1938.*)

Figure 7.14 Granite boulders in regolith, softened by chemical decay and deformed into pancake shapes by creep. In the lower part of the exposure (the wall of an artificial cut) boulders retain their original form. Nearer the surface of the ground the section of each deformed boulder appears as a long, thin, white streak. (*S. R. Capps, U.S. Geol. Survey.*)

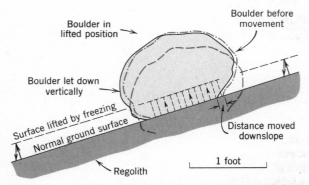

Figure 7.15 A boulder moved downslope by alternate freezing and thawing. Diagram explains mechanism of one step in the movement.

at rates that are as high as 12cm per year (really, per summer season). This *imperceptibly slow, downslope flow of water-saturated regolith* is **solifluction.**

As might be expected, the process is widespread in Arctic regions (Fig. 7.16). Usually the floor beneath the flowing regolith is tightly frozen and therefore impermeable, but solifluction does sometimes occur on a base that is not frozen.

The essential difference between solifluction and

Table 7.2
Causal Factors in Creep of Regolith

Frost heaving	Freezing and thawing, without necessarily saturating the regolith, cause lifting and subsidence of particles
Wetting and drying	Causes expansion and contraction of clay minerals; creation and disappearance of films of water on mineral particles causes volume changes
Heating and cooling without freezing	Causes volume changes in mineral particles
Growth and decay of plants	Causes wedging, moving particles downslope; cavities formed when roots decay are filled from upslope
Activities of animals	Worms, insects, and other burrowing animals, also animals trampling the surface, displace particles
Solution	Solution of mineral matter creates voids, which tend to be filled from upslope
Activity of snow	Where a seasonal snow cover is present, it tends to creep downward and drag with it particles from the underlying surface

most creep lies in the amount of water the regolith contains. For solifluction the material requires a very high water content, but creep can occur in regolith that is completely dry.

Water content large enough to eliminate capillary tension lowers the resistance of regolith to the pull of gravity enough to cause solifluction.

Flow of Rock Glaciers. In mountains where temperatures are very low, remarkable tonguelike bodies, some of them exceeding 1km in length and 3km in width, consisting of angular, blocky boulders as well as finer particles, occur at the bases of cliffs and extend outward into valleys (Fig. 7.17). The kind of rock in the boulders is identical with that which forms the cliffs. These bodies are not the taluses described in a following section; they are too extensive and their lobate form, steep fronts, and multiple concentric ridges suggest that they are flowing or at least have flowed. Because they resemble glaciers in form, they have been given the name *rock glacier,* which we can define as *a lobate, steep-fronted mass of coarse, angular regolith that extends out from the base of a cliff in a cold climate, and that moves downslope aided by interstitial water and ice.*

In rock glaciers studied in Alaska, ice occupies openings between the rock particles. Measurements made during several successive years show these masses are moving outward at rates as great as 1m per year. Boulders are fed to the rock glaciers by frost wedging, water accumulates between the boulders and in the finer material freezes, and the whole mass begins to flow like a true glacier. Nevertheless, despite its resemblance to flow as in a true glacier, the movement in a rock glacier is best classed as a process of mass-wasting.

Sediment Deposited by Mass-Wasting

Colluvium. *Sediment deposited by any process of mass-wasting or by overland flow (Chap. 8) is **colluvium**.* Whether the depositing process is rapid or

Figure 7.16 On this slope, actively moving by solifluction, a line of cones, each with a peg extending down into the regolith, was set up two years earlier along a straight line marked by string. The targets have been displaced in the downslope direction, through a maximum of 25cm during the two-year period. Mesters Vig, northeast Greenland. (*John Scully, from A. L. Washburn.*)

Figure 7.17 Rock glaciers at the base of a cliff, Silver Basin, near Silverton, Colorado. Lobate form, steep fronts, and concentric ridges demonstrate that flow has occurred, but the bodies are not moving now. They are believed to have been active late in the last glacial age. The cliff is the headwall of a cirque (Fig. 11.10). (*Whitman Cross, U.S. Geol. Survey.*)

slow, the particles in a body of colluvium tend to lie in a chaotic jumble (Figs. 7.6,*B*, 7.12) because they were moved by falling, sliding, and similar activities. Also they tend to be angular in shape, because while moving, they did not collide frequently. These characteristics make it possible to distinguish colluvium from sediment deposited after transport in flowing fluids such as water (streams, surf) and air wind). Such sediment tends to consist of rounded particles, sorted and deposited in layers.

Sliderock; Taluses. In areas where steep cliffs prevail and weathering is dominantly mechanical, accumulations of weathered particles usually mantle the bases of the cliffs. The particles, commonly angular and ranging in diameter from sand grains to boulders, are loosened from the bedrock of a cliff by mechanical weathering, accumulate, and move downward at various rates. *The apron of rock waste sloping outward from the cliff that supplies it is a* **talus;** *the material composing a talus is* **sliderock** (Fig. 7.18). From cliff to talus the movement is chiefly falling, sliding, and rolling; within the talus it consists of creep, generally so slow that the rock particles are weathered en route. Water plays little or no part in the movement. Weathering converts

the sliderock into fine-grained regolith which, with its pores of extremely small diameter, can hold much more moisture than sliderock and thus acquire both vegetation and soil (Fig. 7.19).

The activity in a talus constitutes a local system with input of large particles at its head, output of generally finer particles at its toe, and creep in progress from head to toe. The profile of the talus is concave-up, but with a radius of curvature so long that some profiles seem to be nearly straight. In each short segment down its length, the profile represents the angle of rest for the material in that segment. Because the profile as a whole is the sum of all the short segments, in each of which the sliderock is stable, we call it an *equilibrium profile,* or stability profile. In it the force of gravity and other forces, mainly friction, are balanced nicely. The material added from upslope is just equaled by the material that moves on downslope. The same is true of the soil-covered slope downhill from the talus. That slope receives material from the talus and delivers an equal amount to the stream at its foot. Equilibrium profiles, based on a similar nice balance of forces, characterize other natural slopes such as the beds of streams and the beaches created by wave activity along coasts.

121

Figure 7.18. Aprons of coarse sliderock form coalescing taluses against the base of a cliff in Banff National Park, Alberta. (*Jerome Wyckoff.*)

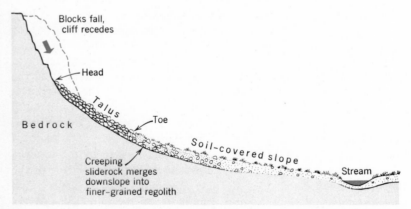

Figure 7.19 Relationship of a talus to bedrock and to fine-grained regolith. The sliderock that forms the talus grades downslope into finer-grained regolith that has been transformed from the sliderock by mechanical and chemical weathering. Also it has been added to by weathering of the bedrock beneath. Fine regolith is slowly fed to the stream, which carries it away. (*After C. F. S. Sharpe, 1938.*)

Practical Applications

Some knowledge of mass-wasting is essential in selecting successful locations for large buildings, bridges, dams, highways, and other engineering works. The St. Francis Dam in southern California, which failed in 1928 with appalling loss of life and property, was built in part on weak, soluble bedrock. This would have been recognized in time to avoid the catastrophe if a competent geologist or an engineer with knowledge of geology had been consulted.

Some engineering works have to be constructed despite troubles with mass-wasting. When work was begun on the deep Culebra Cut for the Panama Canal, slump and debris flow occurred on such a scale that for a time the project appeared to face defeat. It was necessary to reduce the slopes by removing enormous volumes of material beyond the original estimates. Some unstable slopes have been brought under control by building drainage tunnels to carry away water that played an important part in the movement. During construction of the Grand Coulee Dam on the Columbia River, debris flow at a critical location gave serious trouble until engineers conceived the igenious plan of laying pipes through the wet mass and circulating a refrigerant. This created a huge outdoor refrigerator, which froze the ground and kept it stable until the necessary construction was finished.

Private companies and government departments are put to enormous yearly expense in repairing and rebuilding railroads, highways, aqueducts, and other major structures damaged by rapid mass-wasting, especially in country characterized by long, steep slopes. Some of this cost could be avoided by informed planning; the rest is part of the price we pay for civilized living on an unstable Earth.

Despite the spectacularly rapid movements we have described, the part they play in the overall movement of material down the slopes of the lands is minor. Processes such as creep are slow, but they act continuously and they affect nearly all slopes. Hence they do much more work, in the aggregate, as agents of transport. With these almost universal activities in mind we can turn next to running water, the transporting agent that sooner or later receives and handles most of the sediment moved downhill by mass-wasting.

Summary

Downslope Movement

1. Through mass-wasting, residual regolith created by weathering reaches the carrying agencies.

2. Because of the prevalence of slopes, mass-wasting is nearly universal.

3. Some mass-wasting processes are promoted by increased water content of the regolith.

Processes of Mass-Wasting

4. Rock avalanche apparently involves fluidization of a mass of rock particles through compression of air; its motion consists partly of flow of mixed rock and air.

5. Factors that cause creep include cycles of freezing and thawing, wetting and drying, and heating and cooling; also solution, and activities of plants and animals.

6. Although rock avalanche and debris flow are spectacular, imperceptible creep and solifluction, because they are widespread, move more material.

7. Mudflow is particularly common in arid climates where rainfall is sporadic and in areas covered with volcanic ash.

8. Solifluction and the flow of rock glaciers occur chiefly in cold climates, where ice comes into play.

Sediments

9. Sediment deposited by mass-wasting is *colluvium*. It is generally nonsorted and either nonstratified or very poorly so. Although they may show wear, its particles are not rounded. These characteristics distinguish colluvium from sediments deposited in water.

10. Taluses develop at the bases of cliffs; commonly the material composing taluses grades outward into finer, soil-covered regolith.

Selected References

Crandell, D. R., and Waldron, H. H., 1956, A recent volcanic mudflow of exceptional dimensions from Mt. Rainier, Washington: Am. Jour. Sci., v. 254, p. 349–362.

Daly, R. A., Miller, W. G., and Rice, G. S., 1912, Report of the Committee to Investigate Turtle Mountain, Frank, Alberta: Geol. Survey of Canada, Mem. 27.

Eckel, E. G., ed., 1958, Landslides in engineering practice: Highway Research Board, Special Report 29, National Research Council, Washington, D.C.

Howe, Ernest, 1909, Landslides in the San Juan Mountains, Colorado: U.S. Geol. Survey Prof. Paper 67.

Kiersch, G. A., 1964, Vaiont Reservoir disaster: Civil Eng., v. 34, p. 32–39.

Legget, R. F., 1962, Geology and engineering, 2nd ed.: New York, McGraw-Hill, p. 106–128, 385–443.

Rapp, Anders, 1960, Recent development of mountain slopes in Kärkevagge and surroundings: Geografiska Annaler, v. 42, p. 1–200.

Sharp, R. P., and Nobles, L. H., 1953, Mudflow of 1941 at Wrightwood, southern California: Geol. Soc. America Bull., v. 64, p. 547–560.

Sharpe, C. F. S., 1938, Landslides and related phenomena: New York, Columbia Univ. Press. (Reprinted 1960 by Pageant Books, Paterson, N.J.)

Wahrhaftig, Clyde, and Cox, Allan, 1959, Rock glaciers in the Alaska Range: Geol. Soc. America Bull., v. 70, p. 383–436.

Washburn, A. L., 1966, Instrumental observations of mass-wasting in the Mesters Vig district, northeast Greenland: Meddelelser om Grønland, v. 166, no. 4.

Chapter 8

Streams and Sculpture of the Land

Geologic Importance of Running Water

A First Look at a Stream. A good way to understand the important part played by streams in transporting water and sediment down slopes is to sit beside a moderately rapid small stream a few yards wide and watch it. First you see that the water is flowing at different rates at various points. Out in midstream, flow is more rapid than near the banks, where eddies, little whirlpools, are usually in sight. This particular stream is clear, and its bed consists of sand and gravel. You can see pebbles rolling and sliding intermittently along the bed, whereas sand grains now and then take low jumps.

Walk along the stream, and you will note the channel winding from side to side in a rather smooth sequence of curves. At a place where the bank is undercut by the current at the outer side of a curve, slump into the channel is occurring. If the bank is of sand, you can watch the current distributing sand grains along the bed, washed from the slumped mass.

All this activity could have been so interesting that you did not notice the darkening sky, and so you have just time to take refuge under a large tree standing beside a plowed field, as an intense thundershower begins. For a time the tree acts as an umbrella, its leaves and branches holding back the raindrops, which do not reach the ground. But in the field beyond the tree large drops strike the ground and splash up the loose soil, creating tiny craters as they do so. If you were to step out from under the tree, where nothing is happening yet, raindrops would splash soil onto your shoes, a couple of inches above the ground surface.

At first the ground absorbs the water as it falls. But little by little the ground begins to glisten in patches, showing that it is saturated. The excess water starts to flow off over the surface toward the stream. Unless, during plowing, the field was furrowed, the water may be flowing as wide, shallow sheets. It may be slightly turbid (muddied) with fine particles picked up from the soil. Where the water enters the stream, it makes turbid patches that soon blend and disappear. But from the grassy pasture beside the plowed field the water enters the stream through widely spaced grassy gullies, essentially as little, temporary streams of clear water.

The difference in behavior of the water, as it flows across plowed field and pasture, is the difference between overland flow and channel flow, a distinction we will describe in the next section. Meanwhile, observe that the whole stream is swollen, is flowing faster, and has become turbid with silt and clay contributed from plowed fields, roads, and other bare areas unprotected by vegetation.

By now the leaves and branches of the tree can no longer hold back the rain, which is dripping onto the ground, though with less impact than on the unshielded ground outside. The tree being no longer an effective umbrella, you will have to make a dash for better shelter.

Observation and experiment by geologists and engineers indicate that all these stream activities are related to each other in a complex way. These relationships are fundamental to an understanding of geology, because the stream you have been watching is one of the millions of drainageways by which water from rain and snow runs down over the land toward the sea (Fig. 8.1).

Significance of Streams to People. Apart from their strictly geologic importance, streams are significant to society because (1) they are an increasingly important source of water for human and industrial consumption; (2) they are a comparatively small but essential source of industrial energy; (3) many rivers are avenues of industrial transportation; (4) they have great recreational value; and (5) the increase of population in river valleys has created the necessity for protection against damage from stream floods and for controlling pollution created by the discharge of wastes into streams.

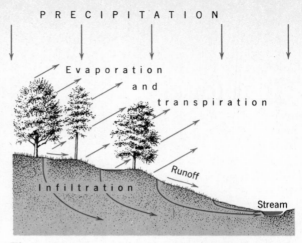

Figure 8.1 Disposition of water precipitated on land. Water returns to the atmosphere directly by evaporation and transpiration. Other water gets into streams by infiltration into the ground and by runoff over the surface.

Streams as Geologic Agents. The significance of streams as geologic agents is threefold.

1. As movers of water from land to sea, streams are an essential part of the water cycle.
2. By far the greatest importance of streams as geologic agents lies in what they carry. From sampling the loads of many rivers in many countries it has been estimated that, every year, streams transport from the lands to the oceans about 1 billion tons of sediment mechanically, plus about 400 million tons of material in solution.
3. Streams, together with mass-wasting, are the prime sculptors of the land. That is to say, the configuration of an enormous part of the Earth's land area consists of stream valleys with hills in between, and is the work of streams and mass-wasting combined. Other agents, such as glaciers (flowing ice) and wind (flowing air), have shaped only minor parts of the lands, and even in many of those parts the principal shapes are the product of streams.

Raindrop Impact and Sheet Erosion

Erosion of the land by water begins even before a distinct stream has been formed. It occurs in two ways: by impact as raindrops hit the ground, and by sheets of water that result from heavy rains.

As raindrops strike bare ground they dislodge small particles of loose soil, spattering them in all directions. On a slope the result is net displacement downhill. One raindrop is very little, but the number of raindrops is so great that altogether they accomplish a large amount of erosion.

What happens to rainwater after it reaches the ground? As Fig. 8.1 shows, rainfall that is not returned to the atmosphere by evaporation or transpiration either infiltrates (sinks into) the ground or flows off over the surface as runoff. We can subdivide the runoff into *overland flow, the movement of runoff in broad sheets or groups of small, interconnecting rills,* and *stream flow, the flow of surface water between well-defined banks.* Stream flow is very obvious. Overland flow is not obvious and commonly occurs only through short distances before ending in a stream valley. But it takes place wherever rain falls in excess of the amount that can be immediately absorbed by the ground. *The erosion performed by overland flow is sheet erosion.*

The erosive effectiveness of both raindrops and overland flow is reduced by vegetation which, where it forms a continuous ground cover, holds erosion to small values. But on bare sloping fields, closely grazed pastures, or areas planted with widely spaced crops such as corn, rates of erosion can be great. Because sheet erosion creates no obvious valleys, the magnitude of its effect on soil was not fully realized until accurate measurements began to be made. Now many experiment stations (Fig. 8.2) maintained by the U.S. Department of Agriculture measure the erosion of soil (Fig. 8.3). To farmers the measurements are of far more than academic interest, for they show that sheet erosion is a menace to soil left unprotected on slopes. In recognition of this fact, wise farmers reduce areas of bare soil to a minimum and prevent the grass cover on pastures from being weakened by overgrazing. When the protective plant cover is weakened, runoff and erosion are increased (Fig. 8.4). If crops such as corn, tobacco, and cotton must be planted on a slope, strips of such crops are often alternated with strips of grass or similar plants that resist sheet erosion.

The examples just illustrated should not, however, lead us to suppose that erosion of slopes results wholly from human activities. This is by no means the case. Under some natural conditions, without any

Figure 8.2 Apparatus for measuring sheet erosion, Marlboro, New Jersey. Runoff, carrying soil particles, flows down a shallow slope; sediment is trapped in containers and measured by weight. A different kind of plant covers each plot. Results from a similar station are shown in Fig. 8.3. (*U.S. Soil Conservation Service.*)

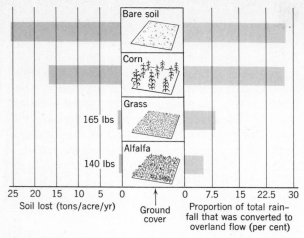

Soil stripped away by sheet runoff

Bedrock exposed

Smooth, mass–wasted slope

Valleys

Figure 8.3 Effect of plant cover on rate of sheet erosion, measured over 4 years at Bethany, Missouri, a station like that shown in Fig. 8.2. Soil is silty, slope is 8 per cent, and annual rainfall is 40 inches. The measurements show that grass and alfalfa, with their continuous network of roots and stems, are nearly 300 times as effective as "row crops" such as corn, in holding soil in place. Erosional loss from bare soil shown here is at a rate of about 1.5 feet/100 years.

agriculture at all, the splash of raindrops and the work of sheet erosion are so effective that they combine to remove large volumes of fine rock particles. For example, in some subtropical grasslands all the rainfall is concentrated within a single rainy season. During the long dry season, evaporation so depletes soil moisture that grass is sparse, covering no more than 40 to 60 per cent of each square meter of ground. Although kept bare by natural causes, the soil is as vulnerable to erosion as soil laid bare by farming.

Although common on slopes, overland flow takes place only through short distances. The flowing water soon concentrates into channels and forms streams. Most sediment transported by running water, therefore, is carried in streams, which likewise deposit sediment. Accordingly streams occupy an important place in the rock cycle, and we must look at them rather closely.

Flowing Water and Movement of Rock Particles

A **stream** is a body of water that carries rock particles and dissolved substances, and that flows down a

Figure 8.4 Too many cattle were grazed on one of these two pastures in northern Kentucky, weakening the grass cover. Erosion did the rest. (*U.S. Soil Conservation Service.*)

slope along a definite path. The path is the stream's channel, and the rock particles are an essential part of the stream itself. Five basic factors of a stream are:

1. **Discharge** (*the quantity of water that passes a given point in a unit of time*). Discharge is usually expressed in cubic feet per second (cfs) or in cubic meters per second (m^3/sec).

2. *Average velocity.*

3. *Size and shape of channel.*

4. **Gradient** (*the slope measured along a stream, on the water surface, or on the bottom*).

5. **Load** (*the material the stream carries*). The load consists of rock particles plus matter in solution. Unlike rock particles that constitute the mechanical load, dissolved matter generally makes little difference to the behavior of the stream.

The Stream's Load

Turbulent Flow. The particles of water in a stream do not move along straight or parallel paths. Their paths form bends and whirls (eddies, Fig. 8.5) so that the particles move in all directions and at different speeds. This kind of motion is *turbulent flow.* Turbulence in fast streams is much greater than in slow ones. Also it is greatest near the sides and bottom of the stream channel, where the flowing water drags against the solid banks and stream bed.

The way in which a stream moves its load of rock particles is easy to observe in an artificial channel with glass sides, an apparatus found in many laboratories. As Fig. 8.6 shows, the stream's load of solid particles consists ideally of *coarse particles that move along or close to the stream bed* (the **bed load**) and *fine particles suspended in the stream* (the **suspended load**.) In addition to these solid particles there is also *matter dissolved in stream water* (the **dissolved load**), chiefly a product of chemical weathering.

Bed Load. As we watch water starting to flow through the channel, at first only a few particles move, by sliding or rolling. Pebbles start to move sooner than sand grains because they project higher into the current and thus into a zone of faster flow (Fig. 8.6).

Now we increase the velocity a little. More particles on the bed start to move and to collide with one another. Some are propelled upward by the collisions and then are pushed forward by the current. Some are sucked upward by the faster current that flows above them, as the wing of an airplane is lifted by the flow of air above it. Gravity pulls the particles down to the bed again, but the result for each particle is a jump forward, downstream.

As we step up velocity just a bit more, particle movement becomes general. Seen from above, the particles travel parallel to the current.

Suspended Load. The experiment we have been watching involved clear water flowing over sand.

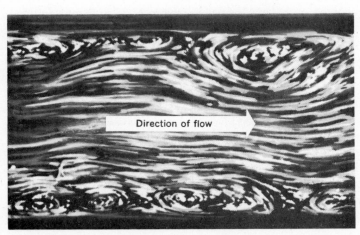

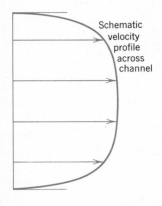

Schematic velocity profile across channel

Figure 8.5 Turbulent flow in an open channel, seen looking down onto surface of stream. Because of drag, turbulence at the surface is greatest near the channel sides. (*From a photograph in Prandtl and Tietjens, Applied Hydro- and Aeromechanics; courtesy Engineering Societies Monographs Committee.*)

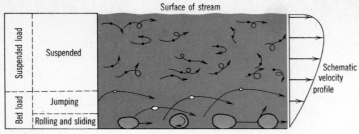

Figure 8.6 Kinds of movement of rock particles carried in a stream; vertical distribution of bed load and suspended load. The velocity profile (right) shows that the stream flows fastest (longest arrow) near the water surface and slowest at the bottom.

Now if we add silt and clay to the water we can see that these fine particles spread quickly throughout the water, making it turbid. If we look very closely we can see that each particle follows a very irregular path. The particles are *suspended* in the water, because upward-moving threads of current within the general turbulence exceed the speed at which each particle settles toward the bed under the pull of gravity.

In any turbulent stream, then, particles of silt size and clay size always remain in suspension; they move continuously, as fast as the water flows. Such particles settle and are deposited only where turbulence ceases; for example, on the floodplain shown in Fig. 8.13, *A*, in a lake, or in the sea. Because of these relationships, the transport of particles in the suspended load differs greatly from that of bed-load particles, which tend to move only intermittently.

Most of the suspended load in streams is derived from the erosion of fine-grained regolith in plowed fields and other areas unprotected by vegetation, and gets into the streams during rains.

Economy of a Stream

The channels of most natural streams consist partly or wholly of sediment. Streams move this material from place to place, and so their channels are continually being altered. Because stream and channel are closely related and are ever changing, we have to examine them together as an interrelated system. We can think of the *economy* of the stream and its channel as *the input and consumption of energy within a stream or other system and the changes that result.*

The annual rainfall on the area of the United States is equivalent to a layer of water 30 inches thick. Of this layer, 21 inches evaporate and 9 inches form *runoff,* defined as *the water that flows over the lands.*

As the average altitude in the United States is about 2500 feet, our 9-inch layer of water falls through that vertical distance as gravity pulls it down to sea level. The potential energy of the water is tremendous; it is equal to the weight of the water times the height of the land. One cubic foot of water weighs 62.4 pounds, and 62.4 times 2500 feet equals 156,000 foot-pounds. This potential energy, multiplied by the enormous number of square feet of area of the United States, is converted into the kinetic energy of flow of U.S. streams. Part of that energy is spent in the geologic work of picking up and carrying rock particles. This kinetic energy makes streams the prime movers of rock waste from lands to ocean. This is why streams occupy a key position in the rock cycle, for they are among the chief agencies by which the sedimentary rocks of the Earth's crust were accumulated.

Some of the sediment carried by streams is deposited on the land, at least temporarily. *Alluvium,* the general name for *sediment deposited in land environments by streams,* exists in great quantity, and still larger quantities of it have been converted into sedimentary rock. Much of the sediment, however, is carried into the sea and deposited there. Once it reaches the sea, it is no longer alluvium but becomes marine sediment.

Factors in the Economy. The economy of the system depends on a continual interplay among the five basic factors mentioned earlier: discharge, velocity, size and shape of channel, gradient, and load. The first factor is calculated from measurements made systematically at selected points along large and small streams for purposes of flood control, irrigation, water supply, and the like. The U.S. Geological Survey, for example, maintains nearly 6500 measurement points, called gaging stations, in various parts of the United States.

The measurements show that as discharge changes, velocity and channel shape must change also. The relationship is definite and can be expressed by the formula:

$$Q = w \quad d \quad v$$

Discharge = width × depth × Velocity
(cubic feet per second) (feet) (feet) (feet per second)

When discharge changes, as it does continually, the product of the other three terms must change accordingly.

When discharge increases, the stream erodes and enlarges its channel, instantaneously if it flows on alluvium, more slowly if it flows on bedrock. The increased load is carried away. This continues until the increased discharge can be accommodated. In contrast, when discharge decreases, some of the load is dropped, making the channel less deep and less wide, and the velocity is reduced by increased friction. In this way width, depth, and velocity are continually readjusted to changing discharge.

Thus, a stream and its channel are related so intimately that we can think of them as a single system. The channel is so responsive to changes in discharge that the system, at any point along the stream, is continually close to a steady state.

Geologic Activity During Floods. Seasonal distribution of rainfall causes many streams to rise in flood seasonally. As discharge increases during a flood, so does velocity. This has the double effect of enabling a stream to carry not only a greater load, but also larger particles of sediment. An extreme example resulted from the St. Francis Dam catastrophe mentioned in Chap. 7. As the dam gave way, the water behind it rushed down the valley, moving as bed load blocks of concrete weighing as much as 10,000 tons through distances of more than 2500 feet.

In most streams more geologic work is accomplished during regular seasonal floods than during intervals of low water. In addition to seasonal floods, rare, exceptional floods (Fig. 8.7) outside the stream's normal economy occur perhaps only once in several decades or centuries. In such floods the geologic work done may be prodigious, but they happen so rarely that their effects probably are less than the aggregate effect of normal activity throughout the very long periods intervening between them. As in the fable, this race, too, is won by the tortoise rather than by the hare.

Changes Downstream. Let us look at the whole length of the stream to see what changes take place. If we go downstream from head to mouth, we see that orderly adjustments take place from point to point. Specifically, four factors change. (1) Discharge increases; (2) width and depth of channel increase; (3) velocity increases slightly; and (4) gradient decreases.

As tributary streams contribute additional water, discharge in the main stream increases in most cases, thus increasing velocity (Fig. 8.8). The demonstration that velocity increases downstream seems to contradict the common observation that water rushes down steep mountain slopes and flows smoothly over nearly flat lowlands. But the physical appearance of the water is not a true measure of its velocity, which increases downstream mainly because channels become deeper and wider in that direction.

Long Profile. The gradient or slope of a mountain stream may be 300 feet per mile or even more, whereas that of the downstream part of a large river may be 6 inches per mile or even less. Gradients of most rivers decrease downstream, and so the *long profile* of a stream (*a line connecting points on the stream surface*) is generally concave-upward (Fig. 8.9).

Base Level. The vertical position of the mouth of a stream is determined by its *base level.* This is *the limiting level below which a stream cannot erode the land* (Fig. 8.10). The *ultimate base level,* for streams in general, is *sea level,*[1] *projected inland as*

---

[1] Although true in principle, this statement is not quite true in detail. A stream confined between channel banks can erode its bed slightly below sea level, as is evident from Fig. 8.14, where depth of base of the topset beds represents depth of stream-channel erosion.

131

Figure 8.7 Flood of Quinebaug River at Putnam, Connecticut in August 1955, the result of extraordinary rainfall. View looking north, upstream. River channel is at left, just out of view; its edge is seen as darker water in lower-left corner of picture. Stream shown is water that overflowed from main channel and is rejoining it at lower left. Although short lived, the overflow eroded a large body of sand and gravel about 25 feet thick (remnants are near flooded railroad tracks) and deposited the resulting load as a series of bars and islands seen in foreground. (*Providence Journal-Bulletin.*)

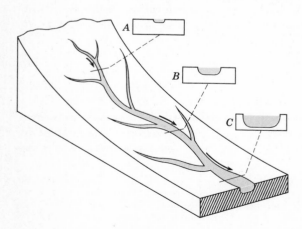

Figure 8.8 Changes in the downstream direction along a stream system. Discharge is increased by entrance of successive tributaries. Width and depth of channel are shown by cross sections *A*, *B*, and *C*. Velocity is shown by relative lengths of three arrows. (*After Leopold and Maddock, 1953.*)

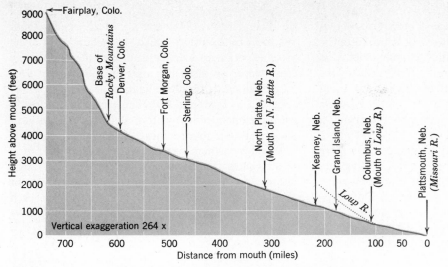

Figure 8.9 Long profile of the Platte-South Platte River in Colorado and Nebraska, illustrating the common concave-up form. Loup River, a tributary (dotted line), joins the main stream smoothly. The vertical scale is greatly exaggerated; if it were not, the upstream end of the curve would not be visible above the base line 0-0. (*After Henry Gannett.*)

an imaginary surface underneath the stream. When a stream cuts down to that surface, its energy quickly approaches zero. For a stream ending in a lake, base level is the level of the lake (Fig. 8.10), for the stream cannot erode below it. But, if the lake were destroyed by erosion at its outlet, the base level represented by the lake surface would disappear, and the stream, having acquired additional potential energy, would deepen its channel. *The levels of lakes and all other base levels that stand above sea level are **local base levels.*** A common kind of local base level is the level of a belt of particularly hard rock lying across the stream's path. Even sea level itself

changes slowly over long periods (Chap. 13) and this too affects the long profiles of streams.

Fans. The building of a fan is a common example. When a stream flows down through a steep highland valley and comes out suddenly onto a nearly level valley floor or plain, it encounters an abrupt decrease of slope. It deposits that part of its load which can not be carried on the gentler slope. The material it deposits takes the form of a ***fan,*** defined as *a fan-shaped body of alluvium built at the base of a steep slope* (Figs. 8.11, 10.4, 10.9). The surface of the fan slopes outward in a wide arc from an apex at the

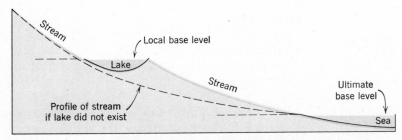

Figure 8.10 Relation between ultimate base level (the sea) and local base levels such as a lake.

mouth of the steep valley. The profile of the fan, from apex to base in any direction, commonly shows the concave-up form characteristic of stream profiles. The exact form of the profile depends chiefly on discharge and on the diameters of particles in the bed load; hence no two fans are exactly alike. A small stream carrying a load of coarse particles builds a shorter, steeper fan than a larger stream carrying a load of finer particles.

Although a fan is originally localized by decrease of slope, as soon as its long profile has become smooth, the chief cause of further deposition on it is the spreading of water through a network of channels, with consequent loss of discharge and velocity in each channel and overall loss of water that percolates down into the underlying sediments.

Unless special circumstances preserve it, the fan will be destroyed piecemeal by continuing erosion downward below the profile *cc'x* (Fig. 8.11). A fan, therefore, is likely to be a temporary deposit.

Meanders. No stream is straight through more than a short distance. The pattern of most streams is a series of bends, and very commonly the bends are smooth, looplike, and similar in size. Such bends are termed ***meanders*** (from a Greek word meaning "a bend"), defined as *looplike bends of a stream channel* (Fig. 8.12). The meanders of the Mississippi and other large streams are under continual study by engineers and geologists concerned with problems such as floods and navigability. These people know that meanders are not accidental, that they occur most commonly in channels with gentle gradients

in fine-grained alluvium, and that they occur even in streams having no load at all. Meanders represent the form by means of which, in making a turn, the river experiences least resistance to flow, does the least work, or dissipates energy most nearly uniformly along its course. The meander form is therefore one of stability. Although stable as a form, a meander changes its position almost continually, as shown by year-to-year measurements of river channels and by artificial streams. The shift or migration of a meander is accomplished by predominant deposition along one bank and predominant erosion of the opposite bank. Along the inner side of each meander loop, the place where speed of flow is lowest, bed load is deposited and accumulates as a distinctive *bar* (Fig. 8.12).

Meanders grow and bars enlarge quite rapidly, as can be shown easily in artificial laboratory streams. Also, because the valley floor slopes downward toward the mouth of the stream, slump is a little more rapid on concave banks that face upvalley than on other banks. Therefore meanders tend to migrate slowly down the valley, subtracting from and adding to various pieces of real estate along the banks, according to location, and causing legal disputes over property lines and even over the boundaries between counties and states.

The behavior of streams in laboratory channels shows that if the bank material is uniform, the meanders are symmetrical and migrate downvalley at the same rate. But, because the material of a bank is usually not uniform, the migration of the downstream limb of a meander can be slowed where it encounters resistant material. Meanwhile the upstream limb, migrating more rapidly, intersects and cuts into the "slow" limb. Thus the channel bypasses the loop between the two limbs and the cut-off loop is converted into a lake (*oxbow* lake, Fig. 8.12).

Since 1776 the aggregate length of channel abandoned by the Mississippi River through cutoffs has amounted to more than 230 miles; yet the river has not been shortened appreciably because the lost mileage has been balanced by the enlargement of other meanders.

Floodplain: Natural Levees. The Mississippi River habitually floods during the spring season. Before the river began to be restrained by artificial flood-control structures, it frequently overtopped its banks and inundated the lower parts of the valley

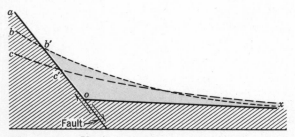

Figure 8.11 Vertical section showing growth of a fan. Bedrock is shaded; alluvium is blue. Line *aox*, profile of surface before deposition of fan. Line *bb'x*, long profile of stream at an early stage of fan building, with apex of fan at *b'*. Line *cc'x* long profile at a later stage, after stream has cut away apex of fan *bb'x*, increasing fan radius while establishing a continuous, concave-up profile.

Figure 8.12 Meanders of Pecos River near Roswell, New Mexico. View looking south, downstream. The areas covered with rows of bushes are bars. At right of center is an oxbow lake. (*Spence Air Photos.*)

floor. *That part of any stream valley which is inundated during floods is a **floodplain;*** the area of the natural floodplain of the Mississippi from Cairo to the delta is 30,000 square miles, but more than half of this area is now protected against floods by dikes and other structures.

The channels of the lower Mississippi and many other meandering streams are bordered by ***natural levees—broad, low ridges of fine alluvium built along both sides of a stream channel by water spreading out of the channel during floods*** (Fig. 8.13). Along the lower Mississippi natural levees are 15 to 25 feet high. The fine alluvium of which they are chiefly built becomes still finer away from the river and grades into a thin cover of silt and clay over the rest of the floodplain. Natural levees were built and are continually added to only during floods so high

that the floodplain is converted essentially into a lake deep enough to submerge the levees. In the water that flows laterally from the submerged channel over the submerged floodplain, depth, velocity, and turbulence decrease abruptly at the channel margins. This results in sudden, rapid deposition of the coarser part of the suspended load (usually fine sand and silt) along the sides of the channel. Farther away from the channel, finer silt and clay settle out in the quiet water. In the vicinity of Kansas City, Missouri, during an exceptional flood in 1952, Missouri River water deposited a layer of silt as much as 6 inches thick over wide areas of the floodplain. In some places fences and other obstacles caused silt and fine sand to accumulate to thicknesses as great as 5 feet. Under ordinary conditions flood-deposited silt is beneficial to agricultural lands, because it contains

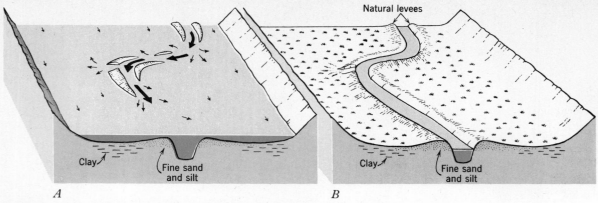

Figure 8.13 Floodplain with natural levees. *A.* During a big seasonal flood much of the valley floor becomes a lake. Water remaining in the channel flows at high speed (large arrows). Water escaping from channel flows with diminishing velocity (small arrows) into adjacent broad, shallow areas. It deposits silt to form natural levees where it leaves the channel and blankets lower land with clay. Highest parts of levees, added to only during still higher floods, form islands. *B.* At times of low water, levees stand as low ridges along sides of channel; beyond them are swamp lands. Vertical scale exaggerated.

organic matter, washed from soils on the watershed, which acts as fertilizer.

Flood Control. The flood deposits of the Mississippi River, including the natural levees, were built up during a long period, but the process has been interfered with by engineering works (mainly dikes built of earth and concrete) designed to prevent the river from generally inundating its floodplain. As described in Chapter 21, the natural levees have been heightened artificially by earth dikes to hold in ordinary floods, and at selected points spillways have been built to allow the water of the highest floods to escape harmlessly into natural channels that parallel the channel of the Mississippi.

Deltas. A *delta* is *a body of sediment deposited by a stream flowing into standing water.* As the water of the stream diffuses into the standing water of sea or lake, its speed is checked by friction, it loses energy, and deposits its load as a delta (Fig. 2.5). Although deltas are of several kinds, the type easiest to recognize and probably most common is shown in Fig. 8.14. Although in form it somewhat resembles a fan, it differs from a fan because of two factors: (1) stream flow is checked by standing water, and (2) the level surface of sea or lake sets an approximate limit to upbuilding of the deposit, the top of which is flatter than the profile of a fan.

The particles in the bed load are deposited first,

in order of decreasing weight; beyond this, the suspended sediments drop out. A layer deposited at any one time (as during a single flood) is sorted, grading from coarse at the stream mouth to fine offshore. The deposition of many successive layers creates an embankment that grows outward like a highway fill made by dumping. *The coarse, thick, steeply sloping part of each layer in a delta* is a *foreset layer.* Traced seaward, the same layer becomes rapidly thinner and finer, covering the bottom over a wide area. This *gently sloping, fine, thin part of each layer in a delta* is a *bottomset layer.*

As successive layers are deposited, the coarse foreset beds one by one overlap the bottomset layers, producing the arrangement seen in Fig. 8.14. The stream gradually extends seaward over the growing delta, erodes the tops of the foreset layers during floods, and at other times deposits part of its bed load in its channel and its suspended load in areas between channels during floods. The channel deposits and interchannel deposits form the *topset layers* of the delta. We can define these deposits as *the layers of stream sediment that overlie the foreset layers in a delta.*

During floods the stream spills out of its channel and forms distributary channels, through which the water enters the sea independently, multiplying the topset deposits. Radiating distributary channels give the delta a crudely triangular shape like the Greek Δ, from which the deposit derives its name.

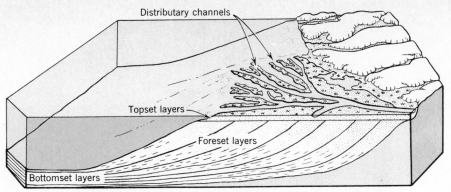

Figure 8.14 Idealized small delta. Foreset layers consist of sand, which grades outward and downward into silt and clay in bottomset layers. Inclination of foreset layers is identical with slope of delta front, shown in phantom view through the water. The less deep the water offshore, the gentler the slope and the less distinct this stratification. The area of such a delta might be 1 square mile or less.

It may seem surprising that the suspended load, much of which has been carried hundreds of miles through the channel of a large river without being deposited, should drop out so abruptly to form part of a localized delta instead of remaining in suspension long enough to be carried far from land. But the salts dissolved in seawater act to coagulate, or flocculate, the suspended fine particles into aggregates so large that they settle to the bottom promptly.

Some of the world's greatest rivers, among them the Nile, the Hwang Ho, the Mackenzie, the Colorado, and the Mississippi, have built massive deltas at their mouths. Each delta has its own peculiarities, and none is so simple as the small delta shown in Fig. 8.14. The Mississippi delta, with an area of 12,000 square miles, not counting the submarine part, is in reality a complex of several coalescing subdeltas built successively during the last several thousand years (Fig. 8.15). Each subdelta was begun by a flood that created a new distributary.

Features of Steep, Rocky Channels. Hitherto we have been discussing streams whose beds consist of sediment. Now let us turn to the erosion performed by turbulent mountain streams whose channels are steep and consist of bedrock. In streams, ***abrasion, the mechanical wear of rock on rock,*** is caused by friction, collisions between rock particles in the mechanical load, and impacts between particles in the load and bedrock in the channel. A stream uses its mechanical load as tools. By rubbing, scraping,

bumping, and crushing, it erodes bedrock and at the same time smooths and rounds the tools. Even where no bedrock is exposed in the channel, rounding of sand and coarser particles goes on, although rounding decreases rapidly as particle diameter decreases.

Under the special conditions in which a stream flows over a rock ledge or cliff, as at Niagara Falls (Fig. 8.16), the increased velocity of the falling water sets up strong turbulence at the base of the falls, and the stream bed is deepened. The cliff is gradually undermined and the falls retreats upstream. The retreat of the Canadian Falls at Niagara Falls has been rapid. Measurements by survey show that between 1850 and 1950 the rate of retreat averaged about 4 feet per year. This rapid rate is favored by the fact that the lip of the falls consists of strong, resistant dolostone beneath which is weak, easily eroded shale. As the shale cliff is eroded back, the dolostone lip is undermined and caves in piecemeal (Fig. 8.17). The long, deep gorge downstream from the falls was created by retreat of the falls, through successive positions, during periods of many thousands of years.

The Steady State in Streams

The characteristics of a stream and its channel are determined by the interplay, at each point along the stream, of factors such as discharge, velocity, channel geometry, and load. These factors continually adjust themselves to one another, and the sum of the adjustments tends toward a condition of steady state

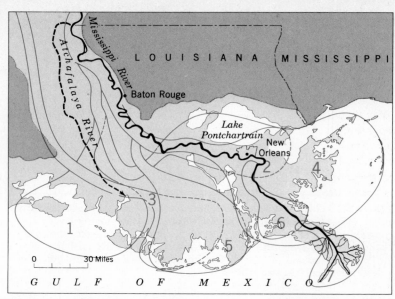

Figure 8.15 The Mississippi River has built a series of overlapping sub-deltas, while occupying successive distributary channels (numbers 1 to 7). Age of subdelta 1 is estimated at 3000 years, of 3, 1500 years, and of 5, 1000 years. Construction of subdelta 7 began more than 100 years ago. Discharge of the Mississippi has been gradually shifting to the Atchafalaya distributary. By 1958, 28 per cent of the discharge was following this new route. Construction of a barrier to stop the diversion was begun in 1955. Had this not been done, the percentage would have increased to 40 by 1975. (*After Kolb and Van Lopik, 1958.*)

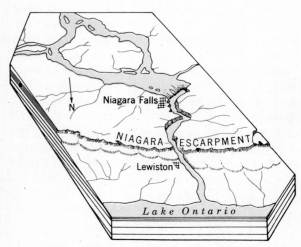

Figure 8.16 Bird's-eye view of Niagara Falls, looking south, showing south-dipping strata and Niagara Escarpment, a great ledge formed by a resistant stratum, from which the Falls originated. By erosion the Falls has retreated through an aggregate distance of about 7 miles. Greatest length of block, 35 miles. Vertical exaggeration 2x.

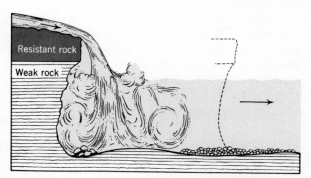

Figure 8.17 A falls similar to Niagara Falls. The stream bed downstream from the falls was created by successive positions of the retreating cliffs; one former position is shown by the dotted line. Turbulence at base of falls keeps bedrock scoured nearly clean of sediment, but in less turbulent water downstream coarse gravel is deposited.

at each point. The conditions at each point, added together, combine to determine the concave-up curve that is the stream's long profile. The tendency toward steady state exists whether the stream's course is steep, narrow, and rocky, or wider and more gently sloping.

Anything that disturbs the steady state sets up a response by the stream, which tends to restore that state. For instance, if velocity is checked by the sea or a lake, the stream responds by building a delta and thereby smoothing the profile at its mouth. If the stream flows over a cliff, it responds by building a fan at the base of the cliff, thereby gradually restoring its smooth profile. Meanders likewise are approaches to a steady state in response to special conditions.

We shall now see how this strong tendency toward the steady state in streams is reflected in the sculpture of the land.

Drainage Systems

Relations of Valleys to Streams. Weathering, mass-wasting, sheet erosion, and stream activities not only create, transport, and deposit sediment; they also shape the land into systems of valleys. Groups or families of valleys are the commonest features of the land. Throughout wide regions the surface consists of little more than a complex of valleys created by erosion, and separated by higher areas that erosion has not yet consumed. Valleys exist in such great numbers that they have never been counted except in sample areas. The enormous number of valleys is commensurate with the huge volume of water runoff over the land. Before the middle of the 18th Century, valleys were generally thought to be the result of catastrophes that somehow broke the Earth's crust and pulled it apart, creating paths for running water to follow. Today we know that with few exceptions, streams themselves make their own valleys. Capacious ones have been created within the last 100 years (Fig. 8.18).

Drainage Basins and Divides. Every stream or segment of a stream has its *drainage basin,* consisting of *the total area that contributes water to the stream. The line that separates adjacent drainage basins* is a *divide* (Fig. 8.19). On a map (Plate B)

we can trace the divide that encloses the huge drainage basin of the Mississippi River, an area that exceeds 40 per cent of that of the conterminous United States.

Spacing of the streams in a drainage basin is orderly. When the streams in a large basin are measured accurately on a map, it appears that the distances between the mouths of tributaries are spaced in an orderly way, and that there is a mathematical relationship between the length of a stream and the area of its drainage basin. This orderliness measured on maps is analogous to the orderliness inherent in a stream's long profile, in which gradient decreases systematically from head to mouth, while discharge, velocity, and channel dimensions increase. All these relationships imply that in response to a given quantity of runoff, stream systems develop with just the size and spacing required to move the water off each part of the land with maximum efficiency.

Surface slopes converge toward the heads of small valleys, and runoff, moving at first as overland flow, soon concentrates in valleys (Fig. 8.20), forming streams. A system of streams does not necessarily require much time to develop, as is indicated by the following example. In August 1959 an earthquake occurred at Hebgen Lake, near West Yellowstone, Montana. The movement tilted the country in such a way that a large area of silt and sand, formerly part of the lake bed, emerged and was subject to runoff. Small-size drainage systems began to develop immediately. Sample areas were surveyed and mapped 1 year and 2 years after the earthquake occurred. The results showed the same basic geometry that characterizes much larger and older systems. The small, newly formed valleys, together with the areas between them, were disposing of the available runoff in a highly systematic way, all within a period of 2 years after the surface had emerged from beneath the lake.

Valleys, Sheet Erosion, and Mass-Wasting. If the sole agency involved in cutting a valley were the stream that flows through it, the valley should be as narrow and steep-sided as the one shown in Fig. 8.21, resembling a cut made by a saw through a block of wood. The shaping of most of the land surface, including the valley sides themselves, is mainly the work of sheet erosion and mass-wasting of weathered rock material.

Figure 8.18 This valley in Stewart County, Georgia is growing headward, swallowing up farm land and threatening a road. Now more than 75 feet deep, it started more than 100 years ago. Many valleys like this one start in a rutted dirt road or in a field with furrows running downslope. (*U.S. Soil Conservation Service.*)

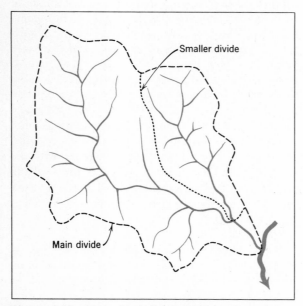

Figure 8.19 Map of a drainage basin, enclosed by a main divide, ending downstream at a large river. A smaller drainage basin, enclosed on one side by a smaller divide, defines a tributary area. Other, still smaller divides within the basin are not shown. Patterns made by streams are like branching trees.

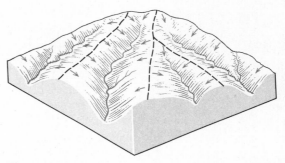

Figure 8.20 Runoff moves over hillslopes as overland flow (suggested by arrows), soon concentrates in small valleys. Broken lines indicate divides.

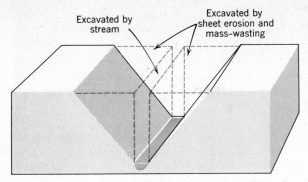

Figure 8.21 In this idealized valley segment the volume of rock excavated by the stream is compared with the much greater volume excavated by sheet erosion and mass-wasting. However, all the waste from the slopes, while being transported out of the area, had to pass through the stream.

Sculptural Evolution of the Land

Sculpture of the land is a grand destructive process in which the forces of erosion are pitted against the resisting forces of cohesion within bedrock, regolith, and plant cover. The erosional forces are limited downward by base level, below which they cannot work. The raising of any land above base level disrupts the relationship in terms of erosion and deposition, between the raised area and the land or sea adjacent to it. Erosion of the raised area and deposition on the adjacent area are increased, in an approach to a new steady state.

In the process the land is *sculptured* into a series of valleys and hills (Fig. 8.23). The sculptured surface, slowly and continuously changing, is the expression resulting from (1) uplift of the crust modified by (2) destructive erosion of the uplifted area. The processes involved form a great interrelated system, in which both (1) water and (2) rock materials are present. Water is continually being lifted from the sea by solar energy and carried again and again over the land (Fig. 5.1). Through weathering, mass-wasting, and stream transport, the water acts to process rock particles and to carry them toward the sea, where they are deposited. The water entering the sea evaporates and is again carried over the land, and so is used again and again. But the rock particles brought from the land move more slowly. They stay in the sea through much longer periods. Eventually, therefore, the system reaches a condition in which so much of the land has been carried away that the remaining land has become low, with very gentle slopes.

Cycle of Erosion. As the system progresses, the land surface undergoes a series of gradual changes somewhat like those which affect an individual organism of any kind, from birth to death, including stages of youth, maturity, and old age. In the stage of youth, stream gradients tend to be steep (Fig. 8.22, *A, a*) and erosion rapid. Valleys are actively deepened and make sharp cuts into the land. But they have not reached their full lengths, so that broad areas of the original surface remain uncut.

In the stage of maturity, valleys have lengthened so that the entire surface is cut up, but stream gradients have become gentler and streams are eroding more slowly. Slopes have developed smoothly curved profiles.

The mature surface is lowered toward base level very slowly, with the rate of erosion becoming increasingly slow as stream gradients and hillslopes become ever more gentle. A surface in a late phase of erosion, lying close to base level, is in the stage of old age. A land surface in a very late phase of old age is a ***peneplain*** ("almost a plain"), defined as *a land surface worn down to very low relief by streams and mass-wasting.* With ever-decreasing energy, erosion of a mature surface becomes so slow that many millions of years are necessary for the creation of an extensive peneplain. Despite its generally low relief, high steep hills can form part of it wherever the rocks are especially resistant to erosion.

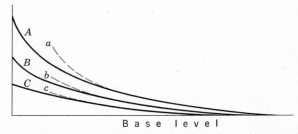

Figure 8.22 Successive long profiles of a main stream (*A, B, C*) and one of its tributaries (*a, b, c*), changing with the passage of time. Although gradients progressively decrease, at all times the tributaries are nicely adjusted to the main stream, which in turn is adjusted to its base level.

We know of few if any peneplains that exist today in an undisturbed state. Numerous peneplains have been submerged beneath the sea, deeply buried by thick layers of sediment and later partly exposed by erosion of their covers of sedimentary rock so that we can see small parts of them. Others have been subjected to uplift and have been dissected so that again only small parts remain; an example is shown in Fig. 8.23. No peneplain, therefore, is seen today lying unaltered in the position in which it was made. This fact implies considerable instability of the Earth's crust during at least the last few million years

Figure 8.23 Result of a sudden increase in rate of erosion. The smooth surface of the area beyond the dotted line is the remnant of a formerly continuous surface, essentially a peneplain. Following a sudden and considerable increase in stream energy, streams are deepening their valleys and are extending them headward; the surface is being rejuvenated. Probably the energy increase resulted from faulting, which lowered the extreme foreground relative to the land next behind, and so increased the fall from stream heads to stream mouths. Streams are creating a new steady-state condition by building fans that help smooth stream profiles. View west, across south end of Inyo Range, toward Sierra Nevada, California. (*John H. Maxson.*)

of geologic time, for if the crust had remained quiet, sculptural evolution of the lands should be more advanced than it is.

Rates of Erosion. How fast is the process of sculptural evolution? It is calculated that the surface of the conterminous United States is being stripped away at an average rate of about 6cm per thousand years. Erosion of 1cm every 166 years may seem a slow rate, but it involves the yearly removal to the sea of nearly 1.5 billion tons of rock material from the area of the United States alone. Since prehistoric, stone-age man hunted big game in the United States 10,000 years ago, rock material equivalent to a layer 60cm thick must have been removed to the sea. The reduction of a broad region to a peneplain may take 15 to 100 million years, depending on height at the start, kinds of bedrock, amount of rainfall, and other factors.

Interruptions in Sculptural Evolution. Our description of a land progressing slowly through a cycle of erosion assumed that the progress of erosion was smooth and uninterrupted by any outside influences. However, when we look closely at valleys we can often see evidence that the stable system has been interrupted. An example of local interruption is seen in Fig. 8.18; a larger and less local one is

that in Fig. 8.23. Land areas in such condition are said to have been *rejuvenated* because, after reaching maturity or old age, they have taken on anew the characteristics of youth.

Rejuvenation, then, is *the development of youthful topographic features in a land mass further advanced in the cycle of erosion.*

On a smaller scale, rejuvenation can result in the creation of stream terraces. A *stream terrace* is *a bench along the side of a valley, the upper surface of which was formerly the alluvial floor of the valley.* In a stream flowing on a broad valley floor sudden increase in rate of erosion results in the cutting of a new valley within the older one. The floor of the older valley is left as a pair of stream terraces (Fig. 8.24), which, of course, will in time be entirely destroyed by continued erosion.

If an area subsides or is tilted so as to decrease the gradient of a stream that flows across it, the stream is likely to have to drop some of its load. The deposited load builds up an *alluvial fill* (*a body of alluvium, occupying a stream valley, and conspicuously thicker than the depth of the stream*), which gradually steepens the gradient to a point at which all the load can be carried (Fig. 8.24).

Causes of Interruptions. Interruptions in the stability of streams, valleys, and hillslopes are traceable to four chief causes.

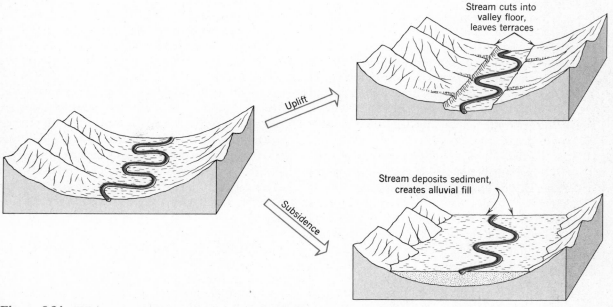

Figure 8.24   Effects of suddenly increased or decreased stream erosion on a valley.

1. Movements of the Crust. The principal cause is probably movement of Earth's crust. If the upstream part of a drainage basin is elevated in relation to the downstream part, energy increases because stream gradients increase and rejuvenation results. The streams flowing down the western slope of the Sierra Nevada in eastern California have been rejuvenated repeatedly by successive uplifts of the mountain range. Conversely, the western part of the same land mass, in central California, has been bent down and buried beneath accumulating sediments.

2. Change of Base Level. Rise and fall of sea level (Chap. 13) change the base level of streams and can cause filling and erosion, respectively, in the segments of valleys that are near the sea. The making of a dam across a valley by a fan, a landslide, a glacier, a lava flow, or even by human construction creates a local base level and can cause alluvial filling in the valley that is dammed. Erosion of the dam then can result in erosion of the fill.

3. Glacial Sediments. A melting glacier commonly delivers so large a load of sediment to a stream that the stream cannot carry it away and deposits it as a fill. The lower Mississippi valley contains an alluvial fill more than 200 feet thick, believed to have resulted partly from the deposition of a copious load of glacial sediment and partly from rise of sea level.

4. Change of Climate. In dry southwestern United States, deepening and headward extension of innumerable small valleys have been going on since about 1880. The valleys are bare, steep-walled canyons cut into soft rock and loose sediment. Some of the canyons have grown headward as rapidly as 1 mile per year and have been eroded to depths as great as 75 feet. Yet before the accelerated erosion began, Spanish settlers found the land surface stable and covered with vegetation.

Study of the valleys and of weather records suggests that the cause lies mainly in very slight changes of climate. During long periods of drought, with few but heavy rainstorms, the grass cover deteriorates and lays the ground bare to erosion. During long periods with more frequent but lighter rains the grass cover improves and protects the ground; valleys tend to fill with alluvium.

A period of few though heavy rains during the last half of the nineteenth century is believed to have

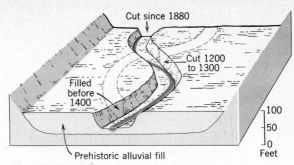

Figure 8.25   Erosion and filling in a small valley in southwestern United States, dated as described in the text. Between A.D. 1400 and A.D. 1880 the stream changed its position, so that the older fill is cut across by the existing valley. (*After Sheldon Judson.*)

caused the erosion now in progress, and overgrazing by cattle and sheep is thought to have accelerated the process. Some of the valleys contain clear evidence of repeated erosion and refilling (Fig. 8.25). Fragments of ancient Indian pottery buried in the alluvium, coupled with records of ancient Indian migrations, give approximate dates of an earlier erosion and filling. A still earlier erosion is probably prehistoric, dating back several thousand years.

Varying Erodibility of Rock

Profiles of Streams and Valleys. The extent to which rocks resist erosion by streams and mass-wasting affects streams and valleys in several ways. When a stream flows over a layer of resistant rock, its long profile is steepened (Figs. 8.26,A; 8.17). A resistant layer exposed along the sides of a valley gives the cross-profile of the valley a steplike form (Fig. 8.26,B). A valley is likely to be narrower where it cuts resistant rock than where it cuts weak rock. A **water gap** (*a pass in a ridge or mountain, through which a stream flows*) is a common feature at such a place (Fig. 8.27). The figure illustrates also that mass-wasting, sheet erosion, and stream erosion have lowered the surface underlain by weak rocks so effectively that the narrow belt of resistant rock has been left standing as a ridge above the general surface.

Stream Patterns. Not only the profiles of streams and valleys but also their patterns, as seen on a map, are affected by the kinds of rock on which they are developed. Patterns are affected also by the history

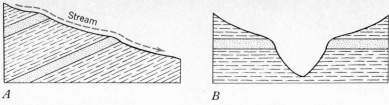

Figure 8.26 Effect of outcrop of resistant strata (stipple) on the long profile *A.* of a stream and on the cross profile *B.* of a valley.

of the areas in which they occur. Three common kinds of stream patterns are shown in Fig. 8.28.

The **dendritic** ("treelike") **pattern** is *a stream pattern characterized by irregular branching in many directions.* This pattern is common in massive rocks and in flat-lying strata. In such situations, differences in rock resistance are so slight that their control of the directions in which valleys grow headward is negligible.

The **rectangular pattern** is *a stream pattern characterized by right-angle bends in the streams.* Generally it results from the presence of joints (Figs. 5.4, 5.5) and faults (Fig. 15.11) in massive rocks or from foliation (Fig. 5.24) in metamorphic rocks. Structures such as these, with their geometrical patterns, have guided the directions of headward growth of valleys.

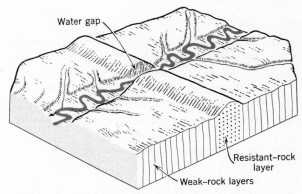

Figure 8.27 Water gap formed where stream has cut through layer of resistant rock. In such rock, a valley is narrower and has steeper sides and gradient than in weak rock. Water gaps are common in the Appalachian region (Fig. 17.13).

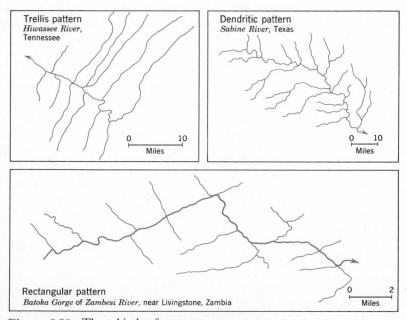

Figure 8.28 Three kinds of stream patterns.

The **trellis pattern** is *a rectangular stream pattern in which tributary streams are parallel and very long*, like vines or tree branches trained on a trellis. This pattern is common in areas like the Appalachian region, where the outcropping edges of folded sedimentary rocks, both weak and resistant, form long, nearly parallel belts.

Classification and History of Streams

Kinds of Streams. On the basis of their patterns and other characteristics, streams are classified into four groups, labeled consequent, subsequent, antecedent, and superposed. The streams in each group have distinctive origins and histories.

A **consequent stream** is *a stream whose pattern is determined solely by the direction of slope of the land*. Therefore, consequent streams generally occur in massive or flat-lying rocks and commonly have dendritic patterns.

A **subsequent stream** is *a stream whose course has become adjusted so that it occupies belts of weak rock*. When such belts are long and straight, subsequent streams constitute the long straight tributaries characteristic of trellis drainage patterns. Figure 8.29 illustrates the difference between consequent and subsequent streams.

An **antecedent stream** is *a stream that has

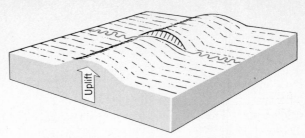

Figure 8.30 This stream has an antecedent relationship to the present surface because of local uplift across its course. Stream has cut a deep gorge across the uplifted belt.

maintained its course across an area of the crust that was raised across its path by folding or faulting (Fig. 8.30). The name comes from the fact that the stream is antecedent to (older than) the uplifting.

A **superposed stream** is *a stream that was let down, or superposed, from overlying strata onto buried bedrock having composition or structure unlike that of the covering strata* (Fig. 8.31). Most superposed streams began as consequent streams on the surface of the covering rocks. The streams' paths, therefore, were not controlled in any way by the surfaces on which they are now flowing.

Stream Capture. When the gradient of one of two streams, flowing in opposite directions from a single divide, is much steeper than that of the other, the

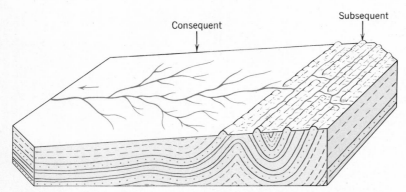

Figure 8.29 Consequent streams contrasted with subsequent streams. On the left the land surface is underlain by flat-lying strata. Drainage has developed under control of the slope of the land (shown by arrow) and is therefore consequent. To the right the same strata are folded. On them tributaries developed most readily along parallel belts of weak rock which determined stream locations. These tributaries are therefore subsequent streams. The main stream crosses ridges of resistant rock through water gaps.

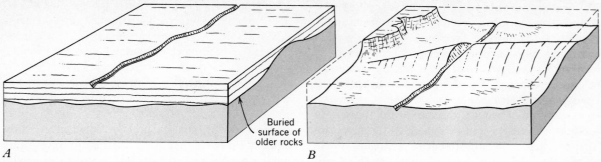

Figure 8.31 Development of a superposed stream. *A.* Stream consequent on strata that bury a former land surface. *B.* After long-continued erosion the stream has become superposed and has cut a water gap through a hill that formed part of the older surface. Overlying strata have been removed by erosion, except for remnant in upper left. Compare Fig. 8.30.

steeper stream can extend its valley headward, shifting the divide against the other stream. In this way the steeper stream can capture the other one little by little. Alternatively, it can capture at one stroke a long tributary of the other stream by intersecting the tributary at its mouth. This process of **stream capture** (or *piracy*), *the diversion of a stream by the headward growth of another stream,* is illustrated in

Fig. 8.32. The Provo River shifted the divide at its head northward and eastward a distance of several miles until the divide intersected and diverted a principal tributary of the Weber River. Evidence of capture is of two kinds: (1) an abandoned segment of the valley of the diverted stream, and (2) tributary streams that are barbed with respect to the new (main) stream they have joined.

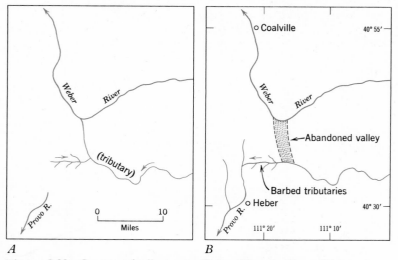

Figure 8.32 Capture of tributary to Weber River by Provo River, upstream from Coalville, Utah. *A.* Reconstructed drainage pattern of an earlier time. *B.* Present drainage pattern. Provo River has extended its valley headward, capturing several small tributaries to Weber River and also the large tributary now part of Provo River. Abandoned segment of valley of former north-flowing stream is floored with stream gravel derived from the territory of the diverted stream. Small tributaries have barbed pattern, showing former flow toward the east. Probable cause of capture: Provo River, shorter and with a steeper gradient than Weber River, had the greater potential for erosion. (*After G. E. Anderson.*)

147

Adjustment of Streams. In a long process of sculptural evolution, a stream system tends to adjust itself to the pattern of the rocks it drains, so that more and more stream segments occupy belts of weak rock or follow joints, faults, and other avenues of easy erodibility. *An adjusted stream system is a system in which most of the streams occupy weak-rock positions.* Through the headward growth of subsequent streams and occasionally through capture, the degree of adjustment of a stream system tends to increase with time and with depth of erosion. The result is a surface that reflects the pattern of the exposed belts of rock. The resistant rocks form hills, ridges, and other highlands, and the weak rocks underlie valleys and lowlands (Figs. 8.27, 8.29, 17.13).

An example of good adjustment is the segment of the Delaware River drainage system shown in Fig. 8.33, in which the principal tributary streams coincide with areas underlain by weak rocks. Even more strikingly, the Delaware River, whose general course lies "across the grain" of the rocks, itself successfully avoids some of the belts of resistant rock and crosses others either at narrow places or at places where the rock is dislocated by faults.

How the tributaries became adjusted is fairly evident. Probably they are the successful competitors among many small streams whose valleys gradually grew headward from the Delaware at an earlier time in history. Those that started on resistant rocks developed very slowly; those on weak rocks lengthened more rapidly and secured most of the potential drainage area. "For he that hath, to him shall be given; and he that hath not, from him shall be

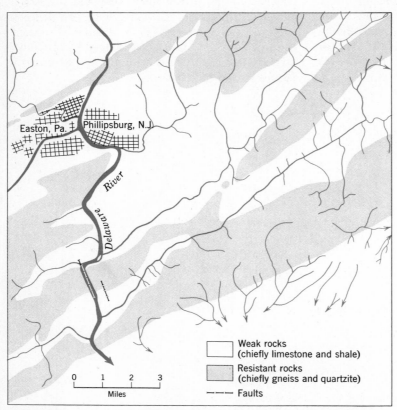

Weak rocks
(chiefly limestone and shale)

Resistant rocks
(chiefly gneiss and quartzite)

—--— Faults

Figure 8.33 Geologic sketch map of area around Easton, Pennsylvania, showing a stream pattern well adjusted to weak rocks. As the gneiss and quartzite resist not only stream erosion but mass-wasting as well, they are left as ridges about 500 feet higher than the weak-rock areas. (*After W. S. Bayley.*)

taken. . . ." Out of context, this is an apt rule for the competition that occurs in a drainage system.

How the Delaware itself became adjusted is less easy to visualize. In every area of complex rocks the locations of streams present puzzles, some of which have been solved with a good degree of probability. But others may never be solved, simply for lack of enough surviving evidence. Erosive forces not only sculpture the land but also tend eventually to destroy their own work and with it the evidence of how the work was done.

Conclusion

In contrast to streams such as the Delaware River, antecedent streams and superposed streams are by definition not adjusted. Whatever their earlier degree of adjustment may have been, it was destroyed at the places where the antecedent or superposed relationship developed.

In the sequence represented by Chaps. 6, 7, and 8 we have followed the creation of rock particles by weathering, their movement down hillslopes into streams, and their transport as sediment toward the sea. These events complete three closely related phases of the rock cycle. Also we have followed the sculpture of the land into characteristic forms, by the same streams as they move and deposit sediment. While a land mass passes through a slowly changing sequence of forms and is gradually eroded to form a peneplain, sediments resulting from erosion are transported and spread out to form new strata, on lower land as basin fills or on the floor of a sea. Thus the cycle of erosion and the rock cycle are complementary. Of course the peneplain can be elevated and destroyed by renewed erosion. But some peneplains sink down and are buried by layers of sediment. The buried surface is preserved until uplift occurs. The uplift renews erosion, which cuts away the cover and exposes the buried surface once more. Ancient peneplains created as long ago as Precambrian time, more than 600 million years ago, are now, after all that time, being again exposed to view.

Summary

Importance of Running Water

1. As part of the water cycle, streams are the chief means of returning water from land to sea. As geologic agents, stream erosion and mass-wasting are foremost among the processes that erode the land and transport sediments from land to sea.

2. Splash of raindrops and sheet erosion effectively erode regolith on bare, unprotected slopes.

Stream Flow and Behavior of Rock Particles

3. Turbulence characterizes the flow of nearly all streams, and is a prime factor in picking up and transporting sediment.

4. A stream's load consists of bed load, suspended load, and dissolved load.

Economy of a Stream

5. Long profiles of streams are concave-up curves. The profiles become gentler with time but are limited downward by base level.

6. Stream discharge, width and depth of channel, and velocity are intimately related, and continually adjust to each other.

7. Because of increased discharge and velocity, a stream can carry a load both coarser and greater in amount during floods, than it can transport at times of low water. Streams do most of their geologic work during seasonal floods.

8. Features of streams in alluvial valleys include meanders, oxbow lakes, bars, natural levees, and floodplains.

9. Fans are built at the toes of steep slopes; deltas are built at the mouths of streams. A common kind of delta consists of foreset, bottomset, and topset layers.

10. Streams tend to maintain a steady-state condition. If this is interrupted, the stream will return to it.

Sculptural Evolution of the Land

11. Most land surfaces consist of complexes of valleys, cut by the streams that flow through them.

12. Uninterrupted sculpture of a land mass follows a broadly predictable cycle of erosion, ending in a peneplain.

13. Weathering, mass-wasting, and sheet erosion together erode more rock material than streams do. The main work of streams is to carry away the material fed to them from slopes.

Interruptions in Sculptural Evolution

14. The orderly progress of land sculpture is commonly interrupted. Among the interruptions are movements of the crust and changes in the position of base level. These can cause rejuvenation or deposition.

Effects of Varying Erodibility of Rock

15. Streams tend to occupy belts of weak rock. Therefore the pattern of rocks exposed at the Earth's surface influences the pattern of streams.

Selected References

Streams

Colby, B. R., 1963, Fluvial sediments—a summary of source, transportation, deposition, and measurement of discharge: U.S. Geol. Survey Bull. 1181, p. A1–A47.

Davis, S. N., and de Wiest, R. J. M., 1966, Hydrogeology: New York, John Wiley.

Davis, W. M., 1899, The geographical cycle: Geogr. Jour., v. 14, p. 481–504.

———, 1902, Base level, grade, and peneplain: Jour. Geol., v. 10, p. 77–111.

Denny, C. S., 1965, Alluvial fans in the Death Valley region, California and Nevada: U.S. Geol. Survey Prof. Paper 466.

Fisk, H. N., 1952, Mississippi River valley geology in relation to river regime: Am. Soc. Civil Engrs. Trans., v. 117, p. 667–682.

Hoyt, W. G., and Langbein, W. B., 1955, Floods: Princeton, N.J., Princeton Univ. Press.

Leopold, L. B., and Langbein, W. B., 1966, River meanders: Scientific Am., v. 214, p. 60–70.

Leopold, L. B., and Maddock, T., Jr., 1953, The hydraulic geometry of stream channels and some physiographic implications: U.S. Geol. Survey Prof. Paper 252.

Leopold, L. B., Wolman, M. G., and Miller, J. P., 1964, Fluvial processes in geomorphology: San Francisco, W. H. Freeman.

Livingstone, D. A., 1963, Data of geochemistry, 6th ed., Chap. G., Chemical composition of rivers and lakes: U.S. Geol. Survey Prof. Paper 440, p. G1–G64.

Shirley, M. L., ed., 1966, Deltas in their geologic framework: Houston Geol. Soc., p. 233–251, maps of existing deltas assembled by A. E. Smith, Jr.

Sundborg, Åke, 1956, The River Klarälven. A study of fluvial processes: Geograf. Annaler, v. 38, p. 125–316.

Motion-picture film: Flow in alluvial channels (16mm color with sound). Shows stream flow and examples of ripples and sand waves formed in a laboratory channel. Available for free loan on application to Map Information Office, U.S. Geological Survey, Washington, D.C.

Sculpture of the Land

Cotton, C. A., 1952, Geomorphology, an introduction to the study of landforms, 6th ed.: New York, John Wiley.

Leopold, L. B., Wolman, M. G., and Miller, J. P., 1964, Fluvial processes in geomorphology: San Francisco, W. H. Freeman.

Morisawa, M. E., 1964, Development of drainage systems on an upraised lake floor: Am. Jour. Sci., v. 262, p. 340–354.

Schumm, S. A., and Lichty, R. W., 1965, Time space and causality in geomorphology: Am. Jour. Sci. v. 263, p. 110–119.

Thornbury, W. D., 1954, Principles of geomorphology: New York, John Wiley.

Chapter 9

Ground Water

The preceding chapter deals with water that flows over Earth's surface, makes clearly defined valleys, and deposits much sediment. In contrast, the present chapter goes below ground in order to examine the less obvious water that exists beneath the surface. Although it stays out of sight, makes no valleys, and deposits little or no mechanical sediment, this **ground water,** defined simply as *all the water contained in spaces within bedrock and regolith,* is important for at least four reasons: (1) it plays an essential part in the water cycle, keeping streams flowing between rains; (2) it performs geologic work by dissolving and depositing substances below ground; (3) it supplies plants and animals (including man) with a sizable fraction of their water requirements; and (4) as a chief factor in controlling the distribution of vegetation, it influences erosion by running water, mass-wasting, and wind.

The questions we must try to answer are these. How is water distributed beneath the ground? How does it get into the ground? How does it move? What geologic work does it do? And last, but not least important, how do we find and develop sub-surface water for economic use, and how does our ever-growing demand for water affect the supply?

Distribution and Origin

Amount. Looking back at our diagrams of the water cycle (Fig. 2.1) and the World's water budget (Fig. 2.2), we see that ground water in the continents acts as a storage reservoir, receiving rainfall at irregular intervals by infiltration from above and transmitting

Table 9.1

Distribution of World's Water. (From data in R. L. Nace, 1967, U.S. Geol. Survey Circ. 536, table 1, and other sources)

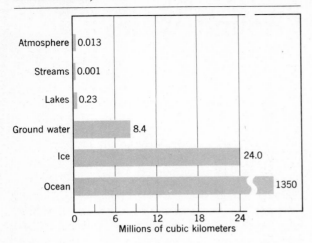

Atmosphere | 0.013
Streams | 0.001
Lakes | 0.23
Ground water | 8.4
Ice | 24.0
Ocean | 1350

Millions of cubic kilometers

this reason some places are much more favorable than others for obtaining ground water.

More than half of all ground water occurs within about 750m of Earth's surface. This probably includes most of the water that is usable, and is estimated to be equivalent in amount to a layer of water some 55km thick, spread over the world's land area. Below the depth of about 750m, water decreases in amount, gradually though irregularly. Holes drilled for oil have found water lying as deep as 9.4km— but everywhere, at some depth, water ceases to be present.

Origin. The ancient Greeks, 2500 years ago, thought that ground water was seawater driven into the rocks by the winds and somehow desalted, or that it was created in some manner from rocks and air deep below the surface. Later it was recognized that rivers are fed, at least in part, by springs emerging from the ground and also that the discharge of rivers does not raise the surface of the sea appreciably. The truth that ground water is derived mainly from rain and snow was recognized by Marcus Vitruvius, a Roman architect of the time of Christ, who wrote a treatise on aqueducts and water supply, a matter of great practical importance to the Romans.

Although true, Vitruvius's statement that ground water comes from rain was qualitative only. Not until the 17th Century was it established on a quantitative basis. Then Pierre Perrault, a French physicist, measured the mean annual rainfall on a part of the drainage basin of the River Seine in eastern France and also the mean annual runoff from it in terms of river discharge. He concluded that the difference between the amounts of rainfall and runoff was ample enough, over a period of years, to account for the amount of water in the ground. Today we accept rainfall as the source of all ground water except for a tiny proportion that comes from magma.

Water Table. Much of our knowledge of groundwater occurrence has been learned the slow way, from the accumulated experience of many generations of people who have dug or drilled millions of wells. This experience (Fig. 9.1) tells us that a hole penetrating the ground ordinarily passes first through a ***zone of aeration,*** *the zone in which open spaces in regolith or bedrock are normally filled mainly with air.*

it downward and outward to streams in a steadier manner. Table 9.1 tells us that an estimated 97.6 per cent of the world's water is in the ocean, the main reservoir, while much less than 1 per cent is ground water. Small though the total volume of ground water is, it is about 35 times greater than the volume of water lying in lakes or flowing in streams on Earth's surface.

True, much of this ground water is not presently of practical use, for at least two reasons. Some of it is firmly held in tiny openings in rock through which it cannot move, and some of it is salty. But so also is much of the surface water not useful, except as a source of salts, for nearly half of it is found in salt lakes.

Depth. One of the reasons why people have been able to establish permanent settlements, not only in well-watered country but also in desert lands, is that few areas exist in which holes, intelligently located and sunk far enough into the ground, do not find at least some water. In a moist country the depth of an adequate well may have to be only a few meters; in a desert it may have to be hundreds. These facts have been learned by experience. Water is present beneath the land nearly everywhere, but whether it is a usable supply depends on depth of occurrence, kinds of rock present, kinds and amounts of substances dissolved in the water, and other factors. For

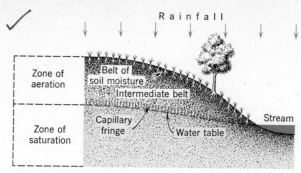

Figure 9.1 Positions of zone of saturation, water table, and zone of aeration. (*After W. C. Ackermann and others, U.S. Dept. of Agriculture.*)

The hole then enters the **zone of saturation,** *the subsurface zone in which all openings are filled with water. The upper surface of the zone of saturation* is the **water table,** which, at any place, normally slopes toward the nearest stream. Ordinarily the water table lies within a few meters of the surface. But along the shore of a lake or the bank of a river it can be at the surface. Again, in some mountainous districts it can lie at a depth of 100m or more. Whatever its depth, the water table is a very significant surface, because it represents the upper limit of all readily usable ground water. We shall return to it shortly.

Movement

Most of the ground water within a few hundred meters of Earth's surface does not lie there inert; it moves. But its movement is unlike the turbulent flow of rivers, measurable in kilometers per hour. It is so slow that velocities are expressed in centimeters per year. To understand why this is so, we must understand the porosity and permeability of rocks.

Porosity and Permeability. The limiting amount of water that can be contained within a given volume of rock material depends on the *porosity* of the material; that is, *the proportion (in per cent) of the total volume of a given body of bedrock or regolith that consists of pore spaces* (i.e., open spaces). So a very porous rock is a rock containing a comparatively large proportion of open space, regardless of the size of the spaces. Sediment is ordinarily very porous, ranging from 20 per cent or so in some sands and gravels to as much as 50 per cent in some clays.

The sizes and shapes of the constituent particles and the compactness of their arrangement affect porosity, as does the degree, in a sedimentary rock, to which pores have become filled with cementing substances. In contrast, igneous and metamorphic rocks generally have low porosity, except where joints and cracks have developed in them.

Permeability is *capacity for transmitting fluids.* A rock of very low porosity is likely also to have low permeability. However, high porosity values do not necessarily mean high permeability values, because size and continuity of the openings influence permeability in an important way. The relationship between size of openings and the molecular attraction of rock surfaces plays a large part. Molecular attraction is the force that makes a thin film of water adhere to a rock surface despite the force of gravity; an example is the wet film on a pebble that has been dipped in water. If the open space between two adjacent particles in a rock is small enough, the films of water that adhere to the particles will come into contact. This means the force of molecular attraction is extending right across the open space, as shown on the left side of Fig. 9.2. At ordinary pressure, therefore, the water is held firmly in place and so permeability is low. This is what happens in a wet sponge before it is squeezed. The same thing happens in clay, whose particles are so tiny their diameters are less than 0.005mm (Table 5.1).

By contrast, in a sediment with grains at least as large as sand grains (0.06mm to 2mm) the open spaces are wider than the films of water adhering to the grains. Therefore the force of molecular attraction does not extend across them effectively, and the water in the centers of the openings is free to move in response to gravity or other forces, as shown at the right in Fig. 9.2. This particular sediment is therefore permeable. As the diameters of the open-

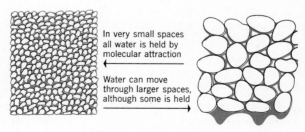

Figure 9.2 Effect of molecular attraction in the intergranular spaces in fine sediment (*left*) and in coarser sediment (*right*). Scale is much larger than natural size.

ings increase, permeability increases. With its very large openings, gravel is more permeable than sand and yields large volumes of water to wells.

Movement above Water Table. Let us return for a moment to Fig. 9.1. Water from a rain shower soaks into the soil, which usually contains clay resulting from weathering of the bedrock. Because of its content of clay the soil is generally less permeable than underlying materials. Part of the water, therefore, is held there by the forces of molecular attraction. This is the belt of soil moisture in Fig. 9.1. Some of the water evaporates directly and much is taken up by plants and is transpired (Fig. 2.1).

In the soil, water that molecular attraction cannot hold seeps downward through the intermediate belt (Fig. 9.1) until it reaches the water table. In fine-grained material a narrow fringe as much as 60cm thick, immediately above the water table, is kept wet wherever open spaces are so narrow that molecular forces can extend across them. Water is drawn upward through these tiny spaces in the same way as ink is drawn upward through blotting paper.

With every rainfall, more water is supplied from above, but, apart from the belt of soil water and the capillary fringe, the zone of aeration is likely to be nearly dry during the times between rains.

Movement Below Water Table: Percolation. In the zone of saturation the flow of ground water is like what occurs when a saturated sponge is squeezed gently. Such movement is called *percolation*. In it, water particles move slowly through small open spaces along parallel, threadlike paths. Movement is easiest through the central parts of the spaces but diminishes to zero immediately adjacent to the sides of each space, because there molecular attraction holds the water in place.

The force of gravity supplies the energy for percolation of ground water. Responding to that force, the water "tends to seek its own level," percolating from areas where the water table is high toward areas where it is lowest; in other words, toward surface streams (Fig. 9.3). Only part of the water travels by the most direct route, right down the slope of the water table. Other parts follow innumerable long, curving paths that go deeper through the ground. Some of the deeper paths turn upward against the force of gravity and enter the stream from beneath. This behavior is explained by the fact that in the zone of saturation the water along any plane of given height, such as h_1 in Fig. 9.3, is under greater pressure beneath a hill than beneath a stream. The water therefore tends to move toward points where pressure is least.

Laboratory models have been made in which dye is injected into the percolating ground water at various depths. Paths followed by the dye resemble those in Fig. 9.3, where they turn upward beneath a model stream at the base of a hill. The rate at which the water moves along these paths decreases sharply with increasing depth. Therefore most of the ground water that enters a stream has traveled to it via shallow paths not far beneath the water table.

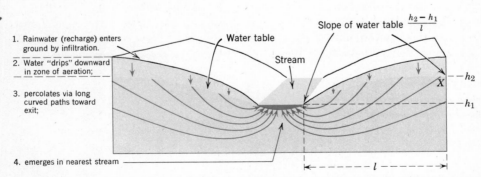

Figure 9.3 Movement of ground water in uniformly permeable rock material. Long curved arrows represent only a few of many possible paths. At any point such as X, slope of water table is determined by $\frac{h_2 - h_1}{l}$ where $h_2 - h_1$ is height of X above point of emergence in surface stream and l is distance from X to point of emergence. (*After M. K. Hubbert.*)

Economy of the Ground Water System

Fluctuation of the Water Table. As can be seen in Fig. 9.3, the water table is a surface consisting entirely of slopes, its form is a subdued imitation of the ground surface above it. It is high beneath hills and low at valleys, where ground water seeps out into rivers. If all rainfall could somehow be stopped permanently, the "hills" formed by the water table would slowly flatten, and would slowly approach the levels of the valleys. Percolation would gradually cease, and the streams in the valleys would dry up. Of course this could not happen, but in times of drought, when rain does not fall for perhaps several weeks, we sense the flattening of the water table in the drying-up of ordinary wells. When that occurs we know the water table has subsided below the bottoms of the wells. It is repeated rainfall, dousing the ground with new supplies of water from above, that prevents the water table from flattening very much.

Velocity of Flow. What, then, determines the steepness of the slopes of a water table and the rate of flow of the percolating ground water? Let us look again at Fig. 9.3. Like the gradient of a stream, the slope of a water table between any point such as X at height h_2 and the point where it emerges at height h_1 is measured by the difference in height ($h_2 - h_1$) divided by the horizontal distance l.

The relationship between slope and velocity of flow was established in the mid-19th Century by Henry Darcy, engineer in charge of public works in the French city of Dijon. In a program of improvement of the city's water supply he experimented with percolation through sand used for filtering the water and, as a result, established the fundamental equation:

$$V = P\frac{h_2 - h_1}{l} \quad \text{or, more compactly,}$$

$$V = P\frac{h}{l}$$

where V = velocity of flow, P = a coefficient representing permeability, and $(h_2 - h_1)/l$ of course is slope. The equation, known as Darcy's law, says that in material of given permeability, velocity increases as slope of the water table increases.

Therefore we can speak of the *economy* of a ground-water system just as we speak of the economy of a stream system. In the ground-water system, however, the terms are simpler because there is no alluvial channel that changes dimensions, as in a stream system. The water is percolating through the openings in a fixed framework of bedrock or regolith, although as the water passes from one kind of rock into another, permeability changes. Apart from changes of permeability, the important variable factor is the slope of the water table, which changes with rainfall as does the discharge of a stream.

Velocity and Discharge. Because of the large amount of friction involved in percolation, velocities are slow, commonly ranging between about 1.5m/day and 1.5m/year. The largest rate yet measured within the United States, in exceptionally permeable material, is only about 250m/year.

Velocity of percolation is measured between pairs of wells. In one method two wells with metal casings are connected to form an electric circuit. A chemical compound that is an efficient conductor and is soluble in water is poured into the upslope well and percolates downslope. On its arrival at the downslope well it creates a short circuit between well casing and electrode; this is recorded on an ammeter. Distance between wells divided by elapsed time gives the velocity.

Another means of measuring rate (and, of course, direction) of percolation is to put a strong dye, such as fluorescein, into a well and then time the appearance of dyed water in neighboring wells.

Of more direct practical importance than velocity, V, is *discharge*, Q, the quantity of ground water that percolates through a given cross-sectional area (A) of rock material in a unit of time. Discharge is more directly important to us because it states the amount available for use.

From its definition given above, we can express discharge thus:

(1) $$Q = VA$$

Now, according to Darcy's law:

(2) $$V = P\frac{h}{l}$$

(Velocity equals permeability times slope of water table.)

Therefore, substituting $P\frac{h}{l}$ for V in equation (1), we get:

$$(3) \qquad Q = P\frac{h}{l}A$$

This says that discharge equals permeability times slope of water table times cross-sectional area. It is through this formula that we can estimate the amount of water that a given well can be expected to deliver.

Ordinary Wells and Springs

There are two classes of wells and springs, determined by the geometry of the rock bodies that supply them. In one class, water movement is unconfined and is therefore comparatively simple. In the other class, movement is confined to a particular part of the rock and in consequence resembles the flow of water through the pipes of a system of plumbing. The wells and springs of the first class we call *ordinary;* the others we call *artesian.*

Aquifers. When we search for a good supply of ground water we search for an **aquifer** (Lat. "water carrier"), *a body of permeable rock or regolith through which ground water moves.* Bodies of gravel and sand are commonly good aquifers and so are many sandstones. But the presence of cement between the grains of sandstone reduces the diameter of the openings and so reduces the effectiveness of these rocks as aquifers.

It might seem that claystones, igneous rocks, and metamorphic rocks should not be aquifers, because in them the spaces between grains are extremely small, and because samples of them, measured in the laboratory, are impermeable. What is true for laboratory samples, however, does not necessarily apply to large bodies of the same material. Many such bodies contain fissures, spaces between layers, and other openings that are too large for water flow to

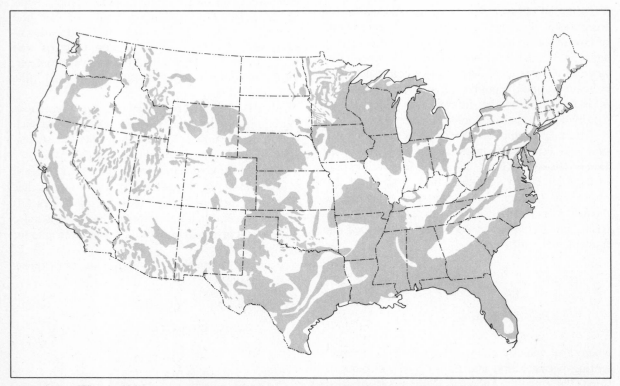

Figure 9.4 The shaded areas are underlain by one or more aquifers that can yield, in individual wells, at least 50 gallons per minute of water containing no more than 0.2 per cent dissolved solids. (*H. E. Thomas, U.S. Dept. Agriculture.*)

be controlled entirely by molecular attraction; these bodies can be aquifers. Even so, they are less effective as aquifers than are coarse-grained sediments of incompletely cemented sedimentary rocks. Whatever their effectiveness, it is in aquifers that we find wells and springs. More than half the area of the conterminous United States is underlain by one or more aquifers (Fig. 9.4).

Ordinary Wells. An ordinary well fills with water simply by intersecting the water table (Fig. 9.5). Lifting water from the well lowers the water level and so creates a **cone of depression,** *a conical depression in the water table immediately surrounding a well.* In most small domestic wells the cone of depression is hardly appreciable. Wells pumped for irrigation and industrial uses, however, withdraw so much water that the depression can become very wide and steep and can lower the water table in all the wells of a district. Figure 9.5 shows that a shallow well can become dry at times, whereas a deeper well in the vicinity may yield water throughout the year.

If rocks are not homogeneous, the yields of wells are likely to vary considerably within short distances. Massive igneous and metamorphic rocks, (Fig. 9.6,*A*), for example are not likely to be very permeable except where they are cut by fractures, so that a hole that does not intersect fractures is likely to be dry. Because fractures generally die out downward, the yield of water to a shallow well can be greater than to a deep one. Again, discontinuous bodies of permeable and impermeable material (Fig. 9.6,*B*) result in very different yields to wells. They also create **perched water bodies** (*water bodies that occupy basins in impermeable material, perched in positions higher than the main water table*). The impermeable layer catches and holds the water reaching it from above.

Gravity Springs. A **spring** is *a flow of ground water emerging naturally onto Earth's surface.* The simplest spring is an ordinary or *gravity spring.* Three common kinds of gravity springs are illustrated in Fig. 9.7.

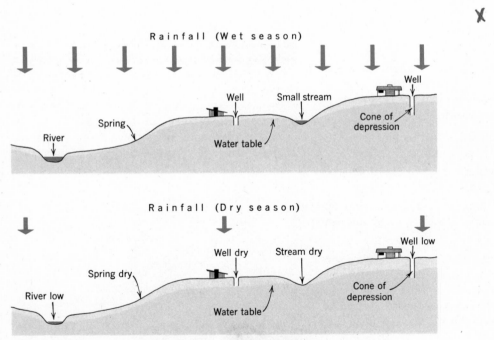

Figure 9.5 Wells and a spring in homogeneous rocks, showing cones of depression and effect of seasonal fluctuation of water table. Note that the slopes of the water table are steeper in the wet season when input of water into the system is greatest than in the dry season when input is least.

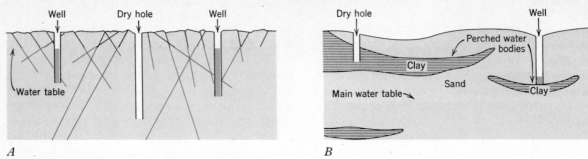

A   *B*

Figure 9.6 Ordinary wells and adjacent dry holes in rocks that are not homogeneous. *A*. In fractured massive rocks such as granite. *B*. In bodies of permeable sand containing discontinuous bodies of impermeable clay. Two perched water bodies are shown.

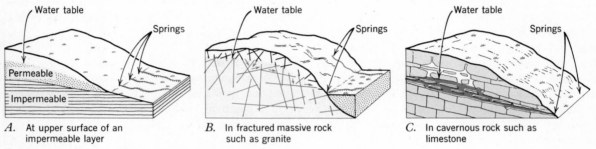

A. At upper surface of an impermeable layer

B. In fractured massive rock such as granite

C. In cavernous rock such as limestone

Figure 9.7 Three common locations of gravity springs.

Artesian Water

Confined Percolation. In some regions the geometry of inclined rock layers makes possible a special pattern of circulation of ground water. Three essentials of the pattern are shown in Fig. 9.8:

1. A series of inclined strata that include a permeable layer sandwiched between impermeable ones.

2. Rainfall, to feed water into the permeable layer where that layer is cut by the ground surface.

3. A fissure or a well so situated that water from the sandstone can escape upward through the impermeable roof.

When these essentials are present, we have a special kind of system. The input consists of rainwater, which enters the permeable layer (now an aquifer by definition) and percolates through it. The output consists of water forced upward through fissures or wells that perforate the roof.

Note, in Fig. 9.8, the position of the water table. Except for a thin zone close to the surface, the whole series of strata is saturated with water. In the claystone the water is held immobile by molecular at-

traction in the tiny spaces between the particles of rock. But in the aquifer it moves, provided only that water can escape through fissures or wells. Percolation in the aquifer, however, is *confined* between the impermeable strata above and below; it moves past the water held immobile in those strata.

If, in the area of *recharge* shown in the illustration, rainfall reaches the ground in greater volume than

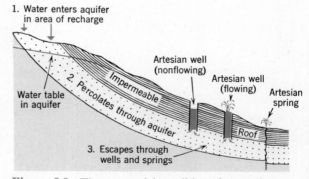

Figure 9.8 Three essential conditions for artesian wells: an aquifer, an impermeable roof, and water pressure sufficient to make the water in any well rise above the aquifer.

that of water output through fissures or wells, only enough water to balance output can enter the system; the excess flows away over the surface. On the other hand, if wells draw out of the system more water than can enter it from the available rainfall on the area of recharge, the yield of the wells will diminish to a quantity small enough to be balanced by the recharge.

The aquifer is like a broad, flat, sand-filled pipe or conduit, holding its ground water confined under the pressure of the column of water that extends up to the water table at its upper end. The water rises, in any well, to the level of the recharge area, minus an amount determined by the loss of energy in friction of percolation. Hence the height to which the water will rise in any well depends on distance from recharge area, height of that area above the well head, and permeability of the aquifer.

A well of this kind is an ***artesian well***, defined as *a well in which water rises above the aquifer*. The name comes from the French province of Artois, in which, near Calais, the first well of this sort in Europe was bored. When the factors listed above are unusually favorable, pressure can be great enough to lift the water above ground, creating fountains as much as 60m high.

Deep wells bored into rock to intersect the water table are popularly called artesian wells, but this is an incorrect use of the term. Such wells are ordinary wells, like those in Fig. 9.5 and 9.6.

Ocala Limestone Artesian System. The Ocala Limestone, a principal artesian aquifer in Florida, is an aquifer because it is full of caverns and smaller openings, intricately interconnected, created by solution. In the central and northwestern part of the peninsula this layer of limestone is exposed at the surface, but eastward and westward it becomes covered by overlying strata as it slopes downward toward both coasts. Impermeable roofs are provided by clayey layers within the limestone; so the Ocala is not merely *an* aquifer; it is a *series* of aquifers one on top of the other.

The age of the water at various places within the system has been determined. This was done by measuring the age of ^{14}C in $(HCO_3)^-$ dissolved in the water, at a series of wells along an 83-mile line generally parallel to the dip of the aquifer. Most of the ^{14}C entered the ground in rainfall on the recharge area and moved through the aquifer in the ground water. The age was found to decrease systematically away from the recharge area. From the sum of the differences in age between samples from pairs of wells in the series is calculated an average velocity of percolation (through the whole distance) of 23 feet per year. According to this velocity, water in the well farthest from the recharge area has been in the ground for nearly 19,000 ^{14}C years.

There is a small error in these calculations. "Young" ground water has been percolating down into the aquifer, and is still doing so, along the entire distance of 83 miles. This "young" water dilutes the older water already in the aquifer, and so the average age of the water sampled in each well is a little younger than it would be if no younger water had been added from above. Hence the travel time calculated at 19,000 years is a little too small.

The ^{14}C dates support the correctness of Darcy's law, which was used independently to calculate the velocity between the same two points. The result was about the same as that calculated from the ^{14}C dates.

Tapping an artesian system is a very ancient art. Four thousand years ago many artesian wells as much as 100m deep were in existence. The well near Calais, in France, was bored in A.D. 1126 and is still flowing today. In that area, at any rate, withdrawal has not seriously exceeded supply.

Artesian systems are not confined to wells. Some are *artesian springs*, in which ground water rises to the surface through a natural fissure rather than through a man-made hole (Fig. 9.8).

Thermal Springs and Geysers

The temperatures of many artesian springs are substantially higher than the local mean annual air temperatures. There are more than a thousand such *thermal springs* in the United States, most of them in western states. Even larger numbers exist in other parts of the world. Ground water can become heated in two ways: (1) by descending so deep that it is warmed by the general internal heat that is measured by the geothermal gradient, and, more commonly, (2) by contact with bodies of igneous rock that are slowly cooling within the crust.

After being heated the water tends to rise, and reaches the surface along a fault or other avenue, forming thermal springs. Water temperatures in such springs range all the way up to the boiling point. Because dissolution is more rapid in warm water than in cold, thermal springs are likely to be un-usually rich in mineral matter dissolved from rocks with which they have been in contact. In some springs the mineral content has medicinal properties.

A hot spring equipped with a system of plumbing and heating that causes intermittent eruptions of water and steam is a **geyser.** The name comes from an Icelandic word meaning *to gush,* for Iceland is the home of many geysers. Most of the world's geysers that are not in Iceland are in new Zealand or in Yellowstone National Park. In all these regions there is evidence of volcanic activity late in geologic time, and the heat for geysers probably consists of masses of hot rock down below the surface.

The feature that marks a geyser is that it erupts not continuously but intermittently. No two geysers behave in exactly the same way, and we cannot observe and study the system of underground pas-sages that supplies any one of them. However, prob-ably they are all alike in that they are fed by ordinary ground water derived from rainfall. The water occu-pies a natural tube, probably crooked, that extends downward from the surface. It is heated by contact with hot rock until its temperature is nearly at the boiling point. In a straight tube convection would occur as it does in a teakettle, and would equalize water temperature throughout the tube. But in a crooked tube convection cannot occur effectively, and so from top to bottom of the tube the water is at its boiling point.

Here we come to a basic principle, which proba-bly explains the on-again off-again character of a geyser: pressure increases with depth and the boiling point rises with increasing pressure. So at the bottom of our tube pressure is greatest and boiling tempera-ture is highest. When this condition is reached, a very slight decrease of pressure or increase of tem-perature will make the bottom water boil. The result-ing steam pushes the overlying water up through the tube and a little is forced out at the top. The loss of water reduces the pressure below, so that through most or all of the tube the water is suddenly converted into steam and a violent eruption occurs at the surface.

Old Faithful in Yellowstone Park, the most fa-mous American geyser, erupts for a few minutes about once an hour, throwing a jet of steam high in the air. During the intervals between eruptions the emptied tube is refilled with water, which is then heated to the critical point at which the next eruption is triggered off.

Water and People

Hitherto we have been dealing with the way ground water moves and the places where it occurs. Now we turn to economic aspects of ground water: finding hidden water, quality of water, and problems of balancing need and supply. Our brief discussion treats surface water together with ground water wherever it seems easier to do so, for the two re-sources are very closely related.

Finding Ground Water. In the days when the population of North America was mostly rural, in any region of fairly abundant rainfall a well could be dug to a few meters' depth with a good chance that it would yield enough water for the use of a family. Sometimes the sites for such wells were located by persons who used forked twigs and other kinds of "divining rods" and who claimed to possess supernatural powers. The search for water by this means, often called "dowsing," dates back at least to the time of Moses and is still widespread. Al-though no scientific basis for this kind of claim is known to exist, use of the divining rod persists, partly because in many areas shallow supplies of ground water are so widespread that successful re-sults would be numerous even though sites were located at random. If the diviner were asked to indi-cate where water is *not* present below the ground and if his predictions were then tested by boring holes, the statistical results would soon reveal the unsoundness of his claims. But, because little money is spent on boring holes in attempts to avoid water, this has never been done.

Today the average depth of wells is deeper, the rock drill has replaced spade and pickax to a large extent, and ground water is being found in aquifers that are well hidden, some of them deep beneath arid regions. Drilling is preceded by detailed study of the local rock or regolith in an effort to picture condi-tions well below the surface, such for example, as those indicated in Figs. 9.6, 9.7, and 9.8.

Recharge. An important aspect of the use of ground water, whether in a gravity system or in an artesian system is *recharge, the addition of water to the zone of saturation.* The Platte River in Nebraska and the Nile in Egypt are examples of rivers that flow from mountains having good rainfall into a much drier region, where the water table lies deep beneath the surface. Water from these rivers leaks downward and recharges the ground water below (Fig. 9.9).

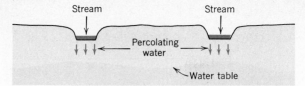

Figure 9.9 Recharge of ground water in a dry region, by infiltration from streams having their sources in mountains with abundant precipitation. The stream channels are actually leaking. Relation of water table to streams is the reverse of that shown in Fig. 9.1, where leaking is impossible.

Although most recharge is supplied directly by rainfall, the intense demand for water in some areas has led to artificial recharging of the ground. One example is the practice of *water spreading* in dry parts of the west. A common way to spread water for recharge is to build a low dam across a valley. This holds back water in a surface stream that would otherwise run to waste and allows it to recharge aquifers beneath the stream bed. The water thereby stored underground is withdrawn through wells as needed.

A chemical-industrial plant in the Ohio River valley, near Louisville, Kentucky, was located on a valley fill of sandy alluvium about 50m thick. Water was needed for cooling. The plant bought city water (purified river water) in winter when the water was cold and fed it into wells in the valley fill, thus recharging the sandy aquifer. In summer the cold water stored in the ground was pumped out for industrial use.

In some districts an aquifer is recharged with used water. This practice has increased with the increased use of air conditioning, which requires a large volume of water. Some cities have laws requiring that certain water that has been used for air conditioning be returned to the ground, where it successfully builds up the water table. This illustrates the basic principle of ground-water conservation: that withdrawal of water must, in the long run, be balanced by recharge; if it is not, then either recharge must be increased or withdrawal curtailed.

Water Quality. The *quality* of a body of water refers to its temperature and the amount and character of its content of mineral particles, solutes, and organic matter (chiefly bacteria), in relation to its intended use. The most common source of pollution of water from wells and springs is sewage, and the

infection most commonly communicated by polluted water is typhoid. Drainage from septic tanks, broken sewers, privies, and barnyards contaminates ground water. If the water contaminated with sewage bacteria passes through material with large openings such as very coarse gravel or the cavernous limestone shown in Fig. 9.10, it can travel for miles without much change. If, on the other hand, it percolates through sand or permeable sandstone, it can become purified within short distances, in some cases less than 30m (Fig. 9.11). The difference lies in the aggregate internal surface area of the material through which it percolates. The large aggregate force of molecular attraction holds the water and promotes its purification by (1) mechanical filtering-out of bacteria (water gets through but most of the bacteria do not), (2) destruction of bacteria by chemical oxidation, and (3) destruction of bacteria by other organisms, which consume and oxidize them. Purification goes on both in the zone of aeration and in the zone of saturation. Because clay

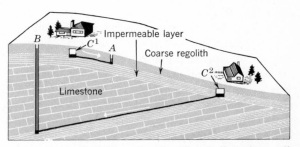

Figure 9.10 Pollution of wells. The shallow dug well, A, was unwisely located a short distance downslope from a septic tank, C^1, and received polluted drainage (black) from it. The owner then drilled a deeper well, B. This well tapped layers of cavernous limestone inclined toward it from the lower septic tank C^2. The water flowed through openings in the limestone, and reached the bottom of well B unpurified by percolation. The well owner must relocate his septic tank or dig a shallow well upslope from C^1.

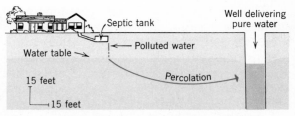

Figure 9.11 Purification of contaminated ground water in sand and gravel during percolation through a short distance.

particles are much smaller than sand particles, we might suppose that clay, with its much larger internal surface area, would be the ideal medium for purification. But it is not, because, as we have seen, it is almost impermeable. Particles of sand are large enough to permit rapid percolation, yet small enough to permit purification within short distances. For this reason treatment plants for purification of municipal water supplies and processing of sewage percolate these fluids through sand.

A substantial proportion of domestic sewage passes through septic tanks (Fig. 9.11) and then mingles with ground water. It percolates into streams, which it generally reaches in a pure condition. On the other hand the domestic waste from many areas, as well as much industrial waste, is dumped unaltered into surface streams. Although purification can be accomplished during stream transport, the distances involved are much greater than those required for the purification of water in the ground, and the amounts of sewage in many rivers are far too great to be dealt with by natural processes. In densely populated industrial countries these facts constitute serious public-health problems.

A dramatic illustration of the difference between surface and underground conditions is this. In some communities much-polluted water that has traveled tens of miles through a river is pumped into the ground, where it becomes a part of the local ground-water supply. In one city, percolation through a horizontal distance of 150m removed impurities from the sewage and made the water fit to drink.

Balance Sheet of Water Supply. Ground water and surface water together are a resource that is an absolute necessity for human use. Table 9.1 shows how much exists, but part of the ground water is not available because it lies too deep, because it lies in rock through which it cannot flow, or because its quality is poor. Table 9.2 shows the four chief uses of ground water in the United States.

Whatever their actual amount, ground water and surface water replenish themselves as the water cycle rolls on and on. Normally, too, the water cycle is in a condition of steady state, so that the supply available to people is nearly constant. In parts of the world where people live in a simple agricultural economy, their withdrawal of water from streams, lakes, and the ground affects the steady state hardly at all because they use little water—possibly less than

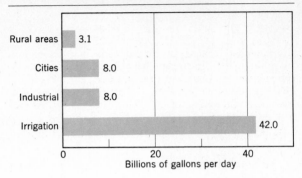

Table 9.2

Use of Ground Water in the United States, 1965. (U.S. Geol. Survey Circ. 556, 1968, p. 12)

10 gallons per day per capita. In contrast, the population of the industrially advanced United States consumes 150 gallons per day per capita in its houses alone. But when we add the per-capita share of the water used industrially, the total jumps from 175 to the enormous figure of 1600 gallons per day—160 times our estimate of consumption by the simplest non-industrial people. Such prodigal use of water, which in the United States doubled during the 15-year period 1950–1965, *does* affect the natural steady-state condition, at least in some areas.

Water use is carried to an extreme in some very dry areas such as parts of Arizona and of Israel. There the carbon-14 ages of some of the water drawn from deep wells are more, perhaps much more, than 10,000 years. We noted similarly "old" water in our discussion of the Ocala Limestone aquifer in Florida. But under Florida's moist climate, recharge from rainfall keeps pace with withdrawal from wells. In Arizona, however, "old" water is relict from much earlier times. It represents rain that fell during the latest glacial age, when climates were cooler and moister. Under today's drier climates rates of recharge are much less than the present rates of withdrawal. Hence the water used today in those areas is being mined and eventually the supply will be exhausted. In some areas, even if all pumping were to cease, it would take more than 100 years to recharge the aquifers.

One of the obvious characteristics of industrialization is the concentration of people in cities. Today more than two-thirds of the people of the United States live in or around cities; only one-third live in the country. Not only do city people use more water for all purposes (including air conditioning) in their buildings, but also the buildings themselves, as well as the streets and roads that give access to them, reduce local additions from rainfall. In cities the ground is covered largely with buildings, concrete, and asphalt, all of which send runoff along the surface (Fig. 1.1) and through sewers, rather than allowing it to soak into the regolith as happens under natural conditions. It is urbanization, more than growth of population overall, that has created a demand which in many cities threatens to exceed supply. Furthermore the available supply is less than it would be were it not for pollution (chemical, physical, bacterial, or thermal) of many streams, some lakes, and some ground water by industrial wastes, agricultural poisons, and sewage.

Figure 9.12 shows that the western part of the United States consumes far more water than does the eastern part, although the eastern part contains more aquifers (Fig. 9.4). The difference is mainly a consequence of the far greater use of irrigation (Table 9.2) in the west, where climates are drier and dependence on agriculture is greater.

Much of the water withdrawn from the ground and from surface sources and used for a variety of human purposes is returned to these natural reservoirs and re-used. As one example, irrigation water may be withdrawn from a river, and the part not evaporated or used by plants is returned through the ground to the stream. This can happen again and again to a parcel of water during its long journey from the middle of a continent back to the ocean.

Many authorities believe that shortage of water in some large cities, although real, is less a matter of inadequate quantity than of inefficient planning and development. They believe that except for the dry southwestern part of the country, the potential supply of water, *if properly managed,* need not be overtaxed for some time to come.

Ground Water in the Rock Cycle

Hitherto we have been dealing with the part played by ground water in the water cycle. Now we can examine the large part it plays in the rock cycle. Ground water dissolves minerals, transports the solutes, precipitates some of the solutes within rocks and regolith, and delivers the rest of them to streams, which carry them to the sea. There some of them are used to build limestone and other marine sedimentary rocks.

We can check on this activity in various parts of the rock cycle. In Chap. 6 we analyzed the decomposition of minerals during weathering. We found that most of this activity is accomplished by ground water, essentially as a weak solution of carbonic acid, and that one of the great results is leaching. In an area in the Piedmont region of North Carolina, underlain by diorite, the discharge of small streams and springs was measured and the total quantity of the many mineral solids in solution in this water was determined. From the data it was calculated that in that area ground water is leaching solids equivalent to a layer of rock 1cm thick every 933 years.

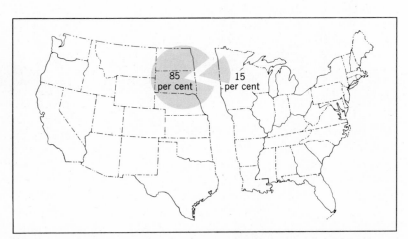

Figure 9.12 Water consumption in the 17 Western States and the 31 Eastern States, 1965. (*C. R. Murray, U.S. Geol. Survey Circ. 556, 1968.*)

Next, sampling the water of big rivers, we find that the Mississippi, down near its mouth, is carrying continuously about 100 million tons of dissolved solids, most of them the result of chemical weathering. Finally, in sampling ocean water, we find the ocean as a whole contains 50 million billion tons of dissolved solids—an amount equal to 3.5 per cent of the weight of the seawater itself. If all this dissolved matter could be extracted and spread over Earth's land area, it would make a solid layer more than 150m thick.

Chemical Composition. Analyses of many wells and springs show that the solutes in ground water consist mainly of chlorides, sulfates, and bicarbonates of calcium, magnesium, sodium, potassium, and iron. We can trace these substances back to the common minerals in the rocks from which they were derived by weathering. As might be expected, the composition of ground water varies from place to place according to the kind of rock in which it occurs. In much of the central United States the water is "hard," that is, rich in calcium and magnesium bicarbonates, because the bedrock includes abundant limestones and dolostones that consist of those carbonates. In some places within arid regions the concentration of dissolved substances, notably sulfates and chlorides, is so great that the ground water is unfit for human consumption. Furthermore, evaporation of water in the zone of aeration precipitates not only calcium carbonate but, in particularly dry regions, sodium sulfate, sodium carbonate, and sodium chloride. Soils containing these precipitates are loosely termed "alkali soils." They are unsuitable for agriculture because crops will not grow in them.

Some ground water is salty. Three causes of saltiness in ground water are:

1. Seawater, trapped in sediment on ancient sea floors, has persisted through the conversion of the sediment to rock.
2. Ground water, circulating slowly through sedimentary rock, has dissolved salts from the rocks themselves and so has become salty.
3. Along coasts, ground water has become mixed with seawater that has percolated into it.

Deposition. The conversion of sediment into sedimentary rock, one of the great transformations that occur as parts of the rock cycle, is primarily the work of ground water. A body of sediment lying beneath the sea is generally saturated with water, as is a sediment lying in the zone of saturation beneath the land. Substances in solution in the water are precipitated as a cement in the spaces between the rock particles that form the sediment. This activity transforms the loose sediment into firm rock. Calcite, silica, and iron compounds (mainly oxides), are, in that order, the chief cementing substances.

Less common than the deposition of cement between the grains in a sediment is *replacement, the process by which a fluid dissolves matter already present and at the same time deposits from solution an equal volume of a different substance.* Evidently replacement takes place on a volume-for-volume basis because the new material preserves the most minute textures of the material replaced. Petrified wood is a common example. But replacement is not confined to wood and other organic matter; it occurs in mineral matter as well.

Caverns. Limestone, dolostone, and marble are carbonate rocks that consist of the minerals calcite and dolomite in various proportions. These rocks underlie millions of square miles of Earth's surface. Although carbonate minerals are nearly insoluble in pure water, they are readily dissolved by the carbonic acid in ground water. As a result the ground water becomes charged with calcium bicarbonate, as shown by these reactions, the first of which we saw in our analysis of chemical weathering in Chap. 6:

$$CO_2 \ + \ H_2O \ \rightarrow$$
(Carbon dioxide) (Water)

$$H_2CO_3 \text{ which ionizes to } H^{+1} \ + \ (HCO_3)^-$$
(Carbonic acid) (Hydrogen ion) (Bicarbonate ion)

The hydrogen ions attack the calcite and dissolve it.

$$CaCO_3 + \ 2H^{+1} \ \rightarrow$$
(Calcite) (Hydrogen ion)

$$H_2O \ + \ CO_2 \ + \ Ca^{2+}$$
(Water) (Carbon dioxide) (Calcium ion)

Such dissolution is a form of chemical weathering just as much as is the decomposition of igneous rocks that contain feldspar and ferromagnesian minerals. In both cases the weathering attack occurs along joints and other partings in the bedrock (Fig. 9.13). But whereas in granite the quartz and other insoluble minerals remain, nearly all the substance of a body of pure limestone can be carried away in solution in the slowly moving ground water. The process of dissolution creates cavities of many sizes and shapes. *A large, roofed-over cavity in any kind of rock is a* **cavern.**

Although most caverns are small, some are of exceptional size. Carlsbad Caverns in southeastern New Mexico include one chamber 1200m long, 190m wide, and 100m high. Mammoth Cave, Kentucky consists of interconnected caverns with an aggregate length of at least 48km.

Some caverns have been partly filled with insoluble clay and silt, originally present as impurities in the limestone and gradually released by solution. Other caverns contain partial fillings of **dripstone** (*material chemically precipitated from dripping water in an air-filled cavity*) and **flowstone** (*material chemically precipitated from flowing water in the open air or in an air-filled cavity*). The "stones" take on many curious forms, which are among the chief attractions to cavern visitors. The most common shapes are **stalactites** (*icicle-like forms of dripstone and flowstone, hanging from ceilings*), **stalagmites** (*blunt "icicles" of flowstone projecting upward from floors*), and **columns** (*stalactites joined with stalagmites, forming connections between the floor and roof of a cavern*) (Fig. 9.14).

As its name implies, dripstone is deposited by successive drops of water. As each drop of water forms on the ceiling of a cavern, it loses a tiny amount of carbon dioxide gas and precipitates a particle of calcium carbonate. This chemical reaction is simply the reverse of the one by which calcium carbonate is dissolved by carbonic acid.

Dripstone can be deposited only in caverns that are already filled with air and are therefore above the water table. Yet many, perhaps most, caverns are

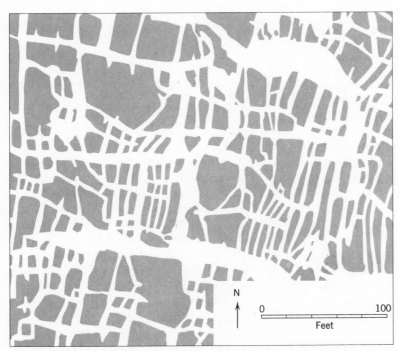

Figure 9.13 Part of a map made by members of the National Speleological Society, of Anvil Cave, near Decatur, Alabama shows how closely solution was controlled by the pattern of joints in limestone of the Gasper Formation. The cavern extends through an area about ten times greater than the part shown. (*W. W. Varnedoe.*)

Figure 9.14 Large cavern in limestone, partly refilled with dripstone and flowstone in the form of stalactites, stalagmites, and columns. Length of view along base of photo, about 20 feet. (*Luray Caverns, Virginia.*)

believed to have formed below the water table, as is suggested by their shapes and by the fact that some caverns are lined with crystals, which can form only in a water environment. How can we reconcile the two conflicting environments? Probably the answer lies in the long process of uplift and erosion of continents, which has been going on for at least hundreds of millions of years. In the process caverns one by one would be lifted above the water table and filled with air, after which they could begin to be filled with dripstone (Fig. 9.15).

Sinks. In contrast to a cavern, a *sink* is *a large solution cavity open to the sky.* Some sinks are caverns whose roofs have collapsed. Others are formed at the surface, where rainwater is freshly charged with carbon dioxide and is at its most effective as a solvent. Many sinks, located at the intersections of joints where movement of water downward is most rapid, have funnel-like shapes. A funnel-shaped sink is shown in Fig. 9.16. Although this sink is small,

some sinks are very large. One, near Mammoth Cave, Kentucky, is 13km² in area.

Karst Topography. In some regions of exceptionally soluble rocks, sinks and caverns are so numerous that they combine to form a peculiar topography characterized by many small basins. In this kind of topography the drainage pattern is irregular; streams disappear abruptly into the ground, leaving their valleys dry and then reappear elsewhere as large springs. This has been termed ***karst topography*** (Figs. 9.17, 9.18) because it is strikingly developed in the Karst region of Yugoslavia, inland from Trieste. It is defined as *an assemblage of topographic forms consisting primarily of closely spaced sinks.* Karst topography is developed through wide areas in Kentucky, Tennessee, southern Indiana, northern Florida, and Puerto Rico.

Sinks and caverns record the destruction of a very large volume of carbonate rock. It is calculated from measured amounts of carbon dioxide in ground

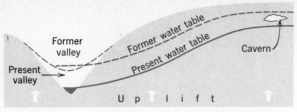

Figure 9.15 Possible history of a cavern containing dripstone. Cavern was excavated below water table. When streams deepened their valleys, water table was lowered as it adjusted to the deepened valleys. This left cavern above water table.

Figure 9.16 Sink in limestone near Sunken Lake, Presque Isle County, Michigan. (*I. D. Scott.*)

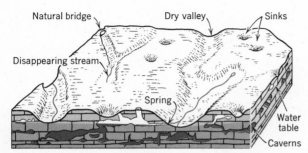

Figure 9.17 Karst topography. Area of block is between 1 and 2 square miles.

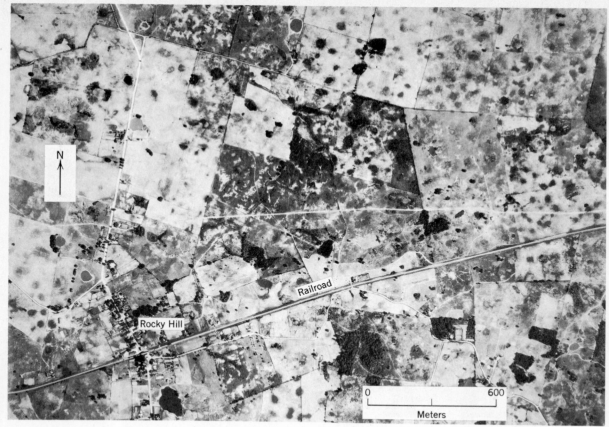

Figure 9.18 Karst topography in area of limestone south of Mammoth Cave National Park, Kentucky, seen in a vertical air photograph. Dark-colored areas with white rims are lakes that occupy some of the sinks. Small gray spots with less definite boundaries are dry sinks. Almost no surface streams are present because most rainfall sinks down into openings in the limestone before it can run off. (*Aero Service Corp., division of Litton Industries.*)

water and from the solubilities of limestones that the amount of precipitation on northern Kentucky is capable, as ground water, of dissolving a layer of limestone 1cm thick every 66 years. This potential is far greater than the average erosional reduction of the surface of the United States by mass-wasting, sheet erosion, and streams. It depends on the presence of exceptionally soluble rocks.

Summary

Geologic Significance

1. In the water cycle ground water plays a part in the return of water from land to sea.

Distribution and Origin

2. Ground water, derived almost entirely from rainfall, occurs nearly universally.

3. The water table is the top of the zone of saturation.

Movement

4. Ground water flows chiefly by percolation, at rates far slower than those of surface streams.

5. Darcy's law states that with constant permeability, velocity of flow of ground water increases as slope of the water table increases.

6. In moist regions ground water percolates away from hills and emerges in valleys. In dry regions it is likely to percolate away from beneath surface streams.

Economy

7. A ground-water system is an open system in each segment of which a state of dynamic equilibrium is approached.

Wells and Springs

8. Ground water flows into most wells directly by gravity but into artesian wells under hydrostatic pressure.

Ground Water and People

9. A basic principle of conservation is that withdrawal of ground water must not exceed recharge.

Ground Water in the Rock Cycle

10. Ground water dissolves mineral matter from rocks; much of the dissolved product eventually gets into the sea.

11. Ground water deposits substances as cement between grains and so reduces porosity and converts sediments into rocks.

12. In carbonate rocks ground water not only creates caverns and sinks by solution but also deposits mineral matter in some caverns.

Selected References

Bretz, J H., 1956, Caves of Missouri: Missouri Geol. Survey, v. 39.

Davis, S. N., and DeWiest, R. J. M., 1966, Hydrogeology: New York, John Wiley.

Ellis, A. J., 1917, The divining rod, a history of water witching: U.S. Geol. Survey Water-Supply Paper 416.

Hubbert, M. K., 1940, The theory of ground-water motion: Jour. Geology, v. 48, p. 785–944.

McGuinness, C. L., 1963, The role of ground water in the national water situation: U.S. Geol. Survey Water-Supply Paper 1800.

Meinzer, O. E., 1923, The occurrence of ground water in the United States, with a discussion of principles: U.S. Geol. Survey Water-Supply Paper 489.

Moore, G. W., and Nicholas, G., 1964, Speleology. The study of caves: New York, D. C. Heath.

Murray, C. R., 1968, Estimated use of water in the United States, 1965: U.S. Geol. Survey Circ. 556.

Todd, D. K., 1959, Ground-water hydrology: New York, John Wiley. U.S. Department of Agriculture, 1955, Water: Yearbook for 1955: Washington, U.S. Gov. Printing Office.

Chapter 10

Deserts
and Winds

Deserts

World Distribution of Deserts

Although the word *desert* means literally a deserted, unoccupied, or uncultivated area, the modern development of artificial water supplies has changed the original meaning of the word by making many dry countries habitable. *Desert* has become a synonym for *arid land,* whether "deserted" or not. But aridity remains the chief characteristic of any desert.

Arid lands of various kinds together add up to 25 per cent of the total land area of the world. In addition there is a smaller though still large percentage of semiarid land, in which annual rainfall ranges between about 10 and 20 inches. These dry and semi-dry areas are seen on a world map in Fig. 10.1, forming a distinctive pattern. The meaning of the pattern is clear as soon as we grasp the general plan of circulation of the atmosphere.

Circulation of Earth's Atmosphere. The atmosphere is continually in motion. It circulates in a definite pattern. The basic cause is that more solar heat is received near the equator than near the poles. The heated air near the equator expands, becomes lighter, and rises, in a process of convection (Chap. 3). High up, it spreads outward toward both poles; and on the way it gradually cools, becomes heavier, and sinks. Meanwhile, beneath it still cooler, heavier air, chilled in the polar regions, forms a return flow toward the equator. The returning air replaces the heated rising air and in turn is heated and rises.

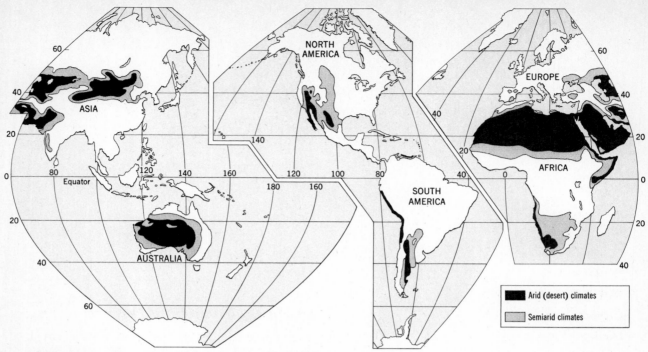

Figure 10.1 Arid and semiarid climates of the world, plotted according to the Köppen-Geiger system. Very dry parts of the polar regions, a special, cold kind of desert, are not included.

This simple circulation, however, is interfered with by Earth's rotation, which sets up a force named the *Coriolis force,* after the 19-Century French engineer who first analyzed it. The effect is to cause any body that moves freely with respect to the rotating solid Earth, to veer toward the *right* in the northern hemisphere and toward the *left* in the southern, *regardless of the direction in which the body may be moving.* Flowing water (such as an ocean current) and flowing air (winds) respond to it, as do smaller bodies such as airplanes, and projectiles of all kinds.

The Coriolis force breaks up the simple general flows of air between equator and poles into sections or belts (Fig. 10.2). Looking first only at the northern hemisphere, we note that at about latitude 30° some of the high-level, north-moving equatorial air is descending toward Earth's solid surface. As it descends it gets warmer; and the warmer it gets the more water it can hold; so it also becomes drier. This means that around that latitude, right around the world, climates are warm and dry, skies are clear, and rain is scarce.

The descending air spreads out along the surface,

toward the north and toward the south. In the northern hemisphere the south-flowing part is twisted toward the *right* (west), so that the south flow becomes a southwest flow. In ordinary language we call such a wind a *northeast* wind, for the direction *from* which it is blowing. As Fig. 10.2 shows, these winds form a belt of *northeast trade winds,* extending around the world.

Returning to the northern edge of this belt where air is descending, we now follow the air that is spreading toward the north. This flow too is twisted toward the right, so that northward flow becomes north*east*ward flow, forming winds that blow from the southwest. These winds are the belt of westerly winds or *westerlies,* again extending around the world.

At higher latitudes, somewhere between 40° and 60°, the westerlies encounter the cold heavy air that flows from the polar region toward the equator. This flow, of course, is also twisted toward the right; so it moves toward the southwest as generally *easterly* winds. Where the warmer and lighter air of the westerlies encounters this cold polar air, some of it

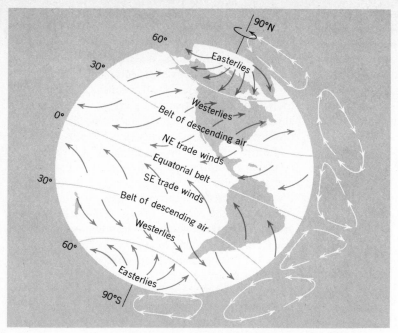

Figure 10.2 Earth's planetary wind belts, shown schematically. At right are vertical cross sections of the six belts. In each cross section the arrows show direction of flow of air.

tends to rise and return toward the equator. The polar air itself becomes warmer, gradually rises, and returns toward the pole.

To see what happens in the southern hemisphere we need only repeat this description, substituting left turns for right turns throughout. The result is a belt of southeast trade winds, a belt of descending air near 30° south latitude, a belt of westerlies, and a belt of polar easterlies, thus completing a neat pattern for the Earth. In this pattern, what is important for our discussion of deserts is that according to Fig. 10.1 most of the world's arid land in both hemispheres is centered between latitudes 15° and 35°. There, because air is descending and becoming warm, skies are clear and rain is scarce. A perfect place for a desert!

Kinds of Deserts. There are three chief classes of arid lands, the most extensive of which is (1) the one determined by Earth's wind belts. Examples of this class are the Sahara and other deserts in northern Africa. (2) A second sort of desert is found in continental interiors, where heating in summer and dry cold continental air in winter prevail. Instances are the deserts of central Asia. (3) Yet a third, more local

kind of arid area is one that lies in the lee of a mountain range. The mountains act as a barrier to rainfall on the desert beyond. Moist air encounters the mountains, rises over their windward slope, is cooled, drops rain. As it descends over the leeward slope the air becomes warmer and drier, creating a dry climate over the country beyond. The high Sierra Nevada in eastern California is the barrier mainly responsible for the arid climate of the country east of it.

Climate and Vegetation

The arid climate of a desert results from the combination of three factors:

1. High temperature. The highest temperature recorded in the United States, 56°C is at Death Valley, a desert area in southeastern California. The world's record, at a place in the Libyan Desert, is 57.7°C.

2. Low precipitation. At Death Valley annual precipitation averages between 2cm and 5cm, and in the Atacama Desert in northern Chile periods of more than 12 consecutive years without rain have been recorded.

3. Great evaporation. The higher the temperature, the greater the evaporation, and therefore the more precipitation an area can receive and still be arid. For if most of the precipitated water evaporates, little is left for streams and for vegetation. In parts of the southwestern United States evaporation from lakes and reservoirs amounts to as much as 250mm annually—10 to 20 times more than the annual precipitation.

Besides aridity resulting from these three factors, deserts are generally characterized by frequent strong winds. These commonly result from convection. During daytime hours air over especially hot places is heated and rises, and this allows surface air to move in rapidly and take its place. Desert winds are effective movers of sediment, as we shall see.

The vegetation in deserts is a direct reflection of dry climate. Usually the vegetation is not continuous. Where grass is present, it is likely to be thin and to grow only in clumps. More commonly the plants consist of low bushes growing rather far apart, with bare areas between them. This pattern of vegetation promotes active movement of sediment by the wind and by running water as well.

Geologic Processes

No major geologic process is restricted entirely to arid regions. Rather, the same processes operate with different intensities in moist and arid regions. As a result, in a desert the forms of the land, the soils, and the sediments show distinctive differences. Let us look at some of the major processes and note the differences.

Weathering and Mass-Wasting. In a moist region regolith is nearly universal, comparatively fine textured, a product chiefly of chemical weathering, in motion downslope mainly by creep, and covered by almost continuous vegetation. Creep fashions hill profiles into a series of curves.

In a desert the regolith, much of it a product of mechanical weathering, is thinner, less continuous, and coarser in texture. Slope angles developed by downslope creep become adjusted to the average diameter of the particles of regolith; the coarser the particles the steeper the slope required to move them. As the particles created by mechanical weathering tend to be coarse, slopes are generally steeper than in a moist region.

Mechanically weathered chunks of rock tend to break off along joints, leaving steep, rugged cliffs. Hills with cliffy slopes, particularly where layers of bedrock are nearly horizontal, are common in dry regions (Fig. 10.3).

Streams. One of the characteristics of deserts is that most of the streams which originate in them soon disappear by evaporation and by soaking into the ground, and never reach the sea. Exceptions are long rivers such as the Nile in Egypt and the Colorado in the southwestern United States, which originate in mountain regions with abundant precipitation. Such rivers carry so much water that they keep flowing to the ocean despite great losses where

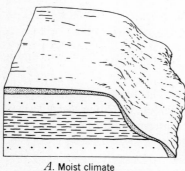

A. Moist climate   *B*. Dry climate

Figure 10.3 Effect of climate on sideslopes of valleys. *A*. In a moist climate resistant strata are partly masked by a creeping mantle of chemically weathered waste. *B*. In a dry climate resistant rocks stand out as broad platforms and steep cliffs, partly concealed by taluses.

Figure 10.4 Fan being built out into Death Valley, California, a down-faulted, desert basin with white, salt-incrusted playas. Highway across fan serves as a scale. (*J. S. Shelton.*)

they cross a desert. In a desert the few plant roots are no great impediment to runoff of rain, and the loose dry regolith is eroded easily. Typical violent rainstorms are therefore likely to be accompanied by "flash" floods that move heavy loads of sediment suddenly and swiftly. The loads are deposited as alluvium, forming fans at the bases of mountain slopes (Fig. 10.4) and on the floors of wide valleys and basins. The deposits of some flash floods are spectacular (Fig. 10.5).

Often, streams in flood effectively undercut the sideslopes of their valleys, causing the slopes to cave. Then, as the flood subsides, the load is deposited rapidly, creating a flat floor of alluvium. The result is a steep-sided, flat-bottomed "box canyon," characteristic of many dry regions (Fig. 10.6).

Playa Lakes. In an arid region, only rarely is water abundant enough to flow into a basin and fill it to overflowing. Streams that flow down from a highland rarely last until they reach the center of the nearest basin. But after an exceptionally large rainstorm some of them discharge enough water to convert the basin floor into a shallow lake that may last a few days or a few weeks (Fig. 10.7). In the dry region of the western United States *an ephemeral shallow lake in a desert basin* is called a ***playa lake;*** when *dry* (that is, most of the time) the *lake bed* is a ***playa*** (Fig. 10.8). Many playas are white or grayish because of precipitated salts at their surfaces. But if the lake water can escape downward through the basin floor before evaporation saturates the water with salts, no salts can be precipitated, and the playa sediments consist mainly of clay.

Ground Water. In a dry region ground water is derived only from the scanty local rainfall plus water that flows in from outside via rivers or via artesian

Figure 10.5 This deposit of boulder gravel resulted from a single flash flood in a mountain valley. Los Angeles County, California. (*U.S. Forest Service.*)

Figure 10.6 Small "box canyon" cut into silty alluvium. Cornfield Wash, Albuquerque district, New Mexico. Compare Fig. 8.25, which shows a valley of a similar kind. (*F. W. Kennon and H. V. Peterson, U.S. Geol. Survey.*)

Figure 10.7 Braun's Playa, near Las Vegas, Nevada, nearly 8km in greatest diameter. Mountain crest is 24km distant. Playa lake on April 10 after an unusually large rainfall. The lake is less than 1m deep. (*C. E. Erdmann.*)

Figure 10.8 Braun's Playa two weeks later, after the water had evaporated. Wind is blowing fine-grained lake sediments (clay and crystals of salts) into dust clouds. Dark spots are desert bushes. (*C. E. Erdmann.*)

aquifers. Useful supplies are therefore small and the water table lies well below the surface. Under these conditions it is especially important that withdrawal of ground water be kept in balance with rate of recharge.

Wind. In dry country, wind is an effective geologic agent. However, contrary to popular belief, deserts are not characterized mainly by sand dunes. Only one-third of Arabia, the sandiest of all dry regions, and only one-ninth of the Sahara are covered with sand. Much of the nonsandy area of deserts is cut by systems of stream valleys or is characterized by fans and alluvial plains. Thus even in deserts more

geologic work is done by running water than by wind.

The way in which wind works in deserts and in moister regions is set forth later in this chapter, after we have summarized desert landscapes.

Land Sculpture

Pediments. In deserts, the land forms sculptured by erosion differ from those in country with more rainfall. When the terrain unit consists of a mountain range and an adjacent basin, two kinds of situations are common. One (shown in Figs. 10.4 and 10.9,*A*) consists of a row of fans along the mountain base;

179

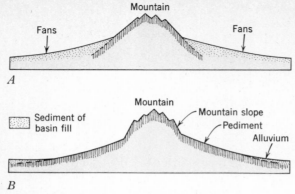

Figure 10.9 Two common relationships at the bases of mountains in desert country. *A*. Fans built by streams at the foot of a mountain. *B*. Pediment eroded across bedrock at the foot of a mountain.

the fans merge outward into a general fill of sediment in the basin. The other situation (Fig. 10.9,*B*), more remarkable and less easy to explain fully, consists of a sloping surface at the mountain base that closely resembles a merging row of fans. But the surface is not that of fans. Instead of being constructional (built of alluvium), it is erosional and cuts across bedrock. Scattered over it are rock particles, some brought by running water from the adjacent mountains and some derived by weathering from the rock immediately beneath. Downslope, the rock particles

gradually merge, forming a continuous cover of alluvium. The bedrock surface, which may be many kilometers long, has passed beneath a basin fill.

The eroded bedrock surface is called a *pediment* because the surfaces adjacent to the two bases of a mountain mass, seen in profile, together resemble the triangular pediment or gable of a roof (Fig. 10.9,*B*). A ***pediment,*** then, is *a sloping surface, cut across bedrock, adjacent to the base of a highland in an arid climate*. The kinds of bedrock on which pediments are cut are those that yield easily to erosion in such a climate. The profile of a pediment (Fig. 10.10), like that of a fan, is concave-up, a form we associate with the work of running water, and the pediment surface is marked by faint, shallow channels. It is likely, therefore, that pediments are the work mainly of running water, flowing as definite streams, as sheet runoff, or as rills, and in any case flowing mainly during rainstorms. Between the infrequent times of runoff weathering occurs.

Beyond this general statement we must admit that the exact way in which a pediment develops is not established and is partly a matter of opinion. But we can see that a pediment meets the mountain slope at its head not in a curve but at a distinct angle (Fig. 10.9,*B*). This suggests that instead of becoming gentler with time, as they would do in a wet region where chemical weathering and creep of regolith are dominant, mountain slopes in the desert seem to

Figure 10.10 Pediment at south base of Little Ajo Mountains, Arizona. The black butte near center is about 1.5km long. Stripelike rows of bushes are growing along faint, shallow channels. The view represents one-half of what is shown in the section (Fig. 10.11,*B*). (*James Gilluly, U.S. Geol. Survey.*)

180   Deserts and Winds

adopt an angle determined by resistance of the bedrock, and to maintain that angle as they gradually retreat under the attack of weathering and masswasting. Retreat of the mountain slope lengthens the pediment at its upslope edge. This growth of the pediment at the expense of the mountain should continue until the entire mountain has been consumed. During the whole time of pediment growth, rock particles are transferred downslope intermittently from mountain and pediment to the basin fill beyond.

Cycle of Erosion in Deserts. The concept of a cycle of erosion (Chap. 8), in which a land having abundant rainfall is gradually reduced to a peneplain, applies equally well to deserts. Figure 10.11 sketches

a desert cycle of erosion in three stages that can be seen today in parts of the western United States.

The stage represented in *A* commonly occurs in northern Nevada, where mountains are being dissected actively. At the mouths of the mountain canyons streams spread out waste in the form of fans that grade outward into playas. The wind picks up fine waste, sorts it, heaps sand-size particles into dunes, and lifts some of the finest particles out of the basins altogether. Meanwhile the rocky mountain slopes are being worn back.

In *B*, a stage occurring in southern New Mexico in the country north of El Paso, Texas, the mountain slopes have retreated, exposing a belt of bedrock at each mountain base to weathering and running water and so creating pediments. The steep mountain

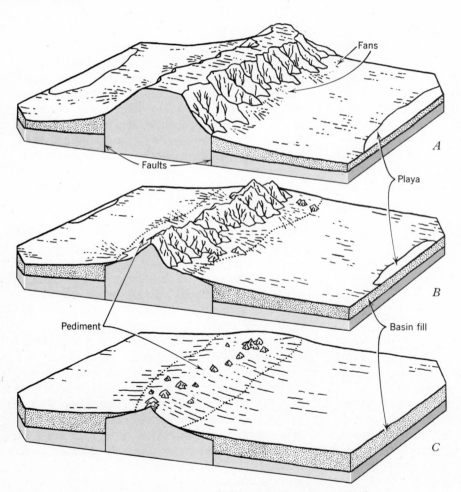

Figure 10.11 Three stages in the sculptural evolution of a mountain range and two basins (originally created by faulting) in an arid climate.

181

slopes, covered sparsely with coarse weathered rock waste moving slowly down them, maintain their steepness instead of becoming gentler with time, as would result from soil creep in a moist region.

As the area of the mountains diminishes, the sediment contributed to the streams during rainstorms decreases also. This reduces the loads of the streams, which begin to erode the heads of the former fans, planing them down. In the process the streams cut sideways into the bedrock at and near the mountain front, planing it off and adding to the area of pediment.

In *C*, a stage seen in the country northwest of Tucson, Arizona, the mountains have been reduced by gradual retreat of their steep slopes to a series of big knobs projecting abruptly above the sloping pediment that surrounds them. Outward beyond the pediment is the surface of the basin fill, beneath which the pediment disappears without any break in the smooth, concave slope. As the basin slopes become gradually gentler, water from the mountains reaches the basin centers more rarely, and increasingly winds, that sweep across the basins, pick up fine sediment and carry it away. The stage shown in *C* is essentially the desert equivalent of a peneplain.

There is reason to believe that in southern Arizona a desert cycle of erosion has been in progress with little interruption for millions of years, and it still has a long way to go.

Wind Action

Transport of Sediment by Wind

We are now in a position to compare flowing air with flowing water as an agent that erodes and deposits sediment. Wind, flowing air, is normally turbulent. Its velocity, like that of a stream, increases with height above the ground. Like a stream, it can carry coarser particles as bed load and finer ones as suspended load. The two loads are clearly visible in a strong wind blowing across a North African desert. The bed load, a layer rarely as much as 1m thick, consists of sand; it grades upward into clouds of suspended silt and clay particles that may reach great heights.

Bed Load. Experiments with sand blown artificially through glass-sided wind tunnels show that sand grains move in long jumps, as they do in a stream of water (Fig. 10.12,*A*; compare Fig. 8.6). The jumps are elastic bounces like those of a ping-pong ball.

A sand grain gets into the air only by bouncing or by being knocked into the air through the impact of another grain. When the wind becomes strong enough, a grain starts to roll along the surface under the pressure of a fast-moving forward eddy. It strikes another grain and knocks it into the air. When the second grain hits the ground, it either splashes up still other grains, making a tiny crater, or bounces into a new jump (Fig. 10.13). In a short time the air close to the ground has become filled with saltating sand grains, which hop and bounce, moving with the wind, as long as wind velocity is great enough to keep them moving.

The jumping sand grains never get far off the ground. They are usually limited to about 10cm, as in the experiment shown in Fig. 10.12,*C*. In desert country they generally jump no higher than about 45cm, as shown by wooden utility poles, which are sandblasted up to about that height but no higher. Even in the strongest desert winds the height of jump rarely exceeds 1m. This explains why windblown sand rarely moves far except on very smooth surfaces. Being always close to the ground, it is easily stopped by obstacles and heaped into dunes.

Suspended Load. To understand the suspended load of fine particles carried by a wind we must note that obstacles on the ground below, whether sand grains, blades of grass, buildings, or trees, create a very thin layer of "dead," motionless air immediately above the surface (Fig. 10.14). The thickness of the dead layer equals $\frac{1}{30}$ the height of the obstacles, regardless of velocity. So over ground that is covered with pebbles 3cm in diameter, the dead-air layer would be 1mm thick. Thin though it is, it plays a controlling part in the movement of all particles finer than medium-size sand grains. The reason why it does not influence the movement of coarser sand grains is that each individual grain constitutes an obstacle that projects far above the dead-air layer. But particles of silt and clay, down within that layer, are crowded so closely together that they do not act as individual obstacles; they present a smooth surface to the wind, which can not lift them off the ground

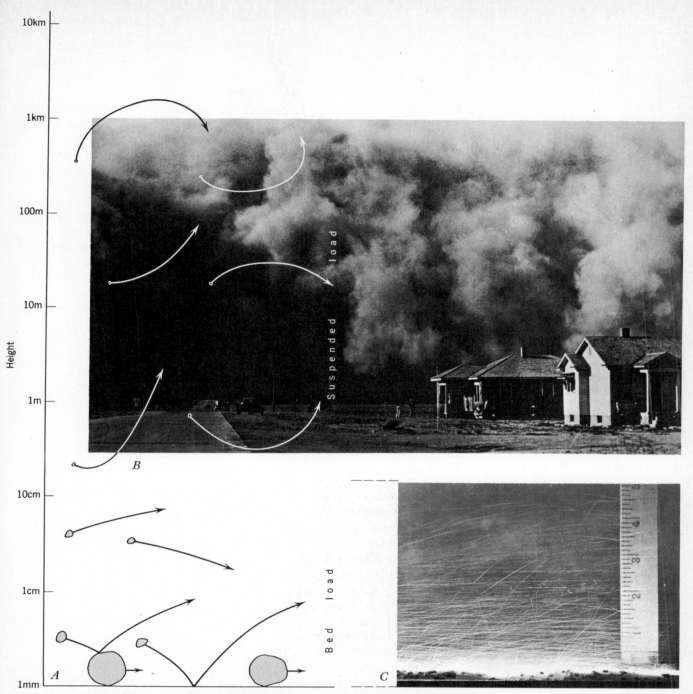

Figure 10.12 Movement of rock particles carried by wind. *A*. Graph showing vertical distribution, comparative sizes, and movement of particles in bed load and suspended load. *B*. The "Dust Bowl" during the 1930s. A dust cloud approaches Springfield, Colorado at 4:47 P.M. on May 21, 1937. Total darkness lasted 30 minutes. (*U.S. Soil Conservation Service.*) *C*. Paths of sand grains being blown through a wind tunnel, photographed in a narrow beam of sunlight. Scale units are inches. Sand grains and small pebbles are visible on tunnel floor. Air current is moving left to right. (*A. W. Zingg, U.S. Dept. of Agriculture.*)

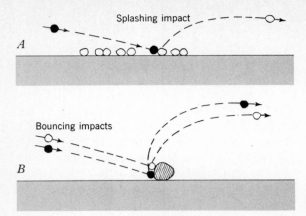

Figure 10.13 *A.* In splashing impact, jumping sand grain strikes one or more other grains and splashes them into air at slow speeds and to low heights. Splashing happens most commonly when all grains are nearly the same size. *B.* In bouncing impacts, sand grains strike pebble or other wide surface and bounce up at high speed and to greater heights. Angle of rise depends on inclination of surface of impact.

directly. It spatters them into the air by the impacts of jumping sand grains.

We can see how this happens by looking at a dusty road in dry country on a windy day. The wind blowing across the road generates little or no dust. But a car driving over the road creates a choking cloud, which is blown a short distance before settling once more. The car wheels have broken up the surface of powdery silt that was too smooth to be disturbed by the wind.

Once in the air, fine particles constitute the wind's suspended load. They are continually tossed up by eddies (again, as in a stream of water) while gravity tends to pull them toward the ground. Meanwhile they are carried forward, and although in most cases suspended load is deposited fairly near its place of origin, strong winds have been known to carry fine particles thousands of kilometers before depositing them. The amount of sediment actually moved by the atmosphere, year in and year out, is probably only a fraction of 1 per cent of capacity, for the air is rarely if ever fully loaded.

During the great windstorms in the dry years of the 1930s, however, loads became abnormally heavy. In a particularly great storm on March 20, 1935, when the sky looked much as in Fig. 10.12,*B*, the cloud of suspended sediment extended 12,000 feet above the ground, and the load in the area of Wichita, Kansas was estimated at 166,000 tons per cubic mile in the lowermost layer 1 mile thick. Sampling of sediment on flat roofs of buildings showed that during that day about 800 tons of rock particles—around 5 per cent of the load suspended in the lowermost 1-mile layer—were deposited on each square mile. Enough sediment was carried eastward on March 21 to bring temporary twilight over New York and New England, 2000 miles east of the principal source area in eastern Colorado. The distance and travel time imply wind velocities of about 50 miles per hour.

Fine-Grained Volcanic Ash. Not all wind-blown sediments originate by being picked up from the ground. Large quantities are shot into the air during explosive eruptions of volcanoes (Fig. 16.3). Although coarse particles fall out quickly, small particles travel long distances. The bulk of the particles that fall out during an eruption tend to form an

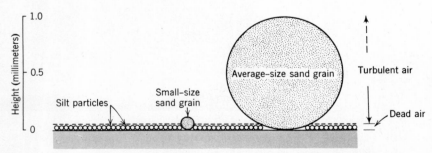

Figure 10.14 Silt particles form a smooth surface that lies within the dead-air layer caused by sand grains and other large obstacles to flowing air. Thickness of dead-air layer is one-thirtieth the diameter of the average-size sand grain.

elongate body of sediment that trails out and thins downwind from the volcano (Fig. 3.9).

Erosion by Wind

Deflation. Flowing air erodes in two ways. The first, *the picking up and removal of loose rock particles by wind,* provides most of the wind's load and is known as **deflation** (from the Latin word meaning "to blow away"). The second, *abrasion* of rock by wind-driven rock particles, is analogous to abrasion by running water.

Deflation on a large scale happens only where there is little or no vegetation and where loose rock particles are fine enough to be picked up by wind. The great areas of deflation are the deserts; others are the beaches of seas and large lakes and, of greatest economic significance, bare plowed fields in farming country during times of drought, when no moisture is present to hold the soil particles together.

In most areas the results of deflation are not easily visible, inasmuch as the whole surface is lowered irregularly. In places, however, measurement is possible. In the dry 1930s deflation in parts of the western United States amounted to 1m or more within only a few years—a tremendous rate com-pared with our standard estimate of rate of general erosion. In Fig. 10.15 the tuftlike yucca plants held the soil in place, but elsewhere it was deflated.

The most conspicuous evidence of deflation consists of basins excavated by wind. These occur in tens of thousands in the semi-arid Great Plains region from Canada to Texas. The lengths of most of them are less than 2km and depths are only a meter or two. In wet years they are clothed with grass and some even contain shallow lakes; an observer seeing them at such times would hardly guess their origin. But in dry years soil moisture evaporates, grass dies away in patches, and wind deflates the bare soil. At the same time, drifting sand accumulates to leeward, especially along fences and other obstructions.

When sediments are particularly prone to deflation, depths of deflation basins can reach 50m, as in southern Wyoming, and even more, as in the Libyan Desert in western Egypt, where the floor of the Qattara basin lies about 125m below sea level. Deflation in any basin is limited finally only by the water table, which moistens the surface, encourages vegetation, and inhibits wind erosion.

A natural preventive of deflation is a cover of rock particles too large to be removed by the wind. Defla-

Figure 10.15 Part of a large area within which a thickness of 1m of deflation can be measured directly.

The plant roots in the hummock, just above the dark band (the remains of an ancient soil) mark the position of the surface before deflation. Note recent ripples in the sand surface. The date is 1936, in the "Dust Bowl" period. (*R. H. Hufnagle, Philip Gendreau.*)

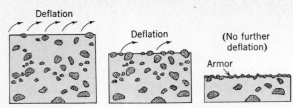

Figure 10.16 Three stages in the development of a deflation armor.

tion of sediment such as alluvium, which consists of silt, sand, and pebbles, creates such a cover (Fig. 10.16). The sand and silt are blown away, and in places are carried off by sheet erosion also, but the pebbles remain. When the surface has been lowered just enough to create a continuous cover of pebbles, the ground has acquired a **deflation armor,** *a surface layer of coarse particles concentrated chiefly by deflation.* Such armors are also called *desert pave-*

A   B

Figure 10.17 Ventifacts.

A. Granite pebble, 6cm in diameter, with three facets. The surfaces are uneven because the quartz grains in the granite, being hard, are cut less rapidly than grains of feldspar and other minerals. The facets do not extend down to the base of the pebble, which probably was embedded in the ground.

B. Basalt pebble, 3cm long, with 2 facets. Either the wind blew from opposite directions at different times, or, more likely, the pebble was rotated after one of the facets had been cut.

ment, because long-continued removal of the fine particles makes the pebbles settle into such stable positions that they fit together almost like the blocks in a cobblestone pavement.

Abrasion: Ventifacts. In desert areas, surfaces of bedrock and of loose stones are abraded by wind-driven sand and silt, which can cut and polish them to a high degree. A **ventifact,** the name given to *a stone the surface of which has been abraded by wind-blown sand,* is recognized by polished, greasy-looking surfaces, which may be pitted or fluted, and by facets separated from each other by sharp edges (Fig. 10.17). Most of the cutting is done by the sand grains in the bed load.

In the laboratory, sand blasting of pieces of plaster of Paris demonstrates that the facets always face the wind (Fig. 10.18). A stone can be worn down flush with the ground by enlargement of a single facet. If the stone is undermined or otherwise rotated or if the wind direction varies from time to time, two, three, or many more facets can be cut on it.

Wind Deposits

In the wind as in a stream of water, the jumping sand grains of the bed load move rather slowly and are deposited early, whereas the finer particles of the suspended load travel faster and much farther before they drop out. Also, winds commonly deposit sand in heaps or hillocks, whereas finer sediment is generally deposited as smooth blankets. Looking more closely at these two kinds of deposits, we begin with the sand that drops out of the bed load, usually forming dunes.

Dunes. A **dune** is *a mound or ridge of sand deposited by the wind.* Generally it is localized by an obstacle small or large (Fig. 10.19), which distorts the flow of air. Velocity within a meter or two of the ground

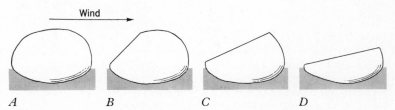

Wind →

A   B   C   D

Figure 10.18 Four stages in the cutting of a ventifact. The pebble becomes a ventifact between stage *A* and stage *B*.

Figure 10.19 Wind blowing from left to right deflated the bare dry field (left), carried away the finer particles in suspension, but dropped sand along the obstacle created by a wire fence with weeds caught in it. The result was the dune along the top of which the man on the right is walking. Hereford, Texas. (*A. Devaney, Inc.*)

varies with the slightest irregularity of the surface. As it encounters an obstacle, the wind sweeps over and around it, but leaves a pocket of slower-moving air immediately behind the obstacle and a similar but smaller pocket in front of it. In these pockets of lower velocity, sand grains drop out and form mounds. Once formed, the mounds themselves influence the air flow. As more sand drops out they generally coalesce and form a single dune.

A dune is unsymmetrical. It has a steep, straight lee slope and a gentler windward slope (Fig. 10.20). The wind rolls or pushes sand grains up the windward slope, and at the crest they drop onto the lee slope, building that slope up to the angle of repose. As noted in Fig. 7.2, angles of repose for loose sand are close to 34°. When more sand drops onto it the slope becomes unstable, sand slides ("slips") downward, and the angle is maintained. For this reason *the straight, lee slope of a dune* is known as the **slip face,** which, as might be expected, meets the windward slope at a sharp angle.

The angle of slope of the windward side varies with wind velocity and grain size, but it is always less than that of the slip face. The asymmetry of a dune with a slip face, then, indicates the direction of the wind that shaped it.

Many dunes grow to heights of 30m to 100m, and some desert dunes reach the great height of 200m. Possibly the height to which any dune can grow is determined by upward increase in wind velocity, which at some level will become great enough to whip sand grains off the top of a dune as fast as they arrive there by creeping up the windward slope.

Stratification and Migration. The dropping of sand grains over the crest onto the slip face of a dune produces cross-strata much like the foreset layers in

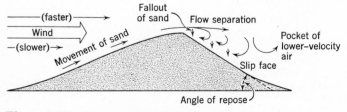

Figure 10.20 Development of windward and lee slopes of a bare sand dune.

a delta (Fig. 8.14). Erosion of the windward slope continually erodes the layers already deposited, as new ones are added to the slip face (Fig. 10.21).

Transfer of sand from the windward to the lee side of a bare dune causes the whole dune to migrate slowly downwind. Measurements on desert dunes of the barchan type (Table 10.1) indicate rates of migration as great as 10 or 20m/year. The migration of dunes, particularly along coasts just inland from sandy beaches, has been known to bury houses and threaten the existence of towns. In such places sand encroachment is countered most effectively by planting vegetation that can survive in the very dry sandy soil of the dunes. A good plant cover inhibits dune migration for the same reason that it inhibits deflation: if the wind can not move sand grains across it, a dune cannot migrate.

A dune covered with grass, however can be reactivated wherever, by drought or trampling by animals, the grass is killed off in patches and allows deflation to start. This converts the bare patches into irregular, shallow basins known as *blowouts*. A *blowout* is merely *a deflation basin excavated in shifting sand*. The view in Fig. 10.15 represents a large blowout.

Shape. The shapes of dunes are varied. Although the factors that determine them are only partly understood, we can classify dunes roughly into five groups based on their form. The groups are shown in Table 10.1.

In the table we see that some dunes, of which the barchan is the best example, are built by winds blowing from one direction, whereas others are built

Table 10.1
Several Kinds of Dunes Based on Form

Kind	Definition and Occurrence	Illustration (Arrows indicate wind directions)
Beach dunes	Hummocks of various sizes bordering beaches. Inland part is generally covered with vegetation.	
Barchan dune	A crescent-shaped dune with horns pointing downwind. Occurs on hard, flat floors in deserts; constant wind, limited sand supply. Height 1m to more than 30m.	
Transverse dune	A dune forming a wavelike ridge transverse to wind direction. Occurs in areas with abundant sand and little vegetation. In places grades into barchans.	
U-shaped dune	A dune of U-shape with the open end of the U facing upwind. Some form by piling of sand along leeward and lateral margins of a growing blowout in older dunes.	
Longitudinal dune	A long, straight, ridge-shaped dune parallel with wind direction. As much as 100m high and 100km long. Occurs in deserts with scanty sand supply and strong winds varying within one general direction. Slip faces vary as wind shifts direction.	

Figure 10.21 If wind had uniform direction and velocity, it would build (*lower left*) simple, parallel cross-strata while dune migrates from AB to A'B'. But variations in direction and velocity are common and result in cross-strata like those at left.

Photograph shows cross-strata in wind-blown sand consolidated to form hard sandstone (Jurassic; more than 150 million years old). Parts of lee slopes of a succession of dunes are exposed. The cavities result from solution at places where calcium-carbonate cement between the grains of quartz is especially soluble. Monument Valley, Arizona. (*Tad Nichols.*)

by winds that shift direction from one time to another. As long as the dune form exists, we can infer wind direction from the position of the slip face. But even after erosion has destroyed the form, and after what is left of the dune sand has been buried, converted into sandstone, subjected to uplift, and reexposed to a new cycle of destruction, we can recover the direction (or directions) of the wind that deposited the sand by measuring the direction of dip of the cross-strata, which are always inclined downwind (Fig. 10.21).

Composition. Virtually all dunes consist of sand-size particles. Since quartz is the most common mineral in sand-size sediments it is not surprising that quartz predominates in most dunes. But where other minerals are abundant, dunes are built from them. The calcite dunes on Bermuda and the gypsum dunes at White Sands National Monument in New Mexico are examples.

Deposits of Silt. *Loess.* Much of the world's regolith includes small proportions of fine sediment deposited from suspension in the air, but so thoroughly mixed with other materials as not to be distinguishable from them. But over wide areas sediment having this origin is so thick and so pure that it constitutes a distinctive deposit. It is known as *loess* (pronounced "less") and is defined as *wind-deposited silt, usually accompanied by some clay and some fine sand.*

Most loess is not stratified, apparently because many grain sizes were deposited together from suspension, and also perhaps because plant roots, worms, and other organisms turned over and churned up the sediment as it was deposited. Where exposed, loess stands at such a steep angle that it forms cliffs (Fig. 10.22) as though it were firmly cemented rock. This is the result of the fine grain size of loess, in which molecular attraction is strong enough to make the particles very cohesive. Also, because the particles are angular, porosity is extremely high, commonly exceeding 50 per cent. Hence loess accepts and holds water and constitutes a basis for productive soils.

Minerals composing loess are chiefly quartz, feldspar, micas, and calcite. The particles are generally fresh, showing little evidence of chemical weathering other than the slight oxidation of iron-bearing minerals that has occurred since deposition and that gives a yellowish tinge to the deposit as a whole.

Origin of Loess. Loess possesses two characteristics that indicate it was deposited by wind: (1) it forms blankets that mantle hills and valleys alike through a wide range of altitudes, and (2) the fossils it contains consist of land plants and animals—mostly air-breathing snails, but mammals too, including elephants.

The world distribution of loess shows that winds carried it from two principal sources, (1) deserts and (2) sediments of streams that originated in the melting of glaciers. Desert loess covers enormous areas that lie to leeward of deserts, an obvious source of mechanically weathered sediments. The loess that covers much of western China, in places as much as 70m thick, was blown from alluvium in the great desert basins of central Asia.

Glacial loess is abundant in the middle part of North America (especially Nebraska, South Dakota,

Figure 10.22 As loess is very cohesive and has vertical joints, it forms cliffs as jointed bedrock does. A typical surface expression of bare loess appears in this cliff, in the east bluff of the Mississippi River valley, Madison County, Illinois. The cliff will stand with little change for many years. (*J. C. Frye.*)

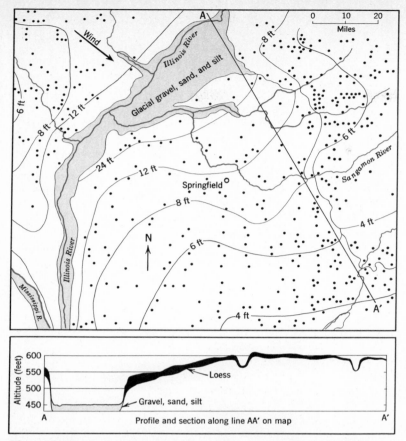

Figure 10.23 Loess in central Illinois. Thicknesses were determined by borings at places shown by dots. Lines are isopachs (App. D); others appear in Fig. 5.9. Thickness decreases away from river valley in both directions, but loess is thickest on southeast side, the leeward side for prevailing winds. Grain diameters also decrease in both directions away from valley. (*Data from G. D. Smith, 1942.*)

Iowa, Missouri, and Illinois) and in the drier parts of Europe. It has two distinctive features: The first is that the shapes and mineral composition of its particles resemble the "rock flour" ground up by glaciers. The second feature is that glacial loess is thickest immediately in the lee of rivers such as the Missouri and Mississippi, which are known to have been swollen with water from melting ice during the glacial ages (Chap. 11). Carbon-14 dates, fossil plants, and other kinds of evidence confirm the relation of loess to great quantities of melting ice. In glacial ages the areas just outside the margins of broad glaciers were cold and very windy. Rivers that were fed by melting ice were filling their valleys with

gravel, sand, and silt so rapidly that plants could not gain a foothold, from one year's flood to the next, on valley floors. So the floors remained bare and were easily deflated. Silt particles, splashed into the air by jumping sand grains, were carried to leeward. The silt settled out, forming blankets 8 to 30m thick near the source valleys and thinning downwind to thicknesses of 1 to 2m, spread over thousands of square kilometers (Fig. 10.23).

Why was the silt not picked up again and again and carried even farther by the wind? One reason lies in the stability of a silt surface, which results from its fine grain size (Fig. 10.14). Another reason is that the silt settled out chiefly on grassland and to some extent on woodland. Once on the ground in environments such as these, silt would be in no danger of further deflation.

Environmental Aspects: Soil Erosion

Climates fluctuate continually. It therefore happens that regions ordinarily suitable for agriculture experience dry periods during which soil erosion by wind reaches tremendous proportions. As we have noted, during the 1930s an enormous volume of soil was blown away from parched, unprotected plowed fields in the "Dust Bowl" region of the Great Plains (Figs. 10.12, 10.15). Sand-size particles were piled up along fences (Fig. 10.19) and around farm buildings, and finer particles were blown eastward to be deposited over wide areas; a good deal of such material was dropped into the Atlantic Ocean.

What determined the location of the "Dust Bowl"? That area closely approximates the largest area in the United States in which *average* wind velocity is great enough both to move sand grains and to keep coarse silt grains in suspension. All that was needed further was a long succession of dry years—the drought in the 1930s. With only very slight changes in Earth's climates deserts expand and contract and can be created or disappear.

In the "Dust Bowl" area, good practice includes the planting of windbreaks (Fig. 10.24), consisting of bushes and hardy trees set in strips at right angles to the strongest winds, at intervals of a mile or so. It also involves planting strips of grass alternating with strips of cultivated grain, for the soil must lie

Figure 10.24 Windbreaks on dry sandy farm land in northern Texas. (*U.S. Soil Conservation Service.*)

bare during alternate years in order to accumulate moisture sufficient to grow grain. The principle on which the rows of plants act to retard deflation is based on the presence, just above the ground, of the dead-air layer we noted earlier, the thickness of the layer being equal to $\frac{1}{30}$ the height of obstacles on the ground. A group of trees and bushes 10m high, although somewhat permeable to wind, should create a dead-air layer nearly 30cm thick, thus effectively preventing deflation beneath it and through some distance to leeward.

Summary

Deserts

1. Deserts, constituting about a quarter of the world's land area, are areas of slight rainfall, high temperature, great evaporation, relatively strong winds, sparse vegetation, and interior drainage.

2. No major geologic process is confined to deserts, but in them mechanical weathering, impact of raindrops, flash floods, and winds are very effective. The water table is low.

3. Pediments are a conspicuous feature of many deserts. They are shaped by streams, rills, sheet runoff, and weathering.

4. In the arid cycle of erosion, pediments grow headward at the expense of mountain slopes, which appear to retreat without becoming gentler as they do in a moist region. Wind removes fine rock waste.

Wind Action

5. Wind carries a bed load of jumping sand grains close to the ground and a suspended load of fine particles higher up. Sorting of sediment results.

6. Wind erodes by deflation and abrasion, chiefly in dry regions and on beaches. It creates deflation basins, blowouts, deflation armor, and ventifacts.

7. With long-continued wind activity, sand grains become rounded.

Dunes

8. Many dunes are localized by obstacles. Bare dunes have steep slip faces and gentler windward slopes. They migrate downwind, forming cross-strata that dip downwind.

9. Dunes are classified by form into beach dunes, barchan dunes, transverse dunes, U-shaped dunes, and longitudinal dunes.

Loess

10. Loess is deposited chiefly in the lee of (a) bodies of glacial outwash, and (b) deserts. Once deposited, it is stable and is little affected by further wind action.

Selected References

Bagnold, R. A., 1941, The physics of blown sand and desert dunes, repr. 1954: New York, William Morrow.

Blackwelder, Eliot, 1954, Geomorphic processes in the desert: California Div. Mines Bull. 170, Chap. 5, p. 11–20.

Bryan, Kirk, 1923, Erosion and sedimentation in the Papago Country, Ariz., with a sketch of the geology: U.S. Geol. SSurvey Bull. 730, p. 19–90.

Cooper, W. S., 1967, Coastal dunes of California: Geol. Soc. America Mem. 104.

Finkel, H. J., 1959, The barchans of southern Peru: Jour. Geology, v. 67, p. 614–647.

Hadley, R. F., 1967, Pediments and pediment-forming processes: Jour. Geol. Education, v. 15, p. 83–89.

Hume, W. F., 1925, Geology of Egypt, v. 1: Cairo, Government Press.

McKee, E. D., 1966, Structures of dunes at White Sands National Monument, New Mexico . . .: Sedimentology, v. 7, p. 1–69.

Neal, J. T., ed., 1965, Geology, mineralogy, and hydrology of U.S. playas: U.S. Air Force, Office of Aerospace Research, Environmental Res. Paper 96, [Defense Documentation Center, Alexandria, Va.].

Sharp, R. P., 1949, Pleistocene ventifacts east of the Bighorn Mountains, Wyoming: Jour. Geology, v. 57, p. 175–195.

——— 1963, Wind ripples: Jour. Geology, v. 71, p. 617–641.

Thorarinsson, Sigurdur, 1954, The tephra-fall from Hekla on March 29th, 1947: Societas Scientiarum Islandica, v. 2, no. 3.

Map

Thorp, James, and others, 1952, Pleistocene eolian deposits of the United States . . .: Boulder, Colo., Geol. Soc. America.

Chapter 11

Glaciers
and Glaciation

Glaciers

Glaciers and the Water Cycle. Glaciers, together with a small amount of snow and a little ice floating on the ocean, represent the part of the water cycle in which the water is in the solid state. Glaciers are the water substance that lingers on the land as a solid instead of returning more quickly to the sea as a liquid. The glacier seen in Fig. 11.1 is pretty inconspicuous. Indeed most of the nearly one thousand glaciers in the United States exclusive of Alaska are small, and very likely their combined volume is less than 50 cubic kilometers. But over the world as a whole there are estimated to be about 24 million cubic kilometers of glacier ice, 89 per cent of which lies on the Antarctic Continent. Even that much ice, however, is only 1.7 per cent of the world's water, because more than 97 per cent of the water resides in the ocean.

Even though the ice of glaciers is holding water out of the sea, it is not dead. It is very active, receiving nourishment in the form of snowfall and giving up water by melting. So water is passing through it all the time, as water passes through a lake. The oldest ice in the great glacier that covers Greenland is believed to be more than 25,000 years old; so it has formed part of the glacier for a long time. But as it melts, it is replaced by new ice contributed by the snow that falls on Greenland every year. In other words the glacier is in or near a steady-state condition.

Obviously, then, the fate of this or any glacier depends on how much ice melts and how much is

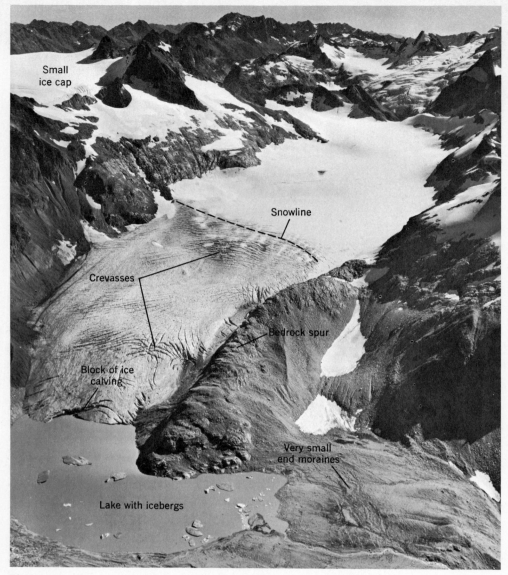

Figure 11.1 South Cascade Glacier, a valley glacier about 3km long, in the North Cascade Range, Washington, September 1960. In view are position of snowline and a small ice cap. (*A. S. Post, U.S. Geol. Survey.*)

replaced by new snow. That, in turn, depends on *climate:* primarily on temperature and secondarily on rate of snowfall. And climate changes continually, although slowly, affecting the sizes and the number of glaciers. Because most of the world's ice lies not in the densely peopled middle latitudes of North America and Europe but in high latitudes far to the north and south, the fact that glaciers fluctuate as

the climate grows warmer or colder may seem to be of no great concern. Yet it does concern us because of the part glaciers play in the water cycle.

The water created by melting ice flows into the ocean. Therefore if world climate becomes warmer, more ice will melt than is replaced by new snowfall, the volume of the ocean will increase, and so the level of the ocean must rise against the continents.

If all the glacier ice that exists today were to melt, returning its substance to the ocean as water, a layer of water more than 60m thick, worldwide, would be added to the ocean, which of course would rise and submerge coastal lands. This unwelcome possibility is sure to make us respect the potential possessed by remote Antarctic ice for influencing life, not just in Antarctica but throughout the world. Between 15,000 and 20,000 years ago the world was in the grip of a glacial age. The volume of ice on the land was more than twice what it is now. Thick ice covered much of northern Europe and the northern half of North America, extending as far south as New York City. The Atlantic Coast of the United States lay some 150 kilometers east of New York because, owing to evaporation of seawater to build the glaciers, the level of the sea was much lower than it is now. Britain was joined to France where the English Channel is now, and where the Bering Strait now separates North America from Asia, Alaska was firmly connected with the USSR.

Apart from the critical role they play in the water cycle, glaciers flow across land areas, erode land, transport eroded rock material, and deposit it elsewhere. As they are conspicuous among the external geologic agents, they are worth examining in more detail.

Kinds of Glaciers. Defined simply, a *glacier* is *a body of ice, consisting mainly of recrystallized snow, flowing on a land surface.*

On the basis of their form, we can readily distinguish four kinds of glaciers. Most numerous and possibly most familiar are *cirque glaciers* (Fig. 11.10), *very small glaciers that occupy cirques.* In the conterminous United States there are about 950 cirque glaciers, and about 1500 exist in the Alps.

A *valley glacier* (Fig. 11.1) is *a glacier that flows downward through a valley.* Such glaciers vary in size from no more than a few acres in extent to mighty tonguelike forms many tens of miles long. In the conterminous United States they number about 50; in the Alps their number is larger.

A *piedmont glacier* is *a glacier on a lowland at the base of a mountain, fed by one or more valley glaciers.* It is shaped like a covered frying pan whose narrow inclined handle represents a feeding valley glacier.

An *ice sheet* is *a broad glacier of irregular shape generally blanketing a large land surface. A small ice*

sheet is an *ice cap;* some ice caps, such as those in Fig. 11.1, are very small. The best-known ice sheet is the Greenland Ice Sheet. Its area is about 1,726,000km^2 and its highest part is more than 3km above sea level. Seismic measurement—that is, measurement by a method using artificial-earthquake waves (Chap. 17)—of the approximate thickness of this glacier was made at intervals along certain routes of travel, one of which is shown in Fig. 11.2. The profile and cross section along it reveal that in places the base of the ice sheet is below sea level and that the ice is very thick; at one point its thickness exceeds 3km. Near its margins the spreading ice sheet encounters mountains, through which it flows along deep valleys to reach the sea. Icebergs floating in the sea are masses broken off from the ends of such glaciers.

The Antarctic Ice Sheet (Fig. 11.3) is far bigger. Its area, including the parts of its margin that float in the ocean, is about 1.4 times the area of the United States, its highest altitude exceeds 4km, its greatest thickness likewise exceeds 4km in one place, and its volume is estimated at around 24,000,000km^3. With this huge volume the Antarctic Ice Sheet alone contains 89 per cent of the world's stock of ice and 66 per cent of all the fresh water in the world.

The Antarctic Ice Sheet overlies a mountainous continent with the highest average altitude, and of course the lowest temperature, of all continents. Centered on the South Pole, it is truly a polar ice sheet. In fact it is the only one, because the North Pole is near the center of the large, deep Arctic Ocean. Consequently there can be no glaciers near the North Pole, only very thin floating ice consisting of frozen seawater and snow.

Valley glaciers, piedmont glaciers, and ice sheets constitute a gradational sequence. Indeed, a great ice sheet like that in Greenland probably began in coastal mountains with the growth of thousands of valley glaciers, which spread out in the interior as piedmont glaciers and gradually thickened to form a body of ice higher in places than the mountain peaks themselves.

Snowline. From what has been said already, evidently two chief requirements for the existence of glaciers are snowfall and low temperature, both of which depend on climate. These requirements are fulfilled in high latitudes and at high altitudes, and occur more commonly in wet coastal areas than in

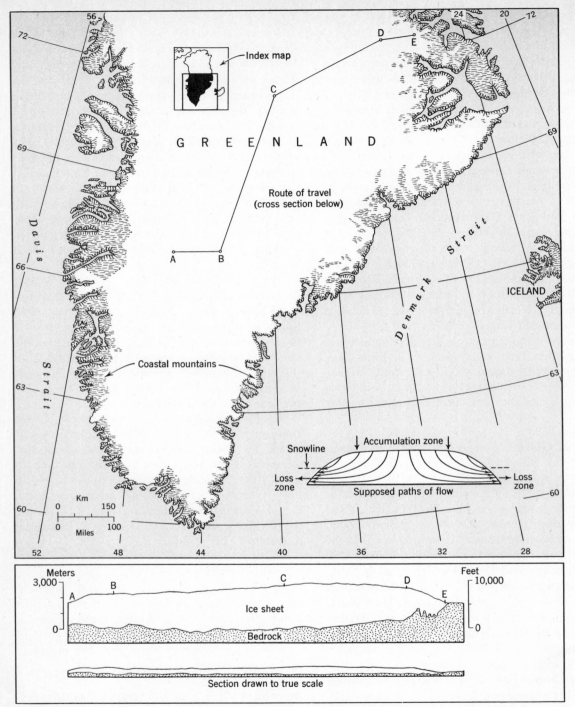

Figure 11.2 Map of south half of Greenland and cross section showing thickness of ice sheet. (*Thickness data from Albert Bauer.*)

Figure 11.3 Part of the Antarctic Ice Sheet, in Marie Byrd Land, with peaks of Edsel Ford Mountains projecting through its surface. In foreground are crevasses, reflecting a change of gradient of the floor beneath the glacier, which in this area is comparatively thin. (*P. A. Siple, U.S. Antarctic Service.*)

the dry interiors of continents. It is not surprising, then, that the world's two largest glaciers, the Antarctic and Greenland Ice Sheets, cover lands that are surrounded by water, and the largest glaciers in Alaska are clustered close to the Pacific Ocean. Altogether, glaciers cover an aggregate area equal to about 10 per cent of the land area of the world.

Glaciers are related to each other by the **snowline,** *the lower limit of perennial snow* (Fig. 11.1). Above the snowline are **snowfields,** *wide covers, banks, and patches of snow,* usually in protected places, *that persist throughout the summer season.* The snowline passes across all active glaciers. It rises from near sea level in polar regions to altitudes of as much as 6000m in tropical mountains and rises also from coasts toward continental interiors.

Conversion of Snow into Glacier Ice. When temperatures fall below 0°C some atmospheric water vapor changes into the solid state, forming the hex-

agonal ice crystals we know as snowflakes. Newly fallen dry snow is feathery, light, and <u>porous</u>. It has much greater internal surface area (Fig. 11.4) than that of ordinary sediment. Air penetrates its large pore spaces. In it, ice evaporates, mostly at the points of crystals, and the water vapor condenses, mostly in the narrow spaces near the centers of the ice crystals. In this way, by evaporation near their edges and condensation near their centers, snowflakes gradually become smaller, rounder, and thicker, and the spaces between them gradually disappear (Fig. 11.4). The whole mass of snow takes on the granular texture we find in old snowdrifts at winter's end. Air is forced out from the spaces between the grains. The snow is being changed from a sediment into a sedimentary rock.

When the specific gravity of the body of granular snow has reached 0.8, the snow has become so compact that it is impermeable to air and is then said to be *ice*. Such ice is a metamorphic rock, for it

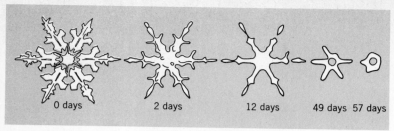

Figure 11.4 Conversion of a snowflake into a granule of ice during nearly two months of evaporation and condensation. (8 times natural size.) (*After Bader and others, 1939.*)

consists of interlocking crystalline grains of the mineral ice. It has been metamorphosed by pressure, the low pressure of the overlying ice itself. Although a rock, such ice has a far lower melting point than any other rock. The specific gravity of glacier ice is about 0.9; so it floats in water. Most of the icebergs in the North Atlantic Ocean consist of ice broken off from glaciers.

Movement. At some time, depending on the steepness of slope of the surface on which it is lying, compacted snow begins to respond to the pressure caused by the weight of snow above it and starts to move, somewhat as regolith starts to creep down a slope in the process of mass-wasting. The movement of ice, generally spoken of as *flow*, takes place because of the pull of gravity on the ice mass. The motion is of two kinds (Fig. 11.5). One kind consists of *sliding* of the whole mass along the ground, as a blanket of snow slides very slowly down the steeply sloping roof of a building (Fig. 17.15). The other kind is *creep*, much of which occurs through slipping on a tiny scale, along flat surfaces within the ice crystals themselves. The crystals are deformed as the body of snow or glacier ice creeps downslope.

The uppermost or surface part of a glacier, having little weight upon it, is brittle. Where the glacier flows over an abruptly steepened slope, such as the downstream side of a rocky "bump," the uppermost ice is subjected to tension and cracks. The cracks open up, forming **crevasses**—*deep, gaping cracks in the upper surface of a glacier* (Figs. 11.3; 11.17). Although sometimes impossible and always dangerous to cross on foot or on a snowmobile, crevasses are rarely as much as 50m deep. At greater depths the glacier flows actively and prevents crevasses from forming. Because it cracks at the surface but flows

at depth, we can liken a glacier to Earth's crust itself, which consists of a surface zone of jointing and fracture, and a deeper zone of flow (Chap. 16).

The surface velocity of a valley glacier is measured by surveying from the sides of the valley, at intervals of time, a line of markers like those in Fig. 7.16, extending from side to side of the glacier. Results show that the central part of the glacier's surface moves faster than the sides, as is true of a stream of water. Velocities normally range from a few millimeters to a few meters per day, about the same slow rates as the rates of percolation of ground water. At such velocities ice takes a long time to travel the length of a glacier. Probably hundreds or even thousands of years have elapsed since the ice

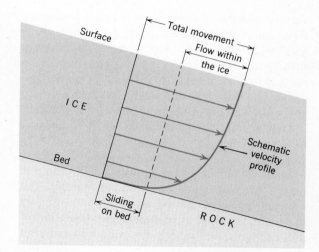

Figure 11.5 The two kinds of motion in a glacier. (1) *Rate of flow* within the ice increases upward while (2) *rate of sliding* on the bed remains the same.

The velocity profile, like that in a stream (Fig. 8.6) shows total movement at any depth.

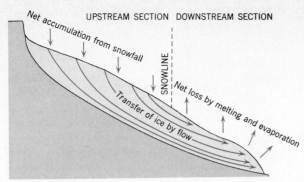

UPSTREAM SECTION DOWNSTREAM SECTION

Net accumulation from snowfall

SNOWLINE

Net loss by melting and evaporation

Transfer of ice by flow

Figure 11.6 Economy of a glacier. Paths of flowing particles of ice are suggested by long, curved arrows somewhat like those of percolating ground water (Fig. 9.3).

now exposed at the downstream end of a long glacier fell as snow upon the upstream part.

Economy. A glacier, like a river, can be thought of in terms of its economy. We see in Fig. 11.6 that a glacier consists of an upstream section that collects snow and a downstream section that wastes ice. The two are separated by the snowline. On the upstream section more snow falls than is removed by melting and evaporation. The weight of the extra snow tends to thicken the glacier and promote movement. In the downstream section the reverse is true. Wastage exceeds snowfall, and this tends to thin the glacier. But the thinning is made up for by the ice that flows down from the upstream section. The lines of flow, represented by arrows in the figure, bend upward to make up for the ice lost at the surface.

So the two sections of a glacier are like a balanced account at a bank. The spendthrift section downstream, a thoroughgoing waster, draws on the ice account and wastes all its assets, while the indulgent section up above underwrites all the losses. It makes continual payments into the account and keeps it balanced. In scientific terms this represents a condition of steady state.

Nevertheless the generous depositor does control the rate of spending, for the account cannot be overdrawn. If resupply is abundant because of increased snowfall upstream, the downstream section thickens and spreads and the snowline shifts downward. With decreased snowfall or increased melting, opposite changes take place. A glacier, like a stream of water, maintains a steady state that is controlled by climate.

When the climate changes, the glacier changes too. Glaciers, therefore, are a sort of yardstick for measuring variations of climate. When the climate in any region becomes cooler and more moist, glaciers expand; when it grows warmer and drier, they shrink. A nearly worldwide shrinkage of glaciers during the first half of the present century implies a general, although slight, warming of climates during at least that span of time (Fig. 11.7).

During the shrinkage of a glacier in a warming climate, the ice of course does not reverse the direction of its movement. The glacier continues to flow in the same direction, but the terminus melts back faster than ice can be supplied to it by flow (Fig. 11.16). Shifting of the terminus, either backward or forward, is merely the adjustment between supply and melting, tending to re-establish a steady-state condition after an interruption caused by a change in the climate.

Surges. Despite the close relation between climate and the behavior of glaciers, from time to time some glaciers experience violent changes, called *surges,* which seem to have little to do with climate. When a surge occurs, a glacier seems to go berserk. It "takes off" and for a few years or decades moves at what, for a glacier, is high speed—in one instance more than 6km per year (nearly 70cm per hour). What makes this happen is still a mystery. It has been thought that perhaps the glacier had been frozen to its bed and then had suddenly melted, "uncoupling" a sizable section of the glacier and allowing it to slide rapidly downslope, overriding its own downstream part, only to freeze up again later on.

Glaciation

Landscapes in Canada, northern United States, and northern Europe are different from those farther south. A principal reason for the difference is that they have been glaciated, and so recently that erosion by weathering, mass-wasting, and running water has not had time to alter them much. Like the geologic work of other external processes, *glaciation, the alteration of a land surface by massive movement over it of glacier ice,* includes erosion, transport, and deposition. To understand how glaciation has affected the landscapes of northern countries we need to examine it more closely.

201

Figure 11.7 Combined Guyot (pronounced Gheé-o) and Yahtse Glaciers, seen from the southeast in 1938. The glaciers flow toward the camera and terminate in Icy Bay, Alaska at West longitude 141°20'. During the last few decades these glaciers, like their neighbors, have been shrinking, principally by calving. Successive positions of the termini in 4 later years, determined by instrumental surveys, are shown. Terminal recession during the 25-year period 1938 to 1963 totaled nearly 10km, a retreat more rapid than would be expected if the glacier had ended on land. By 1963 the two glaciers had become separated by the Guyot Hills. The mountains on the skyline are 190 to 210km distant. (*Photo by Bradford Washburn; data from M. M. Miller.*)

Glacial Erosion and Sculpture

It has been said that a glacier is at once a plow, a file, and a sled. As a plow it scrapes up regolith and gouges out blocks of bedrock; as a file it rasps away firm rock; as a sled it carries away the load of sediment acquired by plowing and filing, plus additional rock waste fallen onto it from adjacent cliffs.

In Fig. 11.8 we see a feature common in northern lands: the result of erosion by a glacier that acted as a plow and a file. The under surface of the ice, studded with rock particles of many sizes, abraded the bedrock over which it moved, and made long scratches (*glacial striations*) and grooves. In places fine particles of sand and silt in the base of the glacier

acted like sandpaper and polished rock to a smooth finish.

At the same time the under surface of the ice drags at bedrock, breaking off blocks of it (usually at joints) and quarrying them out, especially on the downstream sides of hillocks. This process, called *plucking*, creates cliffs facing downstream in contrast to the smooth, abraded upstream slopes of the hillocks. This unsymmetrical form shows which way the glacier was moving (Fig. 11.9). The great length and straightness of some grooves are made possible by the fact that the flow of ice is not turbulent. In contrast, rock surfaces abraded by the turbulent flow of water look different; they consist of a complex pattern of curves.

Figure 11.8 Glaciated surface of sandstone bedrock, showing both striations and grooves. The sandstone contains pebbles and cobbles of quartz, which resist erosion better than the sandstone matrix does. The "tails" of sandstone behind them stream away from the camera; hence the former glacier must have moved in that direction (south).

The straight gouges that parallel the handle of the hammer were made by the teeth of a power shovel. East Haven, Connecticut. (*R. F. Flint.*)

A

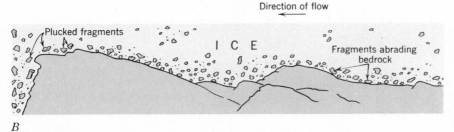

B

Figure 11.9 Results of glacial erosion.
A. Knobs of granitic bedrock abraded and striated on one side and plucked on the other, by a former glacier. Unsymmetrical form shows glacier moved toward southwest. Lake Athabaska, Saskatchewan. (*F. J. Alcock, Geol. Survey of Canada.*)
B. Abrasion and plucking in progress.

Frost Wedging. Although not a process of glaciation, intense frost wedging (Chap. 6) accompanies glaciation because both occur in cold climates. When hills or mountains stand higher than the surface of a glacier, wedged-out blocks of rock roll down onto the surface of the glacier, which carries them away. Intense frost wedging is responsible for much of the detail of the sharp, jagged peaks of glaciated mountains (Fig. 11.12,*D*).

Cirques. Among the characteristic features made by glaciation in many mountain areas is the *cirque* (pronounced *sirk*). This is *a steep-walled niche, shaped like a half bowl, in a mountainside, excavated mainly by frost wedging and glacial plucking* (Figs. 11.10, 11.11). A cirque begins to form beneath a snowbank or snowfield just above the snowline, and

is at least partly the work of frost wedging. On summer days water from melting snow infiltrates openings in the rock beneath the snowbank. At night the temperature drops and the water freezes and expands, prying out rock fragments. The smaller rock particles are carried away downslope by meltwater during thaws. This activity creates a depression in the rock and enlarges it. If the snowbank grows into a glacier, plucking helps to enlarge the cirque still more, but frost wedging continues as water descends the rock wall of the cirque and freezes there. The floors of many cirques are rock basins, some containing small lakes. In summary, small cirques are excavated beneath snowbanks even where no true glacier exists; large cirques are mostly the work of glaciers that continue the excavation process.

Figure 11.10 Six small cirques, containing cirque glaciers (nos. 5 and 6) or the headward ends of small valley glaciers (nos. 3 and 4). Floors of the cirques lie close to a common altitude, that of the snowline. If all the ice and snow were to melt, the snowline could be reconstructed approximately from these floors. Mount Lucania district, St. Elias Mountains, Yukon Territory. (*Bradford Washburn, 1966.*)

Figure 11.11 Glaciated valley, Coast Mountains, British Columbia. A former glacier has ground away projecting spurs to create a U-shaped trough. On the upland, left center, are several cirques containing very small glaciers. (*Canadian Government copyright.*)

Glaciated Valleys differ from ordinary stream valleys in several ways. Their chief characteristics, not all of which are found in every glaciated valley, include (Figs. 11.11, 11.12,*D*): (1) a cross profile that is trough-like (U-shaped), and (2) a floor that lies below floors of tributaries, which therefore "hang" above the main valley. Both (1) and (2) result from erosion by the sides as well as by the base of the glacier, the thickness of which is far greater than the depth of an ordinary stream. In addition (3) the long profile of the floor is marked by step-like irregularities (Fig. 11.17) and shallow basins. Many of these are related to the spacing of joints in the rock, which determine ease of plucking. Finally (4) the valley head is likely to be a cirque or a group of cirques.

Some valleys are glaciated from head to mouth.

Others are glaciated only in their headward parts, and downstream their form has been shaped by streams, sheet erosion, and mass-wasting. This shows that the valleys were there before the glaciers formed and leads to the conclusion that glaciers, unlike streams, do not cut their own valleys but occupy and remodel valleys already made.

Mountain Sculpture. If we examine a mountain area that has been glaciated by a large group of valley glaciers (Fig. 11.12), we find cirques, U-shaped troughs, hanging tributaries, and (in greater detail) striated and polished bedrock. In addition, the forms of the mountain crests are characteristic and are mainly the result of frost wedging in a cold climate. The forms are combinations of three fea-

205

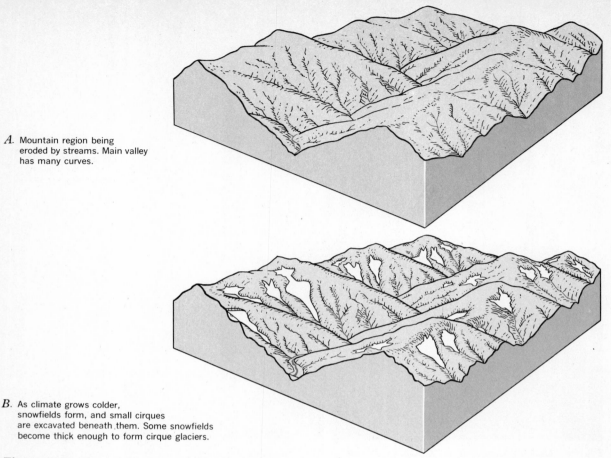

A. Mountain region being eroded by streams. Main valley has many curves.

B. As climate grows colder, snowfields form, and small cirques are excavated beneath them. Some snowfields become thick enough to form cirque glaciers.

Figure 11.12 Erosion of mountains by valley glaciers.

tures, for which we use names given to them by Alpine mountaineers. An **arête** is *a jagged, knife-edge ridge created where two groups of cirque glaciers have eaten into the ridge from both sides.* A **col** is *a gap or pass in a mountain crest where the headwalls of two cirques intersect each other.* A **horn** is *a bare, pyramid-shaped peak left standing where glacial action in cirques has eaten into it from three or more sides.* The Matterhorn is a well-known example.

All these features are primarily the work of frost wedging coupled with glacial erosion and transport. All are shown in Fig. 11.12.

Fiords. The deep, bay-like *fiords* along mountainous coasts like those of Norway, British Columbia/Alaska, and southern Chile are *glaciated troughs partly submerged by the sea.* The form and depths of many fiords imply depths of glacial erosion of 300m or more. The depths of fiord floors below sea level result in part from the rise of sea level since the last glacial age (Fig. 11.25), in part (locally) from subsidence of Earth's crust, and in part from glacial erosion below the sea's surface. Sea level is not a base level for glaciers as it is for streams, because glacier ice 300m thick, with a specific gravity of 0.9, can continue to erode its bed until it is submerged to a depth of 270m, whereupon it floats and bed erosion ends. In contrast to the deep erosion of fiords, some areas that have been glaciated by ice sheets still preserve chemically weathered regolith beneath glaciated surfaces. Because such regolith is ordinarily thin, the thickness of rock removed by glacier ice must have been small, possibly only a meter or two. So the intensity of glacial erosion, like

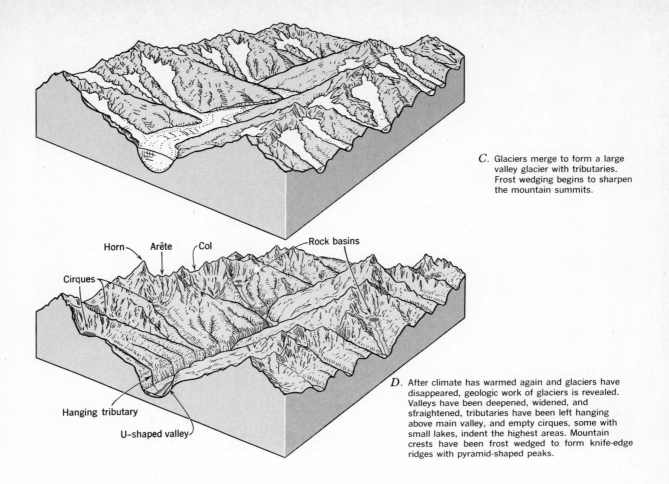

C. Glaciers merge to form a large valley glacier with tributaries. Frost wedging begins to sharpen the mountain summits.

Horn Arête Col Rock basins

Cirques

Hanging tributary

U–shaped valley

D. After climate has warmed again and glaciers have disappeared, geologic work of glaciers is revealed. Valleys have been deepened, widened, and straightened, tributaries have been left hanging above main valley, and empty cirques, some with small lakes, indent the highest areas. Mountain crests have been frost wedged to form knife-edge ridges with pyramid-shaped peaks.

that of stream erosion, varies according to topography, kind of bedrock or regolith, and thickness and velocity of flow.

Glacial Transport

In the way in which it carries its load of rock particles, a glacier differs from a stream in two main respects: (1) its load can be carried in its sides and even on its top, and (2) it can carry much larger pieces of rock; also it can transport large and small pieces side by side without segregating them into a bed load and a suspended load that must be deposited in order of decreasing weight of the individual particles. Because of these differences deposits made directly from a glacier are neither sorted nor stratified.

The load in a glacier is concentrated in base and sides (Fig. 11.17), because these are the areas where glacier and bedrock are in contact and where abrasion and plucking are effective. Much of the rock material on the surface of a valley glacier got there by landsliding from clifflike valley sides.

A good deal of the load in the base of a glacier consists of small particles such as fine sand and silt. Most of these particles are fresh and unweathered, with angular, jagged surfaces. Such particles are the products of crushing and grinding. *Fine sand and silt produced by crushing and grinding in a glacier are **rock flour,*** a material that differs from the chemically weathered, more rounded particles found in the sediments of nonglaciated areas.

Much of the transfer of rock particles from a glacier to the ground occurs by release of particles

60 cm

Facets and polish on boulders

A

Figure 11.13 Till and a cobble from till.

A. Till exposed in a road cut near Bangor, Pennsylvania. Its nonsorted character is clearly evident. The fine sediment between the stones consists of rock flour. Surfaces of two large boulders have been faceted and polished. (*Pennsylvania Geol. Survey.*)

Compare the till-like sediment in Fig. 7.10. How could one prove that that sediment had not been deposited by a glacier?

B. Cobble made of limestone, collected from till exposed in northern New York State. The broad surface is a facet. The random striations on it are deep because the calcite of which the rock consists is a soft mineral. (*R. F. Flint.*)

0   3 cm

B

as the surrounding ice melts. Therefore most glacial deposition takes place in the downstream part of the glacier, below the snowline, where melting is dominant.

Glacial Deposition

Drift, Till, and Stratified Drift. *The sediments deposited directly by glaciers or indirectly in glacial streams, lakes, and the sea together constitute **glacial drift,** or simply **drift.*** The name *drift* dates from the early 19th Century when it was vaguely conjectured that all such deposits had been "drifted" to their resting places by the Flood of Noah or by some other ancient body of water.

Drift consists of two extreme kinds, *till* and *stratified drift,* which grade into each other. Drift whose constituent rock particles are not sorted according to size and weight but lie just as they were released from the ice (Fig. 11.13,*A*) is known as *till,* a name given it by Scottish farmers long before its origin was understood. ***Till*** is defined simply as *nonsorted drift* and is deposited directly from ice.

Probably most till is plastered onto the ground, bit by bit, from the base of the flowing ice near the outer margins of glaciers. We say "probably" because no one has yet devised a means of getting underneath a glacier to observe the process. The surfaces of pebbles and larger fragments in till include facets (Fig. 11.13,*B*) joining each other along smoothed or rounded edges; some facets are striated. Facets are made by grinding. As pebbles turn in their matrix of ice, new facets are made. The sand and silt particles in till generally consist of rock flour (Fig. 11.14).

Colluvium, described in Chap. 7, is nonsorted, and some colluvium, on first comparison, looks very much like till. However, the presence of faceted, striated stones, rock particles of distant origin, and in some places a grooved or striated pavement underneath help to identify till.

On the other hand, much drift is stratified, indicating that water from melting ice has moved and sorted rock particles carried in the ice and has deposited them in immediate contact with the ice or beyond the glacier itself. ***Stratified drift,*** then, is *drift that is sorted and stratified;* it is deposited not by a glacier but by glacial meltwater.

Streamline Hills. Stratified drift and till are sediments. They occur in various bodies, each having

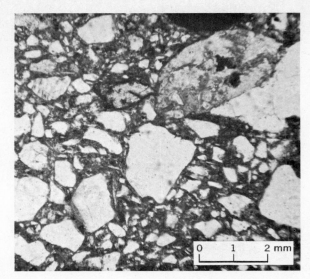

Figure 11.14 Rock flour, forming part of a body of till near Bethany, Connecticut, viewed in a thin section through a microscope. The very angular shapes of particles (mostly quartz and feldspar) are the result of crushing. Particles are not sorted. Note similarity to Fig. 11.13, even though the two scales are very different. (*R. W. Powers.*)

a rather distinctive topographic form. Some of the bodies are described in the following paragraphs.

In many areas drift is molded by flowing ice near the outer edge of an ice sheet into smooth, nearly parallel ridges and furrows that range up to many kilometers in individual length. These forms resemble the streamlined bodies of airplanes and racing cars; they offer minimum resistance to the ice flowing over them. The best-known variety of streamline form is the ***drumlin,*** *a streamline hill consisting of drift, generally till, and elongated parallel with the direction of glacier movement* (Figs. 11.15, 11.23). Not all such forms, however, are built up. Some are shaped by glacial molding of pre-existing drift; these too are drumlins. Others, like the much smaller striations and grooves, are shaped by glacial erosion of bedrock, and even though their form is streamline they are not drumlins. Whether made by building up or cutting out, or both, all these forms reflect streamline molding by flowing ice, and therefore their long axes are reliable indicators of the direction of flow of former glaciers.

Moraine. *Widespread thin drift with a smooth surface consisting of gently sloping knolls and shallow closed depressions is **ground moraine*** (Fig. 11.16).

Figure 11.15 A streamline hill. Drumlin near Madison, Wisconsin. See the map, Fig. 11.23. (*C. C. Bradley.*)

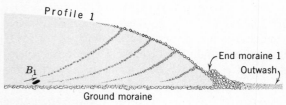

BEFORE RECESSION

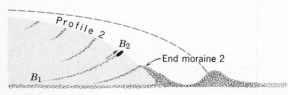

AFTER RECESSION

Figure 11.16 With change to a slightly warmer climate, rate of melting increases and the terminus of a glacier recedes from Profile 1 to Profile 2. Continued flow of glacier during recession moves boulder B from B_1 to B_2. We can expect the boulder to be built into End Moraine 2. Repeated fluctuation can pile up a complicated sequence of end moraines and outwash sediments.

Probably its irregularities result from irregular distribution of rock particles in the base of the glacier.

A ridgelike accumulation of drift, deposited by a glacier along its margin, is an **end moraine** (Figs. 11.16, 11.17). It can be built by snowplow- or bulldozer action, by dumping off the glacier margin as the ice melts, by repeated plastering of sticky drift from basal ice onto the ground, or by streams of meltwater depositing stratified drift at the glacier margin. End moraines range in height from a few feet to hundreds of feet. In a valley glacier the end moraine is built not only at the terminus but along the sides of the glacier as well for some distance upstream. The terminal part is a *terminal moraine;* the lateral part is a *lateral moraine;* but both are parts of a single feature (Figs. 11.1, 11.17, 11.23).

Erratics, boulder trains. Some of the boulders and smaller rock fragments in till are the same kind of rock as the bedrock on which the till was deposited, but many are of other kinds, having been brought from greater distances. *A glacially deposited particle of rock whose composition differs from that of the bedrock beneath it* is an **erratic.** The word means simply "foreign," and the presence of foreign boulders was one of the earliest-recognized proofs of former glaciation. Some erratics form part of a body of drift; others lie free on the ground. Many erratic boulders exceed 3m in diameter, and some are enormous, with weights estimated in the thousands of tons. Such boulders are far larger than those which can be carried by an ordinary stream of water.

In some areas that have been glaciated by ice sheets, erratics derived from some distinctive kind of bedrock are so numerous and easily identified that they can be readily plotted on a map. Generally the plot shows a fanlike shape, spreading out from the area of outcrop of the parent bedrock and reflecting the spreading of the ice sheet. *A group of erratics spread out fanwise* is a **boulder train,** so named in the 19th Century when rock particles of all sizes were called boulders. The boulder train shown in Fig. 11.23 consists of quartzite, very conspicuous in an area in which all the bedrock consists of limestone and sandstone.

Outwash Sediment. On the downstream sides of most terminal moraines is *stratified drift deposited by streams of meltwater as they flow away from a glacier.* Sediments of this kind are **outwash** ("washed out" beyond the ice). *A body of outwash that forms a broad plain beyond the moraine is an* **outwash plain.** In contrast, a **valley train** (Fig. 11.18) is *a body of outwash that partly fills a valley.* Meltwater generally emerges from the ice as one or more swift streams, turbid with a suspended load of rock flour and with an abundant bed load of pebbles, cobbles, and even boulders. The bed load is invisible, but it is there. The deposition of outwash is analogous to the building of a fan (Fig. 8.11). The stream emerges from an ice-walled valley or tunnel onto a broad smooth surface; part of its load therefore becomes excess; it drops bed load, and the profile of

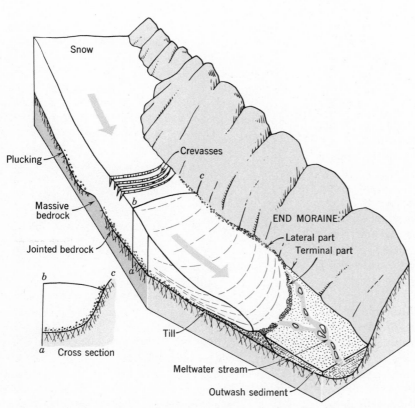

Figure 11.17   Chief features of a valley glacier and its deposits. The glacier has been cut away along its center line; only half of it is shown. Compare Fig. 11.1. The crevasses shown result from steepened slope of the bed beneath the glacier.

Figure 11.18 Valley train being actively built of outwash sediment by streams of meltwater from glaciers upstream. Tasman Valley, Southern Alps, New Zealand. Tasman Glacier (*upper center*), dark with stony debris, is the principal source of outwash. Hooker Glacier (*upper left*) is contributing a fan-like body of outwash that has pushed the streams from Tasman Glacier to one side of their valley. Other, local fans are visible in left foreground. The valley train has little or no vegetation because a new layer of sediment is deposited on it during each melting season. (*V. C. Browne.*)

the entire thick deposit is steep like that of a fan. As the coarse particles are dropped first, the average diameter of particles decreases downstream.

Although the bed load of a meltwater stream is invisible, the suspended load has been measured. The rock flour washed out of the big Muir Glacier in coastal Alaska corresponds to a loss of 2cm of bedrock, from the entire area beneath the glacier, every year. Such a rate of erosion is 332 times faster than the 1cm every 166 years estimated to be lost to the United States by weathering, mass-wasting, and erosion by running water.

Ice-Contact Stratified Drift. When rapid melting and evaporation reduce the thickness of the terminal part of a glacier to 100m or less, movement virtually ceases. Meltwater, flowing over or beside the nearly motionless stagnant ice, deposits stratified drift, which slumps and collapses as the supporting ice slowly melts away. *Stratified drift deposited in contact with supporting ice* is **ice-contact stratified drift.** It is recognized by abrupt changes of grain size, distorted, irregular stratification, and extremely uneven surface form (Fig. 11.19). Bodies of ice-

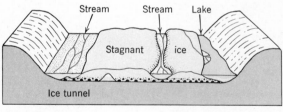

A

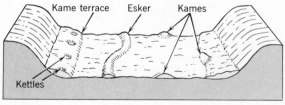

B

Figure 11.19 Origin of bodies of ice-contact stratified drift. *A.* Nearly motionless melting ice furnishes temporary retaining walls for bodies of sediment built chiefly by streams of meltwater.

B. As ice melts, bodies of sediment slump, creating characteristic knolls, ridges, terraces, and closed depressions.

Figure 11.20 This glacial-age esker, overlying ground moraine in Morrison County, Minnesota, consists of gravel and sand deposited in a winding tunnel within an ice sheet but near its edge. When supporting ice melted away, the deposit was left as a curving ridge. (*W. S. Cooper.*)

contact stratified drift are classified according to their shape: short, steep-sided knolls and hummocks are *kames;* terracelike forms along the side of a valley are *kame terraces;* long, narrow ridges, commonly sinuous, are *eskers* (Fig. 11.20); and basins in drift, created by the melting out of masses of underlying ice, are *kettles* and in form are complementary to kames.

The Glacial Ages

History of the Concept. As early as 1821 European scientists began to recognize features characteristic of glaciation in places at considerable distances from any existing glaciers. They drew the inference that glacier ice must once have covered wide regions that are now free of ice. Consciously or unconsciously, they were applying the principle of uniformity of process and materials. The concept of a glacial age with widespread effects was first set forth in 1837 by Louis Agassiz, a Swiss scientist who achieved fame through his hypothesis. Gradually, through the work of many others, information on the character and extent of former glaciation was added to the growing body of knowledge, so that today we have a basic picture of former glacial times, although many important questions still remain unanswered.

Extent of Glaciers. During the second half of the 19th Century, geologists searched for glacial drift and other characteristic features in order to determine the extent of former glaciers. In most mountain regions the characteristics are those shown in Fig. 11.12,*D*. In most lowland regions, however, they consist generally of rolling ground moraine, end moraines, and related features (Fig. 11.23). By 1900 the general extent of glaciation had been learned, and today our information is much more detailed. Figure 11.21 shows areas in the Northern Hemisphere that were covered by glaciers during the glacial ages. Figure 11.22 shows in more detail the extent of glaciation in the northern United States and southern Canada. These maps were compiled from observations by hundreds of geologists on the distribution of features characteristic of glaciation. On a world scale, the areas formerly glaciated add up to the impressive total of more than 44 million square kilometers—about 29 per cent of the entire land area of the world. Today, for comparison, only about 10 per cent of the world's land area is covered with glacier ice, of which 84 per cent (by area) is on the Antarctic Continent. If we neglect that continent and consider only the rest of the world, the glacier-covered area on non-Antarctic lands was more than 13 times larger during glacial ages than it is on the same lands today.

Figure 11.21 Areas (white) in the Northern Hemisphere that were glaciated during the glacial ages. Arrows show generalized directions of flow of glacier ice. Shorelines are shown as they were when sea level was 100m lower than it is today. The gray tone with irregular pattern represents floating ice in the Arctic Ocean, extending south into the Atlantic.

Directions of Flow. In most mountain regions, former glaciers simply flowed down existing valleys (Fig. 11.12,C). In lowland regions streamline forms and end moraines show that ice sheets spread out in a radial manner (Figs. 11.21, 11.23). Flow directions are determined also by tracing erratics of conspicuous kinds to their places of origin in the bedrock. Native copper found as far south as Missouri has been traced to bedrock on the south shore of Lake Superior. In eastern Finland, copper ore traced backward along the line of ice flow led to the discovery of a valuable copper deposit in the bedrock. Several large diamonds of good quality have been found in drift in Wisconsin, Michigan, Ohio, and Indiana. Their source has not yet been found, but

the flow directions of former glaciers suggest it is in central Canada.

Amount of Erosion. The great central parts of the former ice sheets in north-central Canada and northern Scandinavia and Finland removed underlying regolith and bedrock to a depth averaging perhaps 15m to 25m. Much of the load thus acquired by the ice sheets was deposited as drift beneath their broad outer parts. One such belt of drift reaches from Ohio to Montana; another extends from the British Isles to European Russia. In these belts the average thickness of the drift is believed, from well borings and other measurements, to be as much as 12m. Most of the rest of the load went down the Mississippi

River and down various rivers in Europe; some of the fine-grained part was picked up by the wind and blown away.

Where, as in New England and parts of Quebec, the bedrock is resistant and breaks into large chunks, the glacier ice spread boulders liberally over the surface, creating soil that is very difficult for agriculture. Where, as in the southern Great Lakes region and on the Plains, the bedrock is weak and crumbles easily, the glaciers deposited a thick layer of drift, which now supports a rich soil.

Depression of the Crust. We should note at this point the subsidence of Earth's crust beneath the weight of the ice sheets. This effect is described in Chap. 15. Probably the surface of the crust beneath the central part of today's Greenland Ice Sheet would stand as much as 700 to 900m higher if the ice sheet were not there.

Repeated Glaciation. Because most of the drift is fresh and little weathered, it was realized very early that the glacial invasion was recent. But geologists began to find exposures showing a blanket of comparatively fresh drift overlying another layer of drift whose upper part is chemically weathered (Fig. 11.24). This led to the inference that there had been two glaciations separated by a period of time long enough to cause weathering to a depth of one or two meters. Before the beginning of the 20th Century, accumulated evidence of this kind had established the fact of not merely two but several great glaciations, each covering approximately the same areas.

Radiometric dating suggests that the earliest of these great glaciations may possibly have occurred around 2 million years ago. The time that embraced that particular bundle of glaciations is the Pleistocene Epoch (Table 2.1), a time often referred to as the *glacial ages,* notwithstanding the fact that between the glaciations were intervals (*interglacial ages*) in which climates were about as warm as today, and at times a little warmer than existing climates.

Dating by ^{14}C indicates that the last time glaciers in North America and Europe reached a great extent was between 20,000 and 15,000 years ago. The subsequent period of shrinkage to their present condition has been marked by repeated, conspicuous expansions. One of these occurred 2500 to 3000 years ago and another only 200 to 300 years ago, overwhelming small high-altitude farms and villages in the Alps. Some of the dwellings are today still buried beneath ice.

At the height of each one of these glacial ages mean annual air temperatures, averaged over the Earth, were about 5°C less than are those of today; during interglacial ages temperatures were at least 1° more than today's. The mean range was therefore only about 6°. Temperature changes were felt not only in the regions that were glaciated but everywhere, the Equator included, although they varied in amount from place to place.

Glaciers were distributed in the same basic pattern as that of today's glaciers: they originated in the same places but simply were much bigger. Their places of origin are highlands so situated that winds can bring moist air to them. True, the big ice sheets spread far outward over low lands, but they are believed to have originated through the coalescence of many comparatively small glaciers that flowed down from high places.

Each glaciation was not a single event. Instead it consisted of a whole group of minor advances and retreats of glacier margins, as shown by the many end moraines and by other data as well. Radiocarbon dates of fluctuations during the later part of the latest glacial age show that these events have been nearly simultaneous in diverse parts of the world. Although we are not yet completely sure, probably each glacial age affected all the continents at the same time.

Effects of Glacial-Age Climates. The formation and spread of glaciers was the most obvious effect of the relatively cold climates of glacial ages. But there were other effects, apparent in the oceans and in lands not covered with ice.

Change of Sea Level. As we have already said, snowfall that builds glaciers comes through the water cycle from the evaporation of seawater. So a first effect of a glacial age would have been worldwide lowering of sea level, very likely by as much as 100m. As the latest great ice sheets melted away, the level of the sea rose (Fig. 11.25).

Along some coasts there are old beaches and other shore features, high and dry at altitudes of a few meters to more than 35m. Probably these were made during one or more interglacial ages when ice on the lands was less abundant than it is today. Of course, before we fully accept this inference we must make sure the beaches record real changes in the

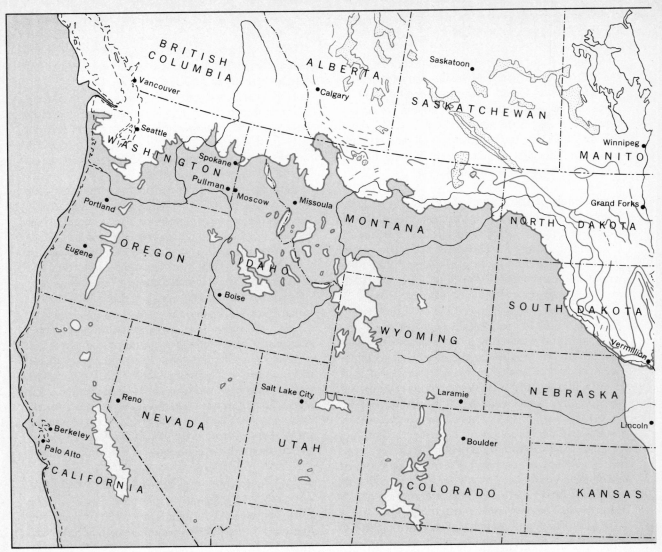

Figure 11.22 Extent of glaciation in northern United States and southern Canada. Inner, ticked line, during latest glacial age; outer line, during earlier glacial ages. (*Compiled from several sources.*)

volume of the oceans, and have not simply been lifted by general uplift of parts of Earth's crust.

Quite likely the shallow channel on the continental shelf, connecting the Hudson River with the Hudson Submarine Canyon, is the work of a lengthened Hudson River during a glacial time of lowered sea level.

Pluvial Lakes. A second effect of glacial-age climates was the creation of lakes in regions that today

are dry or contain saline lakes. Figure 11.26 shows the contrast. The basin of Great Salt Lake, Utah, was occupied formerly by the gigantic Lake Bonneville, more than 300m deep and with a volume comparable to that of Lake Michigan. Beaches, deltas of tributary streams, and bottom sediments are all there to tell the story. Such lakes are ***pluvial lakes.*** None exists today because, by definition, they are *lakes that existed under a former climate, when rainfall in the surrounding region was greater than today's.*

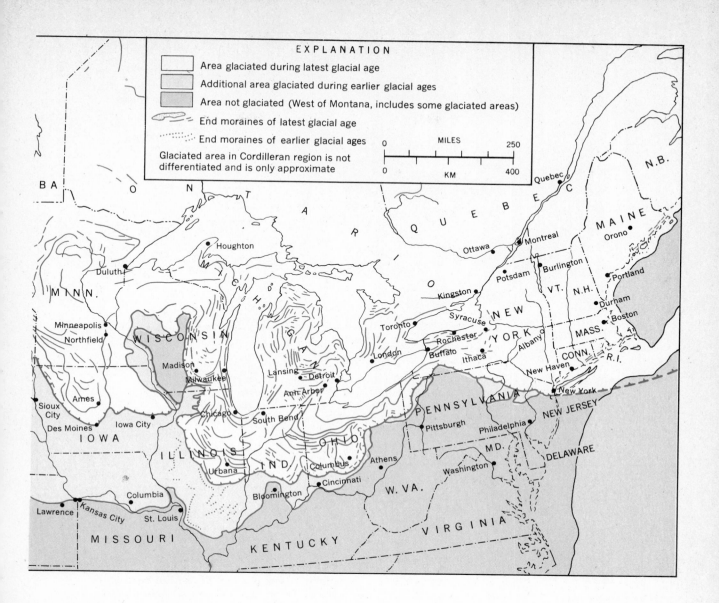

EXPLANATION

Area glaciated during latest glacial age

Additional area glaciated during earlier glacial ages

Area not glaciated (West of Montana, includes some glaciated areas)

End moraines of latest glacial age

End moraines of earlier glacial ages

Glaciated area in Cordilleran region is not differentiated and is only approximate

MILES
0   250

KM
0   400

Although the name implies a pluvial (rainy) climate, we realize that reduced evaporation caused by reduced temperature was a large factor in the changed water economy, but the name *pluvial* still sticks to the ancient lakes. These lakes were, then, the result of (1) lower temperature, and (2) increased rainfall in the regions where they existed.

Direct geologic evidence and ^{14}C dates show that some pluvial lakes were contemporaneous with the most recent of the glacial ages. This does not mean,

though, that such lakes were fed by water derived from the melting of glaciers. Some of them were, but only by coincidence. Most of the lakes occupied basins to which glacial waters had no access.

Pluvial lakes were abundant also in other dry regions of the world. The desert area in the Andes of Bolivia, Chile, and Argentina contained many of them. Numerous pluvial lakes were present also in the Saharan region of North Africa and in western and central Asia. The largest in the world was an

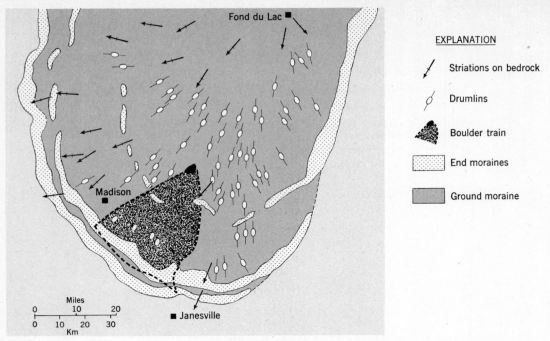

Figure 11.23 Map of lobe-shaped layer of glacial drift in southeastern Wisconsin. Three successive end moraines (both terminal and lateral parts) mark successive positions of the glacier margin. Directions of drumlins, striations on bedrock, and a boulder train all show spreading of ice toward margin of glacier. (*Adapted from Flint and others, 1959, in part after W. C. Alden.*)

EXPLANATION

⟋ Striations on bedrock

⊘ Drumlins

▲ Boulder train

▨ End moraines

▨ Ground moraine

Figure 11.24 Evidence of repeated glaciation. A layer of "fresh" till, weathered only slightly at its surface, overlying a layer of older till that had been deeply weathered before the overlying till was deposited.

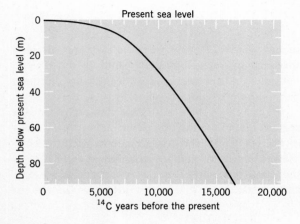

Figure 11.25 Generalized curve showing the rise of sea level relative to the land, through most of the time since the huge ice sheets in North America and Europe last began to melt away. Constructed mainly from the ^{14}C dates of wood, peat, and shells deposited at or near sea level but now submerged. (*After F. P. Shepard, 1963.*)

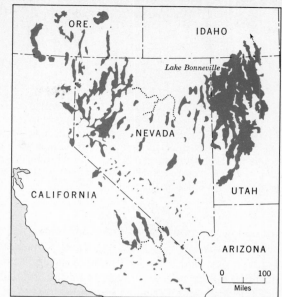

A. Lakes existing today.

B. Lakes that existed about 15,000 years ago.

Figure 11.26 Existing and former lakes in the dry region of western United States. (*O. E. Meinzer, 1922, and other sources.*)

ancestor of today's Caspian Sea; another that is historically famous was an ancestor of today's Dead Sea. Although some of these former lakes probably owed their existence to glacial-age climates, others, especially some of those in Africa, may have originated in very different ways. The problem of their origin is not yet solved.

Spreading of Permafrost. A third effect of the ice-age climates was the spreading of ***permafrost,*** *ground that is frozen perenially.* Actually it is not the ground but rather the ground water that has frozen. It forms a firm cement in the pores of regolith and bedrock, and makes serious trouble where a building is built on it in such a way as to cause the permafrost beneath it to thaw.

Today permafrost extends through a wide belt of far-northern country that adds up to 20 per cent of the land area of the Northern Hemisphere. Over that belt average air temperatures are at or below freezing during most of the year, and loss of ground heat to the atmosphere is the cause of freezing of the ground water. During the short summer melting season only a thin surface zone, usually no more than about 1m thick, thaws out. The thawed layer flows down slopes by solifluction, carrying with it the arctic vegetation that grows at its surface.

Drilling has shown that in places permafrost is

more than 300m thick, representing a long period of freezing. During glacial ages the southern limit of northern permafrost crept southward to much lower latitudes, reaching central Europe and northern North America. Today, under climates that have been growing generally warmer, permafrost is slowly shrinking, as the big ice sheets and the pluvial lakes have done already. But if climates should once more grow colder, permafrost would spread again.

Glacial Ages in Earth's History

Radiometric dates tell us that the bundle of glacial ages whose great ice sheets buried large parts of North America and Europe have occurred apparently within the last 2 million years or so. But other glacial ages, most of them perhaps with less extensive glaciers, seem to have appeared again and again throughout much of Earth's history. Much of the evidence of these earlier cold times consists of ***tillite*** (*till converted into solid rock*) in strata of many different ages. In some places the tillite overlies striated surfaces of still older rock. Tillite and evidence furnished by fossils indicate that the Antarctic Continent and Arctic lands as well have been cold during most of the last 20 million years or so and that probably the Antarctic Ice Sheet has existed

219

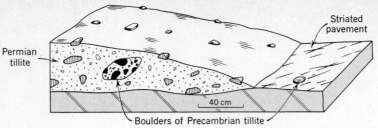

Figure 11.27 Sketch of exposure near Adelaide, South Australia showing a body of Permian tillite about 250 million years old, containing a boulder broken from Precambrian tillite more than 600 million years old. This unusual occurrence demonstrates that Australia has experienced at least two very old glacial ages.

continuously through that time. Still older strata, conspicuously some of those of Pennsylvanian and Permian age, and some of Precambrian age (more than 600 million years old) contain tillites. Evidence like that in Figs. 11.27 and 14.5 implies that glaciation has occurred again and again even in the far-distant past, and has not been confined to a late part of Earth's history.

Causes of Glacial Climates

What, then, is the cause of climatic chilling that recurs, seemingly at irregular intervals, and works great changes in the activities of Earth's external processes? Tens of thousands of pages of reasoning and discussion on this problem have been published, but we are still not certain of the answer. We still have only hypotheses. Let us begin with a check list of the important data on glacial ages:

1. Fluctuation of climate appears to be worldwide and simultaneous. Worldwide, the fluctuation of mean annual temperatures is only a few degrees Celsius. Some fluctuations are of very short duration—recently nearly 1°C in a period of less than 100 years. Fluctuation seems, at least, to be irregular.

2. Glaciations occur on land and begin, at least, on the highest places where snowfall is sufficient and summers are not too warm.

3. Glaciers spread out and sometimes cover great areas of lower land until stopped by warmer climate.

4. In each of the glacial ages within the last 2 million years, glaciers reappeared in the same places that had been occupied by their predecessors.

To answer the requirements of this check list we can suggest three mechanisms, each of which could result in a colder climate at a particular locality on a continent:

1. Slow movement of a crustal plate (Chap. 18) from a lower to a higher latitude, thus carrying a continent into a colder climatic zone than the one it occupied formerly.

2. Uplifts, related to the movement of plates, of parts of continents, creating mountain ranges and other high lands and lifting them high enough to intersect the snowline and thus permit the formation of glaciers.

3. Variation, through time, of the radiant energy (or at least certain wavelengths within it) that reaches Earth's surface from the Sun. Although measurements thus far do not indicate fluctuation in the Sun itself, changes in the composition of Earth's atmosphere (water vapor, carbon dioxide gas, ozone, and very fine volcanic dust have been suggested) might cause changes in the effectiveness of the atmosphere as a filter or insulator.

Although (1) and (2) could explain glaciation of particular continents, those mechanisms are too slow-moving to explain the disappearance and re-appearance of glaciers, and short-term glacial fluctuation on two or more continents at the same time. Item (3) can explain the fluctuation but there is no firm evidence that this received energy does actually fluctuate. So, for the time being, we must live with hypotheses. We can speculate that Earth's surface has been receiving solar energy irregularly throughout a long history, but that only at those times when drifting plates of lithosphere had carried continents into the higher latitudes and/or extensive mountains were present, could low temperatures be translated into glaciers big enough to establish a glacial age.

Summary

Definition and form

1. Glaciers are accumulations of snow and ice; they flow under their own weight. Their surface parts are brittle; below these parts flow occurs.

2. On a basis of form, glaciers include valley glaciers, piedmont glaciers, and ice sheets.

3. Glaciers require low temperature and adequate snowfall. All glaciers are connected with each other by the snowline, which is low in polar regions and high in tropical regions.

Movement

4. The motion of a glacier includes both internal flow and sliding on its bed.

Economy

5. A glacier can be thought of as a system in which thickness, velocity, and other factors adjust to each other to create a steady-state condition.

Glacial Erosion

6. Glaciers erode rock by plucking and abrasion, transport the product, and deposit it as drift.

7. Valley glaciers convert stream valleys into U-shaped troughs with hanging tributary valleys. Cirques form beneath snowbanks, cirque glaciers, and the heads of valley glaciers. Mountain areas that project above glaciers are reshaped by frost wedging into arêtes, cols, and horns.

Glacial Sediments

8. The load, carried chiefly in the base and sides of a glacier, includes particles of all sizes, from large boulders to rock flour.

9. Till is deposited by glaciers directly. Stratified drift is deposited by meltwater. It includes outwash deposited beyond the ice, and kames, kame terraces, and eskers deposited upon or against the ice itself.

10. Ground moraine is built beneath the glacier, end moraines (both terminal and lateral) at the glacier margins.

Glacial Ages

11. During glacial ages huge ice sheets repeatedly covered northern North America and Europe, eroding bedrock and spreading drift over the outer parts of the glaciated regions.

12. The accumulation of glaciers lowered sea level. The cool, moist climates that prevailed in regions now dry created many pluvial lakes. Permafrost developed in areas free from glaciers but with average temperatures that were below freezing.

13. The cause of the glacial climates is not known. Possibly it is related to the drifting of continents into higher latitudes and the creation of mountains and other high land, and, in addition, variations in the receipt of solar energy by the Earth.

Selected References

Agassiz, Louis, 1967, Studies on glaciers (Neuchâtel, 1840), Translated and edited by A. V. Carozzi: New York, Hafner.

Bentley, C. R., and others, 1964, Physical characteristics of the Antarctic Ice Sheet: Am. Geog. Soc. Antarctic Map Folio 2.

Charlesworth, J. K., 1957, The Quaternary Era: London, Edward Arnold.

Crowell, J. C., and Frakes, L. A., 1970, Phanerozoic glaciation and the causes of ice ages: Am. Jour. Sci., v. 268, p. 193–224.

Dyson, J. L., 1962, The world of ice: New York, Alfred A. Knopf.

Flint, R. F., 1971, Glacial and Quaternary geology: New York, John Wiley.

Sparks, B. W., and West, R. G., 1972, The Ice Age in Britain: London, Methuen.

Wright, H. E., and Frey, D. G., eds., 1965, Quaternary of the United States: Princeton, N.J., Princeton Univ. Press.

Glacial-Age Maps

Flint, R. F., and others, 1945, Glacial map of North America: Geol. Soc. America Spec. Paper 60. Scale 1 : 4,555,000.

———— 1959, Glacial map of the United States east of the Rocky Mountains: Geol. Soc. America. Scale 1 : 750,000.

Wilson, J. T., and others, 1958, Glacial map of Canada: Geol. Assoc. of Canada. Scale 1 : 3,801,600.

(*Opposite.*) The Gulf Stream, a current of warm seawater (brown as seen in this picture) passes close to the eastern shore of Florida as it flows north. The picture is not a true photograph because it depicts only differences of temperature, sensed by a Nimbus Satellite on December 24, 1972. Each temperature is represented by a different color. Land surfaces (green) and clouds (blue) are cool compared to normal ocean water (yellow) and to the Gulf Stream. (*NASA.*)

Part III   External Processes in the Sea

Chapter 12

The Ocean and Its Floor

The World Ocean

In driving across the United States or Canada, one is impressed by the vast size of the continent, by the varied topography and the endlessly changing scenery. But unless we happen to be venturesome sailors, we cannot easily get a similar sense of the vastness of the oceans. Nor can we see the irregularity and diverse scenery of the sea floor. Nevertheless, by working together in cooperative research, oceanographers and geologists have found that the sea floor is just as varied as the surface of the land.

Research at sea is difficult and expensive. It requires an array of sensitive instruments and heavy equipment, including specially constructed oceanographic research ships (Fig. 12.1). Until recently, therefore, growth of knowledge of the topography and sediments of the sea floor lagged behind the parallel study of the land. With newly perfected devices for sounding the sea bottom and for sampling its sediment, teams of oceanographers and seagoing geologists have put submarine geology on a basis of knowledge almost as solid as that already established on the land. Diving geologists have visited, photographed, and mapped sea-floor areas at depths as great as 70m, and observers in special submarine chambers have been lowered to the deepest known places in the ocean.

Because of this intensive research, involving many nations, the oceans are gradually giving up their mysteries. The romanticist that is in each one of us cannot help regretting the loss of so much that was built up through more than three thousand years of

Figure 12.1 The oceanographic research vessel *Glomar Challenger,* designed for drilling into the sea floor in the deepest parts of the ocean. The derrick amidships lowers and withdraws drilling rods, a pile of which lie on the rack in front of the derrick. (*Courtesy of Global Marine, Inc., which operates the vessel under a contract with Scripps Institution of Oceanography, University of California, San Diego, California.*)

human legend: the singing mermaids, the strange and threatening gods of the sea, the fabled cities and castles believed to have sunk into watery deeps, the monsters of seafarers' tales. Poetic mysteries are fading away under the hard glare of knowledge. But on the other hand the same knowledge should help us protect the fragile values of an ecosystem that covers more than two-thirds of Earth's surface and that is responsible for a large part of Earth's biological heritage.

Dimensions. The sea covers 71 per cent of Earth's surface and is therefore more than twice as extensive as the land. The 29 per cent that is land is not evenly distributed. In the Northern Hemisphere, often called the land hemisphere, 40 per cent of the surface is land, while in the southern hemisphere only 20 per cent is land. As we shall see, the uneven distribution of land plays an important part in determining the paths along which water circulates in the ocean.

The greatest ocean depth yet measured is slightly more than 11km, near the island of Guam in the western Pacific. If Mount Everest were dropped into the sea at that point, 1.6km of water would cover its peak. The average depth of the sea, however, is about 3.8km, compared to an average height of the land of only 0.75km.

Knowing the area of the sea and its average depth, we can roughly compute the present volume of seawater to be 1350 million km³. We say *present* volume because, as we read in Chap. 11, the volume fluctuates slightly with the growth and melting of glaciers. But if we consider the entire hydrosphere (Chap. 2), the volume of water appears to be constant.

Age and Origin. The world ocean is very old. Earth's oldest rocks include water-laid sedimentary strata similar to those we see being deposited today. Therefore, we are sure that as far back in history as we can see, Earth has been provided with large bodies of water. Indeed, water has probably been present at the surface almost as long as Earth has existed as a solid body.

Where did the water come from? It seems to have been baked and sweated out of minerals in the crust and mantle, arriving at the surface as steam during volcanic eruptions. The process has been gradual; so the world ocean enlarged through time. The enlargement seems now to have stopped. Water is, of course, slowly removed from the hydrosphere by being buried along with sediment, and so eventually becomes incorporated into igneous and metamorphic rock. On the other hand water is added to the hydrosphere by volcanoes and by the slow compaction of sedimentary strata. As the subtraction and addition of water appear to balance out, we can say that the hydrosphere is now in a steady state.

Composition. About 3.5 per cent of average seawater, by weight, consists of dissolved mineral substances (Table 12.1). As everyone knows, this percentage is enough to make the water undrinkable. It is enough also, if precipitated, to form a layer of solid salts about 56m thick, over the entire sea floor.

The measure of the sea's saltiness is termed *salinity.* We commonly express salinity in parts per thousand, rather than in per cent (parts per hundred). Average seawater therefore has a salinity of 35 parts per thousand. The principal elements that contribute to the salinity of the sea are sodium and chlorine; when seawater is evaporated, more than 75 per cent of the dissolved matter is precipitated as common salt (NaCl). But seawater contains all of Earth's other elements as well—most of them in such low concentrations that they can be detected only by super-sensitive analytical devices. As we can see in Table 12.1, more than 99.9 per cent of the salinity is caused by only nine ions.

Where do the dissolved salts come from? Probably many of them have been in the sea from the time water first formed on Earth; sweated out of the crust and mantle along with the water accumulating in primeval oceans. But rivers and streams are still continually adding new materials in solution at the rate of 2.5 billion tons each year. Through millions of years, the amount of dissolved salts added by rivers has far exceeded the salts now dissolved in the sea. Why, then, isn't the sea more salty than it is? The reason is becoming more and more familiar as we deal with geologic processes. The composition of the ocean likewise is in a steady state, so that material is being removed at the same rate at which it is added. Some of the elements, such as silicon, calcium, and phosphorus, are withdrawn from seawater by aquatic plants and animals in order to build shells and skeletons. Other elements, such as potassium and sodium, are removed by clays and other minerals as they slowly settle to the sea floor. Still others, such as copper and lead, are precipitated as sulfide minerals in claystones and mudstones rich in organic matter. The net result of all these processes of extraction, taken together with what is being added, is that the composition of seawater remains constant.

Movements of Seawater

The restless sea is always in motion, at generally slow rates to be sure, though in places at velocities comparable with swift rivers. Although there are several different immediate causes of motion, the chief motive power is solar energy, with some help from the energy of Earth's rotation and from the gravitative pull of Sun and Moon. Motion consists of several kinds of currents and waves. We can conveniently group them in this way:

A. Geologically important along coasts:	1. Waves
	2. Longshore currents
	3. Tidal currents
B. Geologically important in the deep oceans:	4. Surface ocean currents
	5. Density currents

Waves and Longshore Currents. Waves and longshore currents are the movements most easily seen. When we stand on a beach, we are immediately aware of the ceaseless action of waves that wash the shore. If we watch closely, we might also see floating debris moving parallel to the shore, indicating that a longshore current is present.

Table 12.1

The Major Constituents of Seawater (Constituents Are Listed as the Principal Ions in Solution as an Indication of the Kinds of Compounds That Form When Seawater Evaporates)

Ion	Percentage of All Dissolved Matter
Chloride (Cl^{-1})	55.07
Sodium (Na^{+1})	30.62
Sulfate (SO_4^{-2})	7.72
Magnesium (Mg^{+2})	3.68
Calcium (Ca^{+2})	1.17
Potassium (K^{+1})	1.10
Bicarbonate (HCO_3^{-1})	0.40
Bromine (Br^{-1})	0.19
Strontium (Sr^{+2})	0.02
Total	99.97

Waves and longshore currents are the chief agents by which the sea erodes a coast and deposits rock material along the shore. So widespread and distinctive is their geologic work that they are discussed separately in Chap. 13.

Tidal Currents. Tidal currents are caused by the twice-daily tidal bulges that pass around the Earth (Chap. 3). In the open sea, tides are small and do not cause currents, but in bays, river estuaries, straits, and other narrow places, tides can cause rapid currents. As a tidal bulge in the open ocean approaches a restricted inlet, it becomes confined and the water rises rapidly. Tidal flows reach 25km per hour in places, and tidal heights of more than 16m are known (Fig. 12.2). Such fast-moving currents, though restricted in extent, readily move sediment around.

Surface Ocean Currents. *Surface ocean currents* are *broad, slow drifts of surface water*. They are set in motion by the prevailing surface winds. Air that flows across a water surface causes waves, but it also drags the water slowly forward, creating a current of water as broad as the current of air but rarely more than 50 to 100m deep. The marked effect of winds on the ocean is evident when we compare a map of surface ocean currents (Fig. 12.3) with the positions of the belts of prevailing winds (Fig. 10.2). In the low latitudes surface seawater moves westward with the trade winds. The trade winds blow constantly toward the equator, from the northeast in the Northern Hemisphere and from the southeast in the Southern Hemisphere. The overall westerly direction of the North and South Equatorial Currents (Fig. 12.3) is reinforced by Earth's rotation. Both north and south of the equator, the westerly

A

B

Figure 12.2 The Bay of Fundy, between Nova Scotia and New Brunswick, experiences an extreme tidal range. *A.* Hall's Harbor, Nova Scotia, at high tide. *B.* The same place at low tide. (*Russ Kinne, Photo Researchers.*)

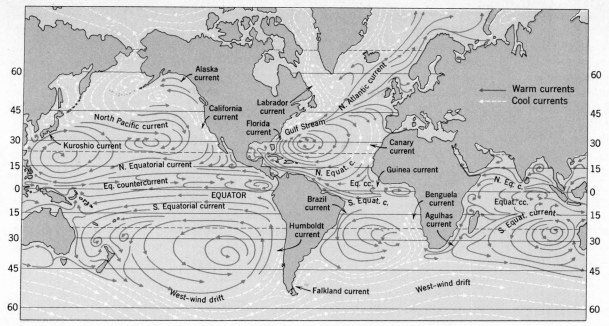

Figure 12.3 The surface currents of the ocean form a distinctive pattern, curving to the right in the northern hemisphere and to the left in the southern. (*Modified from a map by U.S. Navy Hydrographic Office.*)

moving currents eventually become movements of rotation. Rotation is caused partly by deflection when a current encounters a coast, and partly by the tendency of moving objects to veer to the right in the Northern Hemisphere and to the left in the southern.

In the northern Atlantic, then, the North Equatorial Current flows west. Deflected water flows north and is represented by the Florida Current, the Gulf Stream, and the North Atlantic Current. The currents transfer warm equatorial waters to higher latitudes and therefore have considerable effect on climates. Part of the North Atlantic Current eventually moves into the Arctic Ocean, taking heat with it, and part is deflected south along the European-African coast as the Canaries Current. This water has by then lost so much heat that it has become cooler than the surrounding tropical water. Approaching the equator, it has completed its circulation and once more begins to be dragged westward by the trade winds.

We can follow a similar pattern in the northern Pacific Ocean, where we have this sequence: North Equatorial Current, Kuroshio Current, North Pacific Current, and the cool California Current. In the southern hemisphere we find similar great circular movements of surface seawater, but the direction of rotation is counterclockwise, while north of the equator it is clockwise. There is another major difference between circulations north and south of the equator. The far-southern oceans are not impeded by continents, and this makes possible a major globe-circling movement of water, the West-wind Drift. The West-wind Drift is an effective way of moving water from one ocean to another and is therefore important in the mixing of ocean waters, a process that is completed every 1800 years.

Although rates of movement are generally slow, in narrowly confined areas surface ocean currents become rapid. In the narrow strait between Florida and Cuba the rate approaches 3 mph.

Density Currents and Circulation in the Deep Ocean. While the great currents set up by winds are moving slowly through a shallow surface zone of water, deeper—and even slower—circulation of water is also occurring, set up by differences in density. A *density current* is *a localized current, within a body of water, caused by dense water sinking through less-dense water.* The density of seawater increases as the water gets colder or more saline. Dense water tends to sink, displacing less-dense water below.

In polar regions surface water becomes chilled while in contact with the atmosphere. Also it acquires increased salinity by the formation of sea ice. Ice is nearly pure H_2O, and as it forms, the salts remain in solution in the residual seawater, which becomes more saline. As polar water grows denser by cooling and formation of sea ice, it sinks, then slides slowly along the sea floor toward the tropics. Deep, cold polar water may even cross the equator into the opposite hemisphere.

In the Atlantic Ocean four clearly defined masses of cold polar water have been identified (Fig. 12.4). Water in the Gulf Stream and the North Atlantic Current is highly saline as a result of evaporation in low latitudes. When it reaches the Arctic region, it becomes chilled and sinks to the bottom. From the North Atlantic this cold, deep water again flows south, almost into the Antarctic region, before it is obscured by mixing and again approaches the surface.

In the Antarctic region, water is chilled during the winter, and at the same time it becomes saltier through the formation of sea ice. The cold, dense brine sinks to the floor of the ocean and flows northward, crossing the equator and reaching intermediate northern latitudes before being displaced by the cold North Atlantic water. Other water masses form in similar ways. Antarctic Intermediate Water (AIW in Fig. 12.4), forms by chilling of the highly saline waters of the Brazil Current. The Mediterranean Sea is an enclosed basin with a high rate of evaporation. Mediterranean surface water becomes very saline, sinks, and eventually flows, as a deep current (MW in Fig. 12.4), through the Strait of Gibraltar and down into the Atlantic. To counterbalance the flow of deep water out of the Mediterranean, a fast-moving surface current flows in through the Strait of Gibraltar. So, because of differences in density between two great water bodies, the Strait of Gibraltar carries two currents, one above the other, flowing in opposite directions.

Deep circulation in the Pacific and Indian Oceans differs from Atlantic circulation in that all the deep water comes from the Antarctic region. There is no large source of deep, cold water in the northern Pacific, because a shallow barrier at the Bering Strait prevents deep Arctic water from breaking through. It is possible for deep water originating in Antarctica to flow as far north as California and Japan.

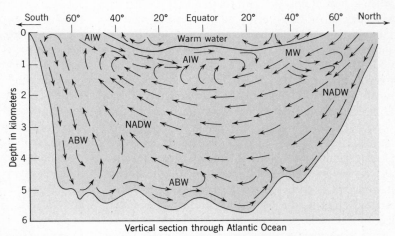

Figure 12.4 Circulation of deep ocean water in the Atlantic Ocean shown along a north-south plot. ABW is Antarctic Bottom Water, dense, cold water that sinks in the Weddell Sea and slowly flows north to form the cold bottom water throughout most of the Atlantic Ocean. NADW is North Atlantic Deep Water, which sinks in the North Atlantic off Greenland, and eventually flows far south, where it meets Antarctic Intermediate Water (AIW), cold, near-surface water flowing north. MW is dense, saline water flowing into the Atlantic from the Mediterranean Sea. (*After K. K. Turekian, 1968.*)

Figure 12.5 Part of the ocean floor in the mid Pacific, as it would appear if the water were drained away. The two flat-topped mountains in the distances are truncated volcanoes called guyots. In the foreground is a valley cut into a seamount, an extinct volcano which differs from a guyot in that it is not flat topped. The top of the seamount is out of view behind the observer. (*From a painting by Chesley Bonestell in E. L. Hamilton, Geol. Soc. America, Mem. 64, 1959, pl. 1.*)

Shape of Ocean Basins

The sea floor is far from smooth. It is just about as irregular as the surfaces of the continents. Long mountain chains, isolated hills including volcanic cones, broad featureless plains, great escarpments, deep basins, and canyons mark the ocean floor as they do the land. Figure 12.5 is an artist's conception of a bit of the Pacific Ocean floor, minus the ocean. The terrain is rugged, the slopes steep, and the mountain relief rather like what we see on land. In Fig. 12.6 we have an ocean basin and a familiar continent shown in profile and on the same scale. The horizontal scale is, of course, compressed and the vertical scale exaggerated, but comparing the profiles, we can see there is not much to choose between them as far as size and number of hills, valleys, and other topographic features are concerned. Nevertheless, although the ocean-floor and land-surface topographies look similar, the two prove to be very different when examined in detail.

Continental Margins. A convenient place to start a discussion of topographic features is the margins of our familiar continents. Continental margins are covered by the sea; in a sense the ocean basins are brimful of water, which slops over and floods the edges of the continents. Looking ahead to Figs. 12.8 and 13.16, we can see examples of *the shallow, nearly flat, submerged borders of the continental masses* which are called the **continental shelves.** The average width of the shelves is about 60km, but there are many local variations, from almost zero to a width of 1300km. Along the Pacific coast of South America a shelf is missing altogether. Off the Pacific coast of North America it is only a few kilometers wide, but off the north coast of Siberia it reaches the maximum known width of about 1300km.

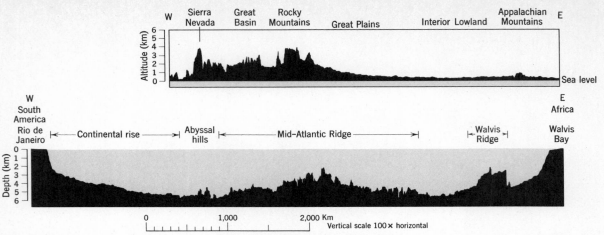

Figure 12.6 Accurate profiles show that the topography of the Atlantic Ocean basin is fully as rugged as the surface of North America. (*After Shepard, 1963.*)

The surface of the shelf is featureless compared to the floors of the deep oceans, but it is not completely smooth. Waves and longshore currents scour shallow depressions in it. Coastal dunes and other shore features have been flooded and submerged by a rising sea level and now form submarine hills. Piles of ice-rafted boulders form undulating surfaces. But the most striking features of the shelf are long valleys, some of them 100km or more in length, that run from the coast out into deeper water. These valleys are seaward extensions of large river valleys on the adjacent continent. They are believed to have been made during glacial ages, when sea level was lowered so much that shelves were exposed to erosion by streams. Off the mouth of the Hudson River, for example, a valley at least 50m deep crosses the shelf.

The origins of continental shelves and of most of the features found on them are closely tied in with wave-, longshore-, current-, and other erosive forces that work on the continents. Therefore we defer further discussion of continental shelves to Chap. 13.

The seaward edges of the shelves are covered everywhere by at least 100m of water, and in places by as much as 600m. We do not fully understand why the depth varies in this way, but the edge of a shelf is not defined by water depth. It is defined by a marked steepening of slope. As we can see in Figs. 12.8 and 13.16, the shelf passes rapidly into the **continental slope,** *a pronounced slope beyond the seaward margin of the continental shelf,* leading down into deep water. Continental slopes mark the outward edge of the continental crust where it abuts against the oceanic crust.

Many continental slopes, like the one shown in Fig. 13.16, grade downward and outward into a region of gentler slopes. This region is a vast, gently sloping pile of sediment, consisting of waste derived from the adjacent continent and from sedimentary debris that slumps down from the shelf and slope above.

The surface of the continental slope, like the surface of the shelf, is rather smooth, with a single exception. Cutting into the continental slopes are many remarkable valleys, some so deep that they have been called submarine canyons. These valleys, as much as 1km deep and with steep side slopes, are widely spread around the world. Their origins are still a puzzle. The valleys extend down to water depths of 3km or more and so could not have been excavated by ordinary rivers, even at times of lowered sea level. Some, such as those off the mouths of the Hudson, Congo, and Ganges Rivers, line up with valleys that cross the continental shelves, while others seem to have no connection with such valleys. One clue to the origin of the canyons consists of vast, fan-shaped piles of sediment on the deep-sea floor at and seaward from canyon mouths (Fig. 12.7). It seems likely that sediment deposited on a shelf sometimes becomes unstable, and gravity causes it to slump, forming a turbid, muddy mass that races down the slope and out onto the deep-sea floor. These masses move so fast that they can erode effectively, in time scouring out great canyons. The tur-

Figure 12.7 A great fan-shaped wedge of sediment lies on the ocean floor off the mouth of the Ganges River in India. Sediment transported by the Ganges is first deposited on the shelf, then transferred to the deeper ocean floor by slump and turbidity currents. (*After B. C. Heezen and M. Tharp, "Physiographic diagram of the Indian Ocean," published by Geological Society of America, 1964.*)

bid masses are believed to be identical with the turbidity currents discussed later in the chapter.

Ocean Basins. Beyond the immediate edges of the continents lie the deep ocean basins. These possess unique topographic forms, of which the most important are these:

Oceanic ridge, a continuous rocky ridge on the ocean floor, many hundreds to a few thousand kilometers wide, with a relief of 600m or more. (Also called mid-ocean ridge and oceanic rise.)

Sea-floor trench, a long, narrow, very deep basin in the sea floor.

Seamount, an isolated volcanic hill, more than 1000m high, rising above the deep-sea floor. Most seamounts have a conical shape.

Guyot, a seamount with a conspicuously flat top well below sea level.

Abyssal plain, a large flat area of deep-sea floor having slopes less than about 1m per km.

These features, together with the adjacent continental margins, are shown in Fig. 12.8. In it, the most striking feature is the Mid-Atlantic Ridge, a sinuous mountain range crossed by great fractures. Many of the fracture zones form deep canyons in the ridge. The Mid-Atlantic Ridge is part of a world-encircling system of oceanic ridges (Fig. 12.9), and as we shall see later, is the site where new magma wells up from the mantle below and continuously forms new oceanic crust.

Sea-floor trenches are the sites of the greatest ocean depths. They are as much as 200km wide and are 25,000km or more in length. Trenches mark the places where moving plates of lithosphere plunge back into the mantle. Most occur in the Pacific Ocean (Fig. 12.10). Some trenches, such as the Aleutian and Kermadec-Tonga Trenches, are far from continental margins. Others, such as the Puerto Rico and Peru-Chile Trenches, are immediately adjacent to continental margins. Sea-floor trenches, as we shall see in Chap. 18, are a vital link in the interactions that continuously occur between Earth's internal processes and her external ones.

Seamounts and guyots (Fig. 12.5) are drowned, isolated submarine volcanoes. Nearly 2000 have been discovered so far, mostly in the Pacific Ocean, but oceanographers believe that as many as 20,000 or more remain to be discovered. Guyots are thought to have had their tops eroded by the action of surf along the coasts of islands; in other words the flat tops are wave-cut benches. Because most of the tops are now covered by at least 1km of water, they must have sunk below sea level after they were eroded.

The deeper parts of the ocean floor are cut by many submarine valleys. Their origins remain mysteries. Some run for more than a thousand kilometers and are so deep that they are spoken of as canyons. One, the Mid-ocean Canyon, which runs down between Canada and Greenland (Fig. 12.8), is more than 2000km long. Others seem to be continuations of valleys on the continental shelf and slope, and fan out onto the abyssal plains. It may be that currents of turbid, muddy water cut valleys farther out on the ocean floor than we thought was possible, or perhaps processes we still know nothing about formed the valleys.

The oceanic ridges, the fracture zones that cut them, and the sea-floor trenches, seamounts, and other topographic features of the ocean basins are

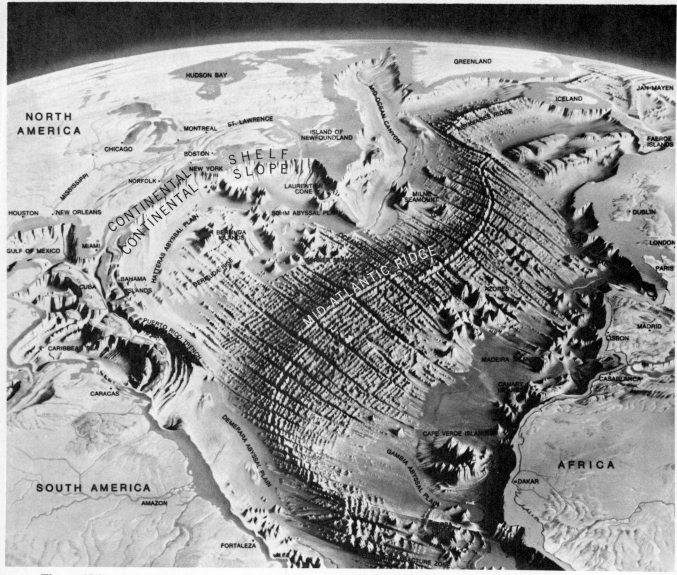

Figure 12.8 The ocean floor is a place of varied topography. This artist's view shows the features we would see in the Atlantic Ocean basin if all the water were removed. Slopes appear steeper than they actually are because the vertical scale has been exaggerated for emphasis.

Along the center of the basin is the Mid-Atlantic Ridge, a great chain of volcanic mountains, in places broken and offset by huge fractures. The fractures are zones of intense breaking of the crust; they form steep-walled valleys. Away from the ridge lie seamounts, a few of which reach up above sea level and form islands. Farther away are abyssal plains, smooth-floored parts of the ocean floors that are bounded by the continental slopes and nearly flat continental shelves. (*From a painting by Heinrich Berann; courtesy of the Aluminum Company of America.*)

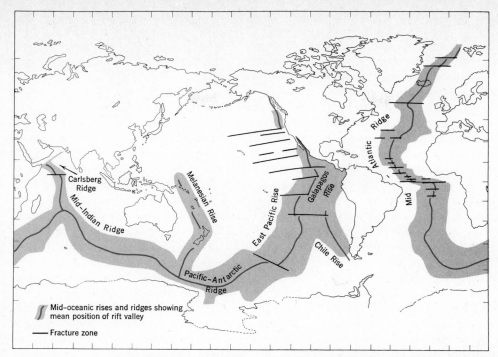

Figure 12.9 The world-encircling system of oceanic ridges have been offset by major fracture zones (black). (*After B. C. Heezen, 1962.*)

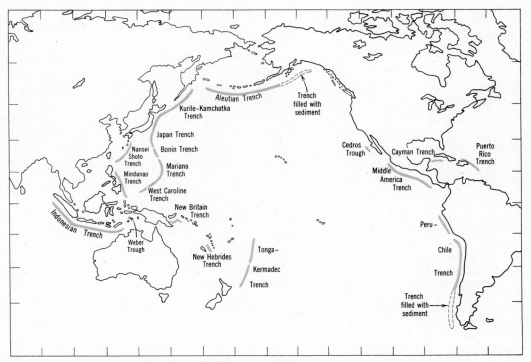

Figure 12.10 Sea-floor trenches are found mainly in the Pacific Ocean. They mark the places where plates of lithosphere plunge back into the mantle.

part of a vast, interconnected system. We will discuss their significance in Chap. 18, together with the connections between major continental and oceanic features. But before we leave them, let us repeat what was said in Chap. 1: oceanic ridges are lines along which new basaltic magma wells up to form new oceanic crust. The magma solidifies into new rock, which is added to the growing edges of plates of lithosphere. To accommodate the continuously added new rock, the lithosphere plates must move slowly away from the oceanic ridges. The farther we are from an oceanic ridge, therefore, the older the igneous rock we find on the sea floor.

But, in addition, the sea floor is mantled by a carpet of sediment that has fallen down through the water or that has slid or flowed down the continental slopes. Deep-sea drilling near an oceanic ridge reveals only young sediments that overlie young igneous rock. Drilling farther away from the ridge reveals thicker and older piles of sediment. Sea-floor sediments therefore tell the same story that is told by igneous rock: the age of the sea floor increases as we progress outward from an oceanic ridge. Because the sediments are such an important part of the ocean basins, we must look at them more closely.

Sediments on the Sea Floor

Sampling. Systematic study of the carpet of sediment that mantles the sea floor dates from the 1870s, when the British scientific ship *Challenger,* on a history-making cruise, collected a vast number of samples at carefully chosen positions. So long and thorough was the resulting study that the published reports were not completed until eighteen years after the cruise ended. Most of the samples were obtained by scoops dragged along the bottom, or by small clamshell buckets that take a bite of sediment and then snap shut.

Since the classic cruise of the *Challenger,* other expeditions of various nationalities have added greatly to the take of samples. Modern improvements in sampling came about through the use of *coring devices.* One of these is a long metal tube, let down on a cable and then forced into the sea floor by one of several mechanisms. The open lower end of the tube closes and brings up a core sample in which all except the topmost layers of sediment are undisturbed. Cores as much as 25m long have been obtained. Another coring device is a drill capable of

cutting into the sea floor at a water depth of 3000m. Within the last few years ships such as *Glomar Challenger* (Fig. 12.1, a second and more modern *Challenger*) using this drilling technique have been able to extract cores as much as 500m long.

Kinds and Distribution of Sediments. Minute analyses of samples brought up by the coring device have made it possible to sort out the various sources from which sea-floor sediment is derived. General sources are listed in Table 12.2: they are the land, the sea, and sources outside the Earth. Table 12.2 shows sources only, not the sediments themselves. This is because study of great numbers of samples indicates clearly that *all* the sediments are mixtures; no one body of sediment comes entirely from a single source. Therefore we have to classify the sediments according to their chief constituents, the predominant kinds of material they contain. Seven principal kinds are described in Table 12.3, but all of them are mixtures and grade from one into another. Figure 12.13 is a map on which the distribution of the seven different kinds of sediment is shown. The scale of the map is so small that the sediment areas are greatly generalized. If we compare the areas of various sediments in the western North Atlantic with the detailed topography in Fig. 12.8, we can see that a detailed map of sediment distribution would be much more complex. Even so, it could not be as accurate as a map of a comparable land area, because the samples on which it must be based are taken from points very far apart.

The sampling apparatus put down by exploring ships sometimes hits not sediment but bare, hard rock. Comparison with topographic records shows that some of the rocky places are cliffs and other

Table 12.2
Origins of Sediment on the Sea Floor

1. Terrigenous sediment (Lat. "derived from the lands"). Sea-floor sediment derived from sources on land. Contributed by (a) rivers, (b) erosion of coasts by waves, (c) wind (clay and silt, including volcanic ash), (d) floating ice.
2. Pelagic sediment (Gr. "belonging to the deep sea"). Sediment, on the deep-sea floor, consisting of material of marine organic origin. Shells and skeletons, mostly microscopic, of marine animals and plants.
3. Sediment derived from submarine volcanoes. Volcanic ash.
4. Extraterrestrial sediment (derived from outside the Earth). Meteorite particles, mostly microscopic.

Table 12.3
Classification of Kinds of Sediment on the Sea Floor

1. Terrigenous sediment. Mainly on the continental shelves, continental slopes, and abyssal plains. Mud, sand, and gravel, varying greatly from place to place.
2. Glacial-marine sediment. Terrigenous sediment, including unsorted mixtures of particles of all sizes, dropped onto the sea floor from floating ice.
3. Sediment displaced by gravity. Mainly terrigenous sediment, originally deposited on the continental shelf and slope, that has moved to the deep ocean floor under the influence of gravity, by gliding, slump, or flowing.
4. Brown clay (also called pelagic clay) (Fig. 12.11,A). Confined to the deep-sea floor, mostly in high latitudes or at depths greater than 4000m. Contains, by definition, less than 30 per cent calcium carbonate. Chief constituents are clay minerals, quartz, and micas. Since these are the sorts occurring in weathered soils, volcanic ash, and fine wind-blown material, they are thought to come from such sources. The clay is brown as a result of gradual oxidation during the very slow process of deposition.
5. Calcareous ooze (Fig. 12.11,B). Contains, by definition, more than 30 per cent calcium carbonate, most of it consisting of shells and skeletons. Confined to regions in which surface water is warm and surface organisms exist in myriads; the resulting shells, falling like snow, accumulate on the bottom more rapidly than do the inorganic-clay particles. Because it contains much carbon dioxide, deep-sea water dissolves calcium carbonate. As the shells drift slowly down, they are gradually dissolved, but only at depths of more than 5000m are they completely consumed. Hence calcareous ooze rarely occurs at such depths.
6. Siliceous ooze. Contains a large percentage of skeletons built of opaline silica. Occurs where organisms with calcareous shells are few in the surface waters and in areas in which such shells are destroyed by solution before they reach the sea floor.
7. Authigenic materials. *Authigenic* means sedimentary deposits *formed in place*. These have not been transported physically. They consist of minerals that crystallized from seawater. The principal authigenic deposits are nodular growths of manganese minerals growing on the sea floor (Fig. 12.12).

steep slopes, from which any accumulating sediment would be expected to slide off. But still others are flat surfaces, and we are not yet sure why they have no cover of sediment. Some such surfaces might be recent lava flows. Others, perhaps, have been scoured bare by currents crossing the floor of the deep ocean.

Sediments on the Continental Shelves. Much sampling has shown that the sediments on the continental shelves are coarser on the average than those on the deep-sea floor. They include much sand and silt, and a large proportion of them have come from the land. Evidently the material was in part contributed by streams and in part worn from coasts by waves, as will be seen in Chap. 13. Theory derived from analogy with the sediment loads of streams and the wind leads us to suppose that sediment is carried offshore in suspension, and is deposited in water so deep that waves and longshore currents cannot pick it up again. Theory suggests also that the diameters of particles should decrease with increasing distance from the shore.

Sampling on some shelf areas confirms the theory. On other areas, however, the distribution of particle sizes is patchy, with coarse sediment, which normally belongs near shore, occurring far offshore near the seaward limits of the shelves. The patchy distribution is believed to be partly the work of localized currents that deposit coarse sediment at some distance from shore. But to a greater extent it is the result of change of sea level. When we remember that during each glacial age sea level was lowered, we can see that the shoreline must have migrated seaward across the shelves, exposing new land. Patches of coarse sediment, deposited near shore or even on the land itself, could then have been submerged as the sea level rose again.

Climate and Chronology Indicated by Deep-Sea Sediments. The sediments recovered in cores from the deep-sea floor reveal an interesting record of changes in the Earth's climate. A common characteristic of many of the cores taken from tropical regions is an alternation of layers of sediment of two kinds. One is calcareous ooze, which contains the shells of minute organisms that today live only in warm surface water. The other is brown clay, which contains fossil organisms that today live in higher, cooler latitudes, and lacks the warm-water kinds. The alternation of "warm" and "cool" sediments has led to the inference that the layers are a record of nonglacial and glacial ages respectively. Again, cores taken right across the floor of the North Atlantic, from Canada to Britain, consist of alternating layers of (1) calcareous ooze, and (2) silt, sand, and small pebbles believed to have been dumped from hosts of melting icebergs. Therefore these cores too suggest alternating nonglacial and glacial ages. So many other cores from widely separated parts of the sea floor contain alternating layers of various kinds that

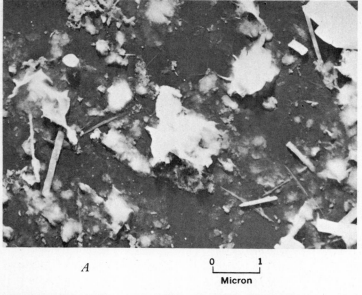

A

0　　　　1
Micron

B

0　1
mm

Figure 12.11 Common materials of deep-sea sediment. *A.* Crystals and particles of various minerals, mostly clay minerals, seen with enlargement about 10,500 times natural size. Brown clay from 1.68m beneath floor of Pacific Ocean, at a depth of more than 4,500m, at a point about 1450km west of Mexico. (*U.S. Geol. Survey.*) *B.* Microscopic calcareous shells of tiny aquatic animals called Foraminifera. Recovered from the floor of the Caribbean Sea 160km west of Martinique, at a depth of 880m, the animals originally lived in surface waters, then sank to the bottom when they died. (*Yale Peabody Museum.*)

fluctuating climate is thought to have been responsible for them. Deep-sea sediments therefore are valuable indicators of world-wide climate changes.

Also it has been found possible to determine the dates of at least the higher, younger layers of deep-sea sediment by measurement of ^{14}C and other radioactive isotopes contained in them. In this way the beginning of a calendar has been established. The record has been extended backward in time by use of the magnetic-field reversals described in Chap. 3. As we read in that chapter, magnetic minerals align themselves parallel to Earth's magnetic field as they settle through the water. Because, in the oceans, sediment is deposited continuously, careful measurements in a core reveal the exact layer at which each magnetic reversal occurs. Since we know the times of reversal (Fig. 3.16), we can readily construct a calendar of the layers in the core.

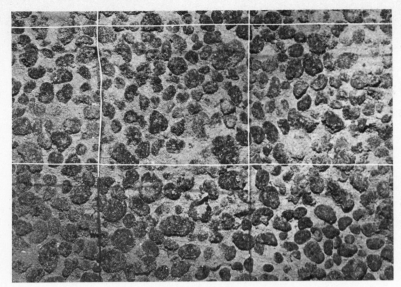

Figure 12.12 Nodular deposits of manganese- and iron-oxide minerals are widely distributed on the deep-sea floor. This photograph, made on the floor of the Pacific Ocean, shows a device used in estimating the abundance and size of nodules for the purpose of mining them. Size of grid 1 foot. (*Photograph courtesy of Global Marine, Inc.*)

As for the sedimentary layers that lie beneath the upper 25m of sediment commonly sampled by core tubes, we have two limited sources of information. Waves generated by earthquakes traverse the crust beneath the sea floor and by suitable recording devices on boats and on land, reveal that sediments up to a thousand meters or more are commonly present. Also direct samples have been obtained from many deep-water sites by drilling conducted from *Glomar Challenger.* In this fashion sedimentary layers as old as the Eocene Epoch have been recovered. But older sediments have not been found. Scientists once thought the oceans contained sediments from Earth's earliest history. Drilling into the deep-sea floor proved this to be incorrect, confirming the theory that plates of lithosphere, together with their mantle of sediment, are continually destroyed when they plunge down through the trenches. The sea floor is therefore being continually renewed and the older sediments continually destroyed.

Special Features and Problems

Some features of oceanic topography and sediments do not fit neatly into the foregoing discussion. They involve both topography and sediments, and because they are among the most unusual features of the sea floor, we discuss them separately.

Turbidity Currents. Among the fascinating problems of sea-floor geology are how coarse sediment can be deposited at depths of 4 to 5km and how abyssal plains are created. The solution to these problems seems to lie, at least in part, in ***turbidity currents,*** *density currents whose excess density results from suspended sediment.* Let us take a look at them.

In 1935, soon after Hoover Dam on the Colorado River, downstream from the Grand Canyon, had been completed, engineers were surprised to find that from time to time the clear water discharging through the lowest outlet pipes became muddy. Also during the 1930s researchers at the California Institute of Technology were pouring streams of water, densified by dissolved salts or suspended silt and clay, into a water-filled tank in their laboratory. They found that the dense water, with kinetic energy acquired by flowing down the front of a delta, possessed enough momentum to travel along the bottom on an extremely small slope, past the clear water above (Fig. 12.14). Energy not expended in friction

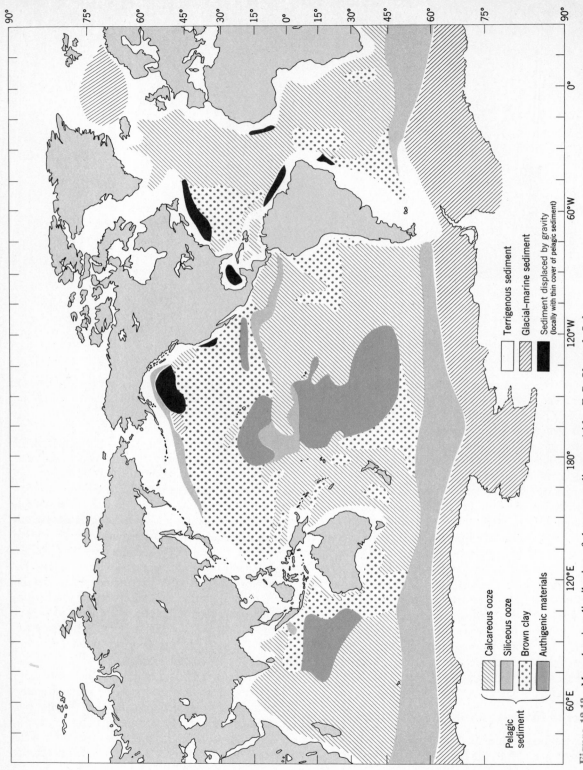

Pelagic
sediment

Calcareous ooze

Siliceous ooze

Brown clay

Authigenic materials

Terrigenous sediment

Glacial–marine sediment

Sediment displaced by gravity
(locally with thin cover of pelagic sediment)

Figure 12.13 Map showing distribution of deep-sea sediments. (*After F. P. Shepard, Sub-marine Geology, 2d ed.: 1963. Fig. 198. With permission of Harper & Row, Publishers, New York.*)

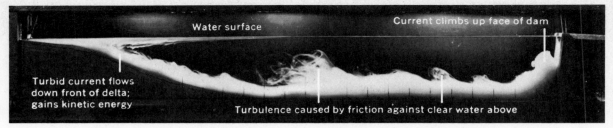

Figure 12.14 Turbidity current flowing from left to right, seen through the glass wall of a water-filled laboratory tank. (*U.S. Conservation Service, California Institute of Technology.*)

en route even enabled the current to climb part way up the face of a dam at the far end of the tank. The current had distinct boundaries, like a cloud of dust; in fact, someone described it as "a dust storm under water." Indeed, it is essentially the same thing, mechanically, as a dust storm.

It was realized that such a current, climbing part way up Hoover Dam, could explain the muddy water appearing in the discharge pipes. Muddy water entering the reservoir, far upstream from Hoover Dam, formed turbidity currents that swept along the floor of the reservoir until they reached the dam, rose up the side, and flowed out through the discharge pipes. Observations in this and other reservoirs confirmed that the suspended loads of rivers were the source of the currents of turbid, dense water, which were then named *turbidity currents.*

As a result of more laboratory experiments, it was learned that turbidity currents could perform a surprising amount of erosion. The densities of natural turbidity currents on gently inclined lake floors are rarely as great as 1.02, and their velocities are less than 30cm per second. But greater densities and velocities have been produced in the laboratory, and a submarine turbidity current, flowing down a continental slope with an inclination of 40m per kilometer, should develop a velocity greater than that of the swiftest streams on the land. In theory, a current with a density of 1.5, moving at about 1.6km per hour, could move a rock particle 14,000 times the diameter of the largest particle that can be moved by clear water at the same velocity.

Turbidity Current on a Continental Slope. Turbidity currents, then, have been observed in lakes, can be created in the laboratory, and, according to theory, have enormous potential for the transport of sediment. But do they occur in the sea? In 1952, re-examination of the record of an old earthquake suggested strongly that they do.

On November 18, 1929, a severe earthquake on the continental slope off Nova Scotia broke 13 transatlantic cables in 28 places. At the time it was supposed that the breaks were the direct result of buckling or other movements of the sea floor, the usual cause of submarine earthquakes. Turbidity currents were then unknown, and the event was lost sight of. But many years later the whole story, supported by a thick file of measurements made by the repair ship, was reviewed and studied. There were two odd things about the cable breaks. First, although all the cables on the continental slope and deep-sea floor were broken, not one of the many cables that crossed the continental shelf was damaged. Second, the breaks occurred in sequence, in order of increasing depth, over a period of 13 hours and through a distance of 480km from the earthquake center. Repair ships found that each cable had been broken at two or three points more than 160km apart. The detached cable segments, between the breaks, had been carried part way down the continental slope or buried beneath sediment beyond the bottom of the slope.

The whole event was like a huge laboratory experiment with times and distances controlled by measurement. Each break was timed by the machines that automatically record the messages transmitted through the cables, and was accurately located after the quake by the electric-resistance measurements always made to enable repair ships to locate breaks and repair damage.

The only hypothesis yet formulated to explain all these facts is that the quake set off great submarine slides on the continental slope. The slides quickly became turbidity currents, which flowed down the slope, breaking each cable as they came to it. Eight cables on the slope were broken instantaneously at the moment the quake occurred, and five others were broken at times ranging from 59 min. to 13 hr. 17 min. after the quake (Fig. 12.15). The area affected was about 320km wide and at least 650km long.

From the times of the breaks and the distances between them the velocities of the inferred turbidity currents could be calculated. Velocities ranged from 93km per hour at the toe of the continental slope down to 22km per hour at the latest and farthest break. Even at the latter point the velocity of the current was four times that of the lower Mississippi River.

Theoretical calculations suggest that near the earthquake area velocities may have approached 135km per hour and that the heights of the turbidity currents may have reached 300m. As the currents lost energy, the sediment eroded from the continental slope was dropped onto the ocean floor, burying some of the long, broken pieces of cable beneath deposits of "sand and small pebbles," the bottom sediment reported by the repair ships. Two core samples collected from the floor, downslope from the cable breaks, show that the floor is blanketed with a layer of silt and muddy sand in the form of a graded layer (Fig. 14.4). In one core the layer is 70cm thick; in the other it is 128cm.

As pointed out in Chap. 14, a graded layer is the result of rapid, continuous loss of energy in the transporting agent. This would occur in a spent turbidity current, but it is not expectable in the other sorts of marine currents we have described.

So, although a turbidity current on the continental slope off Nova Scotia on November 18, 1929, is not proved, it is the only hypothesis that seems able to explain all the facts. The record shows that similar events have occurred at least twice at each of 40 localities around the world within the past 75 years. Some were related to earthquakes; others occurred off the mouths of large rivers, suggesting that the inferred turbidity currents were set off by stream floods. This evidence is impressive. It leads us to believe that turbidity currents are effective agents on continental slopes, despite the fact that no one has yet seen such a current in actual operation in the sea.

Origin of Abyssal Plains. If sediment is carried down a continental slope by turbidity currents, it should be deposited on the ocean floor beyond the slope. Several localized deposits of this kind have been found. They are fanlike in form and occur at the mouths of submarine canyons (Fig. 12.7).

One fan, at the mouth of Hudson Submarine Canyon southeast of New York City, and 5km below sea level, was cored in 1951. Fifteen cores, each 8 to 10m long, show that the deposit consists of layers of clay appropriate to the depth, alternating with layers of sand, silt, and broken shells arranged in graded layers and derived from the continental shelf above. This composition suggests strongly that the deposit was built up by repeated turbidity currents and that during long intervals between the turbid discharges, ordinary fine-grained sediments settled

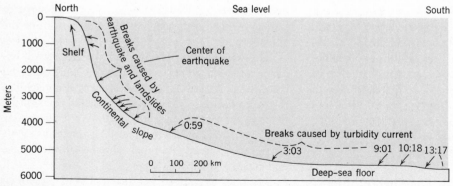

Figure 12.15 Profile of sea floor off Nova Scotia, showing events of the earthquake of November 18, 1929. Short arrows point to locations of breaks in transatlantic cables. Numbers show times of breaks in hours and minutes after earthquake. Vertical scale is exaggerated enormously. (*After B. C. Heezen and M. Ewing, 1952, Am. Jour. Sci., vol. 250, p. 867.*)

out on the sea floor. Several similar fans, one of which has a radius of 120km, occur at the mouths of submarine canyons off the coast of California.

Far more extensive are the sediments of the abyssal plains. The Sohm Abyssal Plain, shown in Fig. 12.8, has been cored. Mud in graded layers has been identified in the cores. The mud is far coarser than are the oozes and brown clay common at great depth. This, and the extremely gentle slope of the plain, have led to the hypothesis that abyssal plains are the deposits of spent turbidity currents, which have gradually buried the hilly sea-floor topography, somewhat as a river makes a valley fill by depositing alluvium. It has even been suggested that some valleys and canyons on the ocean floor, such as the Mid-ocean Canyon in Fig. 12.8, are likewise the work of turbidity currents, but so little is known about the canyons that, as we remarked previously, their origin remains uncertain. We can note again, however, that the Mid-ocean Canyon seen in the northwest corner of Fig. 12.8 is at least 2000km long; so if it was formed by turbidity currents, they must have possessed unusual properties.

Coral Reefs. Related to the supposed history of guyots is that of many coral reefs, which abound in most of the world's equatorial seas. A *coral reef is a ridge of limestone built by colonial marine organisms.* The name is an oversimplification, because most reefs are built by many different organisms, of which coral polyps are only one kind. These animals and plants secrete calcium carbonate as limy skeletons outside their bodies. Each lives in its own tiny chamber like a cliff dweller in a city of stone. Living in colonies of millions of individuals, they combine to build a structure of great size. Because of temperature-, light-, and oxygen requirements of the organisms that inhabit them, reefs are built only at or close to sea level. Therefore they occur along coasts, especially island coasts, and can be classified into three kinds based on their relation to the land. The three kinds are defined in Fig. 12.16.

From that figure it is not difficult to judge that if an island slowly subsided, its fringing reef would grow upward, since the organisms can live only at or near sea level. In time this would convert the reef into a barrier reef enclosing a lagoon. If subsidence exceeded the original height of the island, the island itself would disappear, leaving only the reef, now an atoll. This is believed to be the origin of many

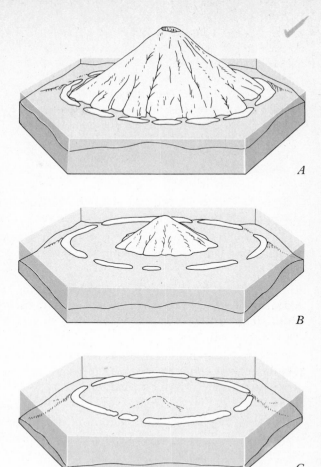

Figure 12.16 Chief kinds of coral reefs. *A.* Fringing reef, attached to the shore of a land mass. *B.* Barrier reef, formed offshore from a land mass. *C.* Atoll. A reef that forms a nearly closed figure within which there is no land mass.

barrier reefs and atolls in the southwest Pacific, although rising sea level, related to glacial ages, may have played a part in altering some reefs. Like guyots, therefore, some reefs owe their origin to gradual subsidence of the crust.

Synopsis

Earth, the Watery Planet, is unique among its sister planets because it has a vast hydrosphere. The ocean basins contain most of the water, within which are dissolved most of the soluble salts. Ocean water moderates the world's climate, provides the starting and ending points for the water cycle, and in one way or another controls or influences all of Earth's external processes. Also the ocean basins are the vast reservoirs into which the products of erosion of the lands are endlessly carried.

The shapes of the ocean basins and the topographic features of their floors are as varied and diverse as those of the land. Topographic features reflect a combination of (1) internal processes, giving rise to features such as oceanic ridges and seamounts, and (2) external features, created by erosion (submarine canyons) or deposition (deep-sea fans). Thus the ocean basins are theaters of great activity, where internal processes wage their endless battle with external ones.

Summary

1. Oceans cover 71 per cent of Earth's surface. Beneath them lies a topography as rugged and diverse as topography on the continents.

2. Seawater is nearly constant in volume, but fluctuates slightly as the amount of glacier ice on land changes. The total amount of water in the hydrosphere is in a steady state.

3. Salts have been in the sea for as long as seas have existed on Earth. They are added continually to the sea by rivers and streams, which derive them from chemical weathering. Salts are extracted by a variety of processes; so the composition of the sea is in a steady state.

4. Surface seawater circulates as currents in a number of huge, circular cells, that rotate clockwise in the Northern Hemisphere, counterclockwise in the south. Surface ocean currents are driven by winds and move warm equatorial water toward polar regions.

5. Ocean circulation takes the form of density currents set up by salty water with increased density caused by chilling or evaporation. The water sinks in polar regions and moves slowly toward, or even across, the equator.

6. Topographic features of the sea floor include continental shelves, continental slopes, submarine valleys and canyons, sea-floor trenches, abyssal plains, seamounts, and guyots.

7. Oceanic ridges occupy all the ocean basins as a world-circling sinuous chain of sea-floor mountains. Oceanic ridges are the lines along which magma from the mantle adds new rock to the edges of growing plates of lithosphere. Trenches are the places where older parts of the lithosphere plunge back into the mantle.

8. Sea-floor sediments are derived from sources on land, in the sea, and outside the Earth.

9. The chief classes of sea-floor sediment are terrigenous sediments (relatively coarse, clastic material), brown clay, calcareous ooze, and siliceous ooze.

10. Both past climates and chronology can be reconstructed from core samples of sea-floor sediment.

11. Turbidity currents probably sweep down the continental slopes and deposit coarse sediment at depth to form abyssal plains.

12. Some submarine canyons are probably sunken land valleys. Others were made beneath the sea, possibly by turbidity currents.

13. Most seamounts are probably volcanic cones. Guyots and some coral reefs indicate subsidence of the crust beneath the sea floor.

Selected References

Carson, Rachel, 1961, The sea around us: rev. ed., New York, Oxford University Press.

Gross, M. G., 1972, Oceanography, a view of the earth: Englewood Cliffs, New Jersey, Prentice-Hall.

Heezen, B. C., and Ewing, Maurice, 1952, Turbidity currents and submarine slumps, and the 1929 Grand Banks earthquake: Am. Jour. Sci., v. 250, p. 849–873.

Heezen, B. C., Tharp, Marie, and Ewing, Maurice, 1959, The floors of the ocean. I. North Atlantic: Geol. Soc. America Spec. Paper 65.

Hill, M. N., ed., The sea. Ideas and observations on progress in the study of the seas: New York, Interscience Publishers, v. 1, 1962, Physical oceanography, Chap. 5, Section III; v. 2, 1963, The composition of sea water, comparative and descriptive oceanography, Chaps. 2, 4, 10, 11, 12, 17, 18, and 23; v. 3, 1963, The Earth beneath the sea. History, Chaps. 4, 5, 12, 14, 17, 19, 20, 25, 26, 27, 28, 30, 31, 33, and 34.

Menard, H. W., 1964, Marine geology of the Pacific: New York, McGraw-Hill, Chaps. 1, 2, 7, 8, 9, and 10.

Scientific American, 1969, The ocean: New York, Scientific American Inc. A series of ten informative and authoritative articles on all aspects of the oceans.

Shepard, F. P., 1963, Submarine geology, 2nd ed.: New York, Harper and Row, Chaps. 1, 2, 11, 12, 13, and 14.

Shepard, F. P., and Dill, R. F., 1966, Submarine canyons and other sea valleys: Chicago, Rand-McNally.

Sverdrup, H. V., Johnson, M. W., and Fleming, R. H., 1942, The oceans, their physics, chemistry, and general biology: New York, Prentice-Hall.

Turekian, K. K., 1968, Oceans: Englewood Cliffs, New Jersey, Prentice-Hall.

Chapter 13

Coasts and
Continental Margins

The coast of a continent is a great boundary between two realms, land and water. Along this, as along other boundaries, two very different realms must adjust to each other, and conflict occurs. At a coast, ocean waves that may have traveled unimpeded through thousands of kilometers suddenly encounter an obstacle to their further progress. They dash against the firm rock, and the long-term results are very great.

Waves, and the currents created by waves, are the agents responsible for most of the erosion of coasts.[1] They are responsible likewise for most of the transport and deposition of the sediment created by wave erosion or washed into the sea by rivers. This sediment is moved outward from the coast and is deposited offshore, mostly in shallow water, in distinct layers. Each layer is a thin fringe around a continent, a fringe continually added to and covered by other layers, to form a thickening continental shelf that adds to the width of its continent, the source from which most of the shelf sediment was derived. The things that happen along a continental coast, therefore, are an essential part of the rock cycle and of the growth of continents. They deserve a closer look.

Waves

Wave Motion. We read in Chap. 12 how ocean waves are generated by winds that blow across the surface. Figure 13.1 shows the significant dimensions

---

[1] The tidal currents mentioned in Chap. 12 likewise erode, but their effect is much smaller.

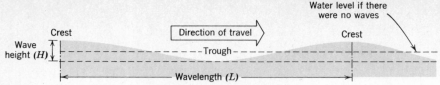

Figure 13.1 Profile of waves in deep water.

of a wave traveling in deep water where it is unaffected by the bottom far below. The motion of a wave is very different from the motion of the particles of water within it. As wind sweeps across a field of grain or tall grass, the individual stalks bend forward and return to their positions (Fig. 13.2), creating a wave-like effect. In similar fashion the *form* of a wave in water moves continuously forward, but each water *particle* revolves in a loop, returning, as the wave passes, very nearly to its former position. This loop-like, or oscillating, motion of the water, first determined theoretically, was later proved by injecting droplets of colored water into waves in a glass tank and by photographing their paths with a movie camera.

Waves receive their energy from wind, and so can receive it only at the surface of the water. Because the wave form is created by loop-like motion of water particles, the diameters of the loops at the water surface exactly equal wave height (Fig. 13.2). But there is progressive loss of energy (expressed in diminished diameters of the loops) downward from the water surface at which energy is received. The rate of decrease of energy downward is so rapid that at a depth equal to only half the deep-water wave length $\left(\text{in other words, at a depth of } \dfrac{L}{2}\right)$ the diameters of the loops have become so small that motion of the water is negligible.

Erosion by Waves. This means that depth $\dfrac{L}{2}$ must be the effective lower limit of wave motion, and therefore also the lower limit of erosion of the bot-

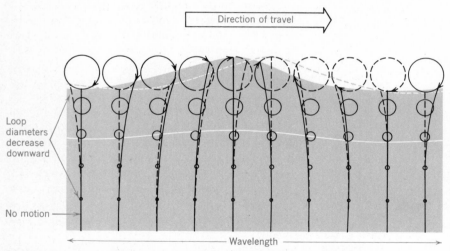

Figure 13.2 Loop-like motion of water particles in a wave of oscillation, in deep water. To follow the successive positions of a water particle at the surface, follow the arrowheads in the largest loops from right to left. This is the same as watching the wave crest travel from left to right. Particles in smaller loops underneath have corresponding positions, marked by continuous, nearly vertical lines. Dashed lines represent wave form and particle positions one-eighth period later. Resemblance to grain stalks bending in the wind is now apparent. (*After Ph. H. Kuenen, 1950, Marine geology: New York, John Wiley and Sons, p. 70.*)

tom by waves. In the Pacific Ocean, wave lengths as great as 600m have been observed. For them, $\frac{L}{2}$ equals 300m, a depth almost twice as great as the outer edge of the average continental shelf. Although the wave lengths of most ocean waves are far less than 600m, we can see it is possible for waves to affect the shelves. What the wave motion does, landward of depth $\frac{L}{2}$, is to lift and drop, endlessly, fine particles of bottom sediment, very slowly moving them seaward along the gently sloping bottom. Such erosion is very slow and not at all spectacular, but by the end of a million years the cumulative result is great.

What happens in the shallow water at the shore is more rapid, sometimes spectacular, and always different in style. When a wave moving toward shore reaches depth $\frac{L}{2}$ it "feels bottom," and when this happens the form of the wave begins to change. The loop-like paths of water particles gradually become elliptical, and velocities of the particles increase. Interference of the bottom with wave motion distorts the wave by increasing its height and shortening wave length. Often the height is doubled. This means the wave is growing steeper. Because the front of the wave is in shallower water than the rear part, it is steeper than the rear. Eventually the steep front becomes unable to support the wave, the rear part slides forward, and the wave collapses or *breaks* (Fig. 13.3).

When a wave breaks, the motion of its water instantly becomes turbulent, like that of a swift river. Such "broken water" is called **surf,** defined as *wave activity between the line of breakers and the shore.*

In turbulent surf each wave finally dashes against rock or rushes up a sloping beach until its energy is expended. Then it flows back. Water piled against the shore returns seaward in an irregular and complex way, partly as a broad sheet along the bottom and partly in localized narrow channels. The returning water is mainly responsible for the currents known to swimmers as "undertow."

Surf possesses most of the original energy of each wave that created it. This energy is quickly consumed in turbulence, in friction on the bottom, and in moving the rock particles that are thrown violently into suspension from the bottom, as in a stream. Ceaselessly the sediment is shifted landward and seaward within the surf zone. Some of the finest particles are carried in suspension into deeper water, where they settle out on the bottom. But they are still lifted and dropped to the bottom again wherever and whenever their depth is less than $\frac{L}{2}$. Most of the geologic work of waves, therefore, is accomplished by surf, shoreward of the line of breakers.

How deep below sea level can surf erode rock and move sediment? The answer depends on the depth at which waves break. Most ocean waves break at depths that range between wave height and 1.5 times wave height. Such waves are rarely more than 6m high; so the depth of vigorous erosion by surf

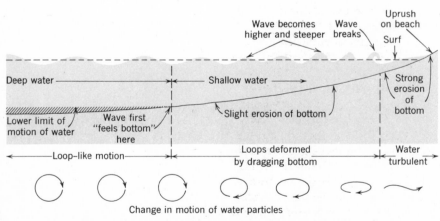

Figure 13.3 Waves change form as they travel from deep water through shallow water to shore. Circles and ellipses are not drawn to scale with the waves shown above. (Compare Fig. 13.2.)

should be limited to 6m times 1.5, or 9m, below sea level. This theoretical limit is confirmed by observation of breakwaters and other structures, which are found to be only rarely affected by surf at depths of more than about 7m. The surf zone, the place where high-energy turbulent water cuts into the land, is limited then to the narrow vertical range extending from sea level down to 7m, or a little more, below sea level.

But during great storms surf can strike effective blows well above sea level. The west coast of Scotland is exposed to the full force of Atlantic waves. During a great storm on that coast a solid mass of stone, iron, and concrete weighing 1350 tons was ripped from the end of a breakwater and moved inshore. The damage was repaired with a block weighing 2600 tons, but five years later storm waves broke off and moved that one too. The pressures applied to such erosion were around 30 tons per square meter. Even waves having much smaller force break loose and move blocks of bedrock from sea cliffs, partly by compressing the air in fissures in the rock; the compressed air pushes out blocks of rock.

The vertical distance through which water can be flung against the shore would surprise anyone whose experience of coasts is limited to periods of calm weather. During a winter storm in 1952, again on the west coast of Scotland, the bow half of a small steamship was thrown against a cliff and left there, wedged in a big crevice, 45m above sea level.

Another important kind of erosion in the surf zone is the wearing down of rock by wave-carried rock particles. By continuous rubbing and grinding with these tools, the surf wears down and deepens the bottom and eats into the land, at the same time smoothing, rounding, and making smaller the tools themselves (Fig. 13.4). But as we have seen, this activity is limited to a depth of only a few meters

Figure 13.4 The tools (rounded pebbles), the rock they have smoothed, and the surf itself are all visible on this beach near Pescadero, California. The wave-cut cliff in the background forms one flank of a rock headland. (*Richard Weymouth Brooks.*)

below sea level. The surf therefore is like an erosional knife edge or saw, cutting horizontally into the land.

Transport of Sediment. The rock particles worn from the coast by surf as well as those brought to the coast by rivers are intermittently in motion. They are dragged or rolled along the bottom, lifted in irregular jumps, or carried in suspension, according to their size and to the varying energy of waves and currents. In the surf, as can be seen on almost any beach, sediment is moved to and fro, shoreward and seaward. But because the bottom slopes down in the seaward direction, the net effect is to carry rock particles from the land gradually out to sea.

Seaward of the surf zone, in deeper water, bottom sediment is shifted by unusually large waves during storms and by currents, again with net movement seaward. Each particle is picked up again and again, whenever the energy of waves or currents is great enough to move it; but as the particle gets into ever-deeper water it is picked up more and more rarely. The result of these movements is that the sediment gradually becomes sorted according to diameter, from coarse in the surf zone to finer offshore.

Wave Refraction. A wave approaching a coast over an undulating bottom can not "feel bottom" along all parts of its crest simultaneously. As each part "feels bottom," wave length at that part begins to decrease and wave height increases. As a result the wave gradually swings around, part by part, to become parallel with the bottom contours; it is said to be *refracted* (Figs. 13.5, 13.9). Thus waves approaching the shore in deep water at an angle of 40° or 50° may, after refraction, reach the shore at an angle of 5° or less. Waves coming in over a submerged ridge off a headland will converge on the headland. Convergence, plus the increased wave height that accompanies it, concentrates wave energy on the headland. Conversely, refraction of waves approaching a bay will make them diverge, diffusing their energy at the shore. Because of refraction, headlands are eroded more rapidly than are bays, so that in the course of time irregular coasts become smoother and less indented (Fig. 13.14).

In summary, *wave refraction* is *the process by which the direction of a series of waves, moving in shallow water at an angle to the shoreline, is changed.*

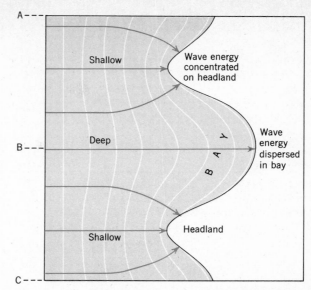

Figure 13.5 Refraction of waves concentrates wave energy on headlands, disperses it at shores of bays. Sketch map showing 8 waves that become more distorted as they approach the shore over a bottom that is deepest opposite the bay.

On- and Offshore Movement and the Shore Profile

To understand the changes made by the sea along a coast, we need to look first at what happens at the surface along a line at right angles to the shore, the shore profile, described in this section. Then in the following section we shall look at the forces that act along, parallel to, the shore. We shall then have a three-dimensional picture of coastal activities.

Elements of the Profile. Seen in profile, the usual elements of a coast (Fig. 13.6) are a wave-cut cliff and wave-cut bench, both the work of erosion, and a beach and wave-built terrace, both the result of deposition.

The **wave-cut cliff** can be defined as *a coastal cliff cut by surf.* Acting like a horizontal saw, the surf cuts most actively at the base of the cliff. The upper part of the cliff is undermined and crumbles, furnishing rock particles to the surf. A cliff in which undercutting keeps well ahead of crumbling has a *notch* at its base. The notch is a concave part of the cliff profile, overhung by the part above. Other minor erosional features associated with cliffs are *sea*

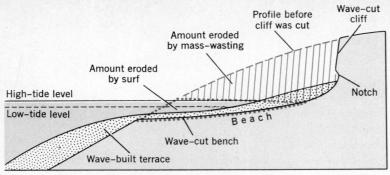

Figure 13.6 Principal features of the shore profile. The comparatively large proportion of erosion by mass-wasting results from notching of the cliff by surf, undermining rock material above the notch.

caves, sea arches, and *stacks* (Fig. 13.7). The cliff as a whole gradually retreats as the surf eats into the land.

The **wave-cut bench** is *a bench or platform cut across bedrock by surf.* It slopes gently seaward and grows wider in the landward direction as the cliff retreats. Some benches are bare or partly bare, but most are covered with sediment in gradual transit from shore to deeper water. This is the condition shown in Fig. 13.6. At low tide the shoreward parts of some benches are exposed (Fig. 13.8). If the coast has been uplifted by movement of the crust, the bench can be wholly exposed (Fig. 13.9). In some areas the bench is concealed and can be inferred only from maps constructed from soundings.

The *beach* is thought of by most people as the sandy surface above water along a shore. Actually it is much more than this. We define a **beach** as *the wave-washed sediment along a coast, extending throughout the surf zone.* In this zone, as we have seen, sediment is in very active movement. The sediment of a beach is derived in part from erosion of the adjacent cliff and from cliffs elsewhere along the shore. As we shall see presently, along coasts in general much more of it comes from alluvium contributed by rivers.

The **wave-built terrace** is *the body of wave-washed sediment that extends seaward from the breakers.* Its extent in the seaward direction is indefinite. "Terrace" is not a very appropriate name for the

Figure 13.7 Minor erosional features along a rocky coast, seen at low tide. Surf hollows out sea cave in more erodible part of bedrock. Cave cut through headland becomes a sea arch. Surf tears away parts of bedrock, leaving isolated stack as an "island" on wave-cut bench.

250   Coasts and Continental Margins

Figure 13.8 Shoreward part of wave-cut bench exposed at low tide, and wave-cut cliff 15m high. Beach here is scanty, consisting mostly of boulders, because high-energy waves carry finer rock particles seaward. North of Bonne Bay, Newfoundland. (*R. F. Flint.*)

Figure 13.9 Emerged and abandoned wave-cut cliff, wave-cut bench, and beach. Portuguese Point, San Pedro, California. Uplift of the crust in this part of Pacific North America has raised these features by 45m. A new set of features are forming along today's shoreline. Note waves being refracted around the headland. (*Spence Air Photos.*)

feature because it is not everywhere a real terrace or embankment. In places it is hardly more than a sort of carpet. Shoreward, it merges with the beach, as shown in Fig. 13.6.

The Steady State Along a Coast. As we noted at the beginning of this chapter, a coast is a scene of conflict between the realm of water and the realm of land. The conflict causes erosion and the creation, transport, and deposition of sediment. As a result the form of the land slowly changes, and the water in motion is obliged to move and to shape sediment derived from the land. The forces that fashion the shore profile—cliff, bench, beach, and terrace—tend to reach and maintain a condition of steady state, a compromise in the water/land conflict. The steady state is analogous to the one that prevails in a stream, where the compromise is between the factors in the stream and the form and character of the stream's channel.

Let us look at some of the ways in which the compromise is reached. On the beach, for instance, the *swash* of a wave running up the beach as a thin sheet of water moves sediment upslope, while gravity pulls it back again. More energy is needed to move pebbles than to move sand grains downslope; so the pebbles brought by the swash remain until the slope becomes steep enough for them to be moved back again. This is why pebble beaches are generally steeper than beaches built of sand.

During storms the increased energy in the surf erodes the exposed part of the beach and makes it narrower. During calm weather, the exposed beach is likely to receive more sediment than it loses and consequently becomes wider. But at all times the beach profile represents an average steady-state condition in the group of forces that shape it.

In the surf zone, on the wave-cut bench, the steady state exists also. Waves tend to move sediment shoreward, in the direction of their travel. But opposed to shoreward movement is the gravitational tendency of sediment to move downslope—seaward. The two tendencies together establish a bottom slope just steep enough to permit slight net movement of sediment down it in the seaward direction. The profile is a product of the steady state among the forces that shape it.

Or consider a delta. The position of the outer limit of a delta, the extent to which it projects seaward from the land, is a compromise between the rate at which the river delivers sediment at its mouth and the ability of surf to erode the sediment and move it elsewhere along the coast. The great size of the Mississippi delta (Fig. 8.15) testifies to the huge volume of sediment carried by its parent river, and to the comparatively small size of the waves of the Gulf of Mexico. Imagine the mouth of a Mississippi-size river on the west coast of Scotland, facing the powerful waves of the Atlantic. The delta would be much smaller.

Whether or not deltas are present, all coasts follow a pattern of evolution that is basically standard. It includes the long, slow retreat of cliffs, the gradual widening of benches, and the transport of sediment seaward, to be deposited on wave-built terraces. These are among the long-term processes by which continental shelves are made.

Movement Alongshore; Depositional Features

Up to now we have been describing mainly the erosional effects of surf and the shaping of the shore profile. But the deposits made by surf and by the currents it sets up are equally important, particularly as they result in large part from movements *along* the shore at right angles to the shore profile. The deposits occur as recognizable forms, which we will now examine.

Longshore Current and Beach Drift. The most common depositional shore features are beaches. Because it shows only two dimensions, Fig. 13.6 may give the impression that the sediment in beaches is derived entirely from the cliff behind it. This is not true. As we indicated earlier, rivers contribute more sediment to most coasts than cliff erosion contributes, but even were there no river-borne sediment, the beach at any point would include more material derived from elsewhere along the shore than particles from the cliff immediately behind it. The same is true of the finer sediment seaward of the beach.

The reason is that despite refraction, most waves reach the shore at an angle, however small. The oblique approach of waves sets up longshore movement of two distinct kinds, both of which move sediment *along* the coast. The first kind consists of a **longshore current,** *a current, within the surf zone, that flows parallel to the shore* (Fig. 13.10). Such currents easily move fine sand suspended in

Surf zone

Line of breakers

Longshore current

Zone of swash and backwash

Landward limit of swash

Figure 13.10 Movement of sediment along a coast as a result of oblique waves. Longshore current moves fine sand in suspension; zigzag path of a typical coarser sand grain along beach is shown by dotted line. (*L. Graff from A. Devaney.*)

the turbulent surf. Meanwhile, on the exposed beach itself, the second kind of movement alongshore is occurring. The swash of each wave is oblique, but the backwash flows straight down the slope of the beach. The result, for sand and pebbles, is *beach drift,* a zigzag movement with net progress along the shore. Both beach drift and a longshore current are shown in Fig. 13.10. The greater the angle of waves to shore, the greater the longshore movement. Pebbles tagged and timed have been observed to drift along a beach at a rate of more than 800 meters per day, transporting a volume of sediment equal to

more than 500 cubic meters in that time. When the amounts of sand moved by the longshore current are added to those moved by beach drift, the total must be very large, and it has important consequences for beaches used by people, as will be seen later in this chapter.

Spits, Bay Barriers, and Other Forms. In addition to beaches, other related forms are conspicuous on many coasts (Fig. 13.11). Common among these is the *spit, an elongate ridge of sand or gravel that projects from land and ends in open water.* Well-

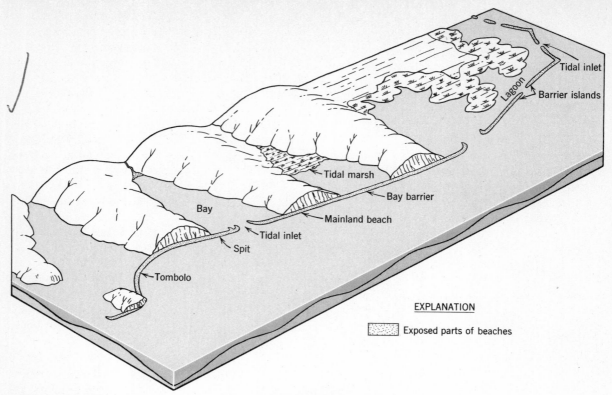

Figure 13.11 This stretch of coast shows several kinds of depositional shore features. Local direction of beach drift is always toward free ends of spits.

EXPLANATION

Exposed parts of beaches

known large examples are Sandy Hook on the southern side of the entrance to New York Harbor, and the northern tip of Cape Cod, Massachusetts. Most spits are merely continuations of beaches, built by beach-drifted sediment dumped at a place where the water deepens, as at the mouth of a bay. When the spit has been built up to sea level, waves act on it just as they would act on a beach. Much of a spit, therefore, is likely to be above sea level, although the tip of it cannot be. The free end curves landward in response to currents created by surf.

A **bay barrier** is *a ridge of sand or gravel that completely blocks the mouth of a bay.* It is believed to be formed by the lengthening of a spit, by beach drift, across a bay in which tidal- or river currents are too weak to scour away the spit as fast as it is built.

A **barrier island** is *a long island of sand, lying offshore and parallel to the coast.* Some such islands are 100km or more in length. Large-size examples are Coney Island and Jones Beach (New York City's shore-playground areas), the long chain of islands on one of which Atlantic City, New Jersey, stands,

the long chain centering at Cape Hatteras, North Carolina, and Padre Island, Texas, 130km long. How barrier islands originate is not known with certainty; very likely they originate in two or more different ways. Some of them are thought to be former spits severed from the mainland by surf erosion during exceptional storms or by submergence of parts of them by general rise of sea level since the latest glacial age.

Barrier islands create lagoons. A **lagoon** is *a bay inshore from a line of barrier islands or from a coral reef.* A **tombolo** (tom′bōlō) is *a ridge of sand or gravel that connects an island to the mainland or to another island.* It forms in much the same way as a spit does.

Sediments and Their Stratification. Now let us examine more closely the sediments themselves. Beaches, barriers, and related features consist of the coarser sizes of whatever range of rock particles is contributed by erosion of cliffs or by rivers. Quartz is the most durable of common minerals in the rocks

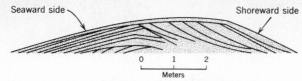

Seaward side   Shoreward side

0   1   2
Meters

Figure 13.12 Idealized cross section through a small beach, spit, or bay barrier, showing stratification of sand. (*Drawn from photographs by W. O. Thompson.*)

of the continents, and in most rocks it occurs as crystals having the diameters of sand grains. It is not surprising, therefore, that the majority of beaches consist chiefly of quartz sand. Bedrock that breaks down into larger pieces, along joints and other surfaces, makes beaches of gravel.

Dragged back and forth by the surf and turned over and over, particles of beach sediment become rounded by abrasion, much as do comparable particles in streams. In fact, we know of no easily recognized difference between the shapes of stream-worn and surf-worn particles.

There are, however, some differences in stratification. Spits, bay barriers, and the exposed parts of beaches, examined where natural erosion or man-made cuts expose them in section, generally are cross-stratified (Fig. 13.12). Their *seaward* parts consist of thin layers, gently inclined at many different angles. By watching calm-weather waves on a beach, we can see that the layers are deposited because part of the uprushing water sinks into the beach, leaving the backwash unable to remove all the sediment brought by the uprush. The varying angles of the layers mostly represent the varying slopes related to sediments of varying diameter. In contrast, the *landward* parts usually consist of foreset layers deposited by high waves that wash entirely over the beach, spit, or barrier and deposit on the far side much of the load carried in the uprush. The result is not unlike the stratification of a sand dune. However, the foreset layers are less variable in direction than those in a dune because, owing to refraction, the angle the waves make with the beach varies less than the angle between wind direction and the crest of a dune.

Sculptural Evolution of Coasts

From what has been said, it is clear that along all coasts erosion and deposition are continually active. Coasts, then, must change with the passage of time. What changes must we expect?

If we examine thoughtfully a number of different coasts we will come to realize that any coast, as it is eroded by surf, will gradually evolve through a series of forms that can be predicted, to an ultimate form that is comparatively smooth and straight. As examples of this process let us take the three short segments of the Atlantic coast of the United States shown in Fig. 13.13,*A*. All are alike in this way: They have been gradually submerged by the great rise of sea level since the latest glacial age, the worldwide event described in Chap. 11. But they differ greatly among themselves as to the extent to which waves and currents have altered them since submergence.

The first segment, the coast at Boothbay Harbor, Maine, (Fig. 13.13,*B*) consists of metamorphic rock, very resistant to erosion, and covered in places with thin glacial drift. Hilltops stand 50m to 100m above the sea. Damariscotta "River," Johns Bay, and the other bays are stream valleys drowned by rise of sea level. Most of the shore consists of bare bedrock, without any beach. In a few minor coves there are beaches so small that they can not be shown on the scale of the map, and there are no obvious wave-cut cliffs and benches. In short, the amount of geologic work performed by waves here is very slight.

The second segment, the coast at Boston, Massachusetts (Fig. 13.13,*C*) consists mainly of tough, compact till and other glacial sediments, all uncemented, and much more erodible than the rocks at Boothbay Harbor. Most of the islands are streamlined, ice-molded hills of till, and the land areas are generally less than 30m above sea level. Like the coast of Maine, that at Boston owes its broad features to the submergence of river valleys. Even though the Boston area is less fully exposed to the open sea than the Boothbay area, changes along the shore, made by surf, are much more evident. Seaward shores of islands have been cliffed, and beaches and spits have been built. Revere "Beach" is a spit grown so long that it is almost a bay barrier. Small former islands have been reduced to shoals (not visible on the map). Rates of erosion by surf here have been measured. During a 48-year period between two surveys, the cliff at Winthrop Head retreated at an average rate of 23cm per year; during two shorter periods, Grovers Cliff retreated 23cm and 30cm per year respectively. The greater amount of wave work accomplished at Boston than near Boothbay must result mainly from the more erodible character of the material exposed to the surf.

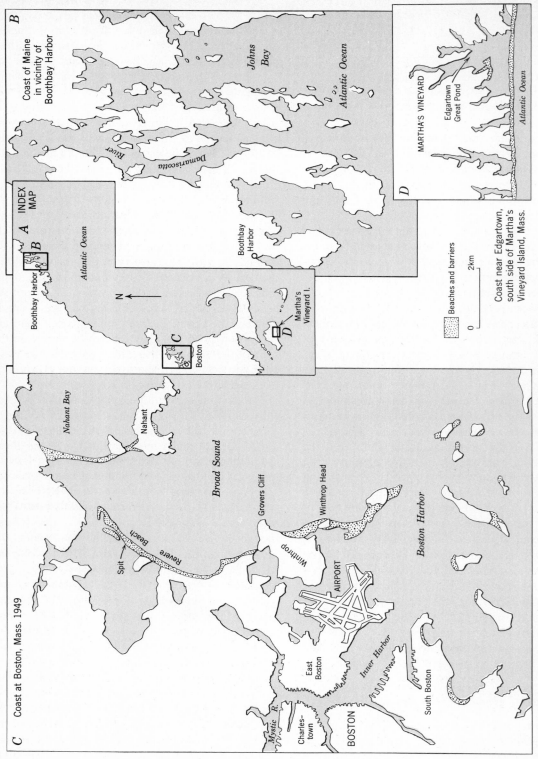

Figure 13.13 Sculptural alteration of coasts by waves and currents. Three New England areas each with a different degree of erodibility. *A.* Index map. *B, C, D.* Local maps, explained in text. (Shore features in *C* after Laurence LaForge, 1932; in *D* after J. B. Woodworth and Frank Wigglesworth, 1934.)

The third segment, the coast near Edgartown, Massachusetts, (Fig. 13.13,*D*) consists of loose, uncemented sand and gravel (glacial outwash) whose surface forms a plain only 3m to 6m above sea level. Also it is more fully exposed to wave attack than the Boston area is. The combination of highly erodible material and full exposure to the Atlantic Ocean has resulted in very rapid erosion by surf. From their shape it is evident that Edgartown Great Pond and its neighbors are small, shallow stream valleys, with tributaries, that have been sumberged. Yet the straight shoreline and the short lengths of the ponds imply that the interstream areas, which originally must have been headlands, have been cliffed and cut back through a considerable distance. During a 40-year interval between two surveys, high cliffs west of Edgartown, cut into till and other non-cemented sediments, retreated at an average rate of 1.67m per year;[2] it is likely, therefore, that the low, sand-and-gravel cliffs in the area of *D* retreated even more rapidly. Whatever the rate, the coast has been simplified to a straight line in which beaches alternate with bay barriers to form a continuous strip of sand.

These three coastal segments constitute a natural sequence, involving progressive retreat of cliffs, creation and widening of a wave-cut bench, and smoothing of the shoreline. The effect is to distribute wave energy more and more uniformly along the shore. By comparing the three segments, we can predict that, barring interruptions, the Boston coast in time will reach a condition comparable to the Edgartown coast and that in a far more distant future the Boothbay coast should do so as well.

A long series of comparisons of this kind lead to the idea, now generally accepted, that a broadly similar evolution should characterize coasts generally. Such evolution is analogous to the sculptural evolution of a land mass as it is eroded by streams and mass-wasting. In other words, just as there are cycles of erosion characteristic of the lands under a moist climate (Chap. 8) and under an arid climate (Fig. 10.11), so also there is a cycle of erosion characteristic of the effect of waves and currents along coasts. As with the stream cycle, the stages described as youth, maturity, and old age represent continuous transformation of the coast and are not sharply marked off from each other. They serve merely to

[2] Data from J. G. Ogden, III.

emphasize the fact of unbroken evolution. A typical sequence is shown in Fig. 13.14.

Could surf, given time enough, eat away the world's lands entirely and convert them into broad benches lying slightly below sea level? This has happened to small islands in Boston Harbor, and also seems to be what planed off the tops of guyots (Chap. 12). Nevertheless it has apparently never happened to whole continents. Continents contain sedimentary strata of all ages, and this fact implies that as far back as we can read Earth's history from rocks there has never been a time when lands have not existed. This is not surprising in view of abundant evidence that sea level keeps changing and that parts of the continental crust are bent upward again and again. Such changes work against long-continued erosion by surf at a single level. For this reason the cycle of erosion along coasts seems never to have destroyed a land of continental size.

Protection of Shorelines Against Erosion

The steady state among the forces that operate on coasts is interrupted temporarily by rare, exceptional storms, which erode cliffs and beaches spectacularly. During a single storm in 1944, cliffs on Cape Cod consisting of compact, non-lithified sediment retreated 15m, more than 50 times the normal rate of retreat per year. The bursts of rapid erosion caused by such storms are insignificant in the overall evolution of a coast. But the spread of people into coastal areas has brought with it the narrower point of view that such erosion causes intolerable damage to property.

To protect a strip of shore that consists of comparatively erodible material like that on Cape Cod, two things can be done. A cliff can be clad with an armor consisting of tightly packed boulders so large that they can withstand the onslaught of storm waves. Or it can be defended by a strong *seawall* built parallel to the shore on foundations deep enough to prevent undermining by surf during storms. Both structures protect cliffs against ordinary storms, at least, but both are expensive.

Because of their great recreational value, beaches in densely populated regions justify greater expense for maintenance than most headlands do. But a beach presents a special sort of problem. Because of beach

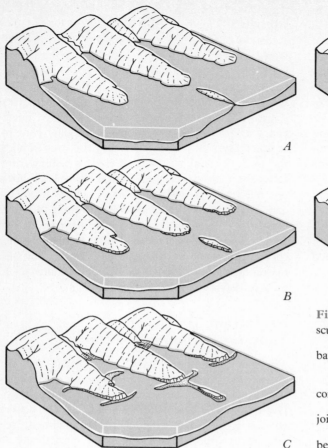

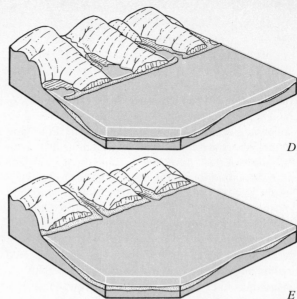

Figure 13.14 Cycle of coastal erosion. Sequence of sculptured forms developed through time.

A. Coast at start, with headlands, islands, and deep bays.

B. Headlands cliffed by surf.

C. Beaches, spits, and a tombolo are built as cliffing continues.

D. Cliffing reduces headlands to short stumps; spits join to form bay barriers.

E. Headlands and bays are eliminated; shoreline has become nearly straight.

drift (Fig. 13.10), what happens on one part of a beach affects all other parts that lie in the downdrift direction. Thus for example a seawall, dock, or other structure built at the updrift end of a beach reduces the amount of sand available for beach drift. The surf becomes underloaded and makes good the loss by eroding sand from along the beach until it becomes loaded again. Small beaches have been completely destroyed by this process in the course of only a few years.

Such erosion can be checked, at least to some extent, by building groins at short intervals along the beach. A **groin** is *a low wall, built on a beach, that crosses the shoreline at a right angle* (Fig. 13.15). Many groins contain openings that permit some water to pass through them. Groins act as a check on the rate of beach drift and so cause sand to accumulate against their updrift sides. Some erosion, however, occurs beyond their downdrift sides.

Another way of protecting a beach that is being eroded is to bring in sand artificially and pile it on the beach at the updrift end. Surf then erodes the pile and drifts the new sand down the length of the beach. In this method the sand that artificially nourishes a beach must be replenished. Both the feeding of a beach and the construction of groins are expensive, and both require maintenance.

In summary, the need for both remedies is caused by human interference with a natural process whose factors have long been in a steady-state condition.

Beaches in southern California are deteriorating for another reason, likewise the result of human interference. As we said earlier, most of the sand on those beaches is supplied, not by erosion of wave-cut cliffs but by alluvium poured into the sea by streams at times of flood. Because the floods themselves cause damage to man-made structures along the stream courses, dams have been built across the streamways in order to control the floods. But the dams also trap the sand and gravel carried by the streams, thus preventing the sediment from reaching the sea. This in turn has affected the steady state among the factors involved in longshore currents and beach drift. Sand is now in short supply, and the shortage is being made good by erosion of beaches. One scientist has estimated that before 1985 most of the beaches in southern California will have been washed away, leaving only the bare rock of the wave-cut benches that underlie them.

A similar situation has developed along the Black Sea coast of the USSR. Of the sand and pebbles that form the natural beaches there, 90 per cent was supplied by rivers as they entered the sea. During the 1940s and 1950s three things happened: Large resort developments including high-rise hotels were built at the beaches. By construction of breakwaters, two major harbors were extended into the sea. Dams were built across some rivers inland from the coast. All this construction interfered with the steady state among supply of sediment to the coast, processes of beach drift and longshore currents, and deposition of sediment on beaches. By 1960, it was estimated, the combined area of all beaches along the coast had

decreased by 50 per cent. Then beach-front buildings began to sag or collapse as the surf ate away the beaches. An ironic twist to the chain of events lies in the fact that contractors removed large volumes of sand and gravel from beaches for use as concrete aggregate, to build not only buildings but also the dams that cut off the supply of sediment to the coast.

Change of Sea Level

We have just mentioned that sea level keeps changing. How do the changes come about? Of course small, temporary, local changes are created by rhythmic rise and fall of the tide and by the piling-up of water against a coast by a big storm. But there are slower, longer-term changes. Measuring them involves special problems because we cannot fix the position of sea level relative to the center of Planet Earth. We can fix it only relative to adjacent land. The vertical relationship of sea to land is somewhat like that between a boat and a floating dock to which it is made fast. Both boat and dock can move. By loading either boat or dock we can alter the vertical position of one relative to the other.

Figure 13.15 Groins on Westhampton Beach, Long Island, New York. Piling up of sand on far side of each groin indicates drift of sand is toward the camera. (*U.S. Army Coastal Engineering Center.*)

Either the surface of the world ocean may rise or fall relative to the continents, or some part of a continent may rise or subside relative to sea level. Or both kinds of movement can occur at the same time. The principal cause of long-term rise and fall of sea level is probably the considerable changes in volume of the world ocean caused by the building and later melting of glaciers during glacial ages, which we described in Chap. 11.

Uplift or subsidence of land, unlike that of sea level, is not universal; it is piecemeal, involving parts of continents rather than a whole continent moving uniformly. Recent piecemeal uplift of land is occurring today at maximum rates of more than 1cm per year, in just those areas that were overspread by thick ice sheets during the latest glacial age. It has been in progress ever since the ice sheets began to melt away, and has lifted some sea beaches more than 200m above sea level, where of course they were formed. We reason that the weight of the ice sheets, in places 3km or more in thickness, caused the crust beneath them to subside, and that when the extra weight was removed by melting of the ice, the crust slowly "rebounded" toward its former position. The glacial age was so recent that the "rebound" is not yet completed.

Another cause of uplift is the slow slipping of a large plate of lithosphere over another (Chap. 1), the edge of the overriding plate lifting as it moves forward. It is thought that such movement is happening in western North America. Wave-cut benches somewhat like that seen in Fig. 13.9, along the coast of southern California, have been lifted as much as 250m above the sea. Along parts of the coast several of them stand one above the other, forming a great flight of step-like terraces. The movements that lifted them out of the sea may be related to the jostling of two moving lithospheric plates where they touch one another (Fig. 18.5).

Inland from the Atlantic Coast of the United States from Virginia to Florida are many marine beaches, spits, and barriers, the highest of which reach an altitude of more than 50m. It is thought that as a group these features owe their present altitude to a combination of two causes. One is uplift of the crust in coastal North America. The other is lowering of the world ocean through the building of glaciers, following periods when climates were warmer than today's, glaciers were smaller, and sea level was therefore higher.

Continental Shelves

Our discussion of the deposition of sediment offshore, the sculptural evolution of coasts, and the repeated fluctuation of sea level together leads us to the continental shelves (defined in Chap. 12) and their history. A continental shelf is a nearly flat surface covered with shallow water. Beyond its seaward edge it is accompanied by a continental slope that descends much more steeply. Thus slope and shelf together resemble the upright face and the tread of a single step in a flight of stairs (Fig. 13.16).

As the maps, Figs. 13.17 and 12.8, show, these treads and faces are very big. If today's sea level were to be lowered enough to expose the shelves and slopes, the area of the world's land would increase by about 15 per cent.

Why does each continent have a step-like pedestal entirely or partly surrounding it? Looking at Fig. 13.6, we might suppose that the shelves are enormous wave-cut benches. Many decades ago, when knowledge was scanty, this origin was thought likely. But today, in the light of many more facts, we realize that bench cutting has played only a minor part in creating the shelves, many of which were built up by deposition of sedimentary layers instead of being carved out of pre-existing rock. Such shelves resemble wave-built terraces (Fig. 13.6) more than they resemble wave-cut benches.

The shelf and slope off North Carolina are a good example of the building-up process of shelf making. A section (Fig. 13.18) running southeast from Raleigh, North Carolina to the base of the continental slope passes close to three deep holes drilled

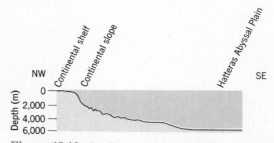

Figure 13.16 Profile across the eastern margin of North America showing the flat, step-like surface of the continental shelf and the steep continental slope. The slope grades gently out into an abyssal plain. The position of this profile, which runs southeastward from Norfolk, Virginia can be identified on Fig. 12.8. (*After Heezen, Tharp, and Ewing, 1959.*)

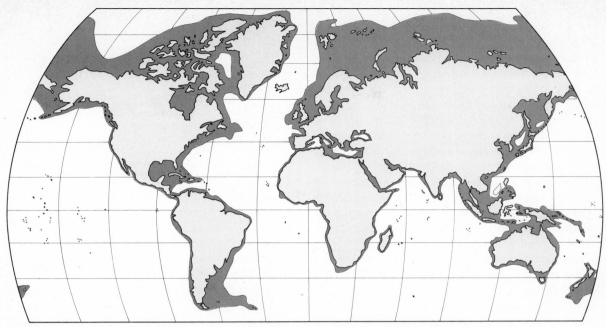

Figure 13.17 The continental shelves and continental slopes, shown together in dark shading, add 15 per cent to the areas of the continents. The elliptical projection used for the base map (from which the Antarctic continent has been left out) exaggerates the shelves in the Arctic region—but they are still very extensive. (*Data from Menard and Smith, 1966.*)

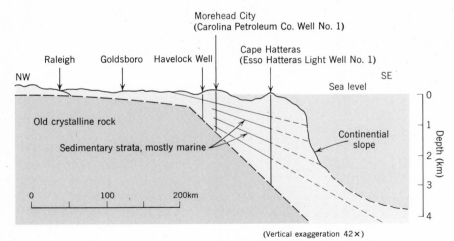

Figure 13.18 Section through coastal North Carolina and continental shelf, showing underlying strata. (*After W. F. Prouty, 1946; B. C. Heezen and others, 1959.*)

for oil. The detailed records of the drill holes show that the shelf consists of a wedge of sedimentary strata, about 3km thick near the slope but thinning landward, overlying a sloping floor of very old crystalline rock. The kinds of sediment and fossils in the drill cuttings show that this great wedge consists mainly of marine sediment that was deposited in shallow water. As the sedimentary layers now lie as much as 3km below sea level, they must have sunk down below the positions at which they were deposited. The layers are in various stages of transformation into rock, some of which is between 70 and 100 million years old. This and similar sections tell us that through long periods the margins of continents have been receiving sediment from rivers and have been gradually subsiding.

The age of the shelf off eastern North America cannot be much more than about 150 million years. Before a shelf could be built, there had to be a North Atlantic Ocean basin into which to build it, and the evidence suggests that the basin began to form at around that time, when the great North American lithospheric plate broke off and drifted slowly northwestward, away from the plates that once joined it on the east. When the break occurred, the severed continental margins—those of the Americas on one side and of Europe and northern Africa on the other—were thinned and bent slightly downward to form a long depression. Seawater flowed into the depression, creating the infant Atlantic Ocean. In this new ocean, sediment brought by rivers that flowed out of the adjacent continents began to accumulate.

As the drifting movement continued at a rate of a few centimeters per year, North American rivers built deltas into the sea and formed a growing shelf. Younger layers extended outward over older ones, and were draped over shelf and continental slope like tablecloths draped over a table, as the margin of the continent slowly subsided.

Looking again at Fig. 13.18, we can see that strata that once formed the continental slope are now cut through by that slope. The most likely explanation is that the slope has been gradually eroded (Fig. 13.19) by repeated slump, turbidity currents, and other processes that were involved in the creation of the canyons that today gash the continental slope.

Other shelves, like the shelf off southern California, are narrow, are irregular in surface form, and apparently consist mainly of bedrock. They seem to have resulted from bending down and faulting of the

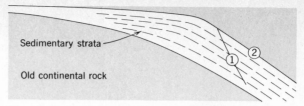

Figure 13.19 Schematic cross section of a shelf like that in Fig. 13.18, showing two continental slopes: (1) A slope resulting from erosion. (2) A slope, resulting from original deposition, on which no erosion has occurred.

western border of the North American Continent, accompanied by repeated chiseling by surf to create wave-cut benches.

Some of the rocky shelves have long, comparatively steep continental slopes that lead directly down (Fig. 13.20) to some of the deepest parts of the ocean floor—into trenches of the kind described in Chap. 12. This relationship of such shelves to trenches is explained in Chap. 18.

Shelves play an important part in the rock cycle because they are the resting places of most of the sediment washed out from continental interiors to continental margins, and so they are likewise the chief places where sediment is gradually converted into rock. Also, this transfer of sediment from centers to margins of continents gradually widens the continents by amounts equal to the widths of the shelves. This process by which continents widen themselves, and the story of what eventually happens to the shelf sediments, are explained in Chap. 18.

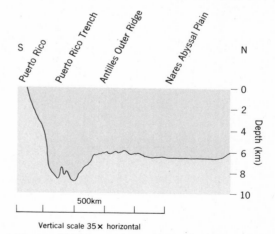

Figure 13.20 The shelf off the north coast of Puerto Rico (shown in Fig. 12.8) is narrow. Its continental slope plunges down to the Puerto Rico Trench 9km below sea level. (*Heezen, Tharp, and Ewing, 1959.*)

Summary

1. In deep water waves have little or no effect on the bottom. At a depth of $\frac{L}{2}$ (usually less than 300m) waves can begin to stir bottom.

2. Most of the geologic work of waves is performed by surf at depths of 9m or less.

3. Wave refraction tends to concentrate wave erosion on headlands and to diminish it along the shores of bays.

4. The shore profile consists of four related elements: wave-cut cliff, wave-cut bench, beach, and wave-built terrace.

5. There is a net gravitational tendency to move sediment seaward, downslope. Sediment becomes finer in the seaward direction.

6. Longshore currents and beach drift transport great quantities of sand along coasts.

7. Depositional shore features include beaches, spits, bay barriers, barrier islands, and tombolos.

8. A coast attacked by waves and currents passes through an orderly sequence of forms, a cycle of erosion, ending with a nearly straight form that permits almost uniform distribution of wave energy.

9. Because of movement of the crust and change of sea level, lands of continental size apparently have never been completely reduced to wave-cut benches.

10. A shore cliff can be protected for a time, at least, by a seawall or an armor of boulders. A beach can be protected by a series of groins or by importation of sand.

11. The level of the sea relative to the land keeps changing, both by rise and fall of the sea surface and by uplift and subsidence of Earth's crust.

12. Continental shelves with their accompanying continental slopes form an area equal to 15 per cent of the combined areas of the continents themselves. Many, perhaps most, shelves consist of thick aprons of sediment washed out from the continents and deposited mostly in seawater. The aprons add to the widths of continents.

Selected References

Bascom, Willard, 1964, Waves and beaches. The dynamics of the ocean surface: Garden City, N.Y., Anchor Books, Doubleday. (Paperback.)

Inman, D. L., 1954, Beach and nearshore processes along the southern California coast: Calif. Div. Mines Bull. 170, Chap. 5, p. 29–34.

King, C. A. M., 1959, Beaches and coasts: London, Edward Arnold.

Shepard, F. P., 1963, Submarine geology, 2d ed.: New York, Harper & Row, Chaps. 4, 6, 7, 8, 9, 12.

————, 1967, The Earth beneath the sea, Rev. ed.: Baltimore, Johns Hopkins Press.

————, and Wanless, H. R., 1971, Our changing coastlines: New York, McGraw-Hill.

Steers, J. A., 1954, The sea coast: London, Collins.

Zenkovich, V. P., 1967, Processes of coastal development: London, Oliver & Boyd. (Transl. from the Russian by D. G. Fry.)

Chapter 14

Sedimentary Strata

Sedimentary Strata in the Rock Cycle

Strata as a Link Between External and Internal Processes. We have postponed discussion of sedimentary strata until all the processes of erosion and deposition that lead to the making of strata had been set forth. Now we can gather together the many threads we followed in studying erosion and deposition, as we observe in the strata features characteristic of the sediments deposited by each of the external processes. We see sedimentary strata as a key phase in the rock cycle. That phase forms a link between the external processes, which break up rock and deposit sediment, and the internal processes that fracture, bend, squeeze, and melt sedimentary rocks and other rocks as well. This is why the present chapter stands between our discussion of Earth's external activities and that of the internal ones.

Sedimentary Strata as Former Sediments. We recall that all sedimentary rock is built of rearranged particles of older rock. It is the conspicuous product of a vast, complex operation, the sorting and transport of sediment. The operation ends with deposition, but is followed by compaction and cementation, which convert the sediment into new, firm rock. We can witness every phase of this segment of the rock cycle: older rocks whose exposed surfaces are being crumbled under the attack of weathering, regolith slowly creeping down hillsides, streams turbid with fine rock particles, sand and silt deposited by a stream and exposed as a flood subsides, sand and

265

mud deposited in shallow water off a sea beach and exposed at low tide, and sediment in all stages of conversion to firm sedimentary rock. Every feature of sediment that geologists have described occurs also in sedimentary strata. This fact is perhaps the most obvious proof of the principle of uniformity of process.

Principle of Original Horizontality. We noted in Chapter 2 that sediment now being deposited in stream valleys and on the floors of lakes and the sea is spread out in layers that are generally almost horizontal. When immersed in water, loose particles are easily moved, and water in motion tends to spread the particles evenly and so to fashion a nearly level surface. In this way, at the end of each stream flood and of each storm offshore, a new sedimentary layer is deposited almost horizontally over the one beneath. Of course some layers of limited extent are deposited in inclined positions (Fig. 8.14), but these represent special conditions and are exceptions to the general rule. It is, then, a principle, recognized as early as the 17th Century, that most layers of sediment are nearly level when formed. We rely on the principle when we analyze sedimentary layers that are no longer horizontal, and conclude that they have been deformed.

Principle of Stratigraphic Superposition. We noted in the preceding paragraph that during each new bout of activity a new sedimentary layer is deposited over the one beneath. This fact embodies the *principle of stratigraphic superposition* (Chap. 2), which says that the order in which strata are deposited is from bottom to top. Together with the fossils the strata contain, this principle is fundamental to the building up of the geologic column.

Stratification

A layered arrangement of the particles that constitute sediment or a sedimentary rock is **stratification.** This is an obvious feature of most sedimentary rocks and is seen also in ancient lava flows. Looking closely at rocks that are stratified distinctly, we can see that the strata differ from one another because of differences in some characteristic of the constituent particles or in the way in which the particles are arranged (Fig. 14.3). Very commonly one stratum consists of particles of different diameter

from those in another. In a clastic rock such changes of diameter result from fluctuations of energy in a stream, in surf, in wind, in a lake current, or in whatever agency is responsible for the deposit. The energy changes, usually small, are not the exception but the rule.

Sorting. A conspicuous result of the transport of particles by flowing water or flowing air is *sorting* in the deposited sediments. Sorting according to specific gravity is evident in mineral placers (Chap. 20). Particles of unusually heavy minerals such as gold, platinum, and magnetite are deposited quickly on stream beds and on beaches, whereas lighter particles are carried onward. Most of the particles carried in water and wind, however, consist of quartz and other minerals with similar specific gravity. Therefore such particles are commonly sorted, not according to specific gravity but according to diameter. In a stream, gravel is deposited first, whereas sand and silt are carried farther before deposition. Thin, flat particles are carried farther than spherical particles of similar weight. Long-continued handling of particles by turbulent water and air results in gradual destruction of the weaker particles. In this way rocks and minerals that are soft or that have pronounced cleavage are eliminated, leaving as residue the particles that can better survive in the turbulent environment. Very commonly the survivor is quartz, because it is hard and lacks cleavage. In this case sorting is based on durability.

Although sorting is the chief cause of stratification, it is not the sole cause. Successive layers that do not differ from each other in grain size, composition, or degree of compaction can still be separated from each other by surfaces of easy splitting representing minor intervals when no deposition occurred. Again two adjacent layers, not otherwise distinct, can differ from each other as to the kind or abundance of cement they contain.

In summary, each stratum nearly always possesses definite characteristics by which it differs from the stratum beneath or above it. With this in mind we can describe two chief kinds of stratification and then examine the particles within a stratum.

Parallel Strata. Layers of sediment fall into two classes according to the geometric relation between successive units. One class consists of **parallel strata,** *strata whose individual layers are parallel*

Figure 14.1 Varves in glacial-lake clay and silt, Uppsala, Sweden. In this case each varve is also a graded layer. Thick, pale silt (summer portion) sharply overlies the layer beneath it and grades upward into thinner, darker silty clay (winter portion), representing an annual cycle. Pencil (placed vertically at base of exposure) is 15cm long. (*R. F. Flint.*)

(Fig. 14.1). Parallelism indicates that deposition probably occurred in water, and that the activity of waves and (except for the special case of graded layers, described in a following section) currents was minimal. Indeed, the sediments of lakes and the deep-sea floor rather commonly occur in parallel layers.

A distinctive variety of parallel strata consists of repeated alternations of layers of unlike grain size or mineral composition. Such alternation suggests the influence of some naturally occurring rhythm, such as the rise and fall of the tide or the seasonal change from winter to summer. *A pair of sedimentary layers deposited during the cycle of the year with its seasons* is a **varve** (Swedish for "cycle"). Varves occur in many glacial-lake sediments (Fig. 14.1) deposited 12,000 years ago or less, during the melting of the latest great glaciers. Radiocarbon dating of related sediments has confirmed that the pairs of layers are true varves.

Paired layers deposited in deep glacial lakes are generally very distinct, because close to an ice sheet the contrast between summer and winter weather markedly affects the rate of melting of ice. Pairs of laminae very similar to these occur in ancient rocks. Certain claystones in South Africa have been inter-preted as varves deposited in glacier-dammed lakes during a glaciation more than 200 million years ago.

Varves of different origin characterize rocks of the Green River Formation, which underlies a vast area in Wyoming, Colorado, and Utah. In each varve one layer consists of calcium carbonate; the other includes dark-colored organic matter. The rhythm is explained as follows: the sediments were deposited in a lake, which warmed in summer, therefore lost carbon dioxide, and precipitated calcium carbonate from solution. During the same warm season floating microscopic organisms reached a peak of abundance. The relatively heavy carbonate sank promptly and formed a summer layer; the lighter organic matter sank much more slowly to form an overlying winter layer. Thus the pair of layers is a varve. It has been estimated from sample counts that between 5 and 8 million varves are present in the Green River Formation; hence we reason that an immense lake occupied the Green River basin for millions of years under remarkably uniform conditions.

Cross-Strata. Very distinct from parallel strata are *strata that are inclined with respect to a thicker stratum within which they occur.* These are **cross-strata.** All such strata consist of particles coarser than silt and are the work of turbulent flow of water or air, as in streams, wind, and waves along a shore. As they are driven forward, the particles tend to collect in ridges, mounds, or heaps in the form of ripples and waves, which migrate forward bodily or simply enlarge in the downcurrent direction. Particles continually accumulate on the downcurrent slope of the pile, forming strata with inclinations as great as 30° to 35°.

Cross-strata are seen in deltas (Fig. 8.14), sand dunes (Fig. 14.2), and beaches (Fig. 13.12). We must keep in mind that although the foreset layers in deltas and in dunes are commonly parallel with *one*

another, nevertheless they are cross-strata because they are not parallel with the larger strata that enclose them. Cross-strata therefore do not argue against our belief in original horizontality. Indeed the larger strata that enclose the cross-strata support that belief.

Under some conditions, however, no larger enclosing strata are present; instead, all the layers deposited at a locality are inclined. For example, volcanic ash and coarser volcanic particles characteristically accumulate in conical piles surrounding their volcanic source and form layers with steep inclinations (Fig. 14.3). Because *all* such layers are

Figure 14.2 These cross-strata are slightly curved. Although the thin individual layers within a larger unit are parallel, the units themselves (separated by dotted lines) cut across each other. The rock consists of wind-blown quartz sand of Jurassic age, cemented to form sandstone. Cavities have formed where cement has been dissolved . Monument Valley, Arizona. (*Tad Nichols.*)

Figure 14.3 These layers of volcanic ash and coarser particles were deposited at a steep angle on the slope of a volcanic cone near Cerro de Camiro, Mexico. Oldest layers are at lower right, youngest at upper left. Each layer represents a single eruption. (*Kenneth Segerstrom, U.S. Geol. Survey.*)

inclined, they are not cross-strata, and so are real exceptions to the idea of original horizontality.

The direction in which cross-strata are inclined is the direction in which the related current of water or air was flowing at the time of deposition. So, by measuring direction of inclination we can learn that one thing about a former environment.

Arrangement of Particles Within a Stratum. In addition to the relationships of layers to each other, several kinds of arrangement of the particles within a single layer are possible. Each kind gives informa-

tion about the conditions under which the sediment was deposited. Chief among these kinds are *uniform layers, graded layers,* and *nonsorted layers.*

Uniform Layers. A layer that consists of particles of about the same diameter is called *uniform.* A uniform layer of clastic rock implies deposition of particles of a single size, with little change in the velocity of the transporting agent. This might occur in the bottomset layers of a delta. A uniform layer of nonclastic rock implies uniform precipitation from solution, which produces crystalline particles of a single size. But a layer that is subdivided into thin layers marked off from each other by differences in grain size suggests a transporting agent the velocity of which fluctuated.

Graded Layers. If we put a quantity of small solid particles of different diameters and about the same specific gravity into a glass jar of water, shake the mixture well, and then let it stand, the solid particles will settle out and form a deposit on the bottom of the jar. The heaviest and largest particles settle first, followed by successively smaller particles; the finest ones, if small enough, may stay in suspension for hours or days, keeping the water turbid, before finally settling out. Thus particle size decreases *gradually* from the bottom upward. This arrangement characterizes a **graded layer,** defined as *a layer in which the particles grade upward from coarse to finer* (Fig. 14.4). The deposit is, as it were, a one-stroke affair. The water receives energy from the shaking of the jar and becomes turbulent; as a result

sediment is lifted above the bottom. When the shaking stops, abrupt loss of energy results in continuous deposition.

Although a graded layer produced by simple shaking in a jar is graded only in the vertical dimension, those produced in nature are graded laterally also, because they are made by currents moving from one place to another. As the heaviest particles settle first, grading is not only from bottom to top but laterally in the downcurrent direction.

Graded sediments are widespread beneath abyssal plains of the deep-sea floor and are present in many deposits on the lands. Some of the processes that cause grading are (1) turbidity currents in the sea and in lakes, (2) streams, as they lose energy and deposit bed-load sediment while floods subside, (3) falls of volcanic ash, and (4) dust storms as they die down. All four processes represent rapid, continuous loss of energy, an essential condition for the creation of graded layers.

Nonsorted Layers. The particles in some sedimentary rocks are not sorted at all. They consist of mixtures of various sizes arranged chaotically, without any obvious order. Processes that create sediments of this class include avalanche, debris flow, mudflow, solifluction, submarine slump, and transport by glaciers and floating ice. Widely recognized among nonsorted sediments is tillite (Fig. 14.5), of glacial origin.

Rounding and Sorting. As we noted in Chaps. 6 and 7, particles broken from bedrock by mechanical weathering and other processes tend to be angular, because breaking commonly occurs along joints, and surfaces of stratification. As they undergo transport by water or air, the same particles tend to become smooth and rounded. Figure 14.6 shows what can happen to pebbles on a beach, and Fig. 14.7 indicates how sand grains become shaped during transport. Degree of rounding therefore gives some idea of the distance or time involved in transport by flowing water or flowing air.

Glaciers, however, tend to make rock particles irregular in shape by crushing and abrading them. The faceted shapes of ideal glaciated coarse particles, and the scratches on them, are illustrated in Fig. 11.13. Flowing air likewise grinds facets on coarse particles, but the facets on ventifacts (Fig. 10.17) are more distinct and meet each other more sharply than

Figure 14.4 Graded layer exposed in a vertical slice through a core taken from sediment beneath the deep-ocean floor at depth 4,074m. The layer, overlain and underlain by nonstratified ooze, extends from about 2 to about 8 on the centimeter scale. (*Lamont Geol. Observatory, courtesy C. D. Hollister.*)

Figure 14.5 Tillite, representing a very ancient glaciation, overlying a smooth surface of diabase on which many glacial striations are visible. The tillite lacks stratification. The pebbles and cobbles in it are mostly diabase, further indicating glacial erosion of the underlying bedrock. View looking southwest, the direction of flow of the glacier. Nooitgedacht, Cape Province, South Africa. (*R. F. Flint.*)

do those made by glaciation. Ventifacts are not rounded because wind does not move them. They lie motionless while fine particles driven by the flowing air abrade them.

We have said that sorting of sediments is responsible for most stratification. Rock particles become increasingly well sorted as they become increasingly rounded. This change with distance makes it possible to draw general inferences about directions and distances of transport of ancient sediments and therefore about geographic relationships at the times when the sediments were deposited.

Figure 14.6 Rounding of pebbles during transport by surf on a beach. These fragments of basalt were collected at random from a talus and an adjacent beach near Clarence, Nova Scotia. They are arranged here to show what can be expected to happen to a plate-shaped piece (*upper left*) and to a spindle-shaped piece (*upper right*) during progressive abrasion. The end product of each could be the same—a spherical pebble. Although this rounding is the work of beach drift, stream transport produces a similar result. (*J. E. Sanders from Longwell and Flint.*)

271

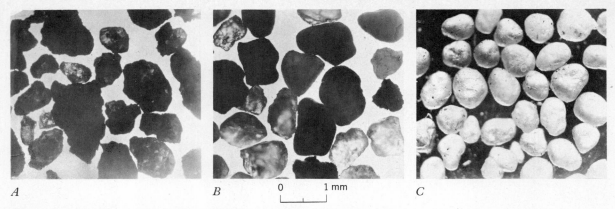

A *B* 0 1 mm *C*

Figure 14.7 Rounding and sorting of mineral grains during transport.

A. Mineral grains loosened and separated from igneous and metamorphic rocks by mechanical and chemical weathering before transport. The angular shapes of the individual grains, slightly altered by weathering, are the forms assumed by the minerals as they crystallized from a magma. The aggregate of grains is a nonsorted sand.

B. Sand carried from an area of rocks similar to those which yielded the sand in *A.* Some of the less-durable mineral grains have been broken up and lost, leaving a larger proportion of the durable mineral quartz. Battering in transit has partly rounded the grains.

C. Sand transported through a long distance. Grains have become well rounded and consist almost entirely of durable quartz. Weaker minerals did not survive transport.

Derivation. Through the kinds of minerals or rocks of which they are composed, sediments reflect the kind of parent rock from which they were derived. An uncomplicated example is a boulder train in glacial drift; another is a placer of gold, diamond, or other economic substance. Many boulder trains and placer minerals have been traced backward (upstream) to their sources; from a boulder train the direction of flow of a glacier can be inferred.

Also we can draw inferences, broad and general at least, as to the kind of rock from which a widespread body of sediment or sedimentary rock was derived. Going back to the chemical weathering of granite and basalt (Table 6.1), we recall that granite yields quartz and clay minerals, but that basalt can yield no quartz. If, however, either kind of rock were weathered mechanically, it would yield bits of the rock itself, and those bits would include feldspar. If transported and deposited quickly, the feldspar would not be destroyed by weathering. The presence of fresh feldspar in sedimentary strata would suggest one of two things about the origin of the sediment. Either a dry or very cold climate with a minimum of chemical weathering prevailed, or the cutting of valleys by streams occurred on slopes so steep that rate of chemical weathering could not keep pace with rate of erosion by streams.

Features on Surfaces of Layers. In foregoing chapters we discussed the making of ripples by streams, by the wind, and by currents and waves in lakes and the sea. Ripples are preserved in some sandstones and siltstones as *ripple marks* (Fig. 14.8).

Some claystones and siltstones contain layers that are cut by polygonal markings. By comparison with sediments forming today, such as those in roadside puddles following a rain, we infer that these are ***mud cracks,*** *cracks caused by the shrinkage of wet mud as its surface becomes dry* (Fig. 14.9). The presence of mud cracks in a rock generally implies at least temporary exposure to air and therefore suggests tidal flats, exposed stream beds, playa lakes, and similar environments. Occurring with some ripple marks and mud cracks, and preserved in a similar manner, are the footprints and trails of animals. Even the impressions of large raindrops made during short, hard showers are preserved in some strata.

Other Features

Fossils. The animals and plants (or parts of them) that were buried with sediments, protected against oxidation and erosion, and preserved through the long process of conversion to rock constitute the fossils (Fig. 14.10) that form a very important part

Figure 14.8 Ancient ripple marks on a big slab of hard sandstone, Capital Reef National Monument, Utah. (*Richard Weymouth Brooks.*)

1 meter

Figure 14.9 Mud cracks preserved in sandstone of Mississippian age near Pottsville, Pennsylvania. This is the underside of a slab turned up on its edge. The sand was deposited over a layer of river mud in which the cracks had formed and then hardened. The claystone (the former mud) has been destroyed by recent weathering, but chips of it litter the foreground. Most ancient mud cracks, like these, are really fillings of cracks by sand that was deposited in them. (*Joseph Barrell.*)

Figure 14.10 Fossils in sedimentary rock. Impressions of the shells of marine invertebrate animals in Paleozoic sandstone from eastern New York. (*C. O. Dunbar.*)

of the record of the strata. Not only are fossils an important clue to former environments, but they are the chief basis for the correlation of strata and the construction of the geologic column.

Color. The colors of sedimentary rocks vary considerably. Some rocks exposed in cliffs are colored only skin-deep by some product of chemical weathering. For example, a sandstone that is pale gray on freshly fractured surfaces may have a surface coating of yellowish-brown limonite, developed during weathering by the oxidation of sparse, iron-rich minerals included with the grains of quartz sand.

The color of fresh rock is the combination of the colors of the minerals that compose it. Iron sulfides and organic matter, buried with the sediment, are responsible for most of the dark colors in sedimentary rocks. Microscopic examination of red and brown rocks shows that their colors result mainly from the presence of ferric oxides, as powdery coatings on grains of quartz and other minerals or as very fine particles mixed with clay.

These oxides get into strata in at least two different ways. In most red and brown rocks probably the colored oxides are secondary, created in the strata by alteration of ferromagnesian minerals such as hornblende and biotite, during weathering and during conversion of sediment into rock. Such chemical alteration can take place in any warm climate, dry or moist.

In some strata the colored oxides are believed to be primary, having been deposited as components of the sediments themselves. Such oxides may have been derived from the erosion of red-clay soils like those forming today in the warm climate of the southeastern United States. Washed down from uplands into rivers, the oxides are deposited in places on swampy flood plains, where decaying plant matter creates strong reducing agents that reduce the red ferric oxide to ferrous oxide, which is not red. Rivers carry the rest of the red sediment into the sea, where organic matter on the sea floor likewise reduces it. Probably, therefore, the red color can be preserved only if the sediments escape reduction after deposition. This they could do if deposited in basins where, for one reason or another, organic matter does not accumulate in amounts sufficient to reduce the ferric iron.

Without very detailed research, therefore, all we can tell about the climate where and when an ancient

274   Sedimentary Strata

red-colored stratum was deposited is that the climate was warm, either there or in the region from which the sediment was derived.

Concretions. Enclosed in some sedimentary rocks are bodies called *concretions.* They range in diameter from a fraction of an inch to many feet and in shape from spherical through a variety of odd shapes, many with remarkable symmetry (Fig. 14.11), to elongate lenses that parallel the stratification of the rock. Concretions are composed of many different substances, including calcite, silica, hematite, limonite, siderite (iron carbonate), and pyrite. Small concretions are dredged up from the sea floor, showing they are forming there today as sediments are deposited. This origin, contemporaneous with the enclosing sediments, is indicated also by the shapes of some concretions and by their relation to the stratification of the surrounding sedimentary rock. Others formed after the deposition of the sediments, as for example those that retain the stratification of the surrounding rock (Fig. 14.12).

The substances of which concretions are made show that these objects are the result of localized chemical precipitation of dissolved substances from seawater, lake water, or ground water. Once precipitation starts around a fossil or other body that differs from the enclosing rock, the concretion thus formed continues to grow. Indeed, in some rocks perfectly preserved fossils are found at the centers of concretions.

Because we do not yet understand concretions fully, the best definition we can suggest is that a **concretion** is *a localized body having distinct boundaries, enclosed in sedimentary rock, and consisting of a substance precipitated from solution, commonly around a nucleus.* Some geologists use the word *nodule* as a synonym for concretion; others restrict it to concretions of small size.

Stratigraphic Interpretation

Environments of Deposition. The study of strata is *stratigraphy.* In the present chapter we have described various aspects of sedimentary rocks. Now we must look at some broader aspects, the first of which concerns our interpretation of the environments in which sediments are deposited.

One of the objectives of research in geology is to reconstruct the *paleogeography* (the ancient geographic relations) of a region at the time when a particular sequence of strata was deposited as sediment. The reconstruction would show the distribution of land and sea, streams with their directions of flow, mountains and lowlands, deserts, glaciers, and if possible, indications of the climate then prevailing. Information of these kinds is commonly assembled in the form of a *paleogeographic map,* representing a series of inferences drawn from the physical characteristics of the sedimentary rocks and from the fossils they contain. The inferences mainly concern the environment in which each kind of sediment was deposited and are based on analogy with the environments of today's sediments, by use of the principle of uniformity of process.

Looking at modern sediments and remembering the slopes shown in Fig. 7.2, we conclude that sediments deposited along streams, in lakes, and in the sea are deposited primarily in basins, where the chances of preservation are better than on high

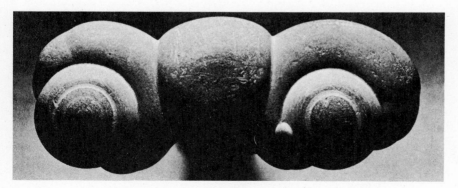

Figure 14.11   Concretion, 7cm long, consisting of calcium carbonate. It was imbedded in claystone. (*Andreas Feininger.*)

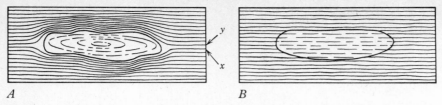

Figure 14.12 Vertical cross sections of two concretions, each several inches long, in claystone, showing different times of origin relative to deposition of the enclosing strata.

A. Concretion formed after layer *x* was deposited but before layer *y* was laid down. Hence the concretion is contemporaneous with deposition of the body of sediment.

B. Concretion transected by layers of the enclosing rock. Hence it was formed after all the layers shown had been deposited.

places and on steep slopes. To distinguish among the various environments in which a group of strata may have been deposited, we must look closely at exposures of the strata, following the west-east line marked by numbers on the map (Fig. 14.13).

1. In the area numbered *1* on the map we see conglomerate consisting of little-rounded and little-sorted cobbles and pebbles, with very irregular cross-strata dipping east. In places the conglomerate is interbedded with thin layers of mudflow sediment. These characteristics suggest a fan or fans at the foot of the steep slope of a highland.

2. Farther east we find better-sorted sandstone with cross-strata associated with elongate bodies of siltstone and broad thin layers of claystone. Fossils might include freshwater mollusks and leaves and branches of trees. This assemblage suggests a floodplain on a lowland, with meanders, natural levees, and overbank-flood sediments.

3. Still farther east is well-sorted, well-rounded sandstone, with long, parallel, gently dipping foreset beds overlain by stream-channel cross-strata. Fossils include plant debris and a few brackish-water mollusks. Evidently we have passed from a former pied-

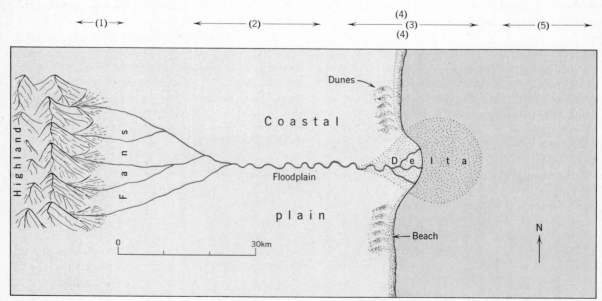

Figure 14.13 Paleogeographic sketch map showing several former environments, reconstructed from characteristics of strata. Numbers indicate positions along a west-east line, described in text.

mont environment, across a coastal plain drained by sizeable rivers, to a low-lying seacoast.

4. Assuming that good exposures of our strata continue eastward, we can see sandy beach sediments (possibly overlain in places by well-sorted sandstone with cross-strata dipping steeply west), suggesting beach dunes of wind-blown sand.

5. Still farther east we find the texture becoming finer, with sandstone grading into siltstone and claystone that contain fossil mollusks of offshore-marine kinds.

From this information, clearly indicating several environments that differ from one another but that are logically related, we can sketch the crude paleogeographic map shown in Fig. 14.13. In a similar manner we can identify former lakes, estuaries, lagoons, and reefs, as well as sandy deserts and country overrun by glaciers, each feature representing a distinctive environment of deposition.

These environments are no longer present in the district we are studying. But they can be recaptured by virtue of the fact that the various sediments formerly deposited in them were preserved and converted into rock, as they subsided and were covered by other sediments. In short, they became a part of the geologic record.

Sedimentary Facies. Implicit in our discussion of environments is the fact that most strata change character from one area to another. In the foregoing discussion we followed a unit or group of strata, all parts of which were deposited in the same interval of time, from the base of a highland across a coastal plain and into a shallow sea that deepened offshore. The changing environments are represented by changing grain size, grain shape, stratification, depositional structures, and fossils in the unit. *A distinctive group of characteristics, within a rock unit, that differ as a group from those elsewhere in the same unit,* is a sedimentary *facies* ("aspect"). Two facies merge laterally into each other either gradually or abruptly, depending on the relations between the two former environments of deposition.

If a sedimentary unit were exposed in section from end to end of its extent, it could be identified as a unit despite changes in its facies. But if, as is usual, only widely separated parts are exposed and if each part represents a different facies, its contained fossils would be needed for correlation. A difficulty

arises here because the assemblages of fossils in two facies may not be exactly the same, even though the organisms they represent lived at the same time. This happens because the environments in which the organisms lived were different. In the same sea, deep-water shellfish are unlike shallow-water kinds; and on land, animals living in deserts are unlike those living at the same time in moist, forested regions. These variations of fossils with varying facies do not, however, make correlation impossible; they only make it more difficult.

Field Study of a Sequence of Strata. Having followed a single unit laterally through several facies, we can now examine a vertical sequence of several units, such as the sequence shown in Fig. 14.14, exposed on a steep slope. We examine the rocks systematically by climbing up the slope, taking notes as we go. We identify the kinds of rock present and determine the thickness of each of the units, of which the figure shows five, lettered *A* to *E*. The surfaces of contact between any two units are not alike. The contacts between *D* and *E* and between *C* and *D* are sharp and distinct, whereas those between *A* and *B* and between *B* and *C* are transitional; that is, they represent a gradation between the layer beneath and the layer above. Looking for fossils to help determine the origins of the rocks, we find fossils of marine animals in units *A*, *B*, *C*, and *D* and fossil bones of land animals in unit *E*.

Examining sandstone *A*, we find it is made up of grains of quartz and other minerals. All the minerals in this group occur in igneous rocks; so we infer that the sandstone was built of the sediment resulting from erosion of igneous rocks. The sediment might have been derived from its parent rocks either directly or remotely, for it could, for a long time, have formed part of one or more sandstones older than the one we are examining. In either case the sandstone shows sorting, for the proportion of quartz to other minerals is far greater than in most igneous rocks. Probably this means that the durable quartz grains survived their trip from the region of erosion to the region of deposition in much greater numbers than did their less resistant associates. However, under a hand lens we can see that even the quartz grains have become moderately well rounded. From this we judge roughly that the journey involved considerable distances, long periods of time, or both.

Stratification of the sandstone is indistinct. The

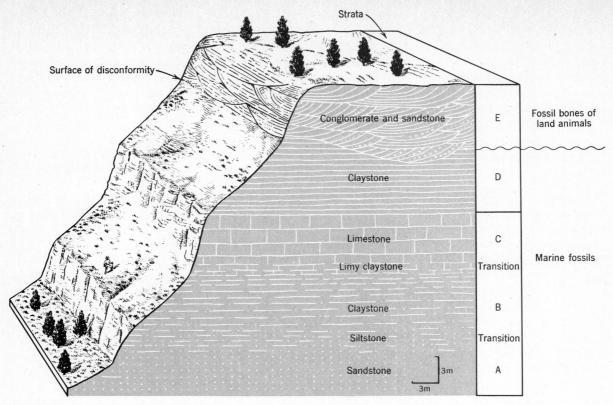

Figure 14.14   Idealized sequence of strata exposed on a steep slope (compare Fig. D-8).

rock lacks well-marked layers because the diameters of all the sand grains are nearly the same; hence there is little distinction between one layer and the next. Yet the faint layers that are present lie flat and nearly parallel. From this we infer the sand was deposited in water deep enough to be beyond the reach of waves and strong currents. We can suppose the sand was either derived from the wave erosion of cliffs of an older sandstone or brought into the sea by rivers that carried some of the products of weathering of inland rocks. In the latter case the streams must have possessed enough energy to enable them to move sand grains; this in turn suggests the slope was fairly steep and therefore that the tributary region was one of hills or mountains. These are only suggestions, which can hardly be verified from a study of this one exposure. Verification could come only from a wide study of the sandstone and of the tributary region.

We note that upward the sand becomes finer and is accompanied by silt. The silt then increases and

is accompanied by clay, so that we pass upward from sandstone through siltstone into claystone (unit *B*). In the claystone we find the same near-parallelism of the laminae and much the same kinds of marine fossils. Hence the only properties that have changed, as compared with the sandstone beneath, are sizes and shapes of grains and also kinds of minerals present, because the fine particles that constitute claystone consist not of quartz but of clay minerals. We can explain the change in one of two ways. Either the water deepened, shifting the zone of sand deposition farther toward shore and causing deposition of mud here or the contributing streams lost energy and deposited their sand before reaching the sea. To determine the cause we would have to study a much wider region.

The upper part of the claystone becomes limy; that is, it begins to include calcite along with the clay minerals. Also it grades upward into nearly pure limestone (unit *C*) containing the same general kinds of fossils as those beneath. On close inspection the limestone is seen to be a mechanical deposit of tiny shell fragments. We can conclude with confidence that, in part of the sea, the water cleared and marine organisms became abundant. This could have been brought about by continued deepening.

Examining unit D, we find no transition from C, but instead an abrupt change from limestone to claystone. Here we find the same kinds of fossils and the same flat-lying, parallel layers; only the composition of the rock has changed. We infer sudden shoaling of the water or sudden increase in stream energy rather than a gradual change.

A summary of our findings on units A, B, C, and D suggests a former sea, deepening gradually and perhaps somewhat irregularly or fed with sediment by rivers with changing characteristics. Hence we are left with uncertainties. However, these are not unexpected, for the nature of the evidence we must rely on often permits us to do no more than judge the relative probability of two possible explanations of the same feature.

However, from our observation that units A, B, C, and D are parallel and show no sign of intervening breaks, we infer with little doubt that together they represent continuous, unbroken sedimentation on the sea floor. Such strata are said to be *conformable*.

Unit E differs from underlying units. It is part sandstone and part conglomerate and is cross-stratified in wedge-shaped units. The repetition of wedge-shaped beds suggests currents that shifted in position and direction; the occurrence of both sand and gravel indicates changes in energy as well; and finally the common direction of inclination of the cross-strata suggests general flow from right to left.

Also unit E contains the fossil bones of land animals. This fact makes a marine origin unlikely; the presence of conglomerate rules out the wind. Accordingly, we conclude that probably E was deposited by a stream or streams, and the dimensions of the wedge-shaped beds give a clue to the size of the stream channels. Streams today are seen to deposit similar beds.

Evidently E overlies D with a contact that is notably irregular and that cuts through some of the uppermost layers in D. This relationship could result only from erosion of D before or during the deposition of E. When two parallel strata are separated by an irregular surface of erosion, the relationship between them is said to be *disconformable*. The irregu-

lar interface is called a *surface of disconformity*.

From these facts we judge that the sea floor emerged for some reason not shown by the evidence at this place. Unit D thereby became land, and a stream flowed over it, eroding an unknown amount of its upper part. To determine how much of D was removed during this erosion and whether to attribute all of the erosion to the stream or part of it to some agency that antedated the stream, we should have to examine D at many other places. At any rate, the stream possessed rather high energy because it carried a bed load of pebbles as well as sand and because the load was so abundant that some was deposited. Evidently the stream was gradually building up its bed because the wedge-shaped layers are piled one on another.

Like most sand grains, those of E are mainly quartz, but the pebbles are samples of various kinds of rock. Here is a clue to the region from which the stream brought the pebbles. To make use of the clue we should have to identify the kinds of rock in the pebbles and to search (probably in the direction to the right of Fig. 14.14) for areas in which those same kinds of rock occur as bedrock. A further clue to the distance the sediment has been transported is found in the degree of rounding of the particles.

This is the way in which sedimentary rocks are commonly studied in the field. From such a study it is possible to re-create a general sequence of events, but, as we have seen, not all our inferences are firm; we can not always rule out other possibilities; and the study leads to questions that can be answered, if at all, only by extending the inquiry into the surrounding region.

Rock Units: Formations. We identified unit A (in Fig. 14.14) as sandstone, but a thorough study must go much further. The unit must be distinguished from other sandstones. One respect in which it differs from all others is its position in the vertical sequence of strata. Hence we give it a designation by which its position is fixed and by which it can be catalogued and referred to. The basic rock unit to which such a designation is applied is the *formation* (App. D). A formation must constitute a mappable unit; that is, a unit (1) thick and extensive enough to be shown to scale on a geologic map, and (2) distinguishable from the strata immediately above and below not just at one exposure but generally wherever the unit is exposed. Within these requirements a formation can be thin or thick, to suit the geologist's convenience. Its thickness is likely to depend on the degree of detail of the field study and correspondingly on the scale of the map to be made. In North America each formation is given a name, typically the name of a locality near which it is exposed (Lexington Limestone, Fox Hills Sandstone, Green River Formation). Not only sedimentary rocks but igneous and metamorphic rocks as well are identified as formations (examples: San Juan Tuff, Conway Schist). Figure D.9 is a series of geologic sketch maps showing three formations.

Formations can be subdivided and also grouped into larger rock units. But because we are concerned here only with the way in which strata are identified, the formation alone serves our purpose. For more about this see Appendix D.

Matching of Rock Units by Physical Characteristics. Once the layered rock units have been identified in vertical sequence, the extent of each (that is, the area it underlies) must be determined as closely as possible. With few exceptions, layers of sediment are deposited in basins located on land, beneath lakes, or in the sea. A layer of sediment may cover all or only a part of a basin. If deposited in the sea or in a lake, the stratum is likely to extend over the whole basin floor. However, during the time, perhaps very long, since conversion from sediment into sedimentary rock, the stratum may have been eroded so much that only parts or remnants of it may now be preserved. An example is the Pittsburgh Coal Seam, an easily recognized formation, whose original area of perhaps 50,000 square kilometers has been reduced by long-continued erosion to approximately 15,000 square kilometers. One

of the tasks of the geologist is to study the remnants so thoroughly that he can determine as nearly as possible the original extent of each formation.

This is done by matching the remnants, preserved in the hills left by erosion, on a basis of physical characteristics such as grain size, grain shape, mineral content, kind of stratification, and color. Matching on this basis is likely to be reliable through short distances, but generally becomes less reliable through longer ones because the physical characteristics tend to change in lateral directions.

Correlation by Means of Fossils. The usefulness of matching by physical characteristics is virtually limited to the area of the basin in which the strata were originally deposited. In order to determine equivalence in the ages of strata in different basins, and even in different continents, geologists compare fossils, the chief basis of stratigraphic correlation (Fig. 2.8).

Correlation by Means of Radiometric Dates. As we learned in Chap. 2, the ages of fossil-bearing strata are fixed by the radiometric dates of igneous bodies closely related to them. Such dates are useful also in correlation. Where, as is not uncommon, sedimentary layers contain no fossils, we can fix their ages approximately through the radiometric ages of related igneous rocks. In some cases this enables us to correlate dated strata that lack fossils with dated strata that contain them, thus gradually enlarging the known extent of strata whose positions in the geologic column have been fixed by fossils.

We have now examined what strata themselves indicate about geologic history. In Chap. 15 we shall see what strata can reveal about geologic structure.

Summary

1. Sedimentary rocks are the product of complex processes of sorting and transport.

2. Most sedimentary strata were nearly horizontal when their sediments accumulated.

3. The most obvious characteristic of sedimentary rocks is that they are stratified. Strata include parallel strata and cross-strata. Various arrangements of the particles are seen in uniform layers, graded layers, and nonsorted layers.

4. Particles of sediment become rounded and sorted in transport by water and air but not in transport by glaciers and by mass-wasting.

5. Ripple marks, mud cracks, fossils, and in some cases red color in sedimentary rocks give evidence of environments of deposition.

6. Stratigraphy is the study of stratified rocks. From it most of our knowledge of Earth's history has been learned.

7. An extensive unit of strata may possess several facies, each determined by a different environment of deposition.

8. The basic physical rock units are formations, each with a locality name.

9. By means of fossils and in some instances by radiometric dates strata can be correlated through long distances.

10. Correlation takes account of the two principles that strata were horizontal when deposited and that they were formed in sequence from bottom to top.

Selected References

Dunbar, C. O., and Rodgers, John, 1957, Principles of stratigraphy: New York, John Wiley.

Hatch, F. H., and Rastall, R. H., 1965, The petrology of sedimentary rocks, 4th ed., revised by J. T. Greensmith: London, Thomas Murby.

Krumbein, W. C., and Sloss, L. L., 1963, Stratigraphy and sedimentation, 2d ed.: San Francisco, W. H. Freeman.

Lahee, F. H., 1961, Field geology, 6th ed.: New York, McGraw-Hill.

Shrock, R. R., 1948, Sequence in layered rocks: New York, McGraw-Hill.

Walker, T. R., 1967, Formation of red beds in modern and ancient deserts: Geol. Soc. America Bull., v. 78, p. 353–368.

Red-hot lava streams out of a volcanic cone during an eruption at Kilauea Volcano, Hawaii. Fountains of lava several feet high are driven by volcanic gases. As the lava cools and forms new rock, the cone grows in size. (*Rick Golt/Rapho Guillumette.*)

Part IV Internal Processes

Chapter 15

Deformation of Rock

How Rocks Are Deformed

The preceding chapters deal largely with Earth's external activities. The next four chapters discuss internal activities such as deformation of rocks, volcanoes, formation of mountains, and earthquakes.

We have seen that the relations between internal and external activities are in a steady state. Continents remain above sea level, indicating that new rocks are brought to the surface as fast as old ones are removed by erosion; new basins form, and catch sediments when old basins fill up. This balance between internal and external activities is indirect evidence that materials of Earth's crust are able to move. But direct evidence of movement exists also. Continents have been rent asunder and the fragments moved thousands of kilometers from their former positions (Fig. 1.10). This proves that great slabs of the lithosphere move horizontally across the mantle at rates as fast as a few centimeters per year. Similarly, layers of sedimentary rock that abound in fossils of marine organisms lie in high mountains and plateaus, far above sea level; large bodies of igneous rock that must have cooled deep underground are widely exposed in high country; and distinctive features that match closely those seen along present seashores are found on high terraces. Far from being quiet and passive, Earth's seemingly solid crust is always in motion, both horizontally and vertically.

Solids cannot flow as liquids do. But they can be deformed, so that their shapes become permanently changed. Change of shape means movement of matter, and we can therefore say that movement within solids indicates deformation. But how is so much movement, and therefore so much deformation, possible in solid rock? Part of the answer lies in fractures. We call *a fracture along which the opposite sides have been displaced relative to each other,* a **fault.** But faults are only part of the answer. We have already seen evidence that rocks can be folded (Fig. 1.8) and sometimes shattered like glass (Fig. 1.9); so it is clear that deformation occurs in other ways too. Let us see how this happens.

What Controls Deformation? Fracturing of rocks, as in the development of faults, is a process easy to understand. In ordinary experience we regard rock as strong but brittle. Building stone, for example, keeps its shape under heavy loads but breaks like glass under sharp blows of a hammer, or with very forcible bending.

How then, can apparently brittle rock be folded, or even show evidence of deformation that looks like flow (Fig. 1.7)? Folding and flowing occur deep in Earth's crust. To understand them we must have in mind some elementary properties of solids. Any solid, including rock, can be deformed as if it were a mass of stiff dough—under certain conditions. The essential conditions are (1) confining pressure, (2) temperature, and (3) time. The higher the confining pressure and the higher the temperature, the weaker and less brittle a solid becomes. We all know, for example, that a rod of iron or of glass is difficult to bend at room temperature. If we try too hard, both will break. But both can be readily bent if we heat them to redness over a flame. The effect of confining pressure is less familiar in common experience. By confining pressure we mean, simply, steady squeezing from all directions. Confining pressure tends to keep a solid mass in the form of a single body and hinders the formation of faults. At high confining pressures, therefore, it is easier for rock to bend and flow than it is for it to break. Weakening of rocks by the high temperatures associated with deep burial, and at the same time loss of brittleness caused by high confining pressures, are therefore part of the reason why solid rock can be bent and folded.

The effect of time on deformation of rock is vitally important, but as with confining pressure, it is not obvious from common experience. When applied to a solid, a force is transmitted by all the constituent atoms of the solid. If the force exceeds the strength of the bonds between atoms, either the atoms must move to relieve the force, or the bonds must break. But atoms in solids cannot move rapidly. So if the force is sharp and quickly applied, the solid breaks. But if the force is applied slowly and is maintained for a long period, the atoms have time to move, and the solid can slowly readjust and change shape by folding and flowing. Forces in the Earth tend to be applied slowly. Earth has had abundant time for even the strongest and most massive rock to be folded and deformed.

Deformation occurs and movement results. These two events are intimately connected. Let us examine first some of the evidence of movement and deformation in the crust today, and then evidence of deformation in the past.

Detection of Movement in Progress

We cannot observe directly events that happen deep inside the Earth, but sometimes we can detect movement of the land surface that is caused by deep-seated events. Apparently many movements happen so slowly that they cannot be measured over a few tens or even a few hundreds of years. Other movements can be detected, however, and for convenience we divide them into two groups: *abrupt movements,* in which blocks of the crust suddenly move a few centimeters or a few meters in a matter of minutes or hours; and *gradual movements,* in which slow, steady motions occur without any abrupt jarring.

Abrupt Movement. Movements along faults tend to be abrupt rather than slow, steady slidings. Forces build up slowly, until friction between the two sides of the fault is overcome. Then abrupt slippage occurs. If the forces persist, the whole cycle of slow buildup, culminating in an abrupt movement, repeats itself. Although movement on a large fault may eventually total many kilometers, it is the sum of numerous small, sudden slips. Each sudden movement may cause an earthquake and, if the movement occurs near Earth's surface, may disrupt and displace surface features, and in doing so, leave clear evidence of the amount of movement.

During the San Francisco Earthquake of 1906, abrupt horizontal movement occurred along a large fault, the San Andreas Fault. Roads and fences that crossed the fault were offset by as much as 7m. In

1940 another earthquake occurred, again with horizontal movement, along the same fault, this time in the Imperial Valley, nearly 800km southeast of San Francisco. The displacement, as much as 5.5m, was registered accurately by offset rows of fruit trees as well as by broken fences. Horizontal movement, then, seems to be a habit of the San Andreas Fault. As we shall see in Chapter 18, this is because that fault marks the boundary along which two plates of the lithosphere are sliding past each other. In this case the surface movement is a record of deformation caused by forces that operate hundreds of kilometers down within the Earth.

Abrupt vertical movements are more obvious than horizontal movements. During the eruption of Kilauea Volcano, Hawaii, in December 1965, underground movement of magma caused abrupt deformation of the surface. Fault cliffs as much as 1m high were formed; one of them cut a main road (Fig. 15.1). Most faults, however, display a combination of horizontal and vertical motion (Fig. 15.2).

The largest known abrupt displacement occurred in 1899 at Yakutat Bay, Alaska, at the time of an earthquake. A stretch of the Alaskan shore, with beach, barnacle-covered rocks, and other telltale features was suddenly lifted as much as 15m above sea level. This visible displacement may be less than the total, because the fault is hidden offshore and the crustal block on the other side of it, entirely beneath the sea, possibly moved downward, thus adding to the total displacement.

Gradual Movement. Movement along faults is not always abrupt, nor is it always accompanied by earthquakes. Measurements along the San Andreas Fault reveal places where slow, steady slipping occurs, sometimes reaching a rate of 5cm a year. Similarly, subterranean movement of magma does not always cause sudden faulting such as that shown in Figure 15.1. Scientists who have studied Kilauea Volcano observe that the top of the mountain slowly rises as magma flows upward prior to eruption. The magma is apparently driven upward under such pressure that it causes the whole mountain to be inflated, as compressed air inflates a balloon (Fig. 15.3). Following eruption, the mountain becomes

Figure 15.1 On December 25, 1965, a fault abruptly cut one of the main roads near Kilauea Volcano, Hawaii. The difference in height of the two sides of the fault is only 1m, but the damage was increased greatly by slump along the fault interface. Faulting occurred when large volumes of magma, previously held in deep-lying magma chambers, were rapidly erupted. (*R. S. Fiske, U.S. Geol. Survey.*)

Figure 15.2 Scarp formed by a fault that crosses 9th Avenue, Anchorage, Alaska. Appearing during the earthquake of March 27, 1964, the fault shows vertical displacement of approximately 2m and horizontal displacement of about 1m. The position of the roadway shows that the upper block has moved from left to right relative to the lower block. The small fault in the foreground likewise shows horizontal movement of about 0.5m. Fault displacement is more recent than the snow. (*Alaska Pictorial Service.*)

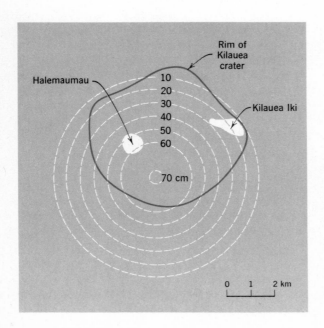

Figure 15.3 Inflation of Kilauea Volcano, Hawaii by magma rising to erupt. Inflation was expressed by a rising land surface during a 22-month period, from January 1966 to October 1967. Contours, showing the total rise in centimeters, are not quite centered over the crater at the summit of Kilauea, nor do they coincide exactly with two of the recently active vents, Halemaumau and Kilauea Iki. From the amount of inflation, and the shapes and positions of the contours, scientists calculate that the magma accumulated in a chamber 2 to 3km below Kilauea Crater, and then moved quickly upward at the time of eruption. (*After R. S. Fiske and W. T. Kinoshita, Science, v. 165, p. 341, 1969.*)

deflated and the surface slowly falls again, but never to exactly the same position it started from.

A classic example of slow changes in the level of the land is seen in the ruins of an ancient Roman marketplace known as the Temple of Serapis, west of Naples. Three columns left standing have been bored by a peculiar marine clam to a height about 6m above the floor (Fig. 15.4), and the shells of the clams still line some of the borings. Along the shore near the ruin is sediment that contains abundant shells of ordinary clams like those now living in the adjoining bay; these deposits are exposed in bluffs as much as 7m above present sea level. At first thought, one might try to explain these observations by a worldwide rise and later sinking of sea level while the land remained stationary. But such fluctuation would have left a record in coastal belts everywhere. As the evidence cited above is found only within a limited area near the old ruin, we conclude that at some time after the Romans built the Temple, this part of the Italian coast slowly sank, and then was re-elevated within very recent time. According to local records, the re-elevation started not long after A.D. 1500.

The examples just cited involve very small portions of the crust. Movement of much larger areas is also well documented. One example is an effect of the latest of the glacial ages (Chap. 11). Any large weight placed on Earth's surface deforms the crust, causing it to sag downward. The great ice masses that covered large parts of continents had just this effect; in some areas they pushed down the crust by at least several hundred meters. When, later, the ice melted, the pressure was relieved; the crust started to rise back to its original position, and although at least 7000 years have passed since the ice masses

Clam borings

High-water mark

Figure 15.4 Columns in a Roman ruin at Pozzuoli, Italy as they appeared in 1828. Borings made by marine clams indicate former submergence. (*Drawing from Lyell, 1875, Principles of Geology, 12th ed.*)

completely disappeared, it is still rising. In north-eastern North America, where the ice was thickest, beaches that were at sea level a few thousand years ago are now hundreds of meters above sea level (Fig. 15.5). The rate of rise near Hudson Bay is still 1m per century. In parts of Scandinavia it is even faster.

It may be that across the entire land surface of the United States there is no spot that is completely stationary. Measurements by U.S. Government surveyors over the past 100 years reveal great areas where the surface is slowly rising, and others where it is slowly sinking (Fig. 15.6). Not all the causes

of these vast, slow movements are well understood, but the movements do prove that the solid Earth is not as solid as it seems at first sight, and that great internal forces are continually deforming its crust.

Evidence of Former Movement

With so much evidence of current movement and deformation of Earth's crust, we might reasonably expect to find a great deal of evidence of former movements. Indeed we do. Studies of land and sea-bottom topography provide abundant evidence of

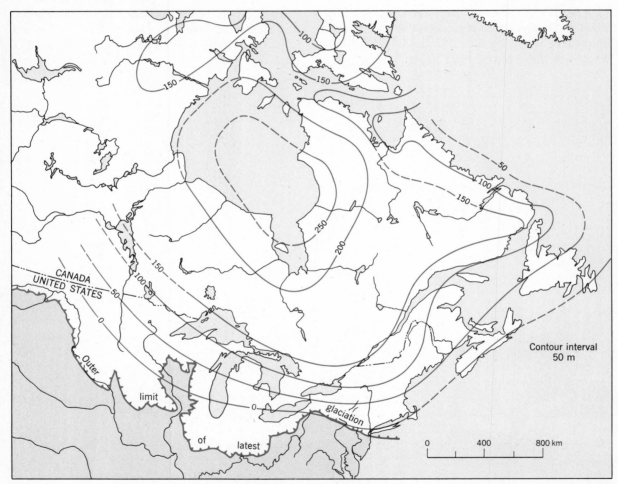

Figure 15.5 Ice, centered on Hudson Bay during the latest glacial age, caused downward buckling of the crust in northeastern North America. When the ice melted, the land surface started to rise again. Beaches that formed at sea level before melting are now many meters above it, and so provide an accurate measure of the amount of uplift. Contours show amount of uplift. Age of beaches, determined radiometrically with ^{14}C, proves rate of uplift is as fast as 1m per century. (*After P. B. King, 1965.*)

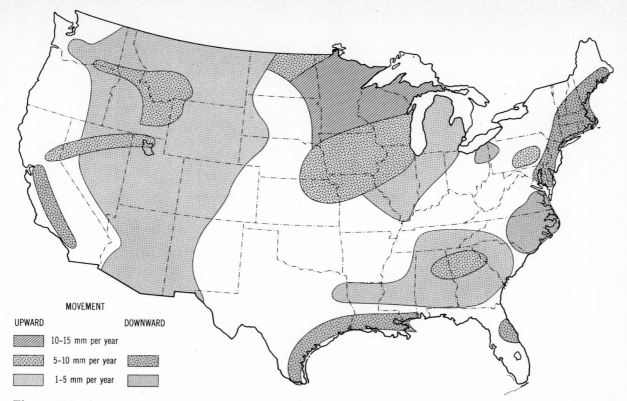

MOVEMENT

UPWARD   DOWNWARD

10-15 mm per year

5-10 mm per year

1-5 mm per year

Figure 15.6 Accurate measurements over a 100-year period show that in large areas of the United States, the surface is slowly moving up or down. Subsidence along the coasts of California and the Gulf of Mexico is believed to have been caused in part by withdrawal of gas, oil and water, which allowed subsurface reservoirs to collapse. Uplift near the Great Lakes is a rebound effect following melting of the last ice sheet (Fig. 15.5).

The causes of movements in other areas are not known with certainty. Those areas in which no movement is shown are not necessarily stationary. They are simply areas in which measurements are very few. (*After S. P. Hand, National Oceanic and Atmospheric Administration, 1972.*)

vertical movements; and in some areas the distributions of various kinds of rock provide clear evidence that horizontal movements through distances as great as several hundred kilometers have occurred. Let us examine a few examples of such evidence. First, topographic evidence.

Topographic Features. In many parts of the world well-developed wave-cut benches stand one above another like stairsteps (Fig. 15.7). Some of the lowest steps terminate inland against typical wave-cut cliffs, where they can be seen still decorated with barnacle shells and standing behind beaches. In higher terraces these critical markings have been dimmed or removed by erosion, but the close similarity in gen-

eral form indicates a like origin for all the steps in a series. Along the California coast the highest recognizable terraces are more than 450m above the sea. Because these terraces are found along only a part of that coastal area, we reason that a segment of the Coast Ranges has risen in a succession of pulses, separated by pauses long enough to allow development of beaches and cliffs. The altitude of a terrace also varies appreciably from place to place, thereby telling us that the amount of uplift was irregular.

Some topographic evidence indicates large-scale subsidence. Accurate surveys of the sea floor north of the Aleutian Islands, Alaska, reveal a submarine topography of high ridges and hills separated by valleys that unite in what appears to be a well-

Figure 15.7 Wave-cut benches far above sea level at Palos Verdes in southern California. The altitude of the lowest wave-cut bench (1), is about 50m, that of the highest, more than 250m. Photo, taken in 1933, shows the ship *Gratia,* stranded during a storm in 1932. (*Aero Service Corp., division of Litton Industries.*)

developed drainage system. The best explanation seems to be the submergence of a wide landscape that was shaped by mass-wasting and stream erosion. Was this drowning of a former landscape caused by rise of sea level or by sinking of the land? We know the level of the sea was raised by the return of water that had been locked up in ice sheets during the glacial ages. The total rise from this cause is estimated to have been about 100m, but the drowned hills and valleys near the Aleutian Islands lie at depths greater than 400m. We infer, therefore, that a large part of the submergence was caused by the sinking of land.

Additional evidence of vertical movements can be found in volcanic islands in the ocean basins. We read in Chapter 12 that the wave-cut tops of many guyots are now covered by seawater more than 1km deep, indicating large subsidence. Some volcanic islands contain evidence of the opposite effect; for example, Aguijan Island in the southwest Pacific has several elevated shore lines, clear evidence of rising land (Fig. 15.8).

Figure 15.8 Emerged shorelines, built by corals and other organisms, crown Aguijan Island, Marianas, in the southwest Pacific. Beneath the corals lies an extinct volcano. Prominent steep slopes are the seaward sides of reefs; their upper edges mark former sea levels; the flat areas are the floors of ancient lagoons. (*U.S. Air Force.*)

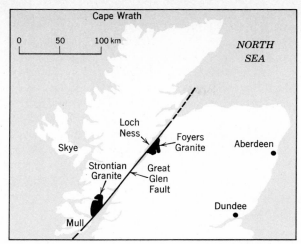

Figure 15.9 The Great Glen Fault in Scotland cuts a granite mass into two separate segments, the Strontian and Foyers Granites. Horizontal movement along the fault has separated the masses by approximately 100km. (*After W. Q. Kennedy, 1946.*)

Bedrock Features. When we examine the distribution of various kinds of rock, we find so much evidence of former movement that one striking example will suffice. A remarkable fault, the Great Glen Fault, crosses Scotland from southwest to northeast. The fault is a line of easy erosion, and a string of lakes now marks its path. One lake is Loch Ness, home of the famous Loch Ness Monster.

When the Great Glen Fault was active, during the Paleozoic Era, it severed a large granite mass into two fragments. The fragments now lie on opposite sides of the fault, approximately 100km apart, and thus provide striking evidence of large-scale horizontal movement (Fig. 15.9).

Not all evidence of movement and deformation observed in bedrock is as striking and obvious as the Great Glen Fault. But once we learn to recognize it, evidence of deformation is seen to be very widespread—so much so that a special branch of geology, *structural geology,* has the study of rock deformation as its primary goal. In order to understand and discuss Earth's internal activities, we must also be able to recognize and evaluate evidence of rock deformation. Let us first examine deformation by fracture.

Deformation by Fracture

We have learned that it is the brittle properties of rock that lead to fracture. Rock in the crust, espe-

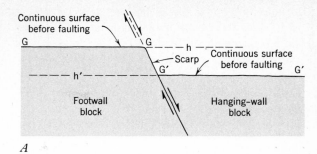

A

B

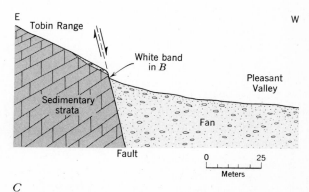

C

Figure 15.10 Fault that displaces surface of ground.

A. Scarp made by displacement of a formerly level surface G-G′ may mean that the hanging-wall block moved up from position h′; or that both blocks moved to some degree, to create the net displacement shown.

B. Scarp at west base of Tobin Range, Nevada, formed in 1915 by abrupt movement along a fault that generated an earthquake. The whitish band of newly exposed rock marks displacement at the top of the fan, several miles from the camera (*Eliot Blackwelder*).

C. Vertical section across lower part of Tobin Range to Pleasant Valley, approximately in the direction of the photographer's line of sight in *B*. Half arrows indicate relative movement of crust blocks.

291

cially rock close to the surface, tends to be brittle; so it is cut by innumerable fractures. The fractures aid erosion, serve as channels for the circulation of ground water, provide entryways along which magma is intruded to form dikes and sills, and in many places serve as the openings in which veins of valuable minerals are deposited.

Most fractures are too small for slipping to have occurred along them. Along many fractures, however, movement has occurred, converting the fracture into a fault.

Faults. Generally there is no way of telling how much actual movement has occurred along a fault. Even if a crystal or a pebble in the rock has been cut through by the fracture and the halves carried apart a measurable distance, we cannot know whether one block stood still while the other moved past it, or whether both sides shared in the movement. Precise surveys of points on the ground before and after movement on an active fault can, of course, indicate which side of a fault is moving. Such a check is made continuously along the San Andreas Fault. But most faults with which we have to deal are old features whose former expression at Earth's surface was long ago destroyed by erosion. In classifying fault movements, then, we can only speak of *apparent* and *relative* displacements.

Block diagram	Name of fault	Definition
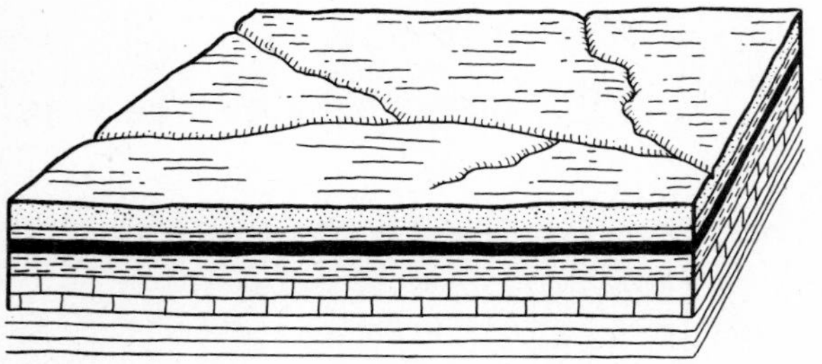		Reference block before faulting. Drainage is from left to right.
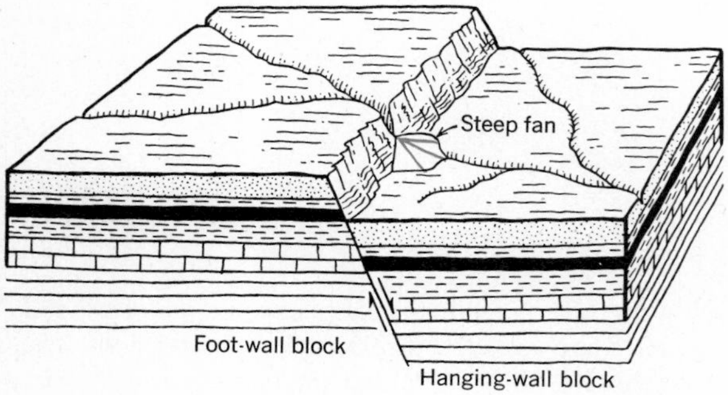	**Normal fault**	A fault, generally steeply inclined, along which the hanging-wall block has moved relatively downward.
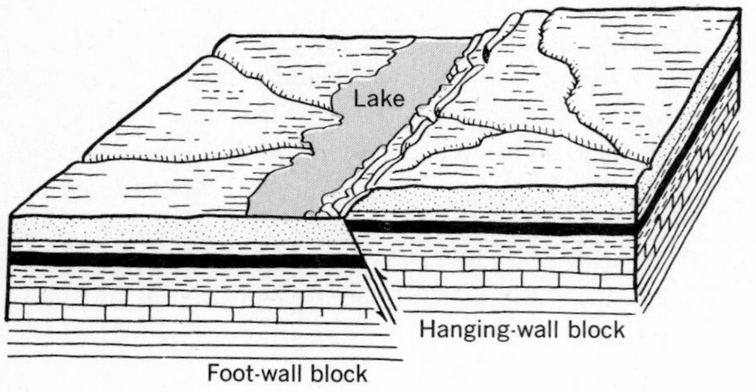	**Reverse fault**	A fault, generally steeply inclined, along which the hanging-wall block has moved relatively upward. A normal or reverse fault on which the only component of movement lies in a vertical plane normal to the strike of the fault surface is a *dip-slip fault*.

Figure 15.11 Principal kinds of faults and some of the topographic changes they cause.

Most faulting occurs along fractures that are inclined. Many veins of metallic ore lie along faults, and we have inherited some terms used by miners digging out the veins. From the miner's viewpoint, one *wall* of an inclined vein overhangs him, the other is beneath his feet. The ***hanging wall*** *is the surface of the block of rock above an inclined fault; the surface of the block of rock below an inclined fault* is the ***footwall*** (Fig. 15.10). These terms, of course, do not apply to vertical faults.

Faults are grouped into classes according to (1) the inclination of the surface along which fracture has occurred and (2) the direction of relative movement of the rock on its two sides. The common classes of faults, together with the changes in local topography they sometimes create, are listed in Figure 15.11. Our standard planes of reference in classifying faults are the vertical and the horizontal. Along some faults, movement is confined to these two reference planes, although, as we saw in Figure 15.2, along other faults both vertical and horizontal movements occur. Not all movement on faults is in straight lines. Sometimes fault blocks are rotated, and, as seen in Figure 15.11, we classify such faults as *hinge faults*.

Normal Faults. Normal faults are caused by tensional forces that tend to pull the crust apart, and

Block diagram	Name of fault	Definition
	Strike-slip fault	*A fault on which displacement has been horizontal.* Movement of a strike-slip fault is described by looking directly across the fault and by noting which way the block on the opposite side has moved. The example shown is a *left-lateral fault* because the opposite block has moved to the left. If the opposite block has moved to the right it is a *right-lateral fault.* Notice that horizontal strata show no vertical dieplacement.
	Oblique-slip fault	*A fault on which movement includes both horizontal and vertical components.* See also Fig. 15-2.
	Hinge fault	*A fault on which displacement dies out (perceptibly) along strike and ends at a definite point.* Figure 15-2 shows a small example.

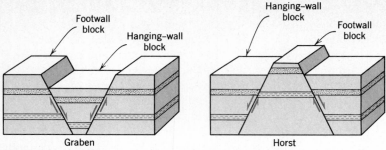

Figure 15.12 *Graben, a trench-like structure bounded by parallel normal faults,* formed when the hanging-wall block that forms the trench floor moves downward relative to the footwall blocks. *Horst, an elevated elongate block bounded by parallel normal faults,* formed when the elevated footwall block moves upward relative to the hanging-wall blocks.

also by forces tending to expand the crust by pushing it upward from below. In the Earth's crust are many zones that have been deformed repeatedly by normal faulting. Commonly two or more similarly trending normal faults enclose an elongate segment of the crust that moves upward or downward. As shown in Figure 15.12, a down-dropped block is a *graben,* an upthrust block a *horst.* The world's most famous graben is the African Rift Valley, which runs north-south through more than 6000km. Along parts of it are volcanoes where magma has formed channel-ways upward along the fault surfaces. The Rift Valley is important for many reasons, but it has a unique claim to fame. In Kenya and Tanzania the earliest human and prehuman fossil bones have been found in sediments deposited in old Rift Valley lakes.

Normal faults are innumerable. Horsts and grabens are also very common, although none is as spectacular as the Rift Valley. The valley in which the Rhine River flows through western Europe follows a series of grabens, and Lake Baikal in central Asia, Earth's deepest lake, is located in a very deep graben. A spectacular example of normal faulting is found in the Basin-and-Range province in Utah and Nevada. There, movement on a series of parallel and sub-parallel normal faults trending north-south, has formed horsts and grabens that are now mountain ranges and sedimentary basins. The province starts with the Wasatch Range in the east and continues westward to the eastern edge of the Sierra Nevada.

Reverse Faults. Reverse faults arise from compressive forces. Movement on reverse faults tends to push older rocks over younger ones, thereby shortening and thickening the crust.

A special class of reverse faults, called **thrust faults** and generally known as **thrusts,** are *low-angle reverse faults with dips generally less than 45°.* Such faults, common in great mountain chains, are noteworthy because along some of them the hanging-wall block has moved many kilometers over the footwall block. In most cases the hanging-wall block, thousands of meters thick, consists of rocks much older than those adjacent to the thrust on the footwall block (Fig. 15.13). The strata above some thrusts lie nearly parallel to those beneath, and so may appear deceptively, to represent an unbroken sequence.

Strike-Slip Faults. We have already mentioned two strike-slip faults, the San Andreas Fault and the Great Glen Fault (Fig. 15.9). The San Andreas is a right-lateral strike-slip fault, and it is believed movement has been occurring along it from at least as long ago as Cretaceous time to the present day—in other words through at least 60 million years. The total movement is not known, but some evidence suggests it amounts to more than 600km. The Great Glen Fault is a left-lateral strike-slip fault; it is no longer active.

Strike-slip faults appear to be widespread in the oceanic crust. For example, many of the great fracture zones that cut the mid-ocean ridge system are strike-slip faults. In Figure 12.8, each of the numerous fracture zones that offset the Mid-Atlantic Ridge is a strike-slip fault. Indeed, so common are

A

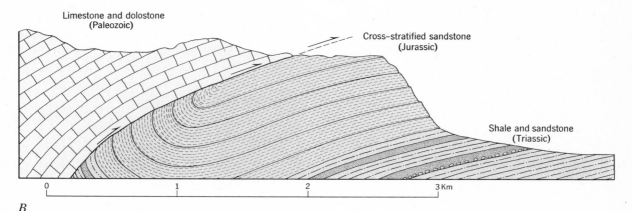

B

Figure 15.13 Keystone Thrust, west of Las Vegas, Nevada.

A. Air view northward shows the fault clearly defined by a color contrast in the strata adjacent to it. Light-colored Jurassic sandstone, forming a cliff nearly 2000 feet high (right), lies below the fault; dark-colored Paleozoic limestones and dolostones (left) lie above it. (*J. S. Shelton.*)

B. Section drawn along white line in photograph and extending somewhat farther east and west. Canyons crossing the fault reveal that it steepens downward toward the west and crosses overturned layers of the sandstone. Farther east the fault becomes essentially parallel to the sedimentary layers, both below and above (*After C. R. Longwell.*)

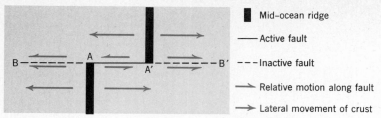

■	Mid-ocean ridge
—	Active fault
- - -	Inactive fault
→	Relative motion along fault
→	Lateral movement of crust

Figure 15.14 Transform faults, a special class of strike-slip faults. In the plan above, a mid-ocean ridge (black) is offset by a transform fault. Crust on both sides of the ridge segments is moving laterally away from the ridge. Between segments of the ridge, along *AA'*, movement on the two sides of the fault is in opposite directions. Beyond the ridge, however, along segments *AB* and *A'B'*, movement on both sides of the fault is in the same direction. A transform fault does not, therefore, cause the ridge segments to move continuously apart.

they that it has been suggested that the four major structural forms of Earth's crust are mountains, deep-sea trenches, mid-ocean ridges, and strike-slip faults. This idea was first proposed by J. T. Wilson, a Canadian scientist, who pointed out that these four major structural features link together and seem to form continuous networks encircling the Earth. When one feature terminates, another commences; their junction point is called a *transform*. Wilson proposed that *a special class of strike-slip faults that links major structural features* occurs at many junctions; he called them **transform faults.** Close study of the strike-slip faults that offset the mid-ocean ridges proved Wilson's suggestion correct. As seen in Figure 15.14, movement along transform faults cutting the ridges is a consequence of the continuous addition of new crustal material and the lateral movement of older crust away from the ridge.

Evidence of Movement Along Faults. Often we find fractures in rock but cannot tell at first glance whether or not movement has occurred along them. For example, in uniform, even-grained rock such as granite, or in a pile of thinly bedded strata, no one of which is unique or distinctive, we would not see displacement of any obvious features. However, examination of the fault surface, or of rock immediately adjacent to it, commonly reveals signs of local deformation, indicating the movement has occurred. Under special circumstances, even the direction of movement can be deciphered.

Adjacent to some faults, bending of strata or other internal features can be seen. Large and small struc-

tures created by bending adjacent to faults are known collectively as *fault drag* (Fig. 15.15).

Movement of one mass of rock past another can cause the fault surfaces to be smoothed, striated, and grooved. *Striated or highly polished surfaces on hard rocks, abraded by movement along a fault,* are **slickensides.** Parallel grooves and striations on such surfaces record the direction of latest movement (Fig. 15.16).

Not all fault surfaces are slickensides. In many instances, fault movement crushes rocks adjacent to a fault into a mass of irregular pieces, forming *fault breccia.* More intense grinding breaks the fragments into such tiny pieces they may not be individually visible under a microscope.

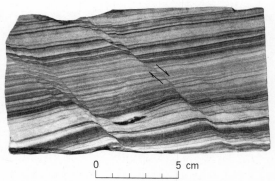

| 0 | 5 cm |

Figure 15.15 Fault drag. Near small faults that cut sandstone, thin layers have been bent during movement. Direction of bending indicates that the direction of last movement was as shown by arrows. The drag in this case indicates these are normal faults. (*B. M. Shaub.*)

Figure 15.16 Slickensides in diabase, Mount Tom Range, Westfield, Massachusetts. The direction of the striations, and the presence of rough steps breaking off the striations, indicates that the missing fault block moved past the one in the picture in the direction of the arrow. (*B. M. Shaub.*)

Deformation by Bending

Bending may consist of broad, gentle warping that extends over hundreds of kilometers, or close, tight flexing of microscopic size. Regardless of the volume of rock involved, or the degree of warping, we refer to the bending of rocks as *folding*. In order to discuss folds and folding, we must become familiar with the few terms used to describe them.

The simplest fold is a *monocline, a one-limbed flexure, on both sides of which the strata either are horizontal or dip uniformly at low angles* (Fig. 15.17). An easy way to visualize a monocline is to lay a book on a table, and drape a handkerchief over one side of the book and out onto the table. So draped, the handkerchief forms a monocline.

Most folds are more complicated than monoclines. *An upfold in the form of an arch is an **anticline*** (Fig. 15.18). *A downfold with a troughlike form is a **syncline***. As we see in Figure 15.18, *the sides of a fold* are called the ***limbs,*** and *the median line between the limbs, along the crest of an anticline or the trough of a syncline is called the **axis** of the fold. A fold with an inclined axis is said to be a **plunging fold,** and the angle between a fold axis and the horizontal is called the **plunge** of a fold. An imaginary plane that divides a fold as symmetrically as possible, and that passes through the axis,* is called the ***axial plane.***

Many folds are nearly symmetrical, similar to the ones shown in Figure 15.18. Others, however, are not symmetrical, and strong deformation may create very complex shapes. The common forms of folds are shown in Figure 15.19. Figure 15.19 shows the difficulty we might encounter in deciding whether a given fold is overturned or not. If we have only fragmentary exposures of bedrock, we must know whether a given layer is right-side up or upside down in order to decide which limb of a fold it is in. This is not always possible, but in some cases sedimentary structures, such as mud cracks and graded layers, do record this. In other examples only careful, thorough mapping of all bedrock exposures can provide the answer.

Figure 15.17 A monocline in southern Utah that interrupts the generally flat-lying sedimentary strata of the wide Colorado Plateau. On both sides of the monocline the exposed layers are nearly horizontal. View looking south. (*J. S. Shelton.*)

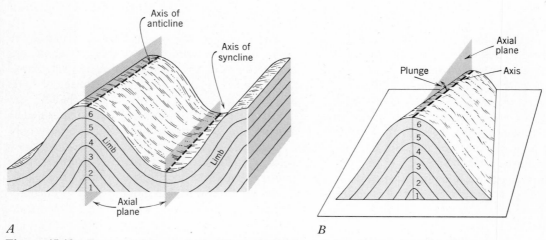

A   *B*

Figure 15.18 Features of simple folds. The upper surface of the youngest layer (6) slopes *toward* the axis of the syncline but *away from* the axis of the anticline. *A.* Fold axis horizontal. *B.* Fold axis plunging.

Name	Description
Symmetrical	Both limbs dip equally away from the axial plane. 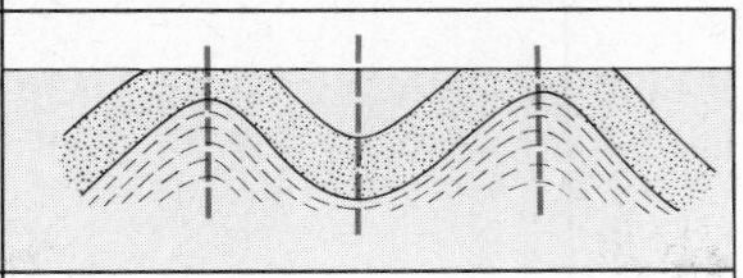
Asymmetrical	One limb of the fold dips more steeply than the other. 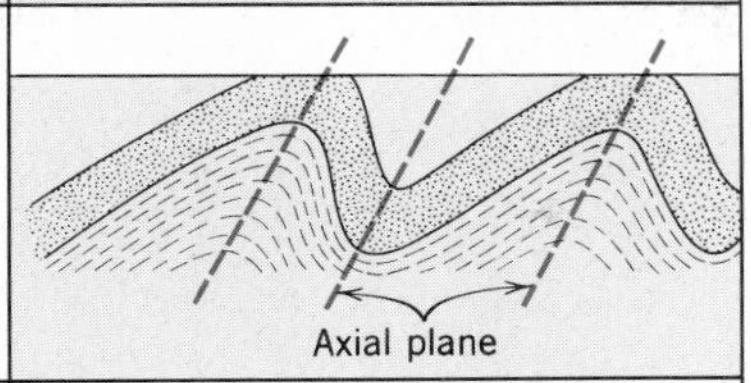
Overturned	Strata in one limb have been tilted beyond the vertical. Both limbs dip in the same direction, though not necessarily at the same angle. 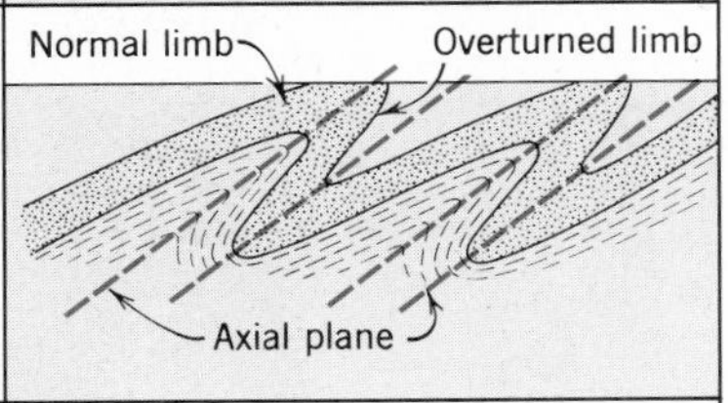
Recumbent	Axial planes are horizontal. Strata on the lower limb of anticline and upper limb of syncline are upside down. 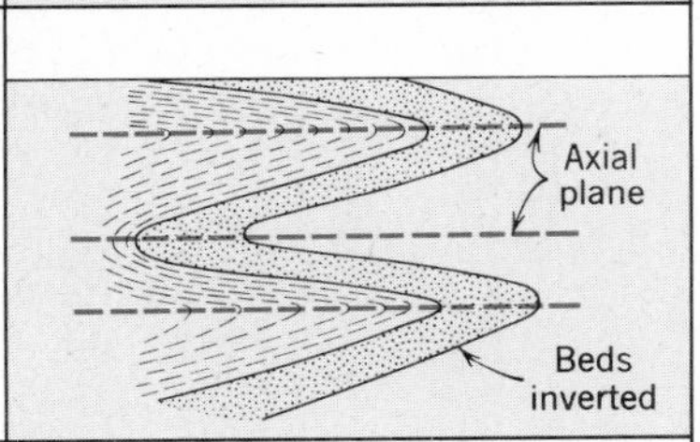
Isoclinal	Both limbs are essentially parallel, regardless whether the fold is upright, overturned, or recumbent. 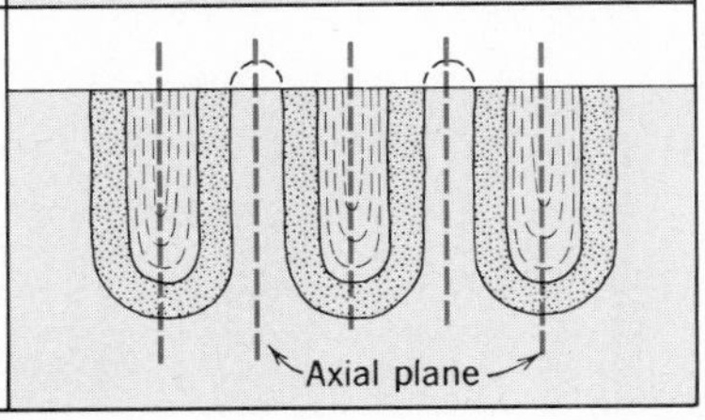

Figure 15.19 Kinds of folds.

Folds are commonly so large that when we examine a single exposure we are not even aware we are seeing folded rock. Nevertheless, when all the exposures of a particular rock body are plotted, and a geologic map is prepared (Appendix D), folds can be recognized from the distribution of the various rock types. Differences of erodibility in adjacent strata can also lead to distinctive topographic forms by which we can recognize the presence of folds (Fig. 15.20).

Gentle warping of the crust, upward or downward, which is really gentle folding on a very large scale, is common. It is likely to be most evident in arid regions where, because of the absence of concealing regolith and the presence of deep dissection by streams, wide areas of rock layers are exposed. Most commonly, however, vegetation and regolith prevent direct observation, and the presence of warping may not even be apparent from examination of several exposures. Inclinations of strata may only be 1° or 2°, and such small slopes are not conspicuous. Yet if a stratum has a persistent dip of 2°, the altitude of the layer changes by about 300m in a horizontal distance of 8km. Careful mapping therefore reveals even the most gentle warping. Upwarping forms a dome, and downwarping creates a basin. Domes are particularly important because some of them are the sites of large accumulations

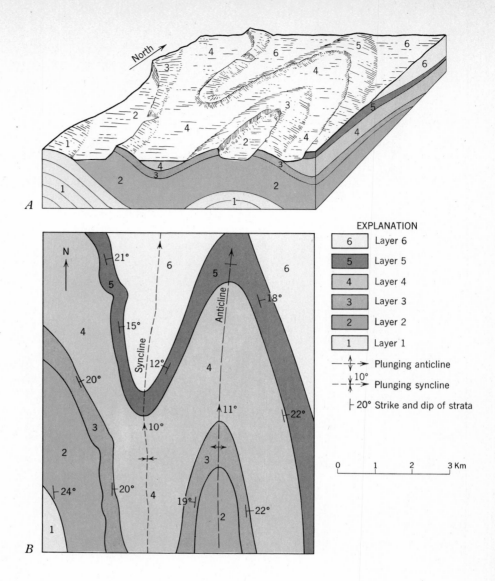

EXPLANATION

6	Layer 6
5	Layer 5
4	Layer 4
3	Layer 3
2	Layer 2
1	Layer 1

Plunging anticline

Plunging syncline

20° Strike and dip of strata

0   1   2   3 Km

of oil and gas. A few domes, especially in desert regions, form distinctive topographic features (Fig. 15.21), but most of them are discovered only by accurate measurements of strata.

Relations Between Faults and Folds. Folds and faults can be related in various ways. As we have just seen (Fig. 15.15), strata near active faults may be folded by the effects of frictional drag along the planes of movement. A steeply inclined fault may pass, either upward or laterally, into a fold and thus die out. The hinge fault shown in Figure 15.22 is an example of how a high-angle fault can become

a monocline. Typically, the projected continuation of the fault surface coincides with the axial plane of the related fold.

Many thrusts in sedimentary rock form after the strata have been compressed and folded into overturned folds. Continued compression finally causes the fold to break, generally near the axial plane, and the former fold becomes a thrust fault. Apparently the Keystone Thrust (Fig. 15.13) was formed in this fashion.

Not only have strata been deformed as the result of active forces of compression, but some thrusts have themselves been folded as if they were strata.

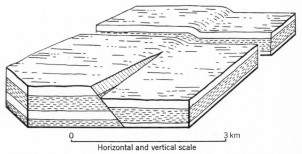

C   **Figure 15.20** Distinctive topographic forms and distinctive patterns in the distribution of various kinds of rock reveal the presence of plunging folds.

A. Block diagram showing topographic effects.

B. Geologic map. (Appendix D gives further details on the preparation and reading of geologic maps.)

C. Air view of a doubly plunging anticline. Sheep Mountain, a high ridge 24km long in the Bighorn Basin, Wyoming, is formed by resistant older strata, which lie along the axis of the anticline. Younger strata, mostly alternating layers of sandstone and shale, have been eroded from the crest of the fold but make jagged, low ridges along the steep limb. These low ridges curve around each end of the mountain, the curve pointing in the direction of plunge. (*J. S. Shelton.*)

Figure 15.21 A great dome, in Mauretania, formed by broad upwarping of the crust, makes a nearly circular pattern of outcrops. The concentric ridges of low relief are made by sedimentary strata that are less erodible than adjoining strata. (U.S. Air Force.)

Figure 15.22 Hinge fault (front block) passes laterally into monocline (rear block).

thrusts have even been overturned. Deciphering the sequence of events in such a case can be a
challenging and difficult task.

Unconformity

Movement along faults brings rocks of different
kinds into contact with each other. It can also bring
into contact two groups of layered rocks that diverge
widely as to dip. When we see such features exposed,
therefore, we think first of faulting as the cause. But
close examination commonly proves our initial assumption wrong. For example, the contact between
the two formations shown in Figure 15.23 is highly
irregular in detail and lacks evidence of movement
along it. Furthermore, in the lower strata of the
upper group we see pebbles and cobbles consisting
of distinctive kinds of rock derived from the steeply
dipping layers beneath. Marine fossils, abundant in
the tilted layers, show that those layers were formed
on the sea floor during the Cambrian Period. Bones
of land mammals, together with the nature of the
sediments around them, tell us that the flat-lying
layers were deposited by streams in the Pliocene
Epoch. Clearly, then, we are looking not at a fault
but at a feature that was created by both deformation
and the deposition of sediment. At some time after
the thick marine layers were formed, a strong dis-
turbance tilted the layers and raised them above sea
level. After a large part of the uplifted mass had been
destroyed by erosion, a blanket of nonmarine sediment accumulated.

The Pliocene alluvium in Figure 15.23 does not
conform with the lower Cambrian strata beneath it.
The relationship between them is one of ***unconformity,*** meaning there is a *lack of continuity* of
sedimentary deposition *between two units of rock in
contact, corresponding to a gap in the geologic record.*
The example in Figure 15.23 is of ***angular unconformity,*** which is *unconformity marked by angular divergence between older and younger rocks.*

Unconformity between a stratum and older rocks
beneath it is commonly a record of movement of
the crust. Such a record gives a few important details; it tells us that the older rocks were brought
into the realm of erosion (this suggests some kind
of uplift) and that at some later time conditions
became favorable for deposition of new sediments.
The record does not necessarily tell us why conditions became favorable for sedimentation. Possibly
it was because the land sank, or because movement,
lava flows, or landslides obstructed drainage and
started deposition by streams or in a lake. If marine
beds lie above and below the contact, the information is more definite: (1) older beds were lifted up
and eroded, (2) submergence brought back the sea,
and (3) another uplift made possible the erosion
which has exposed both sets of beds to our view.

Kinds of Unconformity. Possible combinations of
crustal movement, erosion, and sedimentation are
almost countless, and there are many variations
among the examples of unconformity we find in
rocks. But we recognize only a few differences of

Pliocene
alluvium

Surface of
unconformity

Cambrian
marine strata

Figure 15.23 Surface of unconformity between nearly horizontal gravelly alluvium containing fossil bones of Pliocene age, and tilted and
eroded carbonate rocks containing marine fossils of Cambrian age. Close
examination shows this is not a fault contact. It was formed by deposition
of Pliocene sediments over an old erosion surface. Meadow Valley Wash,
Lincoln County, Nevada. (*C. R. Longwell.*)

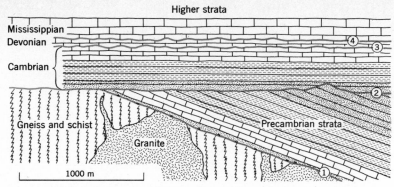

Figure 15.24 Meaning of unconformity. The sedimentary record in the Grand Canyon is broken by large gaps. Angular unconformity indicated at (1) and (2) represents large-scale movements of the crust followed by long intervals of erosion. Surfaces of erosion indicated at (3) and (4), nearly parallel to the sedimentary layers, record broad upwarping followed by erosion, and renewed deposition of sediment.

major importance. Angular unconformity has been presented first because the divergence of beds above and below the surface of unconformity makes it easy to recognize. More common but less conspicuous are examples of unconformity between two beds that are essentially parallel. In the walls of the Grand Canyon is a layer of limestone that contains marine fossils of Devonian age. Through nearly the full length of the canyon, about 320km, this layer overlies another marine limestone that contains fossils of Cambrian age (Fig. 15.24). On close inspection, we find the contact between the two limestones is slightly irregular, with shallow, valley-like depressions cutting into the Cambrian limestone. Hence erosion of the Cambrian layer occurred before the Devonian layer was deposited. There was ample time for erosion, because the combined lengths of the Ordovician and Silurian Periods, which lie between the end of the Cambrian and the beginning of the Devonian, total more than 80 million years. Were sediments deposited in that area during those periods and then completely eroded before the Devonian sea came in? Or was the area above sea level and not receiving deposits during two long periods? In either case the surface of unconformity between Cambrian and Devonian strata in the Grand Canyon represents a gap in history. This sort of unconformity, involving a gap in the record but no obvious deformation, is called *disconformity* to indicate *unconformity that is not marked by angular divergence between two groups of strata in contact.*

Every unconformity testifies to a "lost interval" in the geologic record at the locality, an interval of time in which no deposition of sediments occurred and in which erosion destroyed part of an earlier record. Critical evidence on the basis of which we recognize any unconformity, then, is of two kinds: absence of sedimentary record for a part of geologic time, long or short, and presence of a buried erosion surface. Recognition of this surface is made easier by difference in structure above and below it, but that difference is not essential.

In many places we find sedimentary strata overlying metamorphic and igneous rocks. The contact of sandstone on intrusive igneous rock is recognized as sedimentary if the sandstone fills small valleys and contains pebbles and mineral grains clearly derived from the underlying igneous rock. Rock of any kind, then, may be involved in an unconformity.

Unconformity and Earth History. On the continents the deposition of sediment has been interrupted again and again by uplift and erosion. Indeed such erosional destruction makes the record of former events more complete. In the Appalachians, the Rocky Mountains, and the Alps we see great thicknesses of sedimentary beds folded, faulted, and cut by deep valleys, full of information about geologic history. These three mountain chains were formed at different times, and so the records they reveal partly supplement each other. But every thick sedimentary section includes surfaces of unconformity, each representing a gap in the record, and each recording some important event, such as folding and thrusting in a mountain belt or upwarping of a wide continental area.

Within certain limits we can date the events that create unconformity. If the youngest layers in a sequence of folded strata are Permian and the oldest non-folded beds above are Triassic, we know that the folding occurred near the close of the Paleozoic Era. If the gap in the geologic record is a long one, however, the date may be very imprecise. In central Connecticut the youngest strata in the hanging wall of a great normal fault are of late Triassic age, and the strata deposited across the eroded fault are Pleistocene. Without further information, we cannot fix the date of faulting within the 150-million-year span of post-Triassic, pre-Pleistocene time.

A study of surfaces of unconformity brings out the close relationship between crustal movements, erosion, and sedimentation. All of Earth's land surface is a potential surface of unconformity. Some of today's surface will be destroyed by erosion, but some will be covered by sediment and preserved as a record of the present. Vigorous erosion is now going on where recently there has been uplift. Erosion by streams is laying bare the records in old rocks, and in doing this, it is slowly destroying some of the records. Meanwhile the eroded material is carried away and deposited somewhere else, perhaps on a land surface recently sunk below sea level. And so building up in one place compensates tearing down in another. The many surfaces of unconformity within the geologic column provide records of former surfaces of the Earth and testify that the interactions between internal and external processes have been going on throughout Earth's long history.

Summary

1. Rocks are deformed by fracturing, folding, and flowing.

2. The strength of a rock and the manner in which it is deformed depend on temperature, pressure, and the length of time during which the forces are applied.

3. Evidence of present-day deformation in the crust is found in abrupt movements along faults as well as in areas of the surface that are slowly sinking or rising.

4. Movement on a fault is cumulative, in successive small increments. Some total displacements are hundreds of meters vertically and hundreds of kilometers horizontally.

5. Durable evidence of former crustal movements is registered in deformation of bedrock. The bending and breaking of layered rocks give the clearest record of these movements.

6. Geologic structures, in bedrock, that result from deformation include domes, basins, folds, and faults.

7. Thrust faults, involving great crustal blocks that have moved across low-dipping surfaces, commonly are related to folding in mountain belts.

8. In many areas new rock materials have been deposited as sediment after parts of deformed older bedrock were removed by erosion. The resulting surfaces of unconformity are clear records of crustal movements.

Selected References

Anderson, E. M., 1951, The dynamics of faulting and dyke formation: Edinburgh, Oliver and Boyd.

Billings, M. P., 1972, Structural geology, 3rd ed.: Englewood Cliffs, N.J., Prentice-Hall.

Compton, R. R., 1962, Manual of field geology: New York, John Wiley.

Dennis, J. G., 1972, Structural geology: New York, Ronald Press.

Hills, E. S., 1972, Elements of structural geology: 2nd ed.: New York, John Wiley.

Lahee, F. H., 1961, Field geology, 6th ed.: New York, McGraw-Hill.

Sitter, L. U. de, 1964, Structural geology, 2nd ed.: New York, McGraw-Hill.

Chapter 16

Igneous Activity and Metamorphism

External Processes

Properties of Magma

What Is Magma? Igneous activity and metamorphism seem like strangely different topics to be discussed in the same chapter. Yet they are closely related. Like giant invading armies that change forever the complexion of an invaded country, so great bodies of fluid magma intrude the crust, metamorphosing and changing the rocks they come into contact with. Conversely, under conditions of intense metamorphism, rock may start to melt, forming new magma that rises upward, intruding its parent. Granite batholiths, for example, always intrude metamorphic rocks. All of these activities, plus all other activities of magma on and in the crust, are embraced within the term *igneous activity*.

The best place to commence a discussion of igneous activity is with the properties of magma. This topic leads naturally to volcanoes, because they are the only places we can study the properties of magma at Earth's surface.

Magma is the parent material of igneous rock. In an earlier chapter we defined magma as molten silicate material beneath the Earth's surface, including crystals derived from it and gases dissolved in it. It is a complex mixture of liquid and crystals. As it cools and crystallizes, it reaches a point at which so many crystals have formed, the mixture can no longer flow. A difficult question then arises: when does a magma stop being a magma? When it is half solid, or three-quarters solid? Actually this point varies from magma to magma. The best way to approach the question is to list the essential characteristics of a magma. Magma is characterized by (1) a range of compositions in which silica (SiO_2) is always a predominant constituent, (2) high temperatures, and (3) ability to flow.

Composition. We saw in Figure 5-18 that six minerals or mineral groups—olivine, pyroxene, amphibole, mica, feldspar and quartz—make up the bulk of all igneous rocks. The elements contained in these minerals are therefore the principal elements in magmas. They are Si, Al, Ca, Na, K, Fe, Mg, H, and O. For convenience, we usually express compositional variations in terms of oxide ions, such as SiO_2, Al_2O_3, CaO, and H_2O. The two ions most important for controlling the properties of magma are SiO_2 and H_2O.

The silica content of igneous rock ranges from about 30 per cent to 80 per cent by weight. As we noted in Chapter 5, however, some igneous rock forms when magma cools and the earliest crystals sink and accumulate. We also noted that these earliest-formed crystals differ in composition from the parent magma. Chemical analysis of igneous rock formed in this fashion would provide only the composition of a special fraction of a magma. More accurate indications of complete magma compositions come from lavas that have cooled rapidly and have not separated into different fractions. Such lavas indicate that most magmas fall within the range 45 to 75 per cent SiO_2. They indicate also that three kinds of magma are more common than any others. The first contains about 50 per cent SiO_2 and forms basalt. The second contains about 60 per cent SiO_2 and forms andesite, while the third contains about 70 per cent SiO_2 and forms rhyolite and granite.

Gases dissolved in magma, as we shall see later, are important in determining the properties of magma. Despite their importance, however, the amounts and compositions of dissolved gases are difficult to determine. The principal gas is apparently water vapor, which, together with carbon dioxide, accounts for more than 90 per cent of all gases emitted from volcanoes. Others are nitrogen, chlorine, sulfur, and argon, plus many species present in amounts less than 1 per cent. Although vast quantities of gas are emitted by volcanoes, it is difficult to be sure how much has actually been released by the magma, and how much is merely gas that has leaked out of adjacent rock and has been boiled off. For example, at the height of its activity in May 1945, the volcano Parícutin, in Mexico, released an estimated 116,000 tons of material a day, of which 16,000 tons, or 14 per cent, were water vapor, and 100,000 were lava and ash. But the maximum amount of water vapor it is possible to dissolve in laboratory melts having the same composition as Parícutin magma is only 10 per cent. We must conclude, therefore, that the gases emitted by Parícutin contained not only the gases released from the magma, but also an unknown quantity of ground water heated by the rising magma.

We can never be sure of the exact amount of gas dissolved in any magma. As rising magma nears the surface, pressure decreases and gases gradually bubble off. The problem is rather like estimating how much CO_2 was dissolved in soda water when the only sample we have is a bottle that has been open for an hour. Despite uncertainties, estimates of gases dissolved in magma can be obtained from analyses of glassy volcanic rock. Such rock is erupted and chilled so quickly that dissolved gases cannot escape completely. Typical volcanic glasses have water contents of 0.2 to about 3 per cent, and most scientists believe that the dissolved gas contents of magmas are probably not much greater—perhaps not more than 5 per cent.

Temperature. The temperature of lava coming out of a volcano is likewise difficult to measure. Volcanoes are dangerous places and scientists who study them are not anxious to be roasted alive. Measurements must be made from great distances. Using a pyrometer (an optical measuring device), scientists have obtained a number of measurements. They range from 1040° to 1200°C. Once extruded, however, a flowing lava starts to cool, and temperatures as low as 800°C have been measured in some lava flows that were still barely moving.

Mobility. Although magma may seem to be very fluid, many of its properties are more akin to solids than to liquids; lava is more like asphalt on a road than like water. With rare exceptions, lava flows slowly and with difficulty. *The internal property of a substance that offers resistance to flow* is called **viscosity.** The more viscous a lava, the less fluid it is. Viscosity of a lava depends on composition—especially the SiO_2 content—and on temperature.

The effect of temperature is simple to understand. As with asphalt or thick oil, high temperature leads to greater fluidity. The higher the temperature, therefore, the lower the viscosity and the more readily a magma flows. A very hot magma erupted from a volcano may flow readily, but it soon starts to cool, becoming more viscous and eventually slowing to a complete halt.

The effect of composition on viscosity is not so obvious as the effect of temperature. The same silica tetrahedra that occur in silicate minerals (Chap. 4) occur in magmas also. Just as they do in minerals, the tetrahedra in magmas link together, but unlike those in minerals they form irregular groupings. As the number of tetrahedra in the groups becomes larger, the magma behaves more and more like a solid and becomes more and more resistant to flow. The number of tetrahedra in the groups depends on the silica content of the magma. The higher the silica content, the larger the groups. Magma composition, therefore, exerts a direct and strong control on viscosity (Fig. 16.1). Magma with the composition of basalt sometimes flows readily. Basaltic magma moving down a steep slope on Mauna Loa, in Hawaii, during an eruption in 1850, was clocked at an average speed of 16km per hour. Such fluidity,

however, is very rare. Flow rates are more commonly measured in meters per hour or even meters per day. Magmas with rhyolitic compositions, containing 70 per cent or more SiO_2, are so viscous and flow so slowly that their movement can hardly be detected.

Volcanoes and Volcanism

We noted in Chapter 5 that a volcano is a vent from which molten igneous matter, solid rock debris and gases are erupted. A vent that emits only gases is also called a *fumarole,* and of course does not have an accumulated pile of volcanic debris surrounding it. Where lava, ash, and other rock debris are erupted, a vent is commonly surrounded by either a cone-shaped or sheet-like pile of volcanic material. Most people refer to both the vent and the volcanic pile as a volcano, and we will do so too, although this is not strictly correct in terms of our definition. The shapes of many volcanic piles are so similar, and the processes that build them so nearly alike, that special names have been given to the common types of volcanoes.

Kinds of Volcanoes. The volcano easiest to visualize is one built up by successive flows of lava. Such volcanoes are characteristically built by very fluid lavas, capable of flowing great distances down gentle slopes and of forming thin sheets of nearly uniform thickness. Eventually the pile built up in this fashion develops a shape resembling an inverted shield, convex-side up. A *shield volcano,* then, is *a volcano that emits fluid lava and builds up a broad, dome-shaped edifice, convex upward, with a surface slope of only a few degrees* (Fig. 16.2). Shield volcanoes characteristically erupt basaltic lava; the proportions of ash and other fragmental debris is small. Because basalt is the igneous rock of the ocean basins, shield volcanoes are characteristically oceanic. Hawaii, Iceland, Samoa, the Galapagos, and many other oceanic islands are the upper parts of large shield volcanoes.

Andesitic and rhyolitic lavas are highly viscous and congeal even on steep slopes. Additionally, gas that bubbles out of viscous magma commonly does so with great violence, ejecting vast quantities of pumice fragments, volcanic ash and other fragmental debris (Fig. 16.3). In many instances andesitic and rhyolitic volcanoes erupt larger volumes of fragmental debris than of lavas. A *composite volcano*

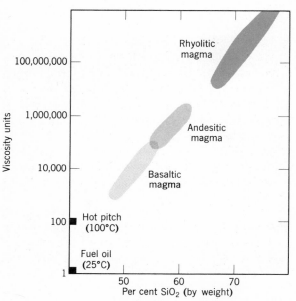

Figure 16.1 Viscosity of magma is strongly controlled by the silica content. The more viscous a magma, the less tendency it has to flow. Viscosities of fuel oil at 25°C and pitch at 100°C are plotted for comparison.

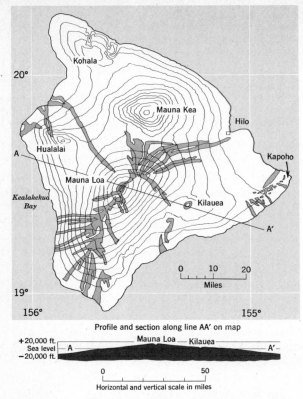

Figure 16.2 The island of Hawaii is composed of five overlapping shield volcanoes, of which the largest are Mauna Loa and Mauna Kea. The topographic map shows the major volcanic centers and the longest lava flows erupted since 1750. The contour interval is 1,000 feet. A profile and section along the line A-A' (lower diagram) illustrates the gentle slope characteristic of shield volcanoes. (*After H. T. Stearns and G. A. Mac-Donald.*)

(also called a ***stratovolcano***) is, then, *a volcano that emits both fragmental material and viscous lava and builds up a steep conical mound.* The surface slope of a composite volcano may reach 30° or more near the summit, decreasing to about 6° near the base. The beautiful steep-sided cones of composite volcanoes are among Earth's most picturesque sights (Fig. 16.4). The snow-capped peak of Mt. Fuji in Japan has inspired poets and writers for centuries. Mount Rainier in Washington and Mt. Hood in Oregon are majestic examples in North America. Andesites and rhyolites are most commonly found on continents; so composite volcanoes are more common on continents than in the ocean basins.

Associated with both shield and composite vol-

canoes are numerous features that give volcanic terrains a special and unique character. Fractures may split the cone so that lava, ash, or both emerge along its flanks. Small cones then develop, peppering the slope of the main mountain like so many small pimples. Gases emerge from small vents, altering and discoloring nearby rocks, and hot springs may form, bubbling off evil-smelling, sulfurous gases. Near the summit of most volcanoes is a ***volcanic crater,*** *a funnel-shaped depression from which gases, fragments of rock, and lava are ejected* (Fig. 16.5).

Many volcanoes are marked, near their summit, by a striking and much larger depression than a crater. This is a ***caldera,*** *a roughly circular, steep-walled basin several kilometers or more in diameter.* The summit of Mauna Loa, in Hawaii, is broken by a spectacular one (Fig. 16.6). Calderas originate principally through collapse because magma has been withdrawn from below. Following eruption, the chamber from which the lava and volcanic ash were emitted becomes empty or partly empty. The now-unsupported roof of the chamber simply collapses under its own weight like a snow-laden roof on a shaky barn, dropping downward within a ring of steep normal faults. Subsequent volcanic eruptions commonly occur along these faults, thus creating roughly circular rings of small cones. Crater Lake,

Figure 16.3 Explosive eruption of volcanic ash and rock fragments from Parícutin Volcano, Mexico. In 1943, when this photo was taken, Parícutin was only a few months old. It was born February 20, 1943, when it suddenly burst into action before the startled eyes of a farmer ploughing his cornfield. (*W. F. Foshag.*)

Figure 16.4 Shishaldin Volcano, a giant composite volcano, towers 9,372 feet above sea level on Unimak Island in the Aleutians. (*U.S. Navy Photo.*)

Oregon occupies an enormous circular caldera 8km in diameter (Fig. 16.7). The caldera formed after a great eruption about 6600 years ago. That eruption pulverized and blew off the top of a composite volcano, posthumously named Mount Mazama. What remained of the summit of Mount Mazama then collapsed into the partly empty magma chamber.

The foregoing volcanic features originate when volcanic materials issue from nearly circular vents. Some lava and ash reach the surface via elongate fractures which, when the walls are spread apart, become fissures. *Extrusion of volcanic materials along an extensive fracture* is a **fissure eruption.** Such eruptions are characteristically associated with fluid basaltic magma, and the lavas that emerge from the fissures tend to spread widely and to build up plains and plateaus, like water from a spring that gradually fills a lake. The fissure eruption of Laki, Iceland, in 1783 took place along a fracture about 32km long. Lava flowed as far as 64km outward from one side of the fracture and nearly 48km outward from the other side; altogether it covered an area of 558km². The volume of lava extruded has been estimated at 12km³, making this the largest single eruption observed during recorded history.

Fissure eruptions are not confined to oceanic islands such as Iceland. They are well known on all continents. *Fissure basalts that form widespread,*

Figure 16.5 Vertical air photo of Parícutin Volcano, Mexico, showing the crater through which most of the ash and other fragmental volcanic material was erupted. Former lava flows (dark areas with lobate margins) issued not from the crater, but from the base of the cone, approximately 1km in diameter. (*Servicio Aerotécnico, Mexico, courtesy Wild Heerbrugg Instruments, Inc.*)

nearly horizontal layers on continents are called **plateau basalts.** One vast tract of plateau basalts covers 500,000km² in Idaho, Nevada, California, Oregon, and Washington (Fig. 16.8). There, in deep canyons cut by the Snake and Columbia Rivers, former lava flows with an aggregate thickness of at least 1300m are exposed. The average thickness of each flow is about 30m, and one, the Roza flow in Washington, has been traced over 22,000km², and has a volume of 650km³. By comparison, the entire volume of volcanic rock in the great composite cone of Mt. Rainier is only about 90km³.

Frequency of Eruption. Some volcanoes give off steam and gases almost continuously. Others lie dormant, seemingly devoid of all volcanic activity through periods of hundreds or even thousands of years; then suddenly they erupt with violent activity. When we consider the frequency of volcanism, two questions arise. How long does activity last at any volcanic center, and how frequent are eruptions? Neither question can be answered with certainty. Perhaps the best way to attempt an answer is to describe the behavior of three well-known volcanic areas.

First, the Hawaiian volcanoes. The Hawaiian Archipelago, a group of many small volcanic islands, spreads in a straight line for nearly 2500km (Fig. 16.9). At the southeast end of the chain are eight large volcanic islands each built by innumerable flows of basalt. To the northwest are many smaller islands, each a coral atoll that caps a deeply eroded volcanic pile below. Geologic study shows that the islands become progressively younger toward the southeast, that each is a shield volcano, and that each ceased to be active before the next-younger one appeared above sea level. K/Ar dates indicate that the exposed parts of most islands were constructed in less than 500,000 years, and suggest that each volcano grew from the sea floor upward to its maximum height in little more than a million years.

The island of Hawaii, most southeasterly of the chain, has five volcanoes, three of which have been active in historic times (Fig. 16.2). Kilauea, most active of the three, has been emitting steam and other gases almost continuously since systematic observations began in 1912. Frequency of lava eruptions also is high. Eruptions have occurred on an average of about two years, although periods of almost ten years have sometimes passed without eruption. Offsetting these quiet periods are times of great activity. From 1968 to 1973 Kilauea was in a state of nearly continuous eruption. Another example of a basaltic volcano in nearly continuous eruption is Stromboli, a volcano in the Tyrrhenian Sea between Sicily and Italy. Since the beginning of recorded history, Stromboli has been hurling clots of red-hot lava upward at 10- to 15-minute intervals. These fall back into the crater and mingle with a pool of molten lava. Sometimes the pool overflows, and lava cascades gently down the mountain, but at all times the luminous masses thrown up by the crater can be seen from great distances by men at sea, thus earning for Stromboli the friendly title, "Lighthouse of the Mediterranean."

An entirely different kind of eruptive style was observed at Mt. Katmai on the Alaskan Peninsula. Until 1912 Katmai, rhyolitic volcano, was not even recognized as being active. In that year it exploded with great violence, blasting the top off its steep-sided cone and leaving a caldera 3.5km long and 2.5km wide. At the same time, from a vent on its lower flank, Katmai erupted a great sheet of pumice, glassy volcanic ash, and pulverized rock fragments. When they had settled, the fragments formed a red-hot blanket of tuff 33km long, up to 5km wide, and as much as 30m thick. It took many years for this hot blanket to cool, during which time rain falling on it caused it to fume and steam from innumerable cracks, leading to the now-famous name, "Valley of Ten Thousand Smokes." Katmai has again lain dormant for 60 years. When and if it will erupt again are unanswered questions. Why it erupted in 1912 is also an unanswered question.

Volcanoes and Man. Some aspects of volcanism are helpful to mankind. For example, weathering converts volcanic ash to exceptionally fertile soils with great rapidity. In some parts of the world crops can be grown as soon as one year after eruption. In other places, such as Italy, New Zealand, Iceland, and California, volcanic steam is tapped by deep drill holes and is used to drive electrical generators. Volcanic power of this sort is called *geothermal power*.

Figure 16.6 Mokuaweoweo Caldera, a huge, steep-walled basin, breaks the summit of Mauna Loa Volcano, Hawaii. This air photograph, taken after a light snowfall, also reveals several smaller calderas in the foreground, each more than a kilometer in diameter. Dark sinuous areas are lava flows. Snow-capped peak in background, projecting through the cloud layer, is Mauna Kea. (*U.S. Army Air Corps photo.*)

A

Despite these helpful aspects, volcanoes are most commonly thought of as danger spots. Underground movement of magma can trigger destructive earthquakes. Flowing lava destroys everything in its path, riding over fields and villages alike (Fig. 16.10). Yet flowing lava is not very dangerous to people. It moves so slowly that one can easily get out of its way and in spite of the frequency of lava flows on Hawaii, loss of life is exceedingly rare.

Unfortunately, during the violent eruptions associated with andesitic and rhyolitic volcanism, a different situation exists. Loss of life cannot always be avoided. Great clouds of volcanic ash and incandescent rock debris erupt explosively, and buoyed by superheated steam and other volcanic gases, form hot, deadly avalanches that roll down steep mountain slopes at high speed. Such an eruption occurred on Mont Pelée, Martinique, in 1902. After 50 years of quiescence, Mont Pelée burst into life, emitting steam and erupting ash for several months. On May 8, a hot cloud of ash roared down the mountainside at 60m/sec., searing everything in its path. Eight kilometers away from the mountain lay St. Pierre, capital of Martinique, a community of 30,000 souls. The fiery cloud engulfed St. Pierre, killing all but two persons, one a prisoner in a dungeon. All buildings were destroyed (Fig. 16.11), ships in the harbor

capsized, and the sea converted to a boiling mass in which fish and other creatures were scalded to death.

Another tremendous and destructive volcanic explosion, the largest of modern times, occurred in 1815 when Tamboro, an Indonesian volcano, decapitated itself. After a long period of apparent dormancy, Tamboro sprang to life, and with a cataclysmic eruption blew 150km^3 of volcanic ash and pulverized rock high into the air. Debris was thickly scattered over a roughly circular area 550km in diameter, the mountain lost 1300m in height and after the top of the mountain collapsed into the space left by eruption of magma, a caldera nearly 11km in diameter remained as mute evidence of the event. The vast number of people killed in heavily populated Indonesia has never been estimated closely. But people all around the world felt the effects of Tamboro. Dust from the eruption was blasted high into the outer atmosphere; some of it remained suspended and circled the Earth for years. Many authorities believe that suspended volcanic dust of this sort prevents so much sunlight from reaching the surface that it provokes world-wide changes in the weather.

The geologic record is filled with evidence of eruptions of the Pelée and Tamboro types. During the last few million years, for example, such erup-

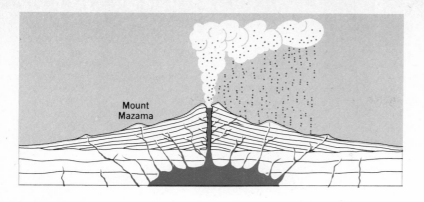

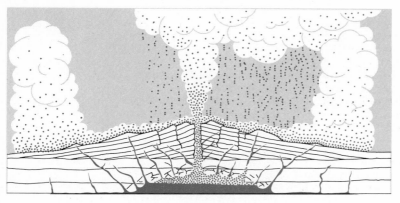

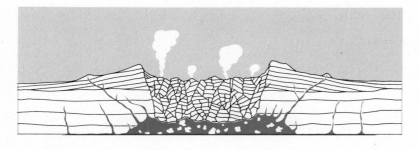

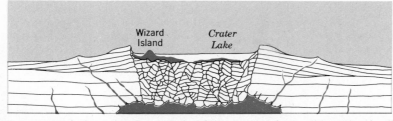

B

Figure 16.7 *A.* (*opposite*) Crater Lake, Oregon, occupies a caldera 8km in diameter, which crowns the summit of a once-lofty composite volcano, posthumously named Mount Mazama. Wizard Island (in the foreground), a small cone of volcanic cinders, capped by a circular crater, formed after the collapse. (*Washington Air National Guard, H. Miller Cowling.*)

B. Following a violent eruption 6,600 years ago, the caldera formed when what remained of the summit of Mount Mazama collapsed into the partly evacuated magma chamber. (*After H. Williams, 1942.*)

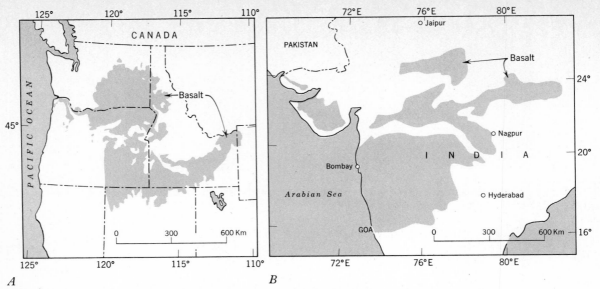

Figure 16.8 Maps of two great basalt plateaus. Areas of basalt (gray) are only remnants of the original flows, reduced by erosion and covered by younger sediments.

A. Columbia Plateau, northwestern United States. (*After A. C. Waters, 1955, and Geologic Map of the United States, U.S. Geol. Survey.*)

B. Deccan Plateau, India. (*After H. D. Sankalia, 1964.*)

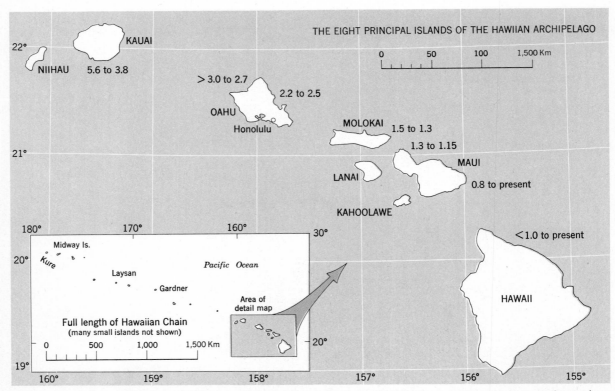

Figure 16.9 In the Hawaiian Archipelago volcanism has ceased on all islands except the two most southeasterly ones, Maui and Hawaii. Activity on Maui is very infrequent. On Hawaii, three volcanoes are still active (Fig. 16.2). All the islands are shield volcanoes. The tops of the most northwesterly islands, such as Midway and Laysan, have been eroded to positions below sea level and are now crowned by coral atolls. Numbers beside larger islands are K/Ar dates (in millions of years) of basalts that form the cones. Volcanism has progressed steadily from northwest to southeast, each island apparently taking little more than one million years to grow from the sea floor to its ultimate height. (*Data from Ian McDougall, 1963.*)

Figure 16.10 Flowing from Parícutin Volcano, Mexico (seen erupting steam and a little volcanic ash in the background) has completely covered the village of San Juan Parangaricutiro, except for the towers of the church. (*Tad Nichols.*)

Figure 16.11 Devastated remains of St. Pierre, Martinique, after a fiery cloud of volcanic ash engulfed it. On May 8, 1902, Mont Pelée, 8km away, erupted a fast-moving sheet of hot volcanic ash and gases. The sheet removed all vegetation, reduced the town to utter ruin, and killed 30,000 people. (*Underwood and Underwood.*)

tions have occurred repeatedly in California, Oregon, Washington, Nevada, New Mexico, and Alaska. What would be the magnitude of the catastrophe if Lassen Peak in California, or one of the supposedly dormant volcanoes in the Cascade Range in Oregon or Washington, should suddenly erupt with the violence of Tamboro?

During the last thousand years about 520 volcanoes have been observed in eruption. They are widely distributed around Earth's surface, and somewhere on Earth a volcano is always erupting. Many others that have not erupted during the past thousand years will almost certainly erupt during the next thousand. The timing and frequency of eruptions are erratic, making predictions uncertain. Yet a promising new development may help. Scientists have noted that prior to many eruptions, slight warming of the ground occurs in the immediate vicinity of the volcano. With films sensitive to radiation in the infra-red region, cameras mounted on orbiting satellites can post a continuous watch for such developing hot spots, warning scientists of areas of likely activity so that closer appraisals can be made and endangered populations warned.

Internal Processes

Properties of Magma

How Rock Melts. We learned in Chapter 5 that two different effects combine to produce a wide range of igneous rocks. One is *crystallization* and it occurs during the cooling and solidification of magma. It includes the many ways that crystals can form and be separated from a parent liquid, and so produce separate solid and liquid fractions with different compositions. The other effect is *partial melting*, which embraces all the complex phenomena involved when a rock composed of many different types of minerals becomes hot enough to start melting. First one mineral melts, then another, so that the melting process is a long-drawn-out affair during which a mixture of liquid and mineral grains exist together. If the liquid is squeezed out from the mixture, its composition differs from the composition of the mixture, just as the composition of water squeezed from a sponge differs from the overall composition of the wet sponge. In Chapter 5 we were concerned only with igneous rock, so we con-

centrated discussion on minerals rather than magma. But magma must precede igneous rock; so the discussion there gives us a starting point for considering the origin of magma.

Making a magma is like cooking in the kitchen. Both ingredients and a source of heat are necessary. Let us first consider the source of heat. In the kitchen it is the stove. In Earth it is the great heat store of the mantle and deeper regions. But temperatures have to be very high indeed before rock starts to melt—so high, in fact, that we usually think of rock as being fireproof and indestructible. How, then, does temperature increase with depth and at what point does rock start to melt? To answer these questions we must first consider the **geothermal gradient,** *the rate of increase of temperature downward in the Earth*. Gradients beneath continental crust differ from those beneath oceanic crusts, because the distribution of radioactive elements, which are the principal sources of heat, differs in the two areas. Nevertheless, as we see in Figure 16.12, temperatures in both cases rise above 1000°C at rather shallow depths. We already know that some lavas are fluid at 1000°C. Why then isn't Earth's mantle entirely liquid? The answer is that pressures are too great. As pressure rises, the temperature at which a compound melts also rises. For example the mineral albite, which melts at 1104°C at Earth's surface where pressure is 1kg per cm^2, must be raised to 1440°C at a depth of 100km below the surface be-

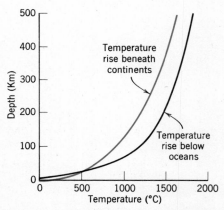

Figure 16.12 Geothermal gradients, the rates at which temperature rises with depth, differ beneath oceanic and continental crusts. The differences arise because naturally radioactive elements, the principal sources of Earth's heat, are distributed differently in the two regimes.

fore it will melt, because the pressure is 35,000 times greater. Therefore, whether a given rock melts and forms magma at any specified depth in the Earth depends on the geothermal gradient and the effect of pressure on its melting temperature.

The effect of pressure on melting seems straightforward and easy to understand. And so it is, provided rocks are dry. When water or water vapor is present, however, a complication enters and another effect occurs, an effect similar to that of salt on an icy road. Salt causes ice to melt by forming a salty solution that can freeze only at temperatures below the freezing point of pure water. We say that salt *depresses* the freezing point of water. Similarly, the presence of water dissolved in magma depresses the freezing point of magma. Or, to say it another way, wet rock will start melting at lower temperatures than dry rock of the same composition. Furthermore, as the pressure rises, the effect of water also rises. Increasing pressure therefore decreases still further the temperature at which a wet rock starts to melt. This is exactly opposite to the effect of pressure on the melting of a dry rock.

Now that we have discussed heat source and melting, we are ready to return to our analogy between magma formation and cooking. A stove or oven is only the first necessity. The second need is cooking ingredients. In the same way that different ingredients produce different cakes, so different source materials give rise to different magmas. We shall consider each of the three common magma types separately.

Basaltic Magma. Basalt is the characteristic igneous rock of the thin oceanic crust. Immediately below the oceanic crust lies the mantle; available evidence points to the mantle as the source of basaltic magma. Basalt contains water-free minerals such as pyroxene, olivine, and feldspar. This fact, plus observations that basaltic magma in Hawaii and elsewhere contains little water, suggests that basalt is essentially a dry or water-poor magma. All evidence suggests that the water content of basaltic magma rarely exceeds 0.1 per cent. Basaltic magma must therefore originate by some sort of dry-melting process. From such considerations as Earth's total mass we know that the mantle does not have the composition of basalt. We surmise, therefore, that basalt must originate by some sort of dry partial melting within the mantle. Much scientific debate

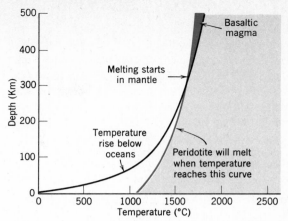

Figure 16.13 Depth temperature curve for the beginning of melting of peridotite (blue). The composition of peridotite is believed to be similar to the composition of the upper mantle. Where the melting curve crosses a curve (black) outlining the geothermal gradient below oceanic crust, a small amount of liquid would develop by partial melting of the peridotite. At a depth of 350km, therefore, dry partial melting of peridotite will develop magma in the mantle. Laboratory experiments show that the magma so developed would have the composition of basalt.

has centered on the exact composition of the mantle and what fraction of that composition melts to form basalt. One possibility is that the composition of the upper part of the mantle resembles that of peridotite (Fig. 3.18). At a depth of about 350km the temperature on the geothermal gradient just reaches a point where the effect of pressure on the dry melting of peridotite is overcome (Fig. 16.13) and a small amount of liquid with basaltic composition could develop. The melting process may not be completely dry, because trace amounts of water vapor may indeed occur in the mantle. The presence of even tiny amounts of water would allow melting to start at even lower temperatures and shallower depths. We can therefore take 350km as a maximum depth at which basaltic magma is formed. Some evidence suggests that small amounts of basaltic magma can form even at depths of about 100km. Although opinions differ as to the exact composition of the mantle, and how much must melt to form basalt, it is generally agreed that the formation of basalt commences with partial melting of dry or nearly dry rock in the mantle.

Once a small amount of liquid is formed in the mantle, how does it ever reach the surface? Most liquids are less dense than the solids from which they form. A small amount of liquid developed by partial melting, therefore, will slowly float and force its way up toward the surface. Furthermore, as the liquid rises, the pressure on it decreases continually. As can be seen in Figure 16.13, therefore, a basaltic magma remains liquid as it approaches the surfaces even though it may be cooled a little in its upward passage. Most basaltic magma actually reaches the surface and forms lava. Basalts are common rocks but gabbros, their deep intrusive equivalents, are rare.

Andesitic Magma. The chemical composition of andesite is close to the average composition of the continental crust. Andesite and its equivalent intru-sive rock, diorite, are commonly found on the continents. From these two facts we might therefore suppose that andesitic magma simply forms by the complete melting of a portion of continental crust. Some andesitic magma may indeed be generated in this way, but this cannot be the origin for all such magma. Andesitic magma is extruded from some oceanic volcanoes that are far from continental masses. It must somehow be developed from the mantle or from oceanic crust, therefore. Laboratory experiments provide a possible answer to the dilemma. Partial melting of a wet basalt yields, under some conditions, a magma of andesitic composition. An interesting hypothesis has been suggested to explain how this might happen. When a moving

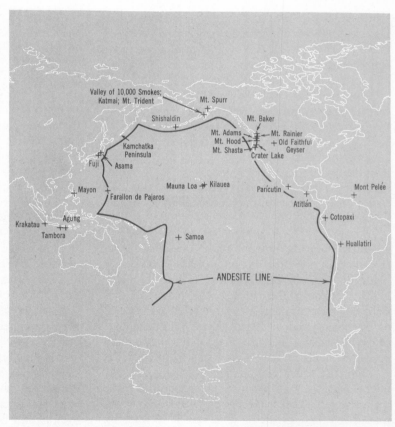

Figure 16.14 The Andesite Line, surrounding the Pacific Ocean basin, separates areas within the basin where andesitic rocks are not found, from areas where andesites are common. Volcanoes such as Mauna Loa, that are inside the line, erupt basaltic magma but not andesitic magma. Those outside the line may erupt basaltic magma too, but they also erupt andesitic magma.

plate of lithosphere plunges back into the mantle, it carries the wet oceanic crust with it. The plate will heat up and eventually the wet rock will melt. A small degree of wet partial melting would produce a liquid having the composition of andesitic magma.

The validity of this interesting suggestion remains to be proved, but one impressive piece of evidence suggests it may be correct. In Figure 16.14, the locations of many active volcanoes in and around the Pacific basin are plotted, together with a well-defined line that separates regions where andesite occurs from regions where it does not occur. The line is called the *Andesite Line*. Inside this line and inside the main ocean basin, andesite is unknown. All active volcanoes inside the line erupt basaltic magma, and all the volcanic rock associated with dormant volcanoes is basalt. Outside the line, andesite is common. The Andesite Line coincides closely with the distribution of deep-sea trenches (Fig. 12.10), which, as we learned in Chapter 12, mark the very places where plates of lithosphere plunge back into the mantle. If, therefore, andesites form by partial melting of wet basalt, the process occurs at the very spot on Earth where wet basalt is probably present in the mantle. The coincidence provides strong support for the hypothesis.

Rhyolitic Magma. Two important observations suggest a unique origin for rhyolitic magma. First, modern volcanoes that extrude rhyolitic magma are confined to regions of continental crust. Similarly, the distribution of ancient rhyolites and their equivalent intrusive rocks (granites and granodiorites) is confined to continents. Second, rhyolitic volcanoes give off a great deal of water vapor, and granitic rocks contain significant amounts of water-bearing minerals such as mica and amphibole. These two points of evidence strongly suggest that the source of rhyolitic magma lies within the continental crust, and that its origin involves some sort of wet melting. Laboratory experiments bear this suggestion out. When, in the laboratory, water-bearing rocks of the continental crust start to melt, the composition of the first liquid that forms is that of granite. As seen in Figure 16.15, the wet-melting curve for granitic magma intersects the geothermal gradient at a depth of 35km to 40km, a depth near the base of the continental crust. Ordinary water-bearing crustal rocks, buried to these depths, will start to melt and the liquid that forms is granitic.

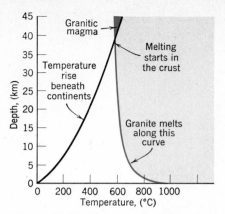

Figure 16.15 Depth temperature curve for the beginning of melting of granite in the presence of water. At a depth of 35km to 40km below the surface of a continent, the temperature on the geothermal gradient is just sufficient to cause granitic magma to start forming from average crustal rocks. Compare Figure 16.13.

Although granite and granodiorite are very common, rhyolite is not a common rock. Why should extrusive rock be rare but intrusive rock of the same composition be common, exactly opposite to the relation between gabbro and basalt? Scientists have long been puzzled by this observation. Figure 16.15 illustrates a possible explanation. Once a granitic magma has formed, it starts to rise. But as it rises, the pressure on it decreases and, as we learned earlier, the effectiveness of water in reducing the melting temperature is diminished. A rising magma formed by wet partial melting must therefore increase in temperature or it will solidify, forming an intrusive igneous rock. There is no way for a rising magma to get hotter, so the generation of magma by partial melting of wet rocks leads naturally to the making of large volumes of intrusive igneous rock and only small volumes of volcanic rock.

One further observation is explained by the origin proposed for granitic magma. As rock is buried deep in the crust, it is metamorphosed. When melting starts near the base of the crust, newly formed magma rises and invades the metamorphic rock above. We should therefore expect to find that bodies of granitic rock are closely associated with metamorphic rock. This indeed is the case. Although granitic batholiths may be thousands of kilometers in individual length (Fig. 16.16), they are invariably intruded into highly metamorphosed rock.

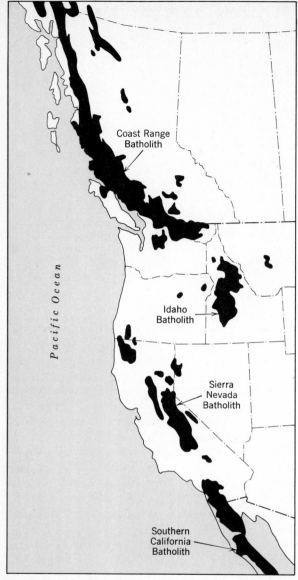

Figure 16.16 The Idaho, Sierra Nevada, and southern California batholiths, largest in the United States, are dwarfed by the Coast Range batholith in British Columbia and Southern Alaska. Each of these gigantic batholiths intrude highly metamorphosed rocks.

Metamorphism

We learned in Chapter 5 that metamorphic rock is formed within the crust by the transformation of pre-existing rock. Metamorphism is apparently a widespread and common process among Earth's in-

ternal activities, for metamorphic rock constitutes roughly 15 per cent of the crust. Metamorphism is like cooking, but with the added wrinkle that pressure as well as temperature can cause transformation. The changes are as dramatic as those occurring in a mixture of flour, salt, yeast, and water to give bread. They involve both mineral compositions and textures and they may be so profound that it is sometimes very difficult to guess what a rock was before metamorphism.

We have already seen that intense metamorphism can lead to melting and the generation of magma. Long before this stage is reached, however, less dramatic changes can be seen. The two processes effecting the changes are *mechanical deformation* and *chemical recrystallization.*

Mechanical deformation includes grinding, crushing, and the development of new textures such as rock cleavage and foliation. Chemical recrystallization includes all the changes in mineral composition, in growth of new minerals, and in loss of H_2O and CO_2 that occur as rock is heated. Deformation and chemical recrystallization generally proceed simultaneously, as in slate, when mica and other new minerals grow as parallel grains and so produce cleavage. Purely mechanical effects do sometimes occur, however, without any changes in mineral chemistry. But they are rare and usually very localized. For example, adjacent to faults, coarse-grained massive rock such as granite may be broken, and individual mineral grains may be shattered and pulverized (Fig. 16.17,*A*). This sort of deformation occurs in brittle rocks. When confining pressures are high, so that brittle properties are suppressed, flattening and elongation without associated chemical recrystallization can occur also (Fig. 16.17,*B*). But chemical recrystallization is so much more common than simple mechanical deformation that our discussion must center on chemical effects. Thus, we shall now examine the two principal kinds of metamorphism: contact metamorphism and regional metamorphism.

Contact Metamorphism. Contact metamorphism occurs adjacent to hot bodies of igneous rock intruded into cooler rock of the upper part of the crust. It occurs in response to a pronounced increase in temperature but virtual absence of mechanical deformation. The temperature of the intrusive may be as high as 1000°C, that of the intruded rocks only

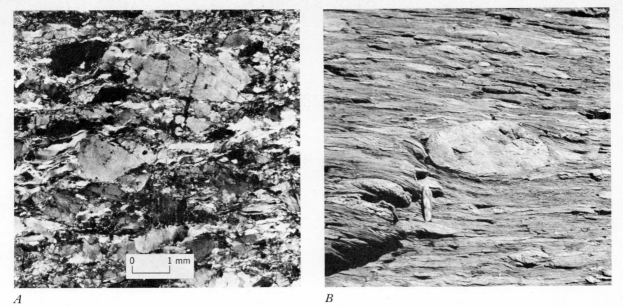

A   *B*

Figure 16.17 Purely mechanical deformation leads to distinctive textures in metamorphic rocks.

A. Thin section of granite from the vicinity of a fault near Branford, Conn. Mineral grains have been shattered, broken, and pulverized. (*B. M. Shaub.*)

B. Deformation under high confining pressure has caused the pebbles in this conglomerate to become flattened and elongate. (*J. W. Ambrose, Geol. Survey of Canada.*)

200 to 300°C. Rock adjacent to the intrusive becomes heated and metamorphosed, developing a well-defined shell, or *metamorphic aureole* of altered rock (Fig. 16.18). The width of the aureole depends on the size of the intrusive body. With a small intrusive, such as a dike or sill a few meters thick, the width of the metamorphic aureole may only be a few centimeters. but when the intrusive is large, perhaps a kilometer or more in diameter, the aureole may reach a hundred meters or more in width. The size of an aureole also depends on how susceptible the rocks are to change, and on how wet the rocks

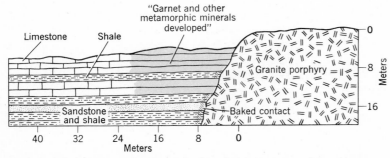

Figure 16.18 Metamorphism around an intrusive mass of granite porphyry near Breckenridge, Colo. Sandstones and shales have been baked to fine-grained hornfels immediately adjacent to the contact. However, in limestone, a much more reactive rock, new metamorphic minerals such as garnet have been developed as far as 20m away from the contact.

are. The presence of abundant water helps many reactions proceed as it helps tenderize tough meat in a stew.

When heated, all kinds of rock react in some fashion, but some are more reactive than others. Those consisting of a single mineral species, such as pure quartz sandstone, or limestone made of pure calcite, are simply recrystallized to quartzite and marble respectively. Most commonly, however, two or more minerals are present and can react to form new minerals. For example, grains of quartz in limestone might react with calcite to form wollastonite, $CaSiO_3$, releasing CO_2 in the process.

The cook who puts a cake in either a cool oven or a very hot oven knows that the final result will probably not be edible. For correct results the ingredients have to be cooked at the right temperature. The final result of rock cooking depends on temperature too. Close examination of a contact-metamorphic aureole usually reveals that several different and roughly concentric zones of mineral reactions can be identified. Each zone is characteristic of a certain temperature and pressure. Immediately adjacent to the intrusive we find water-free minerals such as garnet, pyroxene, wollastonite, and andalusite. Beyond them we find water-bearing minerals such as epidote and amphiboles, and beyond this in turn micas and chlorites. The exact assemblage of minerals we find in each zone depends, of course, on the chemical composition of the intruded rock as well as the temperatures and pressures reached during metamorphism. Because temperature and pressure are the important variables in metamorphism, we refer to *contrasting assemblages of minerals that reach equilibrium during metamorphism within a specific range of physical conditions* as belonging to the same **metamorphic facies.** To continue our analogy with cooking, consider a large roast of beef. When it is carved, one sees that the center is rare, the outside well done, and in between there is a region of medium-rare meat. The differences occur because temperature was not uniform throughout. The center or "rare-meat" facies, is a low-temperature zone; the outside, or "well-done" facies is a high-temperature zone.

Clearly, then, each of the zones observed within a contact-metamorphic aureole belongs to a specific metamorphic facies. Furthermore, whenever a rock of specific composition occurs within a given metamorphic facies, the same mineral assemblage is observed. We can therefore reason in reverse and study (or examine) mineral assemblages, determine the distribution of metamorphic facies, and then reconstruct the temperatures and pressures at the time of metamorphism. This is done most conveniently by selecting key minerals such as andalusite, sillimanite, garnet, or biotite, and noting where they first appear in the aureole. *A line on a map, connecting points of first occurrence of a given mineral in metamorphic rocks,* is called an **isograd.**

Regional Metamorphism. Contact metamorphism is always local and therefore does not affect large portions of the crust. Nevertheless, study of contact metamorphism is very helpful because it allows us to decipher the general processes involved in other kinds of metamorphism. The most common metamorphic rocks occur through areas of thousands of square kilometers and are therefore called *regional metamorphic rocks.* Unlike contact metamorphism, regional metamorphism involves a considerable amount of mechanical deformation in addition to chemical recrystallization. As a result, regional metamorphic rocks tend to be strongly layered and distinctly foliated. Slate, phyllite, schist, and gneiss are the most common varieties, and as we learned in Chapter 5, each has a characteristic texture.

Regional metamorphism occurs when large volumes of Earth's crust are buried during mountain building, and are subjected to the high temperatures and pressures that characterize deep parts of the crust. How such burial occurs is discussed in Chapters 17 and 18.

What happens when a wedge of rock, formerly at or near the surface, becomes deeply buried? Strata that were originally horizontal become faulted, folded, and buckled. In short, mechanical deformation commences. As depth of burial increases, the strata are subjected to increasing pressure and temperatures. New minerals start to grow. But rocks are poor conductors of heat; so the heating-up process is very slow. The temperature reached by a buried pile of strata depends on both depth and duration of burial. If burial is very slow, heating of the pile keeps pace with the temperature of adjacent parts of the crust; that is, a normal continental geothermal gradient is maintained. But if burial is very fast, the pile has insufficient time to heat up; so conditions of high pressure but rather low temperatures prevail. The minerals that grow are controlled both by temperature and pressure; so the mineral assemblages we observe depend on rates of burial.

In a buried pile of strata both temperature and pressure vary. The larger the pile, the greater the variations. Regional-metamorphic rocks, like their contact-metamorphic equivalents, develop zonal sequences of minerals and textures in response to the variations of temperature and pressure (Fig. 16.19). Unlike metamorphic aureoles, however, zones of regional metamorphism tend to be broad and undulating, showing both horizontal and vertical changes. Where aureoles resemble a series of concentric cylinders surrounding an intrusive, zones of regional metamorphism are analogous to a domed pile of blankets, each blanket representing a specific metamorphic facies. Both the concept of facies and the concept of isograds apply equally well to contact- and regional metamorphism. but the sequence of index minerals that develops differs from case to case, reflecting the many different geothermal gradients that are possible.

Role of Volatiles in Metamorphism. So far we have discussed metamorphism as if the only important variables were temperature and depth of burial. But, yet other variables also play a part. These are the amounts of H_2O and CO_2 present, and the ease with which these substances can move around in rocks.

At depths greater than about 10km, the only significant amounts of H_2O and CO_2 in rocks are combined in mineral structures. But the kinds of reactions that take place during metamorphism commonly involve the release of H_2O and CO_2. If they can not escape after release, the reaction cannot proceed. In a sense, this is analogous to saying you can't dry wet clothes if you lock them up in a tight closet. How a regional-metamorphic reaction occurs,

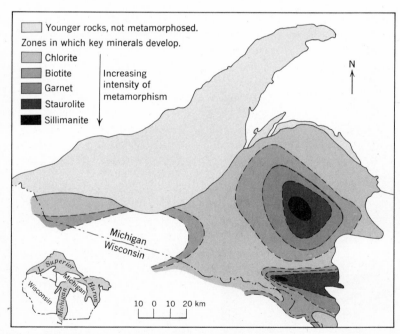

Figure 16.19 Zonal sequences of minerals and textures developed in regionally metamorphosed rocks reflect gradients of temperature and pressure during metamorphism.

Metamorphic zones in an area of Michigan are reflected by the first appearance of key metamorphic minerals. Metamorphism occurred about 1.5 billion years ago. Rocks that were deposited later than the metamorphism are unchanged. Two clearly defined centers of intense metamorphism are identified by the presence of sillimanite. (*After H. L. James, Geol. Soc. America Bull., v. 66, p. 1455, 1955.*)

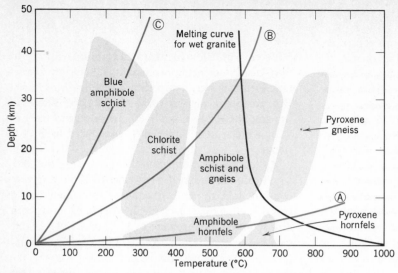

Figure 16.20 The kind of rock that develops from metamorphism of shale depends on temperature and depth of burial. Curve (*A*) is a typical thermal gradient around an intrusive igneous rock that is causing contact metamorphism. Hornfels, a dense, nonfoliated, even-grained rock, is created. Curve (*B*) is a normal continental geothermal gradient and it indicates the rock types developed during slow burial of a pile of strata. Curve (*C*) is the geothermal gradient developed in a rock pile that is buried so rapidly it cannot maintain thermal equilibrium.

Figure 16.21 Complexly folded vein of granite intruding intensely metamorphosed rock. The granite supposedly represents the fraction of the rock that melted. Skeena River, British Columbia. (*B. M. Shaub.*)

therefore, depends on whether the H_2O and CO_2 can escape.

Physical Conditions of Metamorphism. Despite the influence that H_2O and CO_2 must exert, it is observed that common metamorphic rock types and common mineral assemblages are closely related to temperature and depth of burial. One attempt to systematize common metamorphic rocks in this fashion is shown in Figure 16.20. In this figure also is plotted the melting curve for wet granite that occurs in Figure 16.15. Clearly, if sufficient water is present in the rock down near the base of a pile that is being metamorphosed, magma can start to form while amphibole schists and gneisses are forming also. In many places it is possible to see evidence of the phenomenon (Fig. 16.21). Still further melting causes large volumes of magma to develop and to rise through the pile of metamorphic rock above. As discussed earlier, this event leads to the close connection that exists between metamorphic rocks and granitic intrusives.

We have discussed the sort of changes that occur during metamorphism, and how certain igneous activity is connected with metamorphism. But an important part of the story concerning both metamorphic and igneous rock remains to be discussed. Igneous and metamorphic rocks are not randomly distributed like raisins in a pudding. They occur together in elongate belts, and are closely associated with mountainous regions and zones of intense deformation of Earth's crust. In the next chapter we shall see how this association comes about.

Summary

1. The principal controls on the compositions and physical properties of magmas are their content of SiO_2 and H_2O.

2. High temperature and low content of SiO_2 result in fluid magma, such as basaltic magma. Lower temperature and high content of SiO_2 result in viscous magma, such as andesitic and rhyolitic magma.

3. The sizes and shapes of volcanic edifices depend on the kind of material extruded, viscosity of the lava, and explosivity of the eruptions.

4. Volcanoes that dispense sluggish lava, generally rich in silica, tend to be explosive; those that extrude fluid lava, generally low in silica, erupt less violently.

5. Widespread sheets of basaltic rock have resulted from fissure eruptions of fluid lava.

6. The composition of lavas extruded from volcanoes within ocean basins is not the same as that of lavas extruded from volcanoes lying at the margins of ocean basins.

7. Basaltic magma, characteristic of oceanic crust, is formed by the partial melting, under dry or nearly dry conditions, of rock in the mantle.

8. Andesitic magma is believed to be formed by partial melting of water-saturated basaltic rock that is being carried downward to the mantle by descending plates of lithosphere.

9. Rhyolitic and granitic magma are formed by partial melting of crustal rock in the presence of water.

10. Heat given off by intrusive igneous rock creates contact-metamorphic aureoles.

11. Mechanical deformation and chemical recrystallization are the two processes that affect rock during metamorphism.

12. Rocks having the same chemical composition, and subjected to identical metamorphic environments, always react to form the same mineral assemblages.

Selected References

Igenous Activity

Bullard, F. M., 1962, Volcanoes in history, in theory, and in eruption: Austin, Texas, Univ. Texas Press.

Cotton, C. A., 1952, Volcanoes as landscape forms, rev. ed.: New York, John Wiley.

Ernst, W. G., 1969, Earth Materials: New Jersey, Prentice-Hall. Chapter 5.

Hamilton, W. and Myers, W. B., 1967, The nature of batholiths: U.S. Geol. Survey, Prof. Paper 554-C, p. C1-30.

Keefer, W. R., 1971, The geologic story of Yellowstone National Park: U.S. Geol. Survey Bull. 1347.

MacDonald, G. A., 1972, Volcanoes: New Jersey, Prentice-Hall.

Smith, R. L., 1960, Ash flows: Geol. Soc. America Bull., v. 71, p. 795–841.

Tabor, R. W. and Crowder, D. F., 1969, On batholiths and volcanoes—Intrusion and eruption of late Cenozoic magmas in the Glacier Peak Area, North Cascades, Washington: U.S. Geol. Survey Prof. Paper 604.

Thorarinsson, S., 1967, Surtsey: The new island in the North Atlantic: New York, Viking Press.

U.S. Geol. Survey, 1971, Atlas of Volcanic Phenomena: Folio of 20 sheets.

Wyllie, P. J., 1971, The dynamic earth: New York, John Wiley.

Metamorphism

Ernst, W. G., 1969, Earth materials: New Jersey, Prentice-Hall. Chapter 7.

Fyfe, W. S., Turner, F. J., and Verhoogen, J., 1958, Metamorphic reactions and metamorphic facies: Geol. Soc. America Mem. 73.

Turner, F. J., 1968, Metamorphic petrology: New York, McGraw-Hill.

Winkler, H. G. F., 1967, Petrogenesis of metamorphic rocks: rev. second ed., New York, Springer-Verlag.

Chapter 17

Earthquakes and Mountains

Earthquakes

Earth is like a restless sleeper. Most of its internal activities happen slowly, regularly and almost imperceptibly. Yet others, such as movement along faults and violent volcanic eruptions (Chap. 16), are like a sleeper turning over. They happen suddenly and, like a sleeper falling out of bed, sometimes dramatically. All activities, fast or slow, involve transfer of energy. If transfer happens suddenly, the effect on Earth is like striking it with a giant hammer. Earth vibrates, or as we say more familiarly, an *earthquake* occurs.

Bomb blasts cause earthquakes. Sudden volcanic explosions cause earthquakes too, sometimes very strong ones. Most earthquakes, however, seem to be associated with movement along faults. It is still uncertain, though, how a sudden release of energy occurs during movement along a fault. One hypothesis states that release results from elastic deformation of the lithosphere. According to this idea, energy can be stored in bodies of rock that are being elastically deformed, just as in a spring that is compressed. The word *elastic* indicates that when the force is removed and the energy released, the deformed body returns to its original shape. Forces creating movement along a fault must first overcome friction along the fault surface. When friction is exceeded, the opposite walls move suddenly, the stored energy is released, and an earthquake occurs. This theory, sometimes called the elastic-rebound theory, derives support from evidence gathered by study of the San Andreas Fault in California. During field studies in central California, beginning in 1874, scientists from the U.S. Coast and Geodetic Survey determined the precise positions of many points both adjacent to and distant from the fault. As time passed, movement of the points revealed that the crust was slowly being bent. On April 18, 1906, the two sides of the fault shifted abruptly. The stored energy was dissipated rapidly as the bent crust apparently snapped back to a former position, thereby creating a violent earthquake. Repetition of the survey then revealed that the bending had disappeared (Fig. 17.1).

The great majority of all earthquakes originate within the lithosphere, where rocks are brittle and where elastic phenomena can occur. However, earthquakes also occur in localized regions of the upper mantle, where temperatures and pressures are too high for elastic phenomena to operate. Under such conditions deformed bodies undergo permanent changes of shape, even after the deforming forces are removed. Deformation of this kind is *plastic* deformation. Through most of Earth's mass, rock can be deformed plastically, and slowly flows under pressure. Under such circumstances energy is not stored elastically, and the elastic-rebound theory cannot explain the origin of earthquakes. Although

earthquakes do occur at great depths—sometimes as great as 700km—we are still uncertain how they happen. Regardless of how earthquakes happen, let us next see what happens to the energy they release.

Seismic Waves. *The point of first release of the energy that causes an earthquake* is called the **earthquake focus.** The focus generally lies at depth; so for convenience we define *the point, on Earth's surface, that lies vertically above the focus of an earthquake* as the **epicenter** (Fig. 17.7). A convenient way to describe the location of an earthquake focus, therefore, is to state its epicenter and its depth.

How is energy transmitted from the focus to other parts of Earth? As with any vibrating body, waves (vibrations) spread outward from the focus. The waves are commonly called *seismic waves* after the Greek word for earthquake. They spread out in all directions from the focus, just as sound waves spread in all directions when a gun is fired. Seismic waves are elastic disturbances; so the rocks through which they pass return to their original shapes after the waves go by. Because of this we cannot tell, by examining a rock, whether seismic waves have passed through it at some time in the past. Seismic waves must therefore be measured and recorded while the rocks are still vibrating. For this reason many continuously-operated recording stations are scattered around the world.

Seismic waves are basically of two kinds. *Compressional waves* deform solids by change of volume in the same way that sound waves do, and consist of alternating pulses of compression and rarefaction acting in the direction of travel (Fig. 17.2). Compressional waves travel more rapidly than other seismic waves, and are therefore called *P* (for *Primary*) waves. *Shear waves* deform solids by change of shape but not change of volume. They are like electromagnetic waves, and consist of an alternating series of sidewise movements, each particle in the deformed solid being displaced perpendicular to the direction of wave travel. Shear waves are analogous to the waves set up when a rope is shaken (Fig. 17.2). Being slower than *P* waves, shear waves are called *S* (for *Secondary*) waves.

Just as the speed of sound waves is different in air from their speed in water, because air and water have different physical properties, so the speeds of *P* and *S* waves vary as the traverse solids having different properties. Furthermore, the way the speed

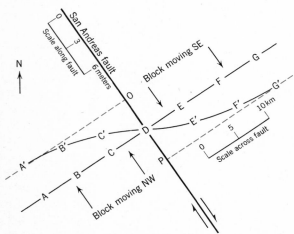

Figure 17.1 Sketch based on detailed surveys near the San Andreas Fault, California, before and after the abrupt movement that caused the earthquake of 1906. The seven survey points, A to G, were originally lined up. Slowly, movement of the two fault blocks bent the crust and displaced the points to new positions, A' to G'. Then suddenly the two sides of the fault moved and the surveyed points lay along the blue lines A'O and PG'. The sudden offset along the fault, distance OP, was 7m.

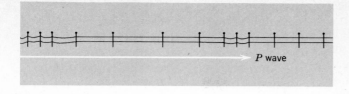

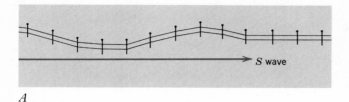

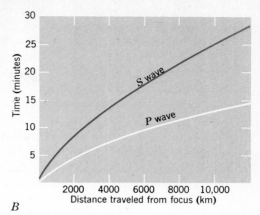

A

B

Figure 17.2 *A.* The difference between seismic waves of the *P* and *S* types is demonstrated by the disturbances they introduce in a fence. *P* waves give rise to alternate compressions and rarefactions, causing the fence to be alternately stretched and squeezed. *S* waves give rise to lateral distortions, causing wave-shaped sideways motions in the fence. Arrows show direction of wave travel.

B. *P* waves travel more rapidly than *S* waves. The curves are drawn for average travel times in the Earth. (*Data from Bullen, 1954.*)

of *P* waves change is different from the way the speed of *S* waves change. For this reason we can use careful measurements of seismic wave velocities to determine how Earth's internal properties change.

But *P* and *S* waves are not the only elastic vibrations generated by earthquakes. Strange as it may seem, a large earthquake can make Earth shake and oscillate freely like a giant bell or a huge tub of jelly. As seen in Figure 17.3, the free oscillations that occur when this happens are of two kinds. *Spheroidal* oscillations involve mainly a change in Earth's total volume, while *torsional* oscillations involve only a change in Earth's shape. The properties of free oscillations are therefore similar to those of *P* and *S* waves, and to an observer at the surface, the oscillations indeed appear very similar to ordinary *P* and *S* waves. The spheroidal oscillations are commonly called Rayleigh waves, and the torsional oscillations are called Love waves, after the English scientists who first recognized them.

Earthquakes and Man

Before going on with our discussion of Earth's structure as deciphered from seismic waves, we need to consider some of the dangers posed by earthquakes. The dangers are profound and the havoc

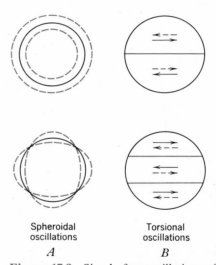

Spheroidal   Torsional
oscillations   oscillations

A   B

Figure 17.3 Simple free oscillations of a vibrating sphere, such as Earth, are of two kinds: spheroidal oscillations, in which both volume and shape may vary, and torsional oscillations, in which the volume is unchanged.

A. The simplest spheroidal oscillations occur when Earth simply expands and contracts radially, and when the sphere becomes deformed.

B. The simplest torsional oscillations occur when the northern hemisphere is twisted in the opposite direction from the southern hemisphere. The next-simplest oscillation divides Earth into three twisting slices.

caused by earthquakes is sometimes catastrophic. The effects are of four principal kinds.

1. *Ground motion* that results from the passage of *P* and *S* waves through surface rock layers and regolith can damage and sometimes completely destroy buildings (Fig. 17.4,*A*). Proper design of buildings can do much to prevent such damage, but in a very strong earthquake no structure is safe.

2. But a greater hazard than moving ground is *fire*. The movement displaces stoves, breaks gas lines, and loosens electrical wires, starting fires. In the disastrous earthquakes that struck San Francisco in 1906, and Tokyo and Yokohama in 1923, probably more than 90 per cent of the total damage was caused by fire.

3. In regions of hills and steep slopes, earthquake vibrations may cause regolith to slip and start *landslides*. This is particularly true in Alaska and parts of southern California (Fig. 17.4,*B*). Houses, roads, and other structures are destroyed by the sliding regolith.

4. Finally, there are *seismic sea waves* (sometimes called by their Japanese name, *tsunami*) that occur following violent movement of the sea floor. Seismic sea waves, often incorrectly called tidal waves, have been particularly destructive in the Pacific Ocean. About $4\frac{1}{2}$ hours after a severe submarine earthquake near Unimak Island, Alaska, in 1946, such a wave struck Hawaii. With a crest 18m higher than normal high tide, this destructive wave demolished nearly five hundred houses, damaged a thousand more, and killed 159 people.

Every year Earth experiences many thousands of earthquakes. Fortunately only one or two are sufficiently strong, or are close enough to major population centers, to cause serious loss of life. The most disastrous earthquake on record occurred in 1556, in the Shensi Province of China, where an estimated 800,000 people died. These people lived in cave dwellings excavated in loess (Chap. 10), which collapsed as a result of the quake. In 1920 another quake, in the Kansu Province of China, killed 180,000 people, and the Japanese earthquake of 1923 killed 143,000 in Tokyo and Yokohama. Since 1900 there have been 26 earthquakes in each of which 500 or more people lost their lives.

Is any place on Earth completely safe from earthquakes? No locality is free from earthquakes, but in some regions the quakes that do occur are weak and consequently are not dangerous to man or his dwellings. In southern Florida, southern Texas, and parts of Alabama and Mississippi, for example, scientists believe that the probability of damaging earthquakes is almost zero. All other parts of the country have experienced damaging quakes in the past, and more can be expected to occur in the future (Fig. 17.5).

Earth's Structure

Detailed measurements of seismic-wave velocities reveal a great deal about Earth's internal structure. Because we cannot see or sample the interior directly, earthquakes serve as a sort of global x-ray, by means of which we are able to examine the interior of our giant patient. Three observations are of particular interest to us. They are (1) Earth's layered structure, which we learned earlier is like the successive skins of an onion; (2) the thickness of the crust; and (3) the existence of a zone of low seismic-wave velocities in the upper mantle.

Earth's Layered Structure. If Earth had a uniform composition, velocities of *P* and *S* waves would increase smoothly with depth because increasing pressure increases the elastic properties of rock. For an Earth of uniform composition, therefore, it would be easy to predict how long it would take seismic waves to pass through it. But when we measure the travel times of waves, we find they differ greatly from such predictions. The only way we can account for the discrepancy is to suppose that velocities change because the Earth's composition changes with depth.

Indeed the composition does change. And another property of waves provides us with exact information on the depths at which the changes take place. Another way in which seismic waves are like light waves and sound waves is that wherever they encounter a boundary between two different substances, the waves are reflected or refracted. For example, light waves are reflected by a mirror and are refracted when they pass through a glass of water. Both *P* and *S* waves are strongly influenced by a pronounced boundary at a depth of 2900km. When *P* waves reach that boundary, they are reflected and refracted so strongly that the boundary actually casts a *P*-wave shadow over part of the Earth (Fig. 17.6). The boundary is pronounced be-

A

Figure 17.4 *A*. Destruction of school in Long Beach, California by ground motion during earthquake of 1933. Fortunately the school was not in session. (*Wide World Photos.*)

  B. A landslide triggered by the earthquake of March 27, 1964 destroyed these houses near Anchorage, Alaska. (*Steve McCutcheon, Alaska Pictorial Service.*)

B

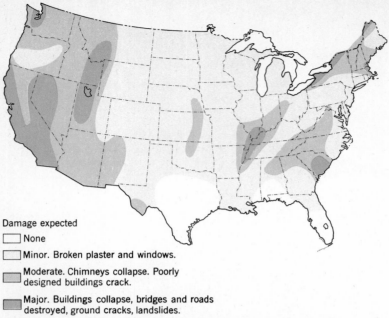

Damage expected

☐ None

☐ Minor. Broken plaster and windows.

☐ Moderate. Chimneys collapse. Poorly designed buildings crack.

☐ Major. Buildings collapse, bridges and roads destroyed, ground cracks, landslides.

Figure 17.5 Seismic-risk map of the United States. Zones refer to maximum earthquake intensity and therefore maximum destruction that can occur. The map does not indicate frequency of earthquakes. For example, the frequency of earthquakes in southern California is high, but in eastern Massachusetts it is low. Nevertheless, earthquakes in eastern Massachusetts, when they do occur, can be just as severe as the more frequent quakes in southern California. (*After Algermissen, 1969, Fourth World Conf. on Earthquake Eng., Proc., v. 1, p. 14–27.*)

cause it is the place where the comparatively light silicate material of the mantle meets the dense metallic iron and nickel of the core. The same boundary casts an even more pronounced *S*-wave shadow, but the reason is not reflection or refraction. Shear waves cannot traverse liquids. The huge *S*-wave shadow therefore proves that the outer core is liquid.

Seismic waves indicate many other boundaries within the Earth. *P*-wave reflections indicate that an inner core of solid iron and nickel exists inside the liquid outer core. Both *P*- and *S*-wave records reveal several boundaries within the mantle. Reflection by boundaries deep within the mantle are not as pronounced as reflection by the core/mantle boundary. The significance of some of the boundaries is still unknown. At the top of the mantle, however, a clearly marked boundary occurs. This is discussed in the following section.

The Crust. Early in the 20th Century the reality of Earth's crust was demonstrated by a scientist named Mohorovičić, (Mō-hō-rō-vĭtch′-ĭck) who

lived in what today is Yugoslavia. He noticed that in measurements of seismic waves arriving from an earthquake whose focus lay within 40km of the surface, stations within 800km of the epicenter recorded *two* distinct sets of *P*- and *S*-waves. He concluded that one pair of waves must have traveled from the focus to the station by a direct path, whereas the other pair represented waves that had arrived slightly later because they had been refracted. Evidently these later waves had penetrated a deeper zone of higher velocity, had traveled within that zone, and had then been refracted upward to the surface (Fig. 17.7). From his conclusions he hypothesized that a distinct boundary, strongly influencing seismic waves, separates the crust from an underlying zone. Scientists now refer to this boundary as the ***Mohorovičić discontinuity*** and recognize it as the *base of the crust.* But as the name is a tongue twister, the feature is commonly called the ***M-discontinuity,*** and in conversation is shortened still further to ***moho.***

By seismic methods, therefore, we can determine

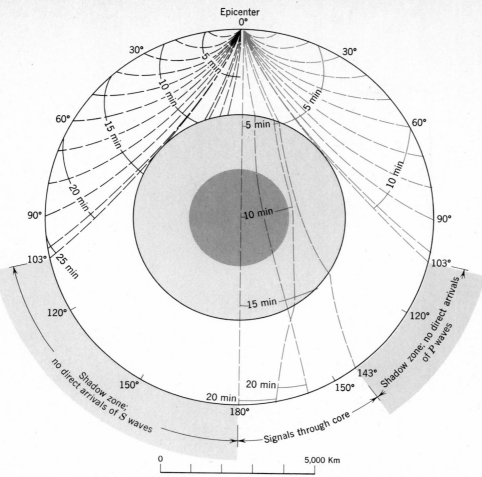

Figure 17.6 Section through Earth showing mantle (white) outer core (light gray) and inner core (dark gray). Paths of *P* waves (blue) from an earthquake focus with epicenter at 0° (top), shown in right half only. Paths of *S* waves (black) shown in left half. Distances reached by waves at 5-minute intervals are indicated. Reflection and refraction of *P* waves at the core/mantle boundary create a *P*-wave shadow zone from 103° to 143°. Because *S* waves cannot pass through a liquid, an *S*-wave shadow exists from 103° to 180°. (*After B. Gutenberg, Internal Constitution of the Earth, 1950. By permission of Dover Publications, Inc., New York.*)

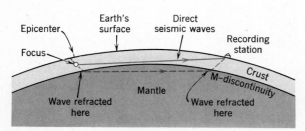

Figure 17.7 Travel paths of direct and refracted seismic waves from shallow-focus earthquake to nearby recording station.

the thickness of the crust. Also from wave velocities and the elastic properties they indicate, we can estimate its probable composition. Beneath ocean basins the crust is thin, in most localities averaging only 10km. The elastic properties of the rock of the oceanic crust are those characteristic of basalt. But in the continental crust both thickness and composition are very different. From 20km to nearly 60km thick, the continental crust tends to be thickest beneath major mountain masses (Fig. 17.8)—a fact discussed further on in this chapter. Velocities in the conti-

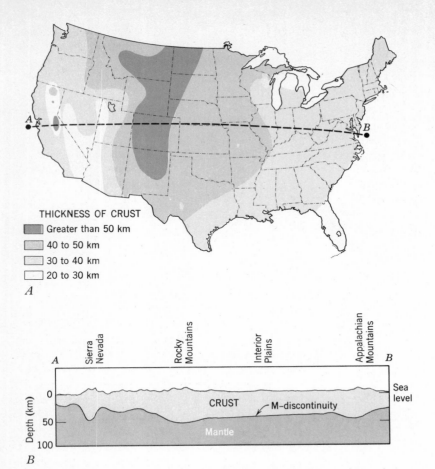

Figure 17.8 *A.* Thickness of crust beneath United States territory, determined from measurements of seismic waves.

B. Section through the crust in North America along the line AB (above). In a general way the crust tends to thicken beneath major mountain masses such as the Sierra Nevada, the Rocky Mountains, and the Appalachians. (*After Pakiser and Zietz, 1965.*)

nental crust are distinctly different from those in the oceanic crust. They indicate elastic properties like those of rocks such as granite and andesite, although at some places just above the M-discontinuity, velocities close to those of oceanic crust are observed. Below the M-discontinuity, however, wave velocities increase sharply. For example, whereas in the continental crust the average *P*-wave velocity is about 6.5km per second, immediately below the M-discontinuity the velocity is about 8.2km per second.

Low-Velocity Zone in the Upper Mantle. Recent studies have confirmed an interesting idea, proposed many years ago but at first not widely accepted.

Within the mantle, at depths ranging from 100km to 350km, a zone of low seismic-wave velocities exists. Presumably this zone of slow wave travel represents a layer having greater plasticity, meaning it flows and is deformed more readily, than material above or below it. Following the realization that plates of lithosphere apparently slide over a somewhat plastic zone in the mantle, the possible significance of the low-velocity zone has increased greatly. The top of the low-velocity zone coincides with the base of the lithosphere. Furthermore, as we read in Chapter 16, basaltic magma apparently originates by partial melting of rock in the mantle at depths of 100km to 350km; this suggests that the low-velocity zone is a place where liquids sometimes form. The low-velocity zone, therefore, seems to be a key to the developing picture of Earth's internal processes. The picture has been clarified by measurements of Earth's free oscillations.

How this is done can be explained by analogy with bells. A bell made of rigid iron is very elastic and rings for many minutes. A bell made of lead or copper—metals that are softer and more plastic—will vibrate only for a few seconds. By analysis of the sound of a bell (which means analysis of the way it vibrates), it is possible to determine its elastic and plastic properties. The same can be done with Earth when an earthquake sets it vibrating.

Although after a large earthquake, Earth may continue to vibrate slightly for days, it is more similar to a bell made of lead than to one made of iron. By analyzing how the different free oscillations die out, scientists can also identify which regions are most plastic and which are most elastic. The oscillations die out in a manner that indicates a zone of considerable plasticity in the upper mantle, at about the depth of the low-velocity zone. Although large earthquakes may have severe effects on people, they nevertheless provide vital scientific information when they start Earth "ringing."

Distribution of Earthquakes

Although no part of Earth's surface is exempt from earthquakes, several *large tracts,* or **seismic belts,** are *subject to frequent earthquake shocks* (Fig. 17.9). Of these the most obvious is the *Circum-Pacific belt,* in which about 80 percent of all earthquakes originate. It follows the western highlands of the Americas from Cape Horn to Alaska, and crosses to Asia, where it extends southward down the coast, and finally loops far southward to New Zealand. Next in prominence, giving rise to 15 per cent of all earthquakes, is the *Mediterranean-Asiatic belt,* extending from Gibraltar to southeast Asia. Lesser belts follow the mid-ocean ridges.

Seismic belts are places where a lot of energy from internal activities is released. We might therefore expect a correspondence with other products of internal activities. Indeed seismic belts correspond closely with some features. Both volcanoes (Fig. 16.14) and young mountain chains (Fig. 18.11) are closely related. Again almost all deep earthquakes that originate in the mantle at depths greater than

100km (Fig. 17.9) are associated with sea-floor trenches (Fig. 12.10). Finally, when we look at the distribution of earthquake foci below trenches, we find they fall within a well-defined region that slopes downward at approximately 45 degrees (Fig. 17.10). These remarkable observations suggest that earthquakes originate on lines along which comparatively cold plates of lithosphere are plunging back into the mantle. As we noted in Chapter 16, there is a close association between andesitic volcanism, oceanic trenches, and downward-moving plates of lithosphere. We can add deep earthquakes to the same association.

Topography and the Principle of Isostasy

We found in Chapter 15 that the weight of a large continental glacier can depress the surface, but that when the mass is removed, the surface slowly rises up again (Fig. 15.5). Evidence of this kind leads us to conclude that the lithosphere can "bob" up and down like an iceberg, although more slowly, and that lithosphere must be "floating" on an easily deformed substratum. *The ideal property of flotational balance among segments of the lithosphere* is known as **isostasy.** This property is so important that it demands more discussion.

Suppose that the lithosphere were an unbendable, rigid sheet and that mountains were nothing more than extra rock piled on its top. If we then measured Earth's gravitational pull at some point close to a mountain, we would record an increased pull caused by the added attraction of the mountain mass. During the last century, surveyors working in India first confronted this problem while they were working near the great Himalaya Mountains. Archdeacon Pratt, a British cleric whose hobby was mathematics, analyzed the measurements and found that the increase in gravitational attraction was actually very much less than would be expected if the mountains simply sat on top of the lithosphere. He quickly recognized that the problem could be explained if the extra mass of the mountain above ground was offset by the presence of rock of low density in the crust below the mountains. Instead of being uniform in density, he suggested, the lithosphere actually consists of blocks of different density. He suggested further that the bottom of each block floats, at the same level, on a fluid substratum, and that the height

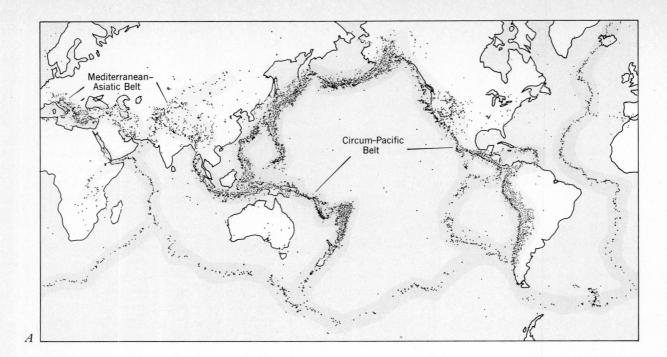

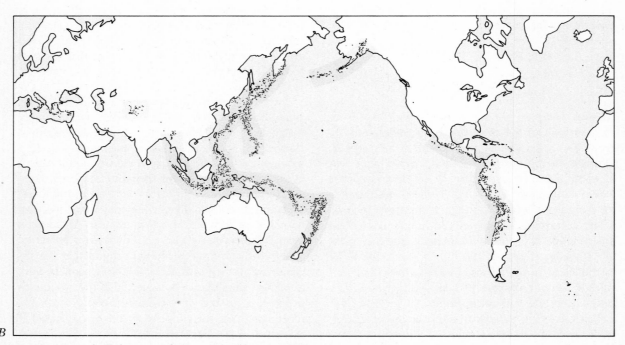

Figure 17.9 *A.* Epicenters of all earthquakes recorded by the U.S. Coast and Geodetic Survey between 1961 and 1967. Each dot represents a single earthquake. The epicenters fall into well-defined seismic belts (blue shading). In the Atlantic, Pacific, and Indian Oceans, seismic activity coincides with mid-ocean ridges.

B. Epicenters of earthquakes having foci deeper than 100km, that occurred during the years 1961 through 1967. These deep earthquakes coincide closely with the deep-sea trenches in Figure 12.10. (*After Barazangi and Dorman, Bull. Seismol. Soc. America, v. 59, p. 369, 1969.*)

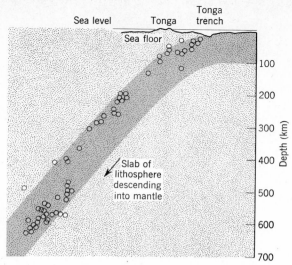

Figure 17.10 Earthquake foci beneath the Tonga Trench, Pacific Ocean, during several months in 1965. Each circle represents a single earthquake. The earthquakes are believed to be generated by downward movement of a comparatively cold slab of lithosphere that is plunging slowly back into the mantle. The slab is represented by the blue band. (*After Isacks, Oliver, and Sykes, Jour. Geophys. Res., v. 73, p. 5855, 1968.*)

of each block is controlled by its density. Blocks of low density stand high and form mountains, while blocks of high density sit low, forming plateaus and ocean basins (Fig. 17.11). The situation suggested by Pratt is similar to a balloon and block of wood floating side by side in a tub of water. The balloon floats high in the water, because it is light. The block of wood sits low because it is dense and heavy. The depth of the fluid substratum, on which Earth's blocks float, was called by Pratt the *depth of compensation*. The actual depth was not specified, but Pratt presumed it to be many kilometers down. Pratt's theory was soon challenged by another English mathematician, Sir George Airy, who pointed out that the gravity observations are also explained if the lithosphere floats as proposed by Pratt, but consists of blocks of the same density but different thicknesses, like a group of icebergs floating in the sea (Fig. 17.11). The biggest icebergs sit deepest in the sea, but also stand highest above the water. According to Airy, the extra mass of a high mountain mass should be offset by deep "roots" of the same material.

We now know that both Pratt and Airy were partly correct. Densities of different parts of the lithosphere do differ. Oceanic crust is about 10 per cent more dense than continental crust, and ocean basins do indeed sit much lower than continents. Furthermore, it is easy to calculate that at a depth of approximately 100km, pressure from the overlying lithosphere regardless of whether oceanic or continental crust caps it, is everywhere the same. This is Pratt's depth of compensation. A depth of 100km puts us at the top of the low-velocity zone—an important piece of evidence in support of the theory that plates of lithosphere float on the upper mantle. In this sense, apparently Pratt was correct. However, the depth to the base of the lithosphere varies from place to place and, as we saw earlier in this chapter, seismic evidence indicates that beneath high mountains there are indeed deep roots of lighter rock. So mountains seem to act very much like icebergs. In this sense Airy was correct. To explain all the details of isostasy, therefore, we must borrow from both the Pratt model and the Airy model.

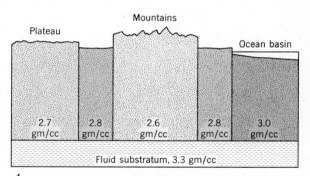

A

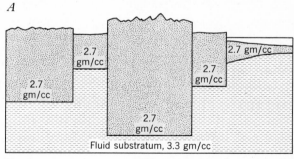

B

Figure 17.11 The two theories of isostasy:

A. Pratt's theory suggests the lithosphere consists of blocks of differing density, all floating on a dense, fluid substratum, so that the height reached by each block depends on its density.

B. Airy's theory suggests the lithosphere has the same density throughout, but varies in thickness. Hence beneath high mountains we should find deep roots. (*After Bowie, 1927 and Longwell, 1925.*)

Isostasy depends on a complicated mixture of both conditions.

The important point for our discussion is that once again the low-velocity zone of the upper mantle apparently acts as a fluid, with the lithosphere floating on it. *Floating* is not exactly the correct word, because the low-velocity zone is really solid rock. Nevertheless the lithosphere can be pushed down into the low-velocity zone. If the pushing force is then removed, the lithosphere will slowly "bob" up again. As rock material is eroded from a mountain, thereby reducing the mass of the mountain, the entire mountain will slowly rise as isostatic adjustment occurs. As we shall see, this is an important feature in the development of mountains.

Characteristics of Mountains

All mountains have one thing in common: they tower above the surrounding terrain. Because of their height, steep slopes, and massiveness, many have been given special names, and by ancient peoples were endowed with mystical properties. The mystery, reverence, and romance that surround mountains still persist; even today we are not exactly sure how some of the greatest mountains formed, and their study attracts many scientists. In many respects, understanding how mountains form is a key to understanding Earth's history.

Earth has always been characterized by mountains. They go through long cycles, during which they form, are elevated, and then gradually waste away and cease to be mountains. But their former presence continues to be recorded in rocks. Granite batholiths and belts of highly deformed metamorphic rock are clear evidence that mountains once existed. Every continent contains such evidence, telling the same story. So widespread is the evidence that there may not be a single part of any continent that has not been, at some time, part of a mountain mass. Today new, young mountains are forming beneath the sea. Other mountains, such as the Himalaya, are middle-aged but still rising; still others, such as the Appalachians, have reached old age and are slowly wasting away.

Mountains differ in age, history, origin, and size. Some, like the volcanic cones Kilimanjaro in East Africa and Mount Ararat in Asia Minor, are single isolated masses. Others form a linear sequence which we call a **mountain range,** meaning *an elongate*

series of mountains belonging to a single geologic unit. Excellent examples are in the Sierra Nevada in eastern California, and the Front Range in Colorado. *A group of ranges similar in general form, structure, and alignment, and presumably owing their origin to the same general causes,* constitute a **mountain system.** Thus, the Rocky Mountain System is a great assemblage of ranges, formed at approximately the same time, extending from near the Mexican border northward through the United States and western Canada. The term **mountain chain** is used somewhat more loosely to designate *an elongate unit consisting of numerous ranges or groups, regardless of similarity in form or equivalence in age.* An example is the gigantic mountain chain that runs along the western edge of the Americas, from the tip of South America to northwestern Alaska, and that includes all the systems and ranges in between. This broad belt of ranges is also called the *American Cordillera.*

Although we can define mountains, it is less easy to classify them because they display a great variety of rocks and structures; no two are identical. If we concentrate on the details we are in danger of seeing only the foliage and missing the forest. The most helpful way to organize our thinking about mountains and to see through the foliage, is to identify some single, most characteristic feature, and use it for classification. On this basis we can identify three principal kinds of mountains:

1. fold mountains
2. volcanic mountains
3. fault-block mountains

Each kind merits separate discussion.

Fold Mountains

Fold mountains are spectacular and complex structures. Occurring in great arc-shaped systems a few hundred kilometers wide, they commonly reach several thousand kilometers in length. The word *fold* clearly indicates their most characteristic feature. Strata have been compressed, are folded, and crumpled, commonly in an exceedingly complex manner. Although folding is the key feature, other kinds of mountain-building processes participate in the making of fold-mountain ranges too: faulting, metamorphism, and igneous activity are always present. Examples are widespread: the Appalachians, the Alps, the Urals, the Himalaya and the Carpathians are all fold-mountain systems. Indeed the Alps, Carpathians and the Himalaya belong to a gigantic fold-mountain chain formed during the Mesozoic and Cenozoic Eras (Fig. 18.11).

Geosynclines. All fold-mountain ranges share another feature related to their folded strata. They develop from exceptionally thick piles of sedimentary strata, commonly 15,000m or more in thickness. Early in the study of mountain ranges, it was realized by the American scientist James Hall that so huge a thickness called for an unusual sort of basin in which the sediments could accumulate. Another famous 19th-Century American scientist, J. D. Dana, discussed this fact in his analysis of the history of the Appalachians. He pointed out that both subsidence of the crust to form an unusually deep basin, and filling of the basin with sediment, must have preceded the final deformation and uplift that created the mountains. Dana coined the name *geosyncline* for the basin in which the Appalachian sediments accumulated. Studies of other fold mountains show that deep, sediment-filled basins preceded all of them. The term **geosyncline** therefore has wide application and describes *a great trough that has received thick deposits of sediment during its slow subsidence through long geologic periods.*

The strata of fold-mountain systems are predominantly marine—a conclusion we can draw from the presence of marine fossils. In systems such as the Alps, the marine strata are mostly of deep-water origin. In others, such as the Appalachians, the sediments apparently accumulated in shallow water. Regardless of water depth, the kinds and thicknesses of sediments found in geosynclines lead us to two important conclusions. First, the fact that most of the sediment is marine indicates that geosynclines are oceanic features. Second, the great thicknesses of sediment, which commonly exceed the greatest depths of the ocean, indicate that the basin *must* have been sinking while it was being filled with sediment.

Some geosynclines occur in pairs. An excellent example is the Appalachians. Two elongate geosynclines, roughly parallel, once occupied the region from Newfoundland to Alabama. One geosyncline lay directly adjacent to the continent and became filled with shallow-water sediment which we can now see as the limestone, sandstone and other rock that are common in most of the Appalachians. The other geosyncline lay farther east and farther off-shore. It became filled with deeper-water sediment of the sort we can observe today at the foot of the continental slope off the east coast of North America. The deep-water geosyncline also contained some volcanic rock.

How and why geosynclines form is still conjectural. In Chapter 18 we shall discuss the possibility that they form as a consequence of movement by plates of lithosphere. But whatever their origin, geosynclines indicate clearly that the life cycle of a mountain system begins with a long period of submarine downwarping. The subsequent deformation and uplift, that resulted in the mountain structures we now see, are simply the latest phases of a long and complicated history that we can describe by means of a real example, that of the Appalachians.

The Appalachians. The Appalachians constitute a fold-mountain system that stretches 2500km from Alabama to Newfoundland and that consists of folded and faulted strata of Paleozoic age. Although sediment apparently accumulated in two geosynclines, only the mountains formed from the western one remain well exposed. Portions of the eastern basin can be observed in Maine, but farther south the equivalent region lies offshore, and so is hidden from view. The shallow-water sediment in the western geosyncline contains mud cracks, ripple marks, fossils of shallow-water organisms, and in places even freshwater materials such as coal. The sediment was deposited on a basement of old metamorphic- and igneous rock, and becomes markedly thicker away from the former western shore; that is, it thickens from west to east.

Today, as we approach the Appalachians from the west, we can still see some of the former sediment, occurring as essentially flat-lying, undisturbed strata (Fig. 17.12). Continuing eastward, however, we see the strata thickening and becoming gently folded and thrust-faulted. In the middle of the elongate Appalachian system, in the region known as the Valley and Ridge Province, the strata along the margin of the Appalachians have been bent into broad anticlines and synclines. The province gets its name because valleys have been developed by erosion in the most erodible strata, such as limestones, dolostones and claystones. The valleys alternate with prominent ridges formed by very resistant strata, chiefly sandstone (Fig. 17.13). At the northern and southern ends of the Appalachian system, strata on

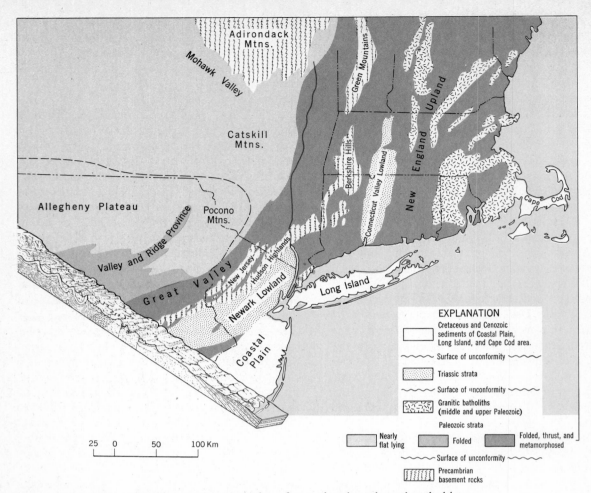

Figure 17.12 Generalized map and cross section of central and northern Appalachians. Approaching the mountains from the west, one passes first through undisturbed strata and then through strata affected by increasing amounts of folding, thrust faulting, and metamorphism. In some areas, where high mountains once stood, rocks of Triassic and younger ages now cover the deformed but deeply eroded strata. (*Map after Tectonic Map of United States; New England, after Richard Goldsmith, 1964; cross section after Erwin Raisz, in Douglas Johnson, 1932.*)

the western margin of the mountains have been subjected more to thrust faulting than to folding. The direction of thrusting is always the same: hanging-wall blocks have been thrust westward, away from the center of the range. The difference in style of deformation between the different regions along the western margin of the Appalachians can be seen by comparing Figure 17.14 with Figure 17.13. Farther eastward, toward the core of the Appalachian system, rocks are increasingly metamorphosed and deformation is increasingly intense. Folds become

isoclinal and then overturned, and faulting is prevalent. In places fragments of the old basement are seen thrust up over younger sedimentary strata. Finally, we reach a region where intense metamorphism has occurred and where granitic batholiths have been emplaced.

Metamorphism and deformation seem to have been most intense where the pile of sedimentary strata was thickest. Because intensity of metamorphism can be equated with depth of burial (Chap. 16), we conclude that the most intense meta-

A

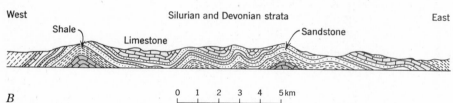

West   Silurian and Devonian strata   East

Shale   Sandstone

Limestone

0 1 2 3 4 5 km

B

Figure 17.13   Valley and Ridge Province in the central Appalachians.

A. View in eastern Pennsylvania, looking southeastward. Resistant strata underlie wooded ridges; more erodible strata underlie cultivated lowlands. Distance between Tuscarora Mountain and Blue Mountain is 37km. (*J. S. Shelton.* From *Geology Illustrated* by John S. Shelton, W. H. Freeman & Co., © 1966.)

B. Section through the ranges having the kind of gentle folding found in the Valley and Ridge Province. Compare Fig. 17.14. (*After U.S. Geol. Survey Folio 59.*)

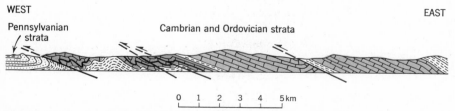

WEST   EAST

Pennsylvanian strata   Cambrian and Ordovician strata

0 1 2 3 4 5 km

Figure 17.14   Thrust faults and folds in the southern Appalachians. Many of the thrusts originated from the stretching and breaking of limbs and folds. Compare Fig. 17.13, where the style of gentle folding found in the central Appalachians is depicted. (*After U.S. Geol. Survey Folio 61.*)

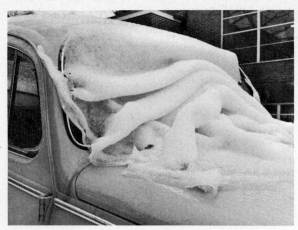

Figure 17.15 Folds in snow that slid down an inclined windshield. Strata that slide because of the pull of gravity display similar folding. (*Yonkers Herald-Statesman, courtesy Westchester-Rockland Newspaper Group.*)

center of the system, corresponding to the Valley and Ridge Province, the sliding strata were somewhat plastic and became folded. A more familiar example of folding by gravity sliding is shown in Figure 17.15. At both ends of the mountain range, strata were apparently brittle rather than plastic; so instead of folding they broke into giant thrust sheets, which slid westward like a series of playing cards in a deck, each one riding up over the next sheet to the west.

We naturally ask how well the Appalachian picture can be applied to other fold mountains. The answer is that similar features are found in all of them. For instance the Jura Mountains, lying along the northwestern edge of the Alps, have the same folded form and the same origin as the Valley and Ridge Province. In the high Alps, which correspond to the now deeply eroded and topographically unimpressive Appalachians we observe in eastern Virginia and Maryland, thrusting appears to have developed on a much grander scale than in the Appalachians (Fig. 17.16).

The Canadian Rocky Mountains, which are a magnificent, much less eroded mountain system than the Appalachians, can also be compared. A section through the Canadian Rockies at about the latitude of Calgary, Alberta (Fig. 17.17) reveals all the features we have discussed for the Appalachians. A central or core zone has been intensely metamorphosed. In the core zone parts of the older basement rocks have been thrust upward, and in the marginal region folding and extensive thrust faulting are evident. The thrust sheets have moved eastward, away from the core zone. It is apparent also (Fig. 17.17) that each sedimentary unit becomes thinner as it is followed from west to east, indicating that the core

morphism, maximum deformation, maximum depth of burial, and maximum uplift all occurred in the same region. This association indicates that the core of a mountain range coincides with what was once the deepest part of a geosyncline. When the unknown force that forms the geosyncline and causes the deformation of the strata is removed, the deeply buried pile of strata starts to rise, and is apparently kept rising until isostatic equilibrium is reached. Upward movement of the mountain core suggests how folding and thrusting may have occurred in the marginal portions. As the core rose, the still undeformed strata on the flanks were tilted. Under the influence of gravity the strata slid westward. In the

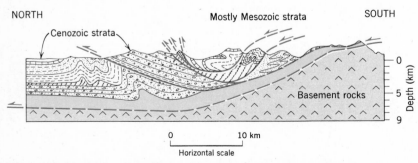

Figure 17.16 Section through the Alps in central Switzerland. Strata have moved northward along great thrust faults (blue lines) which later were themselves folded. Major thrust separates overlying strata from basement rocks. (*After Albert Heim, 1922; Geologie der Schweiz., v. 2, pl. 27.*)

A

Figure 17.17 *A*. Thrust faulting in Canadian Rocky Mountains. View looking northwest. Two thrusts (blue lines) can be seen. Rock underlying Mount Broadwood has moved from left, riding up over steeply dipping strata on right, like a giant sled. Lizard Range, in background, is joined by rock that was thrust many kilometers from the west. (*Photo courtesy of Chevron Standard. Ltd. After Haderson and Dahlstrom, Bull. American Assoc. Petroleum Geologists, v. 43, p. 641-653, 1959.*)

B. Section through the Canadian Rocky Mountains at about the latitude of Calgary, Alberta. The zone of intense metamorphism coincides with the region of maximum uplift and maximum deformation. Further east, where the strata become progressively thinner, the pile has been greatly thickened by movement along thrust faults. Sense of movement on the thrust faults is for each fault-block to move towards the east. (*After Price and Mountjoy, Geol. Assoc. Canada Spec. Paper No. 6, 1970.*)

EXPLANATION

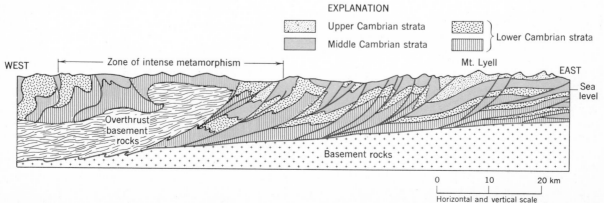

B

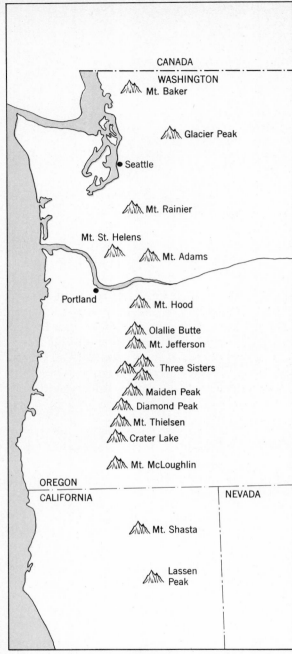

Figure 17.18 Volcanoes of the Cascade Range, each of which has been active during the last two million years. (*After Tabor and Crowder, U.S. Geol. Survey, Prof. Paper No. 604, 1969.*)

zone coincides with the thickest part of the old geosyncline.

Volcanic Mountains

Many fold mountains include a great variety of volcanic rocks. The Andes, for example, contain enormous thicknesses of volcanic rock, yet they owe their origin chiefly to the folding and faulting associated with uplift. But when we say *volcanic mountains* we mean mountains that have been built entirely by volcanic extrusion.

Most volcanic mountains are submarine features. In some of them, as in the Hawaiian chain (Chap. 16), the higher peaks protrude above sea level. In others, as in the East-Pacific Rise, the entire chain is submerged.

A special class of volcanic mountains consists of **island arcs.** These are *great arcuate belts of andesitic and basaltic volcanic islands,* some of them 2000km or more in length. The Aleutian Islands are a conspicuous island arc; another arc runs from Kamchatka through the Kurile Islands and across Japan; yet another consists of the islands of Sumatra, Java, Soemba, and Timor. Each island arc is adjacent to a sea-floor trench—the site, as learned in Chapters 12 and 16, where a plate of lithosphere is slowly plunging downward into the mantle. The volcanism that builds the volcanic mountains is presumably caused by the melting of the downgoing plate (Chap. 16).

Volcanic mountains on land are rare. The only one in the United States is the Cascade Range in Washington and Oregon. The Cascades are a range of huge young, andesitic volcanoes, running from Lassen Peak, California, at the south to Mount Baker, Washington, more than 900km farther north (Fig. 17.18). These snow-covered giants, all apparently active during the past few million years, were erupted onto a platform of older, folded, and deeply eroded rocks. The trend cuts directly across many other geologic features, and for this reason scientists believe that building of the mountains was controlled by deep-seated processes, possibly a downgoing plate of lithosphere.

Fault-Block Mountains

In many parts of the world isolated mountain ranges stand abruptly above surrounding plains. Study re-

Figure 17.19 Frenchman Mountain, east of Las Vegas, Nevada, is a tilted fault block. Steeply inclined marine strata of Paleozoic age, more than 1500m thick, have been elevated by movement along steeply dipping faults that trend north-south. Blue line marks trace of a small east-west fault that offset strata. (*William Belknap, Jr.*)

veals that the ranges are separated from intervening lower lands by normal faults of great displacement, and seem to be giant pieces of the crust punched upward from below. These are fault-block mountains. Rocks within the mountains commonly contain evidence that former fold mountains once occupied the same sites, but that erosion had worn them down before the fault blocks formed.

Basin and Range Province. One of the most extensive fault-block mountain systems occurs in Nevada, western Utah, portions of Oregon, Idaho, Arizona, and California. Known as the *Basin and Range Province*, it contains a spectacular development of uplifted fault blocks. The province is underlain by sedimentary strata deposited during the Paleozoic Era. Following a period of fold-mountain

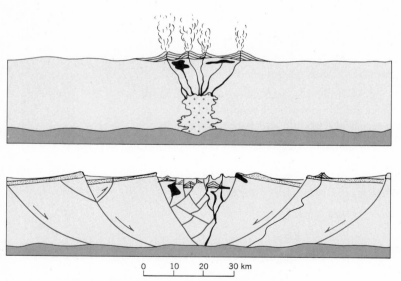

0   10   20   30 km

Figure 17.20 Possible model for formation of structures in Basin and Range Province. Warping of the crust and extrusion of large quantities of volcanic material were followed by collapse along a series of steeply inclined normal faults. (*After Mackin, in Roberts, UMR Journal, No. 1, p. 101–119, 1968.*)

formation, the region was deeply eroded during Mesozoic and Cenozoic times, when it supplied sediment to geosynclines that later developed into the Coast Ranges of California and into the Rocky Mountains. Starting about 25 million years ago, and accompanied by extensive volcanic activity, the region broke into a series of blocks 30km to 40km in width and up to 150km long, bounded by steeply inclined normal faults. Tilting of the blocks is pronounced (Fig. 17.19).

How could faulting and uplift occur on such a gigantic scale? One suggestion is as follows: an initial period during which the crust was stretched by upwarping coincided with the onset of volcanic activity. Later the distended crust collapsed as the volcanism waned. Collapse occurred along a series of steeply dipping, arc-shaped faults (Fig. 17.20), and tilting of the blocks produced the relief we observe today.

Fault-block mountains are common. The Sierra Nevada, and the Grand Teton Mountains in Wyoming are striking examples. Others occur in the Great Rift Valley of East Africa. Here faulting is also associated with tilting of the fault blocks, and this, together with the creation of great horsts and grabens, has produced a chain of fault-block mountains that is several thousand kilometers in length.

The association between mountain ranges of all kinds, batholiths, metamorphic rock, volcanism, and earthquakes has been repeatedly remarked in preceding chapters. The association cannot be accidental; the features must be related and must have a basic underlying cause. In the next chapter we examine possible causes for these intriguing associations and attempt to follow the lithosphere through its long and complicated history.

Summary

Earthquakes

1. Abrupt movement on faults is responsible for most earthquakes, many of which cause destructive damage to dwellings and other man-made structures.

2. Ninety-five per cent of all earthquakes originate in the Circum-Pacific belt and the Mediterranean-Asiatic belt. The remaining 5 per cent are widely distributed.

3. Energy released at an earthquake focus radiates outward as P (compressional) waves, and as S (shear) waves. Earthquake energy also makes Earth vibrate like a giant bell, setting up the free oscillations known as Rayleigh- and Love-waves.

4. From a study of seismic waves scientists infer the internal structure of Earth by locating boundaries or discontinuities in its composition. Pronounced compositional boundaries occur between crust and mantle and between mantle and outer core.

5. Thickness and composition of Earth's crust vary systematically with major relief features. Continental crust, 20km to 60km thick, consists chiefly of granitic rocks. Oceanic crust, only about 10km thick, consists chiefly of basalt.

6. The lithosphere is rigid; it overlies a plastic zone, in the upper mantle, within which seismic waves have low velocities.

Mountains

7. Mountains are of three principal types, fold, volcanic, and fault-block mountains.

8. Fold mountains are elongate systems of ranges within which unusually thick piles of sedimentary strata, many of marine origin, have been extensively folded and faulted.

9. Degree of deformation increases as we pass from the edge to the core of a fold-mountain system. In the core, strata have been metamorphosed, intruded by granitic batholiths, and uplifted.

10. Volcanic mountains occur largely in the ocean basins. Mid-ocean ridges and mountainous island arcs are volcanic-mountain systems.

11. Island arcs are arcuate chains of mountainous volcanic islands associated with sea-floor trenches. They form from volcanic activity generated by a downgoing plate of lithosphere.

12. Fault-block mountains are created when elongate blocks of crust are thrust up, or dropped down, along steeply dipping normal faults.

Selected References

Anderson, D. L., 1962, The plastic layer of the Earth's mantle: Sci. American v. 207, no. 1, p. 52–59.

Andrews, Allen, 1963, Earthquake: London, Argus and Robertson.

Bailey, E. B., 1935, Tectonic essays, mainly Alpine: Oxford, England, Clarendon Press.

Billings, M. P., 1960, Diastrophism and mountain building: Geol. Soc. America Bull., v. 67, p. 1295–1318.

Clark, S. P., 1971, Structure of the Earth: Englewood Cliffs, N.J., Prentice-Hall.

Clark, T. H., and Stearn, C. W., 1968, Geological evolution of North America: New York, Ronald Press.

Davidson, Charles, 1931, The Japanese Earthquake of 1923: London, Thomas Murby.

Gilluly, James, 1967, Chronology of tectonic movements in western United States: American Jour. Sci., v. 265, p. 306–331.

Gutenberg, Beno, and Richter, C. F., 1954, Seismicity of the Earth and associated phenomena, 2nd ed.: Princeton Univ. Press (Facsimile repr., 1963: New York and London, Hafner).

Hodgson, J. H., 1964, Earthquakes and Earth structure: Englewood Cliffs, N.J., Prentice-Hall.

King, P. B., 1959, The evolution of North America: Princeton, N.J., Princeton Univ. Press.

Chapter 18

Dynamics of the Lithosphere

Search for a Solution

Sherlock Holmes was an imaginative detective, able to study a few fragmentary clues and then construct a theory to explain the crime. Usually his theory was correct and the villain was apprehended. Holmes would have been fascinated by the mystery of the lithosphere. In the preceding chapters a lot of clues have been discussed. One of them is the clue that earthquakes, sea-floor trenches, and andesitic volcanoes are closely associated (Chap. 17), and another is the clue, from paleomagnetism, that continents have moved with respect to the magnetic poles (Chap. 3). Now we have to find a unifying thread, or theory, that will tie all the clues together and explain how the lithosphere works.

The search for a unifying theory has occupied scientists for more than a hundred years. During the 19th Century people were inclined to believe that Earth was formerly much hotter and that it is now slowly cooling and contracting. According to that theory, fold mountains and seismic belts are places where most of the contraction occurs. The theory explained some clues but it had some fatal flaws. The most serious flaw, the one that ultimately killed the theory, lies in radioactivity. Earth is *not* cooling down, because the heat generated by radioactive decay keeps it hot.

Theory of Continental Drift. A good detective always has in his mind several theories; when one of them proves incorrect, he moves on to the next. When, early in the present century, the contraction theory collapsed, a suggestion made years earlier was revived and was used as the basis for a fascinating new theory. This one proposed that continents drift slowly across the surface of the Earth, sometimes breaking into pieces and sometimes colliding with each other. This *theory of continental drift* was originally suggested to explain the striking parallelism of the coastlines on the two sides of the Atlantic Ocean. But other bits of evidence were soon found. These supported the idea that the world's land masses had once been joined together in a single great supercontinent, which was dubbed *Pangaea* (pronounced pan-jée-ah), meaning "all lands." According to the theory, Pangaea was somehow disrupted, and its fragments, the continents of today, slowly drifted to their present positions. Proponents of the theory likened the process to the breaking-up of a sheet of ice that floats in a pond. The broken pieces, they argued, should all fit back together again, like pieces of a giant jigsaw puzzle.

However, despite that argument, the shapes of the pieces in Earth's jigsaw puzzle need not be identical with the shapes of today's coastlines. Changes of sea level affect the shapes of coasts, and the parts of the continental crust that form the continental shelves and slopes lie below sea level (Chap. 12). Instead of today's coasts, therefore, the line along which the continental jigsaw pieces should fit most closely is the outer edge of each body of continental crust. The place usually selected for the jigsaw test is a line drawn around continental crust at a depth of about 2000m, halfway down the continental slope. Because we cannot move continents back and forth to see how they fit, we must do the jigsaw with maps or with calculations based on computer data. The reassembly of Pangaea by fitting together today's continents is done most conveniently with the aid of a computer.

All jigsaw puzzles have a few pieces that seem difficult to fit in, and Pangaea is no exception. The smaller pieces can be squeezed into gaps, but no matter how the pieces are fitted a few tiny gaps remain and a few small areas of overlap occur. One of the most successful reconstructions is shown in Figure 18.1,*A*. But rebuilding Pangaea is probably more like reconstructing a broken glass than like doing a jigsaw puzzle, and as anyone who has tried to reassemble broken glass knows, the pieces never seem to fit exactly. After Pangaea broke apart, submarine erosion must have occurred, changing conti-

nental margins to some degree. Additionally, maps covering some parts of the sea floor are so poor that exact reconstruction cannot really be expected. Nevertheless, the fit of the pieces in Figure 18.1,*A* is remarkable. Continents on both sides of the Atlantic fit together so closely that it almost seems a giant zipper could reclose the gap. Australia fits exactly with Antarctica, while India and Madagascar snuggle in between the eastern side of Africa and the western boundary of Australia.

The idea of a former Pangaea explains many clues that have long been known but that, until recently, could not be explained. Some of these are: fossils of identical land-dwelling animals are found in South America and in Africa, but nowhere else; fossils of identical trees are found in South America, India, and Australia; fragments of what is apparently a single mountain system are observed on both sides of the North Atlantic Ocean (Fig. 1.10). But the most impressive clue consists of unmistakable evidence that, about 300 million years ago, a continental ice sheet covered parts of South America, southern Africa, India, and southern Australia (Fig. 18.1,*B*). The ice sheet resembled the one that covers Antarctica today, and evidence of its existence is so well preserved that thousands of tiny glacial striations (Chap. 11) reveal the directions in which the glaciers flowed. But if, 300 million years ago, continents were in the positions they occupy today, the ice sheet would have had to cover all the southern oceans, and in places would even have had to cross the equator! A glacier of such huge size could only mean that the world climate was exceedingly cold. But in that case, why has no evidence of glaciation at that time been found in the northern hemisphere? The dilemma is explained neatly by continental drift. Three hundred million years ago the regions covered by ice lay in very high, cold latitudes. Indeed they were adjacent to the South Pole (Fig. 18.1,*A*). Therefore, at that time the world climate need not have been greatly different from that of today.

Impressive as is the evidence for it, the theory of continental drift does not explain all the clues. One piece of evidence against the theory is the strength of rocks. Because of the great friction it would generate, continental crust simply cannot slide over oceanic crust without both crusts disintegrating. The process is like trying to slide two sheets of coarse sandpaper past each other. In order to overcome objections, many modifications to the theory

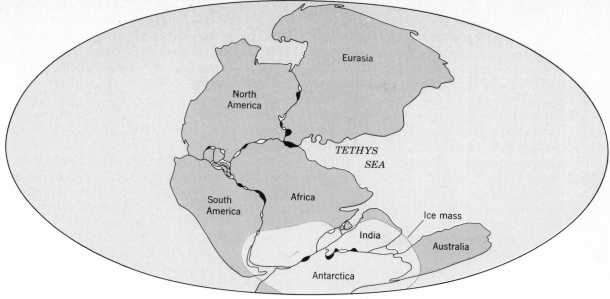

A

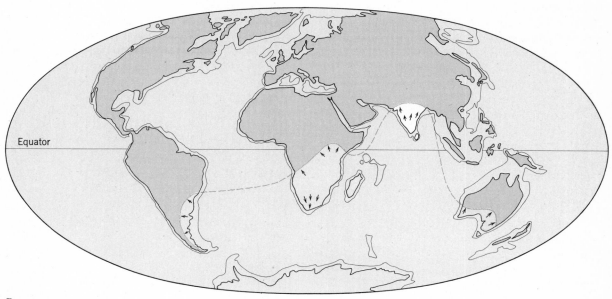

B

Figure 18.1 The continents attained their present shapes when Pangaea broke apart 200 million years ago.

  A. Shape of Pangaea, determined by fitting together pieces of continental crust along a contour line 2000m below sea level. In a few places some overlap occurs (black); in others blank areas are found (white). These are places where existing maps are poor or where later events have modified the shapes of continental margins. Shaded area is the region affected by continental glaciation 300 million years ago.

  B. Present continents and the 2000m contour below sea level. Shading outlines areas where evidence of the old continental glaciation exists. Arrows show directions of ice movement. (*After Dietz and Holden, 1971, and Du Toit, 1937.*)

have been suggested. For example, one suggestion is that Earth is slowly heating up and expanding. According to this modification, the breakup of Pangaea occurred as Earth's surface became stretched so that it cracked. As the cracks widened, the sea penetrated them, and thus today's ocean basins, at first very narrow, began to form. Some expansion of the Earth and cracking of the crust may indeed have occurred, but it seems to be too little to account for the widths of the ocean basins. In addition, the expansion theory does not adequately explain magnetic evidence. Paleomagnetism shows that some continents have twisted and turned as they moved through great distances. Even if expansion of the Earth has happened, therefore, we are forced to conclude that continents have moved about relative to each other.

A scientific Sherlock Holmes, in the guise of Professor H. H. Hess of Princeton University, stepped forward in 1960 and proposed a new theory. He suggested that the sea floor slowly spreads and moves sidewise away from the mid-ocean ridges. The suggestion sounds very simple, and at first glance does not seem to account for moving continents. Nevertheless, it led to a theory that comes

so close to explaining all the clues that it may indeed be the unifying theory everyone has been seeking. That theory is now called the *theory of plate tectonics*. We first referred to it in Chapter 1, and pointed out various aspects of it in subsequent chapters.

Theory of Plate Tectonics

Whereas early theories were proposed to explain clues found on the continents, Hess based his suggestion on new clues from the sea floor. He suggested that features such as mid-ocean ridges and sea-floor trenches could not have been formed by erosion, and proposed instead that all major structures on the sea floor are the direct result of movement within the mantle. A summary of his ideas is seen in Figures 1.11 and 18.2. The basis of his suggestion concerned convection cells. As we saw in Chapter 3, convection is a process that involves fluid flow by which heat is transferred in gases and liquids. Deep in the mantle, however, solid rock becomes so plastic that slow, convective flow can occur even in rock. Hess suggested that hot, solid rock might rise by convection from great depths in the mantle. The convection cells rise beneath mid-

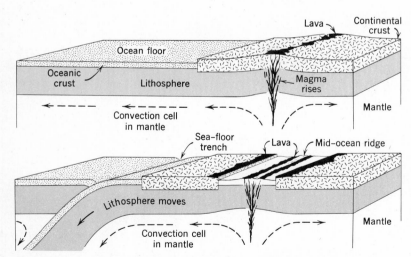

Figure 18.2 Convection cell in the mantle makes plates of lithosphere move. Above the rising convection current, lithosphere and continental crust are split and are dragged apart. Magma rises along the fracture, forming lava and intrusive igneous rock.

As new lithosphere is created, and continents move sidewise, another break develops, and a plate of lithosphere slowly plunges downward where it melts and is absorbed into the mantle. (*After Dietz and Holden, 1971.*)

ocean ridges and move laterally away, carrying with them, as if on a conveyor belt, the cool rigid lithosphere. As the lithosphere is carried in opposite directions away from a mid-ocean ridge, a parallel system of long, thin fractures develops. Basaltic magma, developed by partial melting in the mantle, moves into the fractures. By injection and eruption, the new magma fills the growing fracture and continuously creates new lithosphere.

But rock flowing in a convection cell must eventually cool and flow downward again. Where one sideways-moving current meets another, both currents turn downward. The lithosphere is too rigid for both moving plates to be dragged downward into the mantle by the convective flow. As seen in Figure 18.2, one plate bends and is dragged down, the other remains afloat, but is rammed against the downgoing plate. The places of downward movement are the places where sea-floor trenches and island arcs occur, and, as we see later in this chapter, where fold mountains are made.

The theory of plate tectonics provides explanations for all the problems that troubled earlier theories; for example, it provides a mechanism for continental drift. Continents are simply accumulations of light-weight rock that ride piggy-back on much larger plates of lithosphere. Continents don't drift by themselves; they move when the lithosphere moves, and they break apart when the lithosphere breaks apart (Fig. 18.2). Friction is not a problem after all, because the movement is occurring, not between the continental crust and the oceanic crust, but by flow within the mantle, and specifically within the fluid-like low-velocity zone we describe in Chapter 17.

Tests of the Theory. Unlike Sherlock Holmes, we cannot test theories by catching villains. But we can test them in other ways, and we shall mention three critical tests for the theory of plate tectonics. The first test is based on the data of paleomagnetism, the second on the results of drilling into the deep-ocean floor, and the third on the locations of earthquake foci. But we must remember that we are only testing the theory that plates of lithosphere slide over the top of the mantle. The proposal that convection currents in the mantle are the driving force is really a separate theory, and one it is not yet possible to test. The three tests for the theory of plate tectonics *prove* that lithosphere moves. They *suggest*, but do

not prove, that convection currents cause the movement.

The Paleomagnetism Test. We learned in Chapter 3 that at irregular intervals the polarity of Earth's magnetic field reverses. During the last 4 million years there have been 9 periods of normal magnetic polarity (meaning periods when the magnetic poles were in approximately the same positions as they are today) and 9 periods of reversed polarity (Fig. 3.16). In Chapter 3 we learned also that paleomagnetism in igneous rock provides a record of the magnetic polarity at the time the rock cooled through its Curie point. The first test of plate tectonics makes use of these effects.

Lava extruded at mid-ocean ridges acquires the magnetic polarity existing at the time of extrusion, then is carried sideways like a continuously sliding carpet. When Earth's polarity changes, new lava will acquire the new polarity. The process is endless. Therefore, if we examine the magnetic properties of rock on the sea floor, we should be able to see a pattern in which roughly parallel strips of rock form an alternating sequence of normal and reversed polarity (Fig. 18.3). Indeed magnetic patterns of this sort exist. Southwest of Iceland, strips with constant polarity, each about 30km wide, parallel the mid-ocean ridge. The strips on both sides of the ridge are nearly identical (Fig. 18.4). Similar observations have been made in every ocean of the world. Not only do the observations confirm that the sea floor, and therefore the lithosphere, moves, but also they

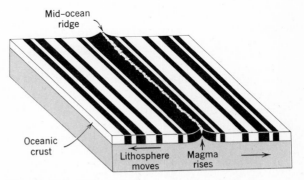

Figure 18.3 Schematic diagram of lithosphere. Lava extruded along a mid-ocean ridge forms a new oceanic crust. As lava cools, it becomes magnetized with the polarity of Earth's field. Successive strips of oceanic crust have alternate normal polarity (black) and reversed polarity (white).

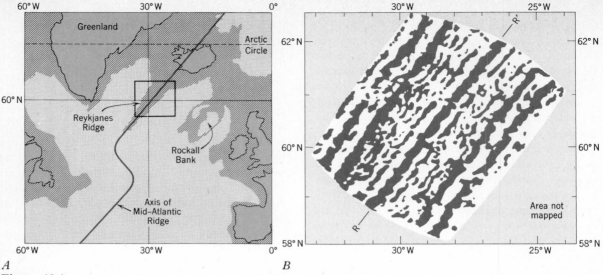

60° W 30° W 0°

Greenland

Arctic
Circle

60° N

Reykjanes
Ridge

Rockall
Bank

Axis of
Mid-Atlantic
Ridge

60° W 30° W 0°

A

30° W 25° W

62° N

60° N

Area not
mapped

58° N

30° W 25° W

B

Figure 18.4 *A.* Index map showing location of Reykjanes Ridge, a portion of the Mid-Atlantic Ridge southwest of Iceland.

B. Map of the magnetic properties of rock on the sea floor. Strips of rock with normal polarization (black) alternate with reversely polarized rock (white). RR' is center line of Reykjanes Ridge. (*After Heirtzler, Le Pichon and Baron, 1966.*)

establish that the lithosphere moves away from mid-ocean ridges in both directions at equal rates. Because we know the times in the past when Earth changed its magnetic polarity (Fig. 3.16), we know also when each of the magnetic strips formed. Knowing that, we are able to say how fast the lithosphere is moving. In the area southwest of Iceland, two plates of lithosphere are moving apart at about 2cm per year. In some parts of the world, rates as high as 18cm per year have been measured. Thus, paleomagnetism confirms that new lithosphere is being created at mid-ocean ridges and is moving sidewise away from the ridge. This first test clearly proves that the lithosphere is moving.

The Sea-Floor-Sediment Test. A second test of the theory of plate tectonics concerns the age of sediments in the deeper parts of the ocean basins. Sediment can be no older than the surface on which it is deposited.

Paleomagnetism suggests that the two continents, Africa and South America, as they are carried along on their conveyor belts of lithosphere, are moving apart at about 4cm per year. If that rate has been constant, the continents must have been in contact about 150 million years ago, and the intervening ocean, the South Atlantic, can be no more than 150

million years old. Drilling conducted from *Glomar Challenger* (Fig. 12.1) has proved that basalt on the Atlantic Ocean floor between Africa and South America is everywhere covered by young strata—that is, strata no older than 150 million years. Drilling has also revealed that the thickness of strata increases away from the mid-ocean ridge; from this we infer that older parts of the ocean floor are farther away from mid-ocean ridges than younger parts of the floor. The older parts of the ocean floor are covered by greater thicknesses of sediment because sediment has been accumulating on them for a longer time. Drilling into other ocean floors—the Indian and Pacific Oceans—reveals that sediment more than 200 million years old does not exist anywhere. All the present ocean basins, therefore, are young geologic features. We infer, therefore, that all parts of ocean floors formed more than 200 million years ago have apparently disappeared down the sea-floor trenches and have been reabsorbed into the mantle.

Moving plates of lithosphere are continuously making new ocean floor, and as they do so they are getting rid of the old. The lithosphere is like a tidy housekeeper; sediment does not accumulate for long before it is swept away. The sea-floor-sediment test confirms that lithosphere is moving, and suggests the process has been going on for at least 200 million years. But a test must still be made to see whether plates of lithosphere are dragged back down into the mantle, and there reabsorbed.

The Earthquake-Focus Test.

The first two tests involved direct examination of the upper surface of a plate of lithosphere. From them we proved that movement of lithosphere is occurring. But how can we test the part of our theory that says plates of lithosphere slide down into the mantle, and are eventually consumed? Studies of seismic waves provide an excellent test.

The theory says that when a moving plate of lithosphere slides downward into the hot, fluid-like mantle, the cold and rigid lithosphere must fracture, much as a sheet of glass plunged into boiling water will fracture. We have learned that fracturing creates earthquakes; so a downward-moving plate of lithosphere should be the site of many earthquake foci. As we learned from Figures 17.9 and 17.10, about 80 per cent of all earthquakes occur beneath sea-floor trenches, the very place where lithosphere is believed to plunge down into the mantle. Furthermore, the upper surface of a plate, which formerly had been sea floor, is colder and more rigid than its lower surface. Therefore, earthquakes should occur most frequently near the *upper* surface of the plate.

Paleomagnetism proves that new lithosphere is created along mid-ocean ridges, while studies of earthquake foci prove that old lithosphere plunges down into the mantle beneath sea-floor trenches. Proof that the lithosphere has been moving for a long time, and that continents move simply as fragments of lithosphere, comes from study of sea-floor sediments. Thus the theory of plate tectonics passes the three tests so far asked of it. Next let us see whether we can decipher some details about the moving plates.

The Plate Mosaic.

The theory of plate tectonics requires that the surface of the Earth consist of a few rigid plates of lithosphere about 100km thick, resting on and sliding over the fluid-like low-velocity zone at the top of the mantle. The moving plates have three kinds of margins:

1. First, along mid-ocean ridges, where new lithosphere and new oceanic crust are being made, two plates are continually being pulled apart. Along such lines, the margin of each of the two adjacent plates is a *spreading edge*. The spreading edge is a line along which the crust is being stretched. The forces that cause the stretching are tensional forces and the margin is therefore called a *margin of tension.*

2. Second, along those lines where two plates are moving toward each other and meet head-on, margins of the plates are *margins of compression.* To relieve the compression, one plate is eventually forced downward into the mantle, where it is reabsorbed. A margin of compression is therefore one along which a plate is being consumed.

3. Third, some margins are formed by the great fracture zones called *transform faults* (Figs. 12.8, 15.4). Along transform faults the plates slide past each other.

The activities that occur along each of these three kinds of plate margins generate earthquakes. Given a particular earthquake, whose focus we have located, we can tell by studying the pattern of seismic waves recorded for that earthquake whether the focus is in a margin of tension, of compression, or on a transform fault. If we plot enough foci on a map, the map will reveal the shapes of plates, and show just how the margin of any plate changes from place to place.

By such plotting we have learned that the lithosphere consists of six large plates, the African, Eurasian, American, Pacific, Australian, and Antarctic plates, plus a number of smaller ones (Fig. 18.5). All of the plates are moving. If we peeled an orange and then replaced the loose peel, the situation would be similar to the fragmented plates of lithosphere that cover the mantle. If we now tried to move one fragment of orange peel, all the other pieces would move too. The same is true for the lithosphere. If one plate moves, all plates must move. We have already discussed paleomagnetic evidence, that proves some plates of lithosphere are moving now; we can infer, therefore, that the other plates of lithosphere must also be moving.

San Andreas Fault.

One consequence of present-day plate motion is of particular concern to residents of California. We have noted that transform faults are one of the three kinds of plate margins. Most commonly these faults offset the spreading margins of plates, and they form while the spreading margins form (Fig. 18.6); as long as the spreading margin continues to be active, the associated transform faults

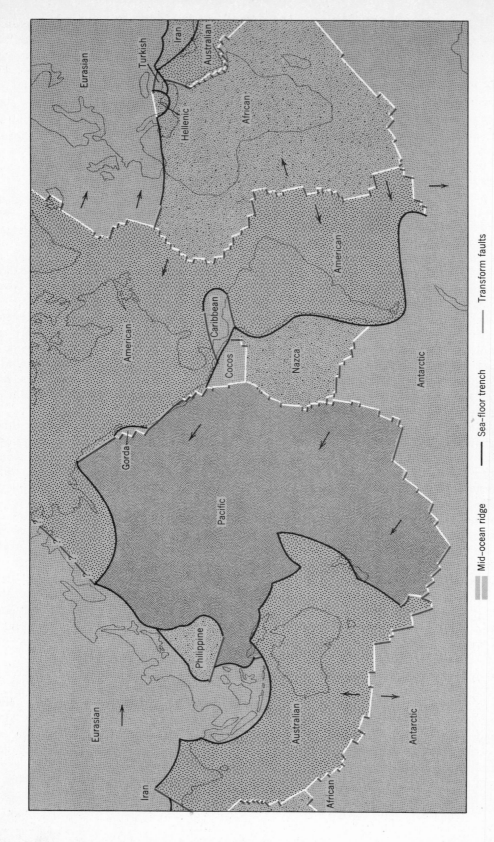

Figure 18.5 Six large plates of lithosphere and several smaller ones cover Earth's surface and move continuously, in the direction shown by arrows. Plates have three kinds of margins: (1) growing or spreading margins, delineated by mid-ocean ridges, (2) margins of consumption, delineated by sea-floor trenches, and (3) transform faults. (*After Dewey, 1972.*)

Mid-ocean ridge   Sea-floor trench   Transform faults

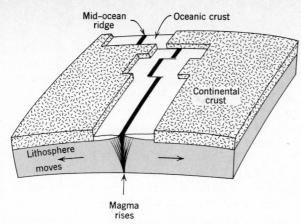

Figure 18.6 Transform faults (blue) form when mid-ocean ridge first forms. Shapes of continental margins reflect faults now found along mid-ocean ridge.

are active. The San Andreas Fault in California is a transform fault, and the many earthquakes that disturb California are caused by movement along it. Some California earthquakes, such as the one that leveled San Francisco in 1906, have been particularly destructive. Possibly others will be just as destructive in the future. As long as the present plates of lithosphere continue to move, movement must occur along the San Andreas Fault, and residents of California can expect more earthquakes.

The San Andreas Fault is one of several faults that offset the East-Pacific Rise. Figure 18.7 shows how the transform faults and segments of the Rise separate the American Plate from the Pacific Plate. Movement along the fault arises from movement between these two giant plates. The peninsula of Baja California, and that part of the State of California that lies west of the San Andreas Fault, are on the Pacific Plate and that plate is moving northwest, relative to the American Plate, at a rate of several centimeters per year. In about 10 million years Los Angeles will have moved far enough north so as to be opposite San Francisco. In about 60 million years, the segment of continental crust on which Los Angeles lies will have become separated completely from the main mass of continental crust that comprises North America.

Fragmentation of Pangaea seems to have commenced when the present plate of lithosphere started moving about 200 million years ago. What preceded Pangaea? Has continental crust always been the same as it now is, and had Earth only one continental mass

until 200 million years ago? Or were there possibly earlier times when plates of lithosphere moved continents around, alternately fragmenting them and welding them together again? Answers to these questions are still very hesitant, and in order to attempt them we must look more closely at the structure of continents.

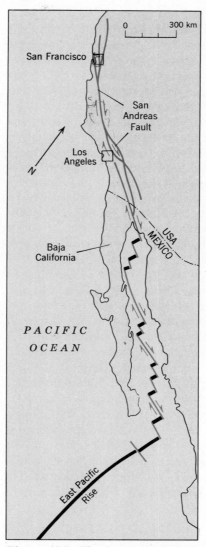

Figure 18.7 The San Andreas Fault is one of several transform faults that offset an ocean ridge (Mid-Pacific Rise). The faults and ridge segments separate the Pacific plate (left) from the American plate (right). The two plates are sliding past each other, causing frequent earthquakes in California. (*After Elders and others, 1972.*)

Origin of Continental Crust

Continental Shields. Every continent has a nucleus of ancient rock. Commonly called a ***continental shield,*** each nucleus is *a mass of Precambrian rock around which a continent has grown.* Within these shields are Earth's most ancient rocks, which, although now changed by metamorphism, were originally lava similar in composition to modern basalt. Textural evidence suggests these earliest lavas were extruded onto an ancient sea floor. When a pile of ancient volcanic rock eventually became large enough to project above sea level, erosion must have commenced and sediment must have formed. Some of the ancient lavas are interlayered with sedimentary strata, and throughout the shields also are deeply eroded granite batholiths. Both the strata and the granites suggest that some form of mountain building must also have occurred very early in Earth's history. But the events that formed the rocks of the shields happened so long ago, and the picture they present is so unclear, that we still cannot be sure exactly what did happen. However, when the ages of igneous and metamorphic rock in the shields are determined by radiometric means, all such rock seems to be older than about 2 billion years. We therefore conclude that the nucleus of each modern continent had formed by at least 2 billion years ago.

We cannot yet say when the earliest continental crust actually formed. As we noted in Chapter 2, the age of the Earth is at least 4.6 billion years. Yet no rock older than about 3.3 billion years has been found. Because we find ancient lavas, 3.3 billion years old, that are interbedded with sedimentary strata, we can be sure that still earlier rock masses must have existed to serve as sources of sediment. Perhaps, then, older rocks await discovery, hidden within the centers of shields. But it is also possible that all traces of the earliest continental crust may have been destroyed by erosion, and that we will never be able to say exactly when the first crust formed.

Structure of Continents. Continental shields today are areas of low relief. Long-continued erosion has worn them down close to sea level, and the ancient mountains have ceased to rise. The shields seem to be both stable (neither rising nor sinking) and in isostatic balance. But continental shields are only small fragments of continents. Somehow, dur-

ing a period starting about 2.5 billion years ago and continuing to the present, continental crust has apparently increased considerably in size. When we look at the structure of North America, we get a clue how this might have happened. In central and northern Canada there is a shield within which all rocks give radiometric dates that are 2.5 billion years or older (Fig. 18.8). Surrounding the shield like a twisted collar is a belt of complexly deformed strata, a belt created during the period 1.0 to 2.5 billion years ago. The collar is not visible everywhere because it is sometimes concealed beneath a veneer of younger, undeformed strata. Nevertheless, all the samples obtained by drilling through the covering strata, and all samples from places where the old deformed collar pokes through, give radiometric dates falling between 1.0 and 2.5 billion years.

The twisted collar consists of many deeply eroded mountain systems. Surrounding it is an even younger belt, somewhat like a showy necklace, of young mountain systems that have not yet been worn down. These are familiar mountain systems, such as the Appalachians, the Rockies, and the Sierra Nevada. They have all been formed during the last 0.6 billion

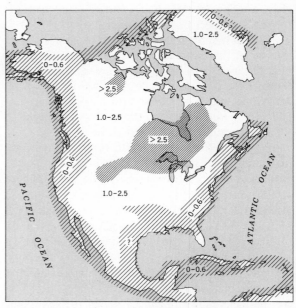

Figure 18.8 Ages of belts of continental crust in North America based on isotopic dates of metamorphic and igneous rock formed during ancient episodes of mountain building. Numbers are ages in billions of years. (*After Engel, 1968.*)

years. The structure of North America suggests, therefore, that an ancient shield forms a nucleus of continental crust, and that the continent has been slowly growing as successive mountain systems were formed along its margins.

How well does the North American pattern fit other continents? The answer is quite well, because all shields are surrounded by belts of younger rock. But some of the patterns are confused; some shields seem to have been broken into two fragments and some old mountain systems seem to be incomplete, as if they had been snapped off. Evidence of this sort suggests that there may have been several periods in the past when plates of lithosphere moved their loads of continental crust around on the surface and that fragmentation of continental shields and truncation of mountain systems occurred when continents were split apart.

The structure of continents therefore lets us infer that continental crust has increased in size through the ages, and that more than one period of movement of plates has occurred. Before we can rely on these inferences, though, we must answer two important questions; first, why do continents seem to grow by the addition of mountainous collars, and second, where does new material come from to account for the increasing volume of continental crust? Possible answers to both questions can be obtained by considering the origin of fold-mountain systems.

Origin of Fold-Mountain Systems. When we discussed fold-mountain systems in Chapter 17, we noted that formation and filling of geosynclines always precede formations of the fold mountains. But we left unanswered all questions of origin.

The Principle of Uniformity (Chap. 2) suggests that processes operating in the past are probably also operating today. To examine the beginnings of new mountain systems, therefore, we should study modern geosynclines. For a geosyncline to be fed enough sediment to fill it up, a continent must be nearby shedding the product of its erosion. It was realized long ago that geosynclines must form along the margins of continents. When we examine the margins of continents, two principal kinds of geosyncline are found: one in which sediment accumulates in shallow water, and one in which sediment accumulates in deep water.

Examples of modern geosynclines have been discovered along the Atlantic coast of North America.

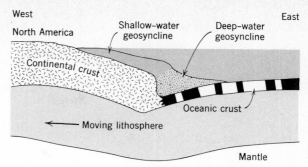

Figure 18.9 Schematic section through lithosphere and upper mantle along the eastern margin of North America. Oceanic crust (showing magnetic strips) joins continental crust at a point where lithosphere is warped slightly downward forming two geosynclines. In one geosyncline, sediment accumulates in shallow waters of the continental shelf; in the other, sediment accumulates in deep water beyond the continental slope. (*After Dewey and Bird, 1970, and Dietz, 1972.*)

One contains a wedge of sediment that accumulated in shallow water over the continental shelf. The other is a great pile of sediment that accumulated in deep water, adjacent to and abutting the continental slope (Fig. 18.9). Basins of accumulation are present because the American Plate is bowed down beneath the edge of the continent; therefore both geosynclines contain thick piles of strata.

The Atlantic Ocean is now slowly growing larger, and the depth of sediment in both the shallow- and deep-water geosynclines lying along its flanks is increasing. The geosynclines that flank the Atlantic coast of North America are 180 million years old, having formed when the North Atlantic Ocean first appeared. The geosynclines in the South Atlantic appear to be somewhat younger, because this part of the Atlantic formed later than the North Atlantic did. But the accumulation of sediment apparently does not go on forever; so eventually all the geosynclines in the Atlantic Ocean will become fold mountains. For some reason still not understood, the lithosphere eventually breaks beneath the deep-water geosyncline and starts to plunge down into the mantle. The downward-sliding plate of lithosphere butts into the deep-water strata, crumples them, and rams them against the edge of the continent (Fig. 18.10). Strata near the bottom of the geosyncline can also be dragged downward by the moving plate. Becoming heated and metamorphosed, the deepest strata may begin to melt. Granitic magma formed

361

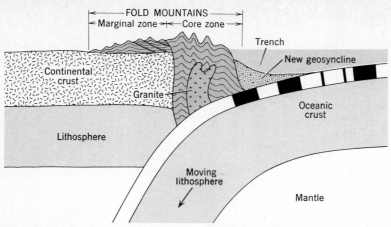

Figure 18.10 Schematic section showing how fold mountains are formed when a plate of lithosphere plunges downwards beneath the edge of a continent. Strata originally accumulated in a deep-water geosyncline are crumpled, metamorphosed and intruded by granite, which forms the core zone of mountains. Strata of shallow-water geosyncline are pushed and slid sideways to form marginal mountain zone. A sea-floor trench forms a new geosyncline on the seaward side of the mountains. (*After Dewey and Bird, 1970.*)

by the melting then rises and intrudes the strata up above. In this way the deep-water geosyncline becomes the core of a newly made chain of fold mountains.

Strata in the shallow-water geosyncline are compressed and pushed aside by the mass of deformed and metamorphosed rock in the new mountain core. The shallow-water strata become folded and deformed by thrust faulting into the marginal ranges of a fold-mountain system.

There are many places in the world where plates of lithosphere are now plunging downward beneath the edges of the plates of continental crust, and where mountain systems are still actively growing. One such place is along the west coast of South America (Fig. 18.10). Imagine we are looking southward at that figure. If we draw a section through the western edge of South America, the downward-plunging plate of lithosphere in Figure 18.10 is equivalent to either the Nazca Plate or the Antarctic Plate, the fold-mountain system is equivalent to the Andes, and the sea-floor trench is equivalent to the Peru-Chile Trench. At some time in the past, the west coast of South America probably resembled the present Atlantic coast of North America. We can infer therefore, that at some unknown time in the future the Atlantic coast of North America

will come to resemble the Pacific coast of South America, and a great new mountain system will start to form.

The Andes and the Pacific coast of South America also suggest a way by which the volume of continental crust might have increased through the ages. As we noted in Chapter 16, when a plate of lithosphere slides down into the mantle, some of the basalt from the ocean floor undergoes partial melting, and yields andesitic magma. The magma is then erupted, as lava, onto the continental crust, and thus increases the volume of the crust. The same process is occurring along island arcs such as the Aleutians, in northern Japan, and in Indonesia. The process of making andesite seems to imply, therefore, that material from the mantle is being slowly added to the continental crust.

According to the inferences we have just drawn, mountain building and the addition of new material to the continental crust should be most active along the edges of continents. But not all mountain systems occur in such positions. Let us, then, consider the distribution of mountain systems to see if the theory of plate tectonics can account for them.

Distribution of Mountain Systems. If the theory of plate tectonics is correct, mountain systems formed during the last 200 million years—during the Cenozoic and Mesozoic Eras—must have been made by movements that are continuing today. Observing the distribution of Cenozoic and Mesozoic mountains (Fig. 18.11), we note that they occur in two sorts of positions. The first position parallels the margins of the existing continents; these are mountains formed along coasts of the kind now seen along the west side of South America, and, as we have seen, they are readily explained by the theory of plate tectonics. The second position is in continental interiors; these are mountains flanked on both sides by continental crust, such as the Himalaya in southern Asia.

Interior mountain systems can be explained also by plate tectonics. For the sake of clarity, let us think of a downward-plunging plate of lithosphere as being a moving plate, and the plate beneath which it is plunging as stationary. A continent on the stationary plate, then, will have a mountain system developed along the margin which overlies the downward-plunging plate. But if a second continent is riding on the moving plate, it will eventually crash into the stationary continent (Fig. 18.12,*A*). The moving continent will have both deep- and shallow-water geosynclines in front of it. The strata in these geosynclines are caught as though in a giant vise, and will become folded, faulted, and elevated (Fig. 18.12,*B*). The original system of fold mountains is thereby increased in size.

A moving continent is too big, and its rocks too light in weight, to be carried down into the mantle with the descending plate of lithosphere. Like a slab of cork, the continent will bob up again. During

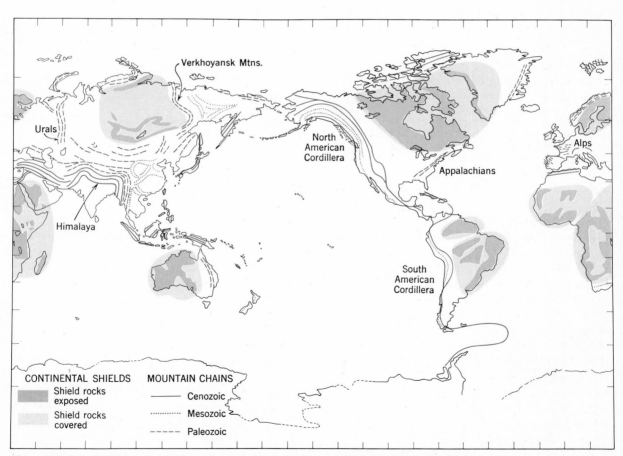

Figure 18.11 Continental shields and mountain chains formed since the beginning of the Paleozoic Era. Each shield itself is a patchwork of ancient mountain chains. (*Shield areas after Dunbar, 1966.*)

363

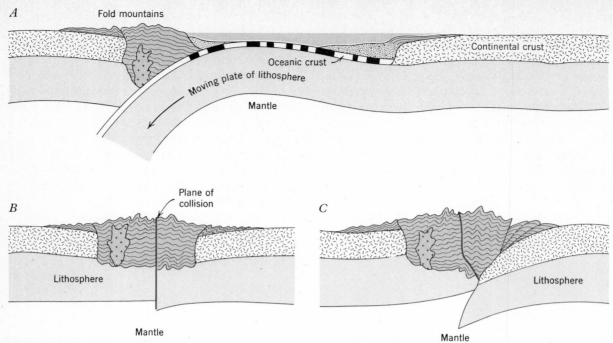

Figure 18.12 Fold-mountain system formed by collision between continents. *A*. Continental margin of the type now found on the western coast of South America forms a fold-mountain chain. Moving plate of lithosphere carries a second continent on a collision course. *B*. Collision between continents crumples geosynclines and increases size of mountain system. *C*. The downward-moving plate of lithosphere breaks off, but the edge of the remaining segment of plate is partly thrust under the edge of the stationary plate, causing further elevation of the fold mountains. (*After Dewey and Bird, 1970.*)

collision, the edge of the moving continent, together with the plate of lithosphere on which it rides, will be partially thrust down beneath a stationary continent (Fig. 18.12,*C*). As the depressed edge of the moving continent tries to bob back up, the edge of the stationary continent, with its fold mountains attached, will also be pushed upward and its mountains elevated to even greater heights. This is believed to be the origin of the Himalaya, a great mountain mass formed by the crumpling of strata that had been deposited in the ancient Tethys Sea (Fig. 18.1). When Pangaea broke up, Asia moved slowly westward, while India, riding on its plate, moved rapidly north. Eventually India rammed into, and partly slid under and lifted, the edge of Asia. The consequent uplift has made the Himalaya the highest mountain range on Earth.

Mountain systems older than 200 million years now occur in the interiors of some continents and provide important clues that plate movements ante-

dating the present episode must have occurred. In the interior of Asia are two mountain systems of Paleozoic age, the Urals and the Verkhoyansk Mountains, both shown in Fig. 18.11. We infer that these systems formed by continental collisions that happened several hundred million years ago. Perhaps this earlier period of movement was the one that brought Pangaea together. A piece of evidence to support the contention lies on the east coast of North America.

Apparently the Appalachians were once joined to mountains of similar age in western Europe (Fig. 1.10). But if North America was joined to western Europe, how did a geosyncline form where the Appalachians now stand? Was there a previous North Atlantic Ocean, one that antedates the present ocean 180 million years old? Evidence strongly suggests there was. The sequence of events that formed the Appalachians could have been as follows. Geosynclines formed in an ancient North Atlantic Ocean were crumpled and elevated when the continental margin of ancient North America looked like the Andean margin of South America does today. This happened many hundred million years ago. The plate of lithosphere on which Europe sat began to move westward. Eventually, perhaps 500 million years ago, Europe and North America collided and a great mountain system was formed. The system included the Appalachians and many of the Paleozoic mountains of western Europe. For a long time—at least 200 million years—the continents remained welded together and the Appalachians were part of a mountain system in the interior of Pangaea. When movement started again, one of the new spreading edges formed along a line close to the center of the system of fold mountains—the same line that earlier marked the collision between Europe and North America.

The evidence from mountain systems strongly suggests, therefore, that at least one episode of plate movement preceded the present one. It is not possible to be sure, but it is tempting to guess, that there may have been several periods during Earth's long history when convection cells formed and when plates of lithosphere slid around on top of the mantle. In the process, continents may have been repeatedly split apart, rammed together, and moved past each other, increasing in size as new mountain systems were created. After each episode of movement and collision, there may have been a period of quiet that ended when new convection cells formed and when plates of lithosphere started to move again.

When will the present episode of movement stop? We cannot be sure. Probably it will continue for a hundred million years or more. In the meantime, continents will continue to ride piggyback on their plates and the map of Earth's surface will continue to change. One prediction of what the map will look like in about 50 million years is shown in Figure 18.13.

A Word of Caution

The theory of plate tectonics seems to provide answers for so many questions that we are tempted to believe it is the long-sought unifying theory that explains the lithosphere. But we must be careful.

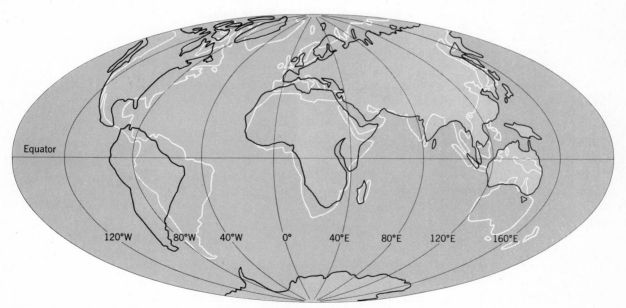

Figure 18.13 If the present motions of plates should continue for 50 million years into the future, the map of the world would show some notable changes. Present map outlines (white) will be strikingly changed in Africa, because a large fragment in the northeast corner will have almost separated from the rest of the continent. Australia and New Guinea will have moved north, many islands in the East Indies will have vanished and open seaways will have replaced both the Panama and Suez Canals. (*After Dietz and Holden, 1971.*)

Other theories, too, have seemed overwhelming in their promise, yet in the long run have proved incorrect. The theory of plate tectonics is still only a theory. Also it is so new that there has not been time enough to examine and test it fully. Perhaps there are clues the theory cannot explain, and perhaps the theory will have to be changed somewhat. A few scientists have already found clues that they believe strain the theory. Only with time and much more work can we decide whether the amazing sequence of events discussed in this chapter actually happened as we described them.

The theory does not explain how the first continental crust formed, nor does it tell us much about the first billion years of Earth's history. Perhaps future studies will show that movement of plates occurred only during the more recent part of Earth's history, and that in earlier times entirely different things happened. We have to face the fact that much of the vital evidence of Earth's earlier history has been obscured by the results of long-continued erosion. But the early histories of all the smaller planets are thought to be similar to that of Planet Earth. We should, therefore, examine Earth's sister planets, where erosion may not have obscured the evidence. Perhaps Moon, Mars, or Venus holds the key to Earth's earliest days. In the next chapter we discuss the many exciting clues that are coming out of our exploration of space.

Summary

1. In the 19th Century the concept of contraction by cooling of a hotter Earth served as a theory to explain the origins of earthquakes, mountains, and continents.

2. The discovery of radioactivity destroyed the contraction hypothesis.

3. The theory that continents drift over the oceanic crust replaced the contraction theory. It was discarded because the friction between continental crust and oceanic crust is too great for movement to occur.

4. The theory of plate tectonics proposes that plates of lithosphere slide over the top of the mantle. Continents are carried piggy-back on the plates.

5. The lithosphere is about 100km thick; it sits on a fluid-like region of the mantle called the low-velocity zone.

6. Six large and several small plates of lithosphere cover the Earth. The plates move away from mid-ocean ridges where new lithosphere is made by intrusion and extrusion of magma from the mantle.

7. Plates move past each other along transform faults. Eventually they slide down into the mantle, where they are reabsorbed.

8. Geosynclines and systems of fold mountains form at the margins of continents as a consequence of the movements of plates.

9. Continents have grown larger through the ages. Each continent has a nucleus or shield at least 2.5 billion years old. Around the nucleus are rings of progressively younger fold mountains.

10. The present period of moving plates of lithosphere commenced about 200 million years ago, disrupting an old supercontinent we call Pangaea.

11. It is still uncertain whether there were periods of plate movement prior to the present one. The distribution of ancient mountain systems suggests that prior movements have occurred.

Selected References

Anderson, D. L., 1971, The San Andreas Fault: Sci. American, v. 225, No. 5, p. 52-68.

Dewey, J. F., 1972, Plate tectonics: Sci. American, v. 226, No. 5, p. 56-68.

Dewey, J. F. and Bird, J. M., 1970, Mountain belts and the new global tectonics: Jour. Geophys. Research, v. 75, p. 2625-2647.

Dietz, R. S., 1972, Geosynclines, mountains, and continent-building: Sci. American, v. 226, No. 3, p. 30-38.

Dietz, R. S. and Holden, J. C., 1970, Reconstruction of Pangaea: Breakup and dispersion of continents, Permian to present: Jour. Geophys. Research, v. 75, p. 4939-4956.

Elders, W. A. and others, 1972, Crustal spreading in southern California: Science, v. 178, p. 15-24.

Engel, A. E. J., 1963, Geological evolution of North America: Science v. 140, p. 143-152.

Hallam, A., 1972, Continental drift and the fossil record: Sci. American, v. 227, No. 5, p. 56-66.

McKenzie, D. P., 1972, Plate tectonics and sea-floor spreading: American Scientist, v. 60, p. 425-435.

Takeucki, H., Uyeda, S. and Kanamori, H., 1970, Debate about the Earth: (revised ed.), San Francisco; Freeman, Cooper.

Tarling, D. and Tarling, M., 1971, Continental drift: Garden City, Doubleday.

Vine, F. J., 1966, Spreading of the ocean floor: new evidence: Science, v. 154, p. 1405-1415.

Wyllie, P. J., 1971, The dynamic Earth: New York, John Wiley.

(*Opposite.*) An area of several thousand square kilometers on the surface of Mars, photographed by Mariner 9 in 1972. Meteorite-impact craters like those on the Moon are clearly visible. Two photographs were taken 37 days apart. Although originally photographed in black and white, the first (larger) photograph is printed in green, the second (smaller) in red. When the two colored images are superimposed, unexplained seasonal changes on the surface of Mars are observed. If the two photos were identical, the area of overlap would have a uniform orange color. But some areas on it appear greenish, indicating that they reflected less light when the red photograph was taken; that is, the areas now greenish had become darker during 37 days. Other areas appear reddish, meaning that they reflected more light after 37 days and had therefore become lighter. The observation of seasonal variations on Mars raises the possibility that processes of erosion similar to those observed on Earth may be operating. (*Jet Propulsion Laboratory.*)

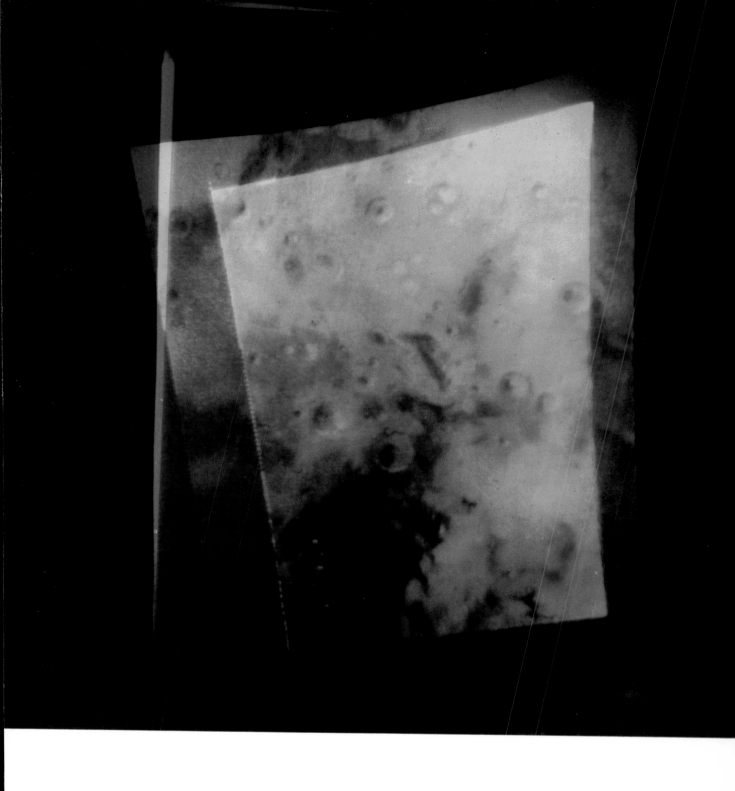

Part V The Earth's Neighbors

Chapter 19

Earth's Neighbors in Space

The Solar System

A President's every action is thought about, talked about, debated, and analyzed. The result is a developing picture of the man, documented by a flood of evidence. But people also ask what he was like during his childhood and his formative years; do the experiences of those earlier days influence his present actions? One way to seek answers is to talk to family members who shared in, and were influenced by, the same early experiences. In some respects Planet Earth is like a President. Through the preceding eighteen chapters we have examined and discussed it in detail. But we now know that little evidence of Earth's birth and childhood remains, because it has been destroyed by erosion. Nonetheless, we ask whether Earth's present activities reflect events that happened during those long-ago times, and as with a President, a way to answer the question is to examine Earth's neighbors in space. This way is promising, because all the planets were born together and have been neighbors a long time. It should be possible to glean from other planets, clues to the childhood of Planet Earth.

Space vehicles are a means of making direct examinations of other planets. Starting in 1964 and continuing to the present, a number of manned and unmanned visits to the Moon and to the nearby planets Mars and Venus have been made. The observations made during these magnificent achievements have created a revolution in our knowledge of Earth's neighbors. To create a convenient framework within which we can discuss the observations, let us begin with the birth of the **Solar System,** *the system comprised of the Sun and the group of planetary bodies that revolve around it, held by gravitational attraction.*

The Birth Event. The birth throes began with space that seems to be empty, yet is not entirely empty. Atoms of various elements are present everywhere, albeit thinly spread; they form a tenuous gas, and a star such as the Sun is formed when the gas thickens by the slow gathering together of all the thinly spread atoms. The gathering force is gravity, and as the atoms slowly move closer together, the gas gradually becomes hotter and denser. Earth and the other planets formed as one part of the gathering process. Exactly how this was accomplished is still being studied, but the principal features of the most likely process can be identified.

More than 99 per cent of all atoms in space are atoms of hydrogen and helium, the two lightest kinds of atoms. Near the center of the gathering cloud of gas, the atoms become so tightly pressed and so hot, that atoms of hydrogen and helium begin to fuse together to form heavier elements. As we learned in Chapter 3, fusion of light elements to form heavier ones causes heat energy to be released; the hydrogen and helium undergo *nuclear burning.* When, in the gas cloud that formed our Solar System, nuclear burning commenced, the Sun was born. The time was about 6 billion years ago. But nuclear burning was confined to the center of the gas cloud. A vast envelope of less-compressed gas still surrounded the Sun, and because of the way in which atoms gather in space, the gaseous envelope rotated. We noted in Chapter 3 that rotation gives rise to centrifugal force. While gravity tended to pull the gaseous envelope in toward the Sun, then, centrifugal force tended to pull it outward. As a result of these two opposing forces, the gas cloud slowly became *a flattened, rotating disk of gas* surrounding the hot Sun. Such a disk is called a **planetary nebula** (Fig. 19.1).

At some stage the outer portions of the planetary nebula became both compacted and cool enough to allow solid objects to condense, in the same way that ice condenses from water vapor to form snow. The solid condensates eventually became the planets. Planets nearest the Sun, where temperature was highest, contain only compounds that are capable of condensing at high temperatures. Those compounds consist of elements such as iron, silicon, magnesium, and aluminum; we call them *refractory elements.* Planets distant from the Sun, where temperatures were lower, contain not only refractory elements but also *volatile elements,* such as hydrogen, helium, and sulfur, that readily form gases even at low temperature.

The size of a planet reflects its distance from the Sun. A large-diameter ring of gas contains more atoms than a small-diameter ring. Planets close to the Sun formed from comparatively small amounts of gas, and are therefore smaller than more distant planets which condensed from large rings of gas (Fig. 19.2). The small planets, Mercury, Venus, Earth, and Mars, are dense, rocky objects and they appear to be similar in composition. We call them the *terrestrial planets.* Planets distant from the Sun are much larger than the terrestrial planets; the masses of Saturn and Jupiter are more than a hundred times larger than the mass of Earth. Because the large planets contain large amounts of the volatile elements, especially hydrogen and helium, their densities are low.

The large planets, then, are very different from the terrestrial planets. We still know little about them because space vehicles have not yet visited them for a close look. Perhaps when we do get close to them, we will learn much more about the birth of the Solar System and the childhood of the planets. But until then we must seek our evidence from closer neighbors. Let us begin by re-examining two of the clues Earth has provided about its earliest history: its composition and its layered structure. Both clues help us understand the process by which the planets were made.

Earth's Composition. The condensation of a planetary nebula is analogous to the cooling of magma. First one solid compound condenses, then another. The earliest compounds gather together to form a core, and later compounds, as they condense out of the nebula, are added to it in successive layers. The earliest compounds to condense, therefore, should be near the center of the planet, and the later ones should be near the outside. A widely held theory states that the earliest materials would have been metallic iron and nickel. This corresponds to Earth's core. As the planetary nebula cooled, other materials should start to condense, mantling the core with compounds that contain, in addition to iron, ele-

Figure 19.1 The gathering of atoms in space created a rotating cloud of dense gas that eventually became the Sun, and a surrounding disk-shaped planetary nebula. The planets formed by condensation in the planetary nebula. (*After Turekian, 1972.*)

ments such as silicon, magnesium, and oxygen. These compounds correspond to Earth's mantle. As condensation proceeded and as the planetary nebula cooled still further, increasing amounts of less-abundant elements such as sodium, potassium, aluminum, sulfur, and carbon should condense. These are elements we now find concentrated in Earth's crust. The structure of our planet and the distribution of elements within it thus reflect strongly the

way the planet formed. But the distribution of elements is probably no longer the same as it was when Earth formed. The heat of Earth's interior has caused redistribution of material, both by the formation of magma and possibly also by convection in the mantle. We know that as radioactive decay caused Earth to heat up, many gases, including water, were given off, because we now find them in both atmosphere and hydrosphere.

Figure 19.2 The planets, arranged in order of their relative positions, outward from the Sun. Each object is shown in correct relative size: the Sun, 1.6 million kilometers in diameter, is thirteen times greater than Jupiter, the largest planet.

But one question we ask about condensation in a planetary nebula cannot be answered from evidence collected on Earth. This question concerns the exact way in which the condensates form. Did Earth start with a single, tiny nucleus and grow larger by addition of layer on layer of atoms, like so many layers of paint? Or did condensation occur like millions of small snowflakes, and did the planetary snowflakes then accumulate, by gravitational attraction, into a planetary snowball? Although Earth does not provide us with answers to these questions, the Moon does. So we will begin our discussion of Earth's neighbors with the Moon.

Earth's Moon

In some ways our Moon is unique in the Solar System. Its diameter is 3476km, only a little smaller than that of Planet Mercury; for this reason Moon is often described as a small terrestrial planet, and the Earth-Moon pair is described as a double planet. The moons of Jupiter and Saturn, although as big as or even bigger than our Moon, are tiny by comparison with the sizes of their giant neighbors. And the moons of the other terrestrial planets are only small bodies, a few kilometers in diameter. One unique feature about the Moon, then, is its large size by comparison with Earth's size.

The exact place of Moon's birth in the planetary nebula is not yet known. Moon may have formed where it now is—close to Earth and attracted to it by gravity. Or Moon may have condensed as a separate planet, moving in its own orbit around the Sun. Then, when the separate paths of Earth and Moon brought them momentarily close together, Earth's gravitational pull may have plucked the Moon from its path, causing it to move into a new path and endlessly to orbit the Earth. We still cannot choose between these two possibilities, but they do not influence an important conclusion we *can* draw: because Moon is a small, dense, rocky planet, it must have condensed in the inner regions of the planetary nebula, just as the other terrestrial planets did. Moon's structure and composition should, therefore, be similar to the structures and compositions of other terrestrial planets. Now that Moon has been visited and examined by astronauts, we can say that in general these statements are true; yet the differences between the structures of Earth and Moon are many.

Structure. Each time astronauts have visited the Moon, they have made measurements that provide clues about Moon's structure. When they departed, they left behind instruments that continued the measurements and that transmitted the results to Earth. The most informative measurements are of four kinds: measurements of seismic waves, of magnetism, of the heat that flows out of the Moon, and of gravity.

Seismic Evidence. Compared to earthquakes, moonquakes are weak; also they are infrequent. Moonquakes large enough to be detected by instruments carried to the Moon by astronauts number fewer than four hundred a year; on Earth the same instruments would record about a million quakes a year. The comparative scarcity of moonquakes, and their weakness, immediately suggest that processes such as volcanism and plate tectonics, the causes of most earthquakes, are not happening on the Moon.

Moonquakes tend to occur in groups; most of them happen when Moon's elliptical orbit brings it closest to Earth. At the point of closest approach, gravitational forces between Earth and Moon are strongest. This suggests, then, that most moonquakes result from the gravitational pull that Earth exerts on the Moon. The pull causes slight movement along cracks, each one causing a tiny moonquake. Foci of moonquakes have been recorded as deep as 800km; this observation too is informative. It means that Moon, unlike Earth, is rigid enough at 800km for elastic effects to happen. This in turn means that Moon's temperature at 800km must be considerably less than Earth's temperature at the same depth.

But regardless of the causes of moonquakes, the seismic waves they generate tell us a good deal about the structure of the Moon. They tell us that on the side of the Moon that faces the Earth—or at least on those parts that have been visited by astronauts—there is a crust about 65km thick. They tell us also that the crust is layered (Fig. 19.3). Covering the surface is a layer of regolith that ranges from a few meters to a few tens of meters in thickness. Below the regolith is a layer, about 2km thick, of shattered and broken rock (how the rock was broken is discussed later in this chapter). Below this, again, is about 23km of basalt, then a further 40km of a rock rich in feldspar. At a depth of 65km, the velocities of seismic waves increase rapidly, indicating that the lunar crust overlies a mantle. The exact composition of the lunar mantle is not known. Wave velocities suggest it is possibly similar to that of Earth's mantle. Wave velocities also suggest that the lunar mantle is considerably colder and less plastic than Earth's mantle.

Whether or not the lunar mantle is layered is not

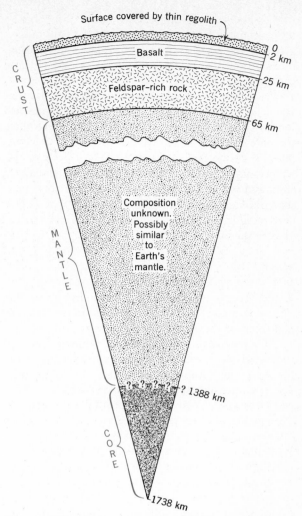

Figure 19.3 Section through the Moon, showing its probable layered structure. The structure of the crust is known with certainty only in the vicinity of the astronauts' landing sites, and the existence of the small core is still uncertain.

yet known. Moonquakes are too weak to set off the strong seismic waves needed to penetrate deeply and resolve the question of layering in the mantle. Likewise they cannot pass through the Moon and test the presence or absence of a core. But the case can be resolved in part by measurements of lunar magnetism.

Magnetism. Although the Moon lacks a magnetic field like that of Earth, it is weakly magnetic. The magnetism comes from a surprising source: igneous rocks, all of which retain ancient magnetism. Therefore, although there is no lunar magnetic field today, a field must have existed at the time the igneous rocks solidified from magma. Unless magnetic fields can be generated in ways that we know nothing about, we can infer that Moon has a core, that the core once created a magnetic field, and that it is no longer capable of doing so. We can infer also that the core must be small. Moon's density is 3.35gm per cm^3, while that of Earth is 5.52gm per cm^3. If we presume that Moon's core, like Earth's, is mainly iron, and is therefore very dense, we can state that its radius can be no larger than about 700km. If the core were larger, Moon's density would have to be greater than it is.

Heat Flow. Seismic waves and magnetism prove that Moon has a layered structure. What can we infer about Moon's internal temperatures and about the chance that convection currents exist in the mantle? Evidence on these matters comes from two sources.

The first kind of evidence of Moon's internal temperatures is obtained with the help of the Sun. We noted in Chapter 3 that the Sun continually emits electrically charged particles which stream away into space. The streams are similar to electrical currents, and as they reach the Moon they cause such currents to flow within it. The currents are weak—so weak that they cannot be felt—but they are detected by sensitive measuring devices. The strength of an electrical current depends on the composition and temperature of the substance in which it flows; for rocks, the principal control is temperature. The strength of Moon's electrical currents indicates that at a depth of 1000km the temperature of the lunar mantle can be no greater than 1000°C to 1100°C. Under such conditions, melting cannot take place, and we can therefore infer that magma is not presently forming in the Moon. Such low internal temperatures let us infer also that in Moon's mantle, convection probably does not happen.

The second evidence of Moon's internal temperatures comes from direct measurement of the amount of heat that flows out from Moon's interior. The flow was measured on the Apollo 15 and Apollo 17 missions, and was found to be high—about half the value of Earth's heat flow. A high rate of flow at the surface and a cool lunar interior may seem contradictory, but the two values let us draw an important inference. High flow at the surface means that

the lunar crust must contain a high concentration of heat-producing radioactive elements. From this we can infer that some kind of magmatic differentiation (Chap. 5) must have been involved in the formation of the lunar crust, as a result of which the radioactive elements became concentrated.

Gravity. We can divide Moon's surface into two general categories: (1) *highlands,* mountainous areas that appear as light-colored patches when we look at the Moon; and (2) *maria* ("seas"), (singular *mare*). These are smooth lowland areas that appear to us as dark-colored regions (Fig. 19.4).

In the highlands, mountains soar tens of thousands of meters above the maria, and in places stand even higher than Earth's mountains. Because lunar mountains are big, one way to obtain clues about Moon's interior regions is to learn whether isostasy operates as it does on Earth, and whether the great lunar mountains have roots. The test is best made by measuring variations in Moon's gravitational pull (Chap. 17). Such measurements are made continually as spacecraft orbit the Moon; they indicate that the mountains must have roots because the highlands are isostatically balanced. We can infer, then, that when the lunar mountains formed, at least one of Moon's outer layers must have been sufficiently fluid-like and plastic to enable the highlands to float.

These measurements of Moon's gravity have given us an unexpected surprise and an additional clue. Some of the mare areas are not isostatically balanced; they have too much mass near the surface, and they produce *an anomalously high gravity pull* called a **mascon** (abbreviated from *mass concentration*). The origin of mascons is still unknown. But mascons prove an interesting point. When the mare formed, the Moon was no longer capable of attaining isostatic balance. As we see later, the highlands formed before the maria did. Thus the Moon had once, but later lost, the capacity for isostatic adjustment.

Astronauts not only measured Moon's structures; they returned to Earth with a treasure-trove of samples of rock and regolith samples. From these, too, we have obtained many clues about Moon's composition and history.

Rock and Regolith. The astronauts brought back three different kinds of lunar material: (1) a variety of igneous rocks, (2) breccias (App. C), and (3) regolith, popularly called *Moon dust* (Fig. 19.5).

The most interesting samples are the igneous rocks; in terms of age and composition, there are three different kinds. The first and oldest kind consist of both feldspar-rich lava and anorthosite, a variety of igneous rock formed by extreme magmatic differentiation, and consisting largely of calcium-rich plagioclase. Radiometric dates of these oldest rocks, which come from the highlands, indicate they were formed more than 4 billion years ago. The second kind of igneous rock is basalt that contains high concentrations of potassium and phosphorus; this too is about 4 billion years old. The potassium-rich basalts likewise come from highland areas, and may possibly be the rocks that cause the high heat flow, because they are richer in radioactive elements than any of the other rocks found on the Moon. The third kind of igneous rock is likewise basalt, but is rich, not in potassium but in iron and titanium. It has been found only in the maria, and is dated radiometrically at 3.3 to 3.7 billion years. We infer it is the material that underlies each mare to a depth of about 25km. Mare basalt, then, formed several hundred million years after the highlands had formed. Even though magmas do not seem to occur on the Moon today, the mare basalt proves that magmas similar to those on Earth formerly existed on the Moon. But they existed a long time ago. The *youngest* mare basalt is as old as the *oldest* rock yet found on Earth.

What clues can be extracted from samples of the other two kinds of Moon rock, regolith and breccia? The lunar regolith is a fine-grained mixture of gray pulverized rock fragments and small dust particles, many of which are glassy. Its composition is essentially that of the lunar igneous rocks. Regolith covers all parts of the lunar surface like a gray shroud (Fig. 19.6), as if a giant hammer had crushed the surface rock. Indeed it *is* apparently the product of hammering—by the steady stream of meteorites, large and small, that crash onto Moon's surface. On Earth, most meteorites burn up in the heat of the friction generated as they come speeding into the atmosphere. But on the atmosphereless Moon even small meteorites reach the solid surface, pulverizing bedrock and continually creating more regolith. Presumably the layer of shattered rock, 2km deep, beneath the regolith likewise was created by similar hammering. The breccias brought back by the astronauts are compacted aggregates of rock fragments, mineral chips, and regolith. They, too, seem to have been formed by the impact of meteorites. Apparently some meteorites compress the regolith and broken rock so greatly that the particles are welded together again to form new breccia.

The samples of regolith and breccia, then, provide clues about erosion on the lunar surface—erosion caused by meteorite impacts. Erosion of the kind we are familiar with on Earth does not exist, because the Moon lacks both an atmosphere and a hydrosphere.

History of the Moon. From the clues we have discussed, let us construct a history of the Moon. The story begins about 4.6 billion years ago. Although one popular theory proposes that the Moon formed when a giant fragment split apart from the Earth, Moon's layered structure makes such an origin very unlikely. Apparently, Moon's origin, like Earth's, was the result of condensation from the planetary nebula. But condensation was not an atom-by-atom and layer-by-layer affair. It involved a gathering together of millions of small solid particles. As particles condensed from the planetary nebula, gravity continually gathered them together. We know this to be the case because the final stages of the gathering process left scars that can still be seen on Moon's surface. These are the countless craters, large and small, left by falling meteorites—for **meteorites** are simply *small solid bodies formed during condensation of the planetary nebula.*

The process by which solid particles gather together to form a planet is **accretion.** By 4.6 billion years ago, the Moon had accreted to about its present size, apparently developing, in the process, a core and a mantle. But as the Moon accreted, there occurred a phenomenon we have not mentioned before. As the Moon grew larger, the strength of its gravitational attraction increased, so that the speeds at which the accreting meteorites reached the lunar surface grew ever greater. Eventually, speeds became so great that each impact generated a large amount of heat. Near the end of the accretion process, so much heat was generated that an outer layer, 150km to 200km thick, of the Moon's body was melted. The Moon thus had a solid interior but a molten outer shell. The period of rapid accretion soon ended, and the outer layer of magma began to cool and crystallize. The earliest crystals that formed in the magma were apparently plagioclase feldspar, and because they were lighter than the parent magma, they

Figure 19.4 *A.* The "front" side of the Moon. Light-colored areas, pitted with craters caused by impacts of meteorites, are lunar highlands. Dark-colored, lowland areas are maria (plural of *mare*), formed when basaltic lava flowed out to fill the craters produced by exceptionally large impacts. Copernicus, Kepler and Tycho are three prominent young meteorite craters formed after the mare basins were filled. The impacts that produced them splashed bright-colored rays of rocky debris over ancient basalt. Sites of the six Apollo lunar landing missions are indicated. (*Mosaic LEM-1, 3rd edition, 1966, made from telescopic photographs by U.S. Air Force.*)

Figure 19.4 *B*. The "back" side of the Moon, invisible from Earth. This photograph was taken by astronauts, as they prepared to leave the Moon after the Apollo 16 mission. Mare Crisium, visible on the "front" side of the Moon, can just be seen at upper left. The back of the Moon is mostly highland. The scarcity of maria on it is one of Moon's unsolved mysteries. (*Courtesy of NASA.*)

A

B

C

Figure 19.5 Three kinds of lunar material brought back by astronauts.

A. Basalt, with numerous vesicles formed when gases escaped during cooling and crystallization of lava. Collected during the Apollo 12 mission, this sample is the kind of basalt found in the mare basins. (*Photograph by W. Sacco.*)

B. Breccia, a rock composed of numerous fragments of igneous rock and glassy fragments similar to those found in the lunar regolith. This sample was collected on the Apollo 15 mission. (*Courtesy of NASA.*)

C. Sample of regolith, a mixture of many rocks and minerals, together with glassy fragments produced by bombardment of the lunar surface by meteorites. Only coarser grains from the regolith are shown in the picture. (*Photograph by J. A. Wood, Smithsonian Astrophysical Observatory.*)

floated in it. Thus magmatic differentiation was operating, and soon a thick crust, rich in feldspar crystals, floated in the remaining liquid. Meteorites fall on the lunar surface today; so we can infer that meteorites must also have been falling, although at a continually declining rate, while the crust was forming. Larger impacts must have broken the crust, letting liquid from below ooze out to form the ancient, feldspar-rich lava flows. When differentiation was far advanced, the ancient potassium-rich lavas probably formed in the same way. Those ancient products of magmatic differentiation are the materials the astronauts found in the lunar highlands. In places the astronauts even saw crude layering created when the ancient lavas flowed out (Fig. 19.7). The

crust has an irregular, mountainous surface because impacts tilted great blocks of it, and possibly also because convection currents piled broken fragments up, much as ice on the sea becomes piled into irregular, tangled masses.

The layer of magma, 200km thick, would have cooled and crystallized within about 400 million years. By about 4 billion years ago, therefore, the crust and mountains had formed, and the major activity on the Moon had become the steady rain of meteorites that pitted and pocked the surface. Some meteorites, larger than others, made exceptionally large impact scars. These huge circular cavities eventually became the mare basins. If we study the photograph of the front side of the Moon closely

Figure 19.6 Photograph of Hadley Rille, a deep graben-like valley, made by Apollo-15 astronauts near their landing site. The valley walls are cloaked with dusty regolith several meters deep. The numerous small pits on the slope, left rear, were caused by meteorite impacts. Diameter of the largest boulder is 15m. (*Courtesy of NASA.*)

Figure 19.7 Silver Spur, a prominent peak in the lunar highlands, about 20km from the Apollo-15 landing site. The pronounced layering in Silver Spur is believed to represent a series of ancient lava flows, tilted to their present positions by the giant meteorite impact that created the basin now occupied by Mare Imbrium. (*Courtesy of NASA.*)

(Fig. 19.4,*A*) we can see that all the mare basins are roughly circular.

We infer that while the Moon's surface was being bombarded by meteorites, its interior regions were slowly heating up. The lunar mantle must contain some radioactive elements, and as time passed these must have heated the interior. If we judge from meteorites (because they too are samples of the condensed planetary nebula), the amount of radioactivity is small, but nevertheless it was sufficient to cause partial melting in the upper mantle, beginning about 3.7 billion years ago. The magma so formed apparently worked its way up to the surface along places of weakness in the crust, that had been caused by the impacts of the largest meteorites. When it reached the surface, the magma flowed out to form the basalt flows now seen in the maria (Fig. 19.8). About 3 billion years ago, the extrusion of mare basalts ceased and we infer that from that date to the present no further magmatic activity has occurred. Except for the continuing rain of meteorites, Moon has remained a dead planet.

By the use of seismic data, scientists of the U.S. Geological Survey have, in effect, "peeled" away the debris from the youngest meteorite impacts and have determined the shapes of the mare basins beneath the lava. In this manner they have prepared reconstructions of the "front" face of the Moon at two stages in its history (Fig. 19.9).

Lessons for the Earth. Did Earth ever look like the Moon? Although we cannot be sure, probably it did. The Moon proves that the terrestrial planets formed by accretion of solid bodies. Earth is larger than the Moon, and its gravitational pull is stronger. Presumably, therefore, an even deeper layer of

Figure 19.8 Photograph, made in 1966, of the surface of Oceanus Procellarum, with the crater Marius (Fig. 19.4,*A*) near the horizon. The dome-shaped objects are believed to be volcanoes; the irregular ridges are ancient lava flows. The craters were caused by meteorite impacts. The area of the photograph is a region as large as the States of Connecticut, Massachusetts, and Rhode Island combined. Each of the two prominent volcanic domes in the central foreground is about 20km in diameter. (*Courtesy of NASA.*)

A

B

Figure 19.9 How the "front" side of the Moon probably looked at two times in the past. The figures, which should be compared with the Moon's aspect today (Fig. 19.4,*A*), were prepared by starting with a photograph of the Moon as it is today, and painting in a surface reconstructed from geologic maps.

A. As Moon probably looked about 3.9 billion years ago, before the mare basalts filled the largest craters. The crater that later became Mare Imbrium looks like a giant bullseye in the upper left-hand corner. Virtually the entire Moon is covered with ancient crustal rock of the highlands. The few dark areas represent old potassium-rich lavas.

B. As Moon appeared 3.0 billion years ago. The basaltic lava flows that filled the mare basins between 3.0 and 3.7 billion years have ceased. Except for the young impact craters, the Moon looked much as it does today. (*Wilhelms and Davis, 1971.*)

magma must have covered the Earth at the end of the accretion process, and as it cooled, an early crust must have formed on Earth. But all signs of Earth's primitive crust have now gone. The reason is not hard to find. Radioactive heating on Earth continued to create magma, and the earliest crust has probably been remelted and reabsorbed. We can infer that radioactive heating would certainly have started to create bodies of magma on Earth about the same time as on the Moon—that is, about 3.7 billion years ago. If this inference is correct, then the familiar magmatic process on Earth, and related tectonic events such as mountain building, might only have commenced about 3.7 billion years ago.

On one point we can be sure Earth and Moon shared a common experience; both were scarred by meteorite impacts. Despite the speed with which erosion removes features from Earth's surface, many ancient meteorite scars remain today (Fig. 19.10). Now let us see whether the other terrestrial planets tell a similar story.

Mars

Mars, with a diameter of 6720km, possesses a mass only one-tenth of Earth's mass. Yet despite its small size, Mars is Earth-like in many ways. It rotates once every 24.5 hours; so the length of the Martian day is nearly the same as the length of an Earth day. Also Mars has an atmosphere, although it is only one one-hundredth as dense as Earth's, and consists largely of carbon dioxide; and Mars has polar ice caps (Fig. 19.11), consisting mostly of frozen carbon dioxide ("dry ice"). Like Earth, Mars has seasons, and the diameters of the ice caps alternately grow and shrink with the coming of winter and summer.

But Mars also has many features unlike Earth's. One difference is the Martian moons, of which there are two, Phobos and Deimos. Both Martian moons are small, and irregular in shape. For example, Phobos, the larger, is roughly oval, about 27km long, and heavily pitted and cratered from innumerable meteorite impacts (Fig. 19.12).

We can infer little about the internal structure of Mars because we have few observations on which to base our inferences. Like the Moon, Mars lacks a magnetic field. It has a density of 3.96gm per cm³, which means that if an iron core is present, it must be small. Some scientists have suggested that condensation in the planetary nebula, at a place as far

away from the Sun as Mars is, should lead to a core of iron oxide rather than one of metallic iron. Perhaps Mars does have an iron oxide core, but if so we must await future measurements in order to find out. From the density of Mars and from the way it wobbles on its axis, we infer that it probably does have a layered structure; but all details of the layering remain a mystery.

Surface Features. Four attempts to send spacecraft to Mars have been successful, but none of the craft have landed. The craft have been aptly named Mariners, and they sailed through distant seas of space to reach their destinations, for when Earth and Mars are closest to each other, a one-way trip lasts about 5 months. The Mariners, and particularly Mariner 9, which started orbiting Mars on November 13, 1971, radioed back a series of measurements and photographs that revealed a number of interesting things about Mars and its history.

Cameras on Mariner 9 could see only those features on the Martian surface that are larger than about 200m in diameter; but even so, Mariner 9 recorded many striking features. A map of Mars, with the locations of the most prominent features, is shown in Figure 19.13.

Most of the southern hemisphere looks very like the Moon, in that it is pitted and pocked by innumerable meteorite craters. Two craters are as large as maria on the Moon, and presumably were created by very great impacts. These are the features called Argyre and Hellas. Because of the large numbers of impact craters, we infer that most of the southern hemisphere is covered by ancient rock.

The northern hemisphere presents a very different picture. Here we see vast areas nearly free of craters; and we therefore infer that the surface there is younger than that of the southern hemisphere. The clue to the origin of the northern hemisphere comes from a series of photographs that reveal the presence of huge shield volcanoes. At least fifteen volcanoes have been discovered, and the giant among them is Nix Olympica (Snows of Olympus). The basal diameter of Nix Olympica (Fig. 19.14) is 540km, approximately the distance from Boston, Massachusetts to Washington, D.C., and it stands an estimated 15km to 20km above the surrounding plain. Nix Olympica is at least twice as wide and twice as high as the largest shield volcano on Earth. Mauna Loa, in Hawaii, one of Earth's largest volcanoes, is only

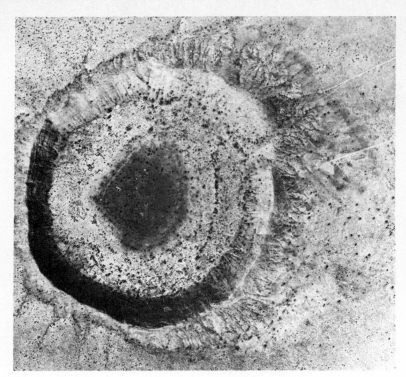

A

Figure 19.10 Scars left on Earth's surface from the impact of meteorites.

A. Wolf Creek Crater in the northern part of Western Australia. The age of the crater is not known exactly, but we infer it is not great because the crater has not been deeply eroded. The size of the crater can be judged from the two single-lane dirt roads that approach it (upper right-hand corner). (*Courtesy of the Pickands-Mather Co.*)

B. Clearwater Lake, west of Hudson Bay in northern Quebec, marks the site where a meteorite struck the surface of the Earth during Precambrian times. The crater must have looked originally like the Wolf Creek Crater (above), but erosion has removed the rim, leaving only the shallow circular feature we see here. (*Photo by Royal Canadian Air Force.*)

B

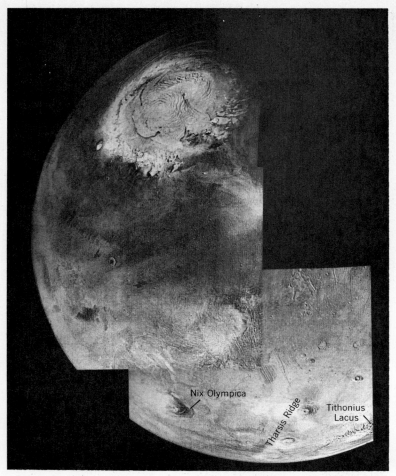

Nix Olympica

Tharsis Ridge

Tithonius
Lacus

Figure 19.11 Mars seen from a distance of 13,700km. Photographed by
Mariner 9, in the late spring of the northern hemisphere, the north polar
cap is plainly visible as a circular, white feature. The Martian polar caps,
believed to consist of solid carbon dioxide, shrink and grow with the seasons. In late spring the cap is greatly shrunken from its winter maximum.
The origin of the spiral-shaped structures in the cap is unknown. Near
the base of the view several shield volcanoes, including the giant Nix
Olympica, are visible. The great canyon-like structure, lower left, is
Tithonius Lacus (see text and Figure 19.13). (*Courtesy of NASA.*)

225km across and stands only 9km above the sea
floor. The summits of all the Martian volcanoes are
huge calderas; that of Nix Olympica is thirty times
larger than Mauna Loa's caldera (Fig. 16.6). Inside
the calderas, numerous young-looking craters are
visible. We cannot be sure whether Nix Olympica
and the other volcanoes are still active, but the almost complete absence of impact craters on their
slopes suggests that they may be.

Adjacent to Nix Olympica, and running generally
parallel to the Martian equator, is a region called
Tithonius Lacus. Here Mariner 9 photographed the
most remarkable canyons yet discovered on any
planet (Fig. 19.15). Approximately 4000km long,
100 to 250km wide and up to 7km deep, the giant
Martian canyons dwarf the Grand Canyon—the entire Grand Canyon of the Colorado River could be
lost in one of the tributaries. The origin of these

Figure 19.12 Photographs of Phobos, one of the two moons that orbit around Mars. The photograph was taken by Mariner 9 from a distance of 5758km. The irregular shape of Phobos is probably caused by all the meteorite impacts it has sustained. The circular markings are impact craters. (*Courtesy of NASA.*)

great valleys is not known, but they are thought to be a series of giant grabens (Chap. 15). Possibly they formed when the crust subsided into openings left empty when magma was extruded to build the giant volcanoes.

Erosion. Remarkable as the giant volcanoes and canyons are, even more remarkable features are seen in some of the Mariner 9 photographs. They seem to have been caused by erosion of unfamiliar kinds.

When Mariner 9 reached Mars, a sandstorm of spectacular proportions was raging. Dust and sand particles were observed as high as 55km above the surface, winds reached velocities of 300km per hour, and the storm raged unabated for several months. Under conditions such as these, sand particles could abrade rock effectively; it seems likely that on Mars, wind-driven sand is the main agent of erosion. Peering into one of the impact craters near Hellas, Mariner 9 photographed a series of spectacular sand dunes (Fig. 19.16). This suggests what probably happens when the great storms abate. Probably all craters and all low places on Mars contain piles of sand. Every time a storm blows up, the sand is swirled into the atmosphere. When the storms die down, gentle winds slowly sweep the sand back into all the sheltered hollows on the surface.

But still more spectacular than the Martian sand is evidence, sent back by Mariner 9, that water or some other liquid has cut stream channels on Mars (Fig. 19.17). There are valleys that look very like those cut by intermittent desert streams on Earth. They meander, they branch, and they have braided patterns and other features characteristic of valleys made by running water. Yet, Mars lacks rainfall, streams, lakes, and seas. The sparse water that does occur on Mars is in the atmosphere, or exists, near the poles condensed as ice. The Martian surface is too cold for water to exist as a liquid. Some have suggested that ice might be present beneath the surface dust as permafrost, and that at times the Martian climate warms up, causing the ice to melt and create torrential floods. Others believe that Mars simply does not have sufficient ice in the solar caps or as permafrost to melt and to cut great stream channels. The origin of the Martian stream channels, then, remains a mystery. Perhaps future spacecraft will help solve it.

History. Can we infer, from our limited clues, anything about the history of Mars, and does this in turn tell us anything about Earth's history? The answer is yes, but not much.

The early history of Mars must have been much like that of the Moon; indeed the terrain of the southern hemisphere, formed in the early days, is probably equivalent to the lunar highlands. Then, like both Earth and Mars, radioactive heating inside Mars started to create magma by partial melting. Because Nix Olympica and its mates are shield volcanoes, we can infer that the magma they extrude is of low viscosity and therefore is probably basalt. This means that magmatic differentiation has occurred, but we cannot yet say when the volcanism started. To be able to do that we must have radiometric-age measurements. But the fresh-looking surfaces of the big cones suggest strongly that the volcanoes are still active and that Mars, like Earth, is still continuing to make magma.

Measurements made from Mariner 9 show that Mars has several elevated plateaus. Gravity measurements, also made from Mariner 9, suggest further that Mars is close to isostatic balance, but that, like Earth, some force may be continually disturbing the balance. These observations have suggested to some that convection in the Martian mantle might be causing plates of lithosphere to move around, and that the elevated plateaus are blocks of low-density

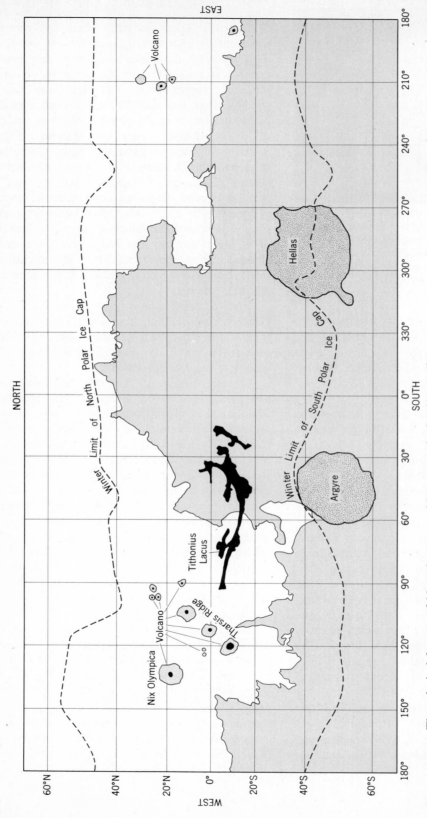

Figure 19.13 The principal features of Mars shown on a Mercator projection (Appendix E). Some features on this map can be located on Figure 19.11. The southern hemisphere is occupied by much-cratered terrain (blue) believed to be ancient crust. The northern hemisphere is much smoother, and the crust there is believed to be younger. Several large shield volcanoes, each crowned by a caldera, occur in the younger crust, while two very large impact structures, Argyre and Hellas, occur in the ancient crust. The region labeled Tithonius Lacus is a region of vast canyons and gorges (Fig. 19.11). The winter limit of the northern polar cap is much farther south than the edge of the cap seen in later spring in Figure 19.11. (*Adapted from a map, based on Mariner 9 photographs, prepared by U.S. Geological Survey.*)

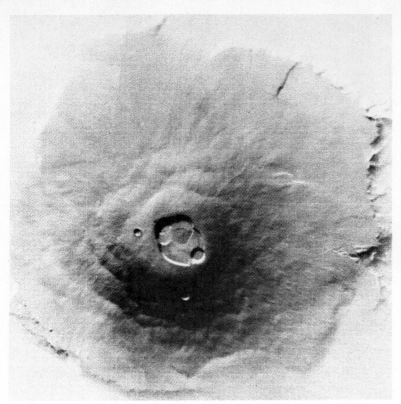

Figure 19.14 Nix Olympica, giant shield volcano on Mars. Its diameter is about 540km, and its base is marked by steep cliffs believed to have been cut by intense wind erosion during fierce Martian sandstorms. Nix Olympica, the largest volcano known on any planet, is capped by a caldera 65km across. Within the main caldera, smaller calderas and a circular crater can be seen. The two small circular structures on the flanks of the volcano, to the left of and below the caldera, are believed to be volcanic craters. (*Courtesy of NASA.*)

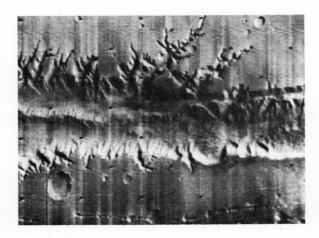

Figure 19.15 Part of a giant canyon in the Tithonius Lacus region of Mars. The segment shown here is 480km long and 7km deep. The canyon is believed to be a huge graben. Although the branching structures that run into the canyon look like water-cut stream channels, they cannot be, because high-resolution photography reveals that some of them are closed basins. Probably the channels were cut by fast-moving winds that are believed to roar up and down the canyon. (*Courtesy of NASA.*)

Figure 19.16 Dark-colored area of sand dunes on Mars, near Helles-pontus. The distance from crest to crest of adjacent dunes is about 1.5km. Numerous dark areas have been observed on Mars, many of them inside large craters, suggesting that dunes may be common features of the Martian landscape. (*Courtesy of NASA.*)

continental crust of the kind we have on Earth. It has been suggested that because Mars is smaller than Earth, its heat leaks away faster, so that Mars may still be heating up. If this suggestion is correct, the time when widespread convection and very active plate tectonics will occur on Mars is still in the future. If this is indeed the case, Mars will provide yet another great source of information about Earth's childhood, because we shall then know that Mars's stage of development lies intermediate between those of Moon and Earth.

Venus

For still more information we can now turn to Venus, the last of the terrestrial planets that has been visited by spacecraft. Venus is the planet most like Earth in size and mass—its diameter is 12,042km and its density 5.25gm per cm^3. But there the similarities stop. Venus is enveloped by a cloudy atmosphere of carbon dioxide, a hundred times more dense than Earth's atmosphere, that hides the solid surface from view.

Russian scientists have sent to Venus several spacecraft, of which two landed successfully. The spacecraft reported that the surface of Venus is as-tonishingly hot, with a temperature of about 500°C. At that temperature metals such as lead, zinc, and tin are in a molten state. The explanation of the high temperature is not hard to find. First, Venus is closer to the Sun than Earth is. But more important is the fact that the carbon dioxide in Venus's atmosphere acts like the glass of a greenhouse: it lets the Sun's electromagnetic rays through to heat the surface, but serves as a barrier that prevents heat from leaving.

Although we cannot see the surface of Venus, we can infer something about it. Visiting spacecraft show that Venus has a weak magnetic field, about three one-thousandths as strong as Earth's. This observation, combined with knowledge that the density of Venus is very close to Earth's density, tells us that Venus probably has an iron core.

Sensitive radar measurements of Venus's shape show that the planet is nearly smooth, with relief apparently much less than that measured on Earth or Mars. Nevertheless, relief as great as 2.5km does occur, and long, arcuate features resembling eroded mountain ranges have been detected. Perhaps Venus once had great mountains, the product of tectonic activity, so that what we now detect are merely the eroded remains of an active past. Even the composition of the atmosphere tells something about Venus.

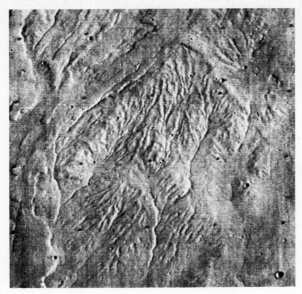

A

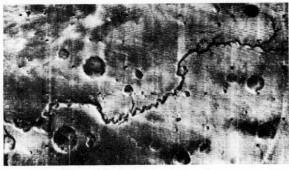

B

Figure 19.17 Part of the surface of Mars, showing features that have been interpreted as ancient stream channels.

    A. Channels that branch in a dendritic pattern. The small impact craters are about 2km in diameter.

    B. Valley, approximately 6km wide, showing meanders and branching tributaries. (*Courtesy of NASA.*)

Because gases react quickly with hot solids, the composition of the Venusian atmosphere must be controlled by chemical reactions with hot surface rock. Measurement of the composition of the atmosphere, then, should indicate the probable composition of the surface. Such reasoning indicates that because the composition of Venusian surface rock seems to be close to that of andesite, Venus has apparently undergone magmatic differentiation.

We can speculate that the history of Venus may be very like Earth's. Apparently magmas were created, perhaps convection occurred in the mantle, and possibly even a period of movement of lithospheric plates transpired, creating the mountain ranges. Some have suggested that Venus is even more advanced in its history than Earth is, and that further study of Venus will help us peer into Earth's future. Perhaps; but such suggestions are at best speculations, and we must await future spacecraft and more reliable observations before we can draw any definite conclusions.

Conclusion

Earth, Moon and the other terrestrial planets seem to have similar histories. Each formed by condensation in the planetary nebula and each bears marks from the rain of meteorites accreting to its surface— even on tiny Mercury, recent photographs have shown this to be the case. Each terrestrial planet seems to have been through—or is still going through—a period of radioactive heating that produced kinds of magma familiar to us on Earth. Other familiar features on Earth, such as mountains and volcanoes, can also be observed. Yet the terrestrial planets have vital differences among themselves. For example, Earth's atmosphere and hydrosphere are unique; and the importance of different kinds of erosion differ from planet to planet. The terrestrial planets, then, are similar in some respects but different in others. By examining the differences we can gain a clearer idea of how Earth's processes occur.

Our research on the similarities and differences between Earth and its neighbors has led us to a conclusion of peculiar importance—Earth's sister planets are not hospitable places for Man. Nor is it likely that they contain great mineral riches such as deposits of gold, copper, lead, and silver. As we shall see in Chapter 20, the mere existence of mineral deposits of various kinds involves the presence of a hydrosphere and an atmosphere. The absence of either or both would make the formation of such deposits unlikely. Earth remains the planet on which we must rely for our food, and for all the other resources we need to keep our civilizations going. For this reason it is appropriate to discuss, in the next chapter, the distribution and abundance of Earth's mineral resources.

Summary

1. The Sun formed when atoms in space became sufficiently compacted for nuclear burning to begin.

2. The planets formed by condensation from a disk-shaped envelope of gas that rotated around the Sun.

3. Planets, close to the Sun, such as Earth and Venus, are small, dense, rocky objects. Planets farther away, such as Saturn and Jupiter, are large, light-weight objects.

4. Earth, Moon, and the other terrestrial planets have a layered structure that probably formed when they condensed from the planetary nebula.

5. Moon probably has a small core surrounded by a thick mantle, and is capped by a crust 65km thick.

6. On the Moon, magma was formed earlier, but is no longer generated. Moon is a dead planet.

7. The highlands of the Moon are remnants of ancient crust built by magmatic differentiation more than 4 billion years ago.

8. The maria (lunar lowlands) are vast basins created by the impacts of giant meteorites, and later filled in by lava flows.

9. Mars seems to be geologically active. Nix Olympica, a shield volcano on Mars, is the largest volcano yet found in the Solar System.

10. The principal eroding agent on Mars is wind-driven dust, but water or some other flowing liquid has cut stream channels at some time in the past.

11. Venus has about the same size and density as Earth. It may once have been tectonically active, but now seems dead.

12. Venus has a dense atmosphere of carbon dioxide and a surface temperature of about 500°C.

13. None of the terrestrial planets have climates hospitable to Man.

Selected References

Solar System

Carr, M. H., ed., 1970, A strategy for geologic exploration of the planets: U.S. Geol. Survey, Circular 640.

Jastrow, Robert, and Thompson, M. H., 1972, Astronomy: fundamentals and frontiers: New York, John Wiley.

Kuiper, C. P., and Middlehurst, B. M., 1961, The Solar System. Vol. III. Planets and satellites: Chicago, University of Chicago Press.

Moon

Kosofsky, L. J., and El-Baz, Farouk, 1970, The Moon as viewed by lunar orbiter: NASA Publ. SP-200.

Levinson, A. A., and Taylor, S. R., 1971, Moon rocks and minerals: New York, Pergamon Press.

Mutch, T. A., 1970, Geology of the Moon: Princeton, New Jersey, Princeton University Press.

Wilhelms, D. E., and McCauley, J. F., 1971, Geologic map of the near side of the Moon: U.S. Geol. Survey Map I-703.

Mars

McCauley, J. F., and others, 1972, Preliminary Mariner 9 report on the geology of Mars: Icarus, v. 17, p. 289–327.

Murray, B. C., 1973, Mars from Mariner 9: Sci. American, v. 228, no. 1, p. 49–69.

Weaver, K. F., 1973, Journey to Mars: Nat. Geogr. Mag., v. 143, p. 231–263.

Venus

Lewis, J. S., 1971, The atmosphere, clouds and surface of Venus: American Scientist, v. 59, p. 557–566.

The Trans-America Highway being cut through virgin forest in Central Brazil. (*Toby Molenaar/Woodfin Camp.*)

Part VI Man and Earth

Chapter 20

Sources of Energy and Minerals

Materials and Energy

When we speak of the present epoch in western civilization as "the industrial age," we mean the age of suddenly and enormously increased use of fuels and metals; another term would be "the age of high-energy industry." Stone-Age man, 30,000 or more years ago, had an industry too: he chipped and flaked mineral substances (mostly pieces of quartz) to make tools and weapons. But his was a low-energy industry, with the energy supplied by human muscle. Although he was limited by this fact and by a very narrow choice of materials to work with, we must not underestimate him. He was as intelligent as we, lacking only our accumulated experience and the skill that comes from experience.

The lack was gradually, though not steadily, overcome by him and by his descendants. Here are a few of their accomplishments, discovered and dated by archeologists:

4000 B.C.: Chaldeans had become skilled workers in metals such as gold, silver, copper, lead, tin, and iron.

3000 B.C.: Eastern Mediterranean peoples were making glass, glazed pottery, and porcelain.

2500 B.C.: Babylonians were using petroleum instead of wood for fuel.

1100 B.C.: Chinese were mining coal and were drilling wells hundreds of feet deep for natural gas.

These accomplishments were arts learned by experience, and they implied the substitution of metals,

393

glass, and other substances for stone. But the making of a copper implement used more energy than the manufacture of a stone tool. Still more was needed to make objects of glass. So with each advance, more energy and therefore more fuel was needed. Most of the energy with which the working of iron, copper, lead, and other materials were pursued came from wood fuel and from the muscles of men and animals. But as time passed, other fuels came to be used: fuels such as coal, oil, and natural gas. The new fuels were more efficient than the old. By using them, each man could produce many more implements than he could have produced with wood fuel or with muscle. The new fuels therefore sparked rapid growth of the amounts of metals used.

Supplies of Minerals. Metals to build machines and the energy needed to power the machines are complementary. The patterns of use of metals apply also to other mineral substances. High-energy industry demands a host of nonmetallic mineral products, such as shale and limestone for making cement, gypsum for making plaster, and asbestos for insulation. The number and amounts of mineral substances we now use are enormous. In the United States, each year, the quantity of iron and steel used is so great that it is equivalent to about 640kg[1] for every man, woman, and child in the country. This already-large figure rises every year and shows every sign of continuing to rise. The rate at which all other mineral products are being used is also steadily increasing. The same pattern holds true for the U.S.S.R., where mineral production of iron and steel now exceeds that of the United States. The total amount of mineral products produced each year is now so great that it is not easy to imagine where it all goes. In the United States the annual use of sand and gravel exceeds 4000kg per person, of copper 10kg, of aluminum 20kg, of coal 2200kg, and of oil 4000 liters.[2] This huge mass of material is used to make roads, to build skyscrapers, to generate electricity, to build automobiles and trucks, and to fashion the countless other things our civilization uses.

The majority of industrialized nations possess a strong mineral base; that is, they are rich in mineral deposits and are exploiting them vigorously. But because no nation is entirely self-sufficient in mineral

[1] One kilogram (kg equals 2.205 pounds.
[2] There are 3.78 liters in a gallon.

supplies, each must rely, to some extent, on other nations to fulfill its needs.

Mineral resources have certain peculiar aspects that differ from those inherent in such resources as agriculture, grazing, forestry, and fisheries. First, occurrences of usable minerals are both limited in abundance and distinctly localized at places within Earth's crust. This is the main reason why no nation is self-sufficient where mineral supplies are concerned. Because usable minerals are localized, they must be searched out, and the search ranges over the entire globe.

Second, the extent of the reserves of a given mineral available in any one country is rarely known with accuracy, and the likelihood that new occurrences will be discovered is difficult to assess. As a result, production over a period of years may be difficult to predict. Thus, a country that today can supply its needs of a given mineral substance may face a future in which it will have become an importing nation. Third, unlike plants and animals that are cropped yearly of seasonally, deposits of minerals are depleted by mining, which eventually exhausts them. Minerals therefore have "one-crop" availability per occurrence; this disadvantage can be offset only by finding new occurrences or by making use of scrap—that is, by re-using the same material repeatedly.

These peculiarities of the mineral industry place a premium on the skill of geologists, prospectors, and engineers, who play essential parts in finding and bringing to the surface mineral substances for use in industry. This task of finding and mining is being accomplished through the application of the basic principles that have been set forth in this book, and by the use of additional specialized knowledge concerning the origin and distribution of mineral deposits. Much ingenuity has been expended in bringing the production of minerals to its present state. Because known deposits are being rapidly used up, while at the same time demands for minerals are increasing, we can be sure that even more ingenuity will be needed in the future.

Supplies of Energy. The chief sources of the energy consumed in highly industrialized nations are few: the fossil fuels (coal, oil, natural gas), hydroelectric power, nuclear power, wood, wind, and a little muscle. Probably nuclear energy will become an important source in the future, but at present it is just beginning to develop. As recently as a century ago wood was an important fuel, but now it is used mainly for space heating in some dwellings. Wood and nuclear power together supply only one per cent of the energy consumed today. Excluding muscle (a resource now getting very little exercise in industrial countries), the energy consumed in the United States in 1972 came from these sources:

Oil	44%
Natural gas	33%
Coal	18%
Water power	4%
Wood plus	
nuclear power	1%
	100%

For western Europe the breakdown is different: coal accounts for about half the energy consumed. For the world as a whole, with its hundreds of millions of agricultural workers, the breakdown is different again; wood is a more substantial source of energy, and fossil fuels, especially oil and gas, bulk less.

Because our ability to produce and use Earth's mineral supplies depends on our having energy available to do so, we first discuss the principal sources of energy, and then the sources of mineral supplies.

Coal

The black, combustible sedimentary rock we call coal is the most abundant of the fossil fuels. These fuels are so called because they contain solar energy, locked up securely in chemical compounds by the plants or animals of former ages. Most of the coal mined is burned under boilers to make steam for electrical generators, and is converted into coke, an essential ingredient in making steel. Varying amounts of liquid fuel can be derived from coal, and because internal-combustion engines are generally more efficient than steam engines, the future will probably see more and more coal used in this way. In addition to its uses as energy, coal is the chief raw material from which nylon, some plastics, and a multitude of chemicals are made.

Origin of Coal. Coal occurs in strata (miners call them *seams*) along with other sedimentary rocks,

mostly shale and sandstone. A look at a piece of coal with a magnifying glass shows that it consists of bits of fossil wood, bark, leaves, roots, and other parts of land plants, chemically altered but still identifiable. This observation leads at once to the conclusion that coal is fossil plant matter (Fig. 20.1), a conclusion that we incorporate in our definition of *coal* as *a black sedimentary rock consisting chiefly of partly decomposed plant matter and containing less than 40 per cent inorganic matter.*

Many coal seams include fossil tree stumps rooted in place in the underlying shale, evidently a former clay-rich soil. Unlike the material in most sedimentary rocks, the sediment we now find in coal seams was not eroded, transported, and deposited, but accumulated right where the plants grew. It was recognized as long ago as 1778 that the places of coal accumulation were ancient swamps, because (1) a complete physical and chemical gradation exists from coal to peat, which today accumulates only in swamps, and (2) only under swamp conditions is the conversion of plant matter to coal chemically probable. On dry land, dead plant matter (composed chiefly of carbon, hydrogen, and oxygen) combines with atmospheric oxygen to form carbon dioxide and water; it rots away. Under water, however, oxygen is excluded from dead plant matter and oxidation is prevented. Instead, the plant matter is attacked by anaerobic bacteria, which partly decompose it by splitting off oxygen and hydrogen. These two elements escape, combined in various gases, and the carbon gradually becomes concentrated in the residue. Although they work to destroy the vegetal mat-

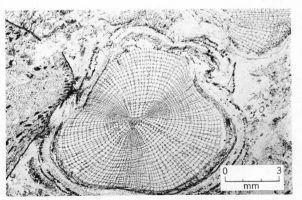

Figure 20.1 Coal of Pennsylvanian age from Illinois, showing cellular structure in fossilized fragment of wood. (*General Biological Supply House; Julius Weber.*)

ter, the bacteria themselves are destroyed before they can finish the job, because poisonous acid compounds they liberate from the dead plants kill them. This could not happen in a stream because the flowing water would dilute the poisons and permit the bacteria to complete their destructive process.

With the destruction of bacteria, the plant matter has been converted to peat. But as the peat is buried beneath more plant matter and beneath accumulating sand, silt, or clay, further changes occur (Fig. 20.2). The peat is compressed, water is squeezed out, and volatile organic compounds such as methane (CH_4) escape, leaving an ever-increasing proportion of carbon. The peat is converted successively into lignite, subbituminous coal, and bituminous coal. These coals are sedimentary rocks, but a still later phase, anthracite, is a metamorphic rock. As anthracite generally occurs in folded strata, we infer that it has undergone a further loss of volatiles and a concentration of carbon, as a result of the pressure and heat that accompany folding. Because of its low content of volatiles, anthracite is hard to ignite, but once alight, it burns with almost no smoke. In contrast lignite, rich in volatiles, ignites so easily that it is dangerously subject to spontaneous ignition (in chemical terms, rapid oxidation), and burns smokily. In certain regions where folding has been intense, coal has been metamorphosed so thoroughly that it has been converted to graphite in which all volatiles have been lost, leaving nothing but carbon. Graphite, therefore, will not burn.

Occurrence. We have said that coal occurs in layers or seams, which are merely strata. Each seam is a flat, lens-shaped body corresponding in area to the area of the swamp in which it accumulated originally. Most coal seams are 0.5m to 3m thick, although some reach more than 30m. They tend to occur in groups: in western Pennsylvania, for example, there are about sixty beds of bituminous coal. The occurrence of coal beds in the rocks of every period later than Devonian indicates that during the last 300 million years or so, swamps rich in vegetal growth have been recurrent features of the land. Peat is accumulating today, at an average rate of about 1m every 100 years, in swamps on the Atlantic and Gulf coastal plains of the United States. The swamps now represented by coal beds must have been much the same as these.

Distribution. Coal is not only abundant, but also rather widely distributed; it is found even in Antarctica. Experts believe that most of the world's coal seams have been discovered, so that a good estimate of the amount of mineable coal can be made. Coal seams thinner than 0.3m are too thin to be mined, and seams deeper than 2000m make mining very dangerous as well as expensive. Using 0.3m as the limit of the thickness, and 2000m as the depth limit for mining, experts estimate that 8415 billion tons of coal can eventually be mined. Most of this reserve is in Asia (Fig. 20.3)—principally in Russia—but North America is well endowed too (Fig. 20.4).

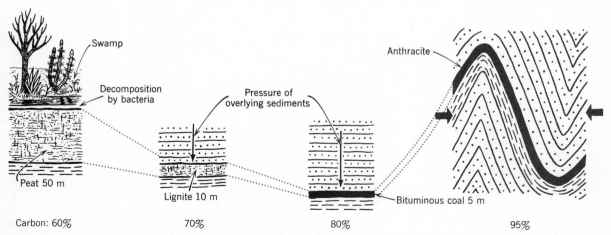

Figure 20.2 Accumulating plant matter is converted into coal by decomposition and pressure. A layer of peat decreases to one-tenth of its original thickness by the time it has become bituminous coal. At the same time the proportion of carbon increases continually.

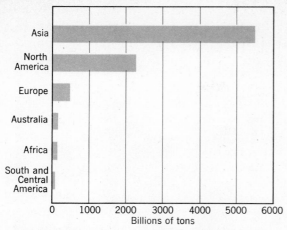

Figure 20.3 Geographic distribution of the world's recoverable coal reserves. (*Data from U.S. Geol. Survey, 1969.*)

Mining. We have noted that thin coal seams and deep lying strata cannot be mined; the average thickness of all seams now mined in the United States is about 1.6m. Increasing costs have stimulated the invention of increasingly efficient methods of extraction. In underground mining the automatic coal-cutting and loading machine, which eliminates drilling and blasting, has increased production per man-hour 10 to 20 times. Coal within 60m of the surface is being mined, without ever going underground, by power shovels that take 10 to 100 cubic meters at a single bite (Fig. 20.5) and by huge augers, 2m or more in diameter, that in a single operation bore horizontally into a seam and move the coal outward between their spiral blades to loading belts (Fig. 20.6).

Experiments are now being made with the burning of coal, while it is still in the ground, to produce gas for industrial use. This technique would eliminate mining altogether.

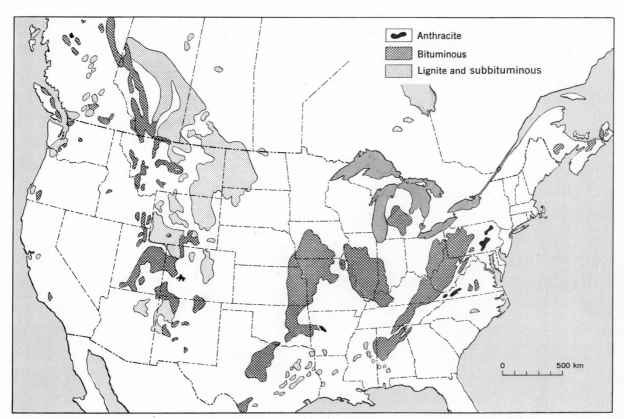

Figure 20.4 Coal fields of United States and southern Canada. (*Adapted from maps by U.S. Geol. Survey and Geol. Survey of Canada.*)

Upper shovel removes overburden

Conveyor belt carries overburden to dump

Lower shovel digs coal

(man shows scale)

Conveyor belt loads coal

Figure 20.5 This coal strip-mine at Wyodak, Wyoming produces, each year, 300,000 tons of subbituminous coal from a seam 20m to 30m thick. (*U.S. Bureau of Mines.*)

Figure 20.6 A huge multiple auger in eastern Kentucky bores into a coal seam from an artificially excavated bench. Coal is cut, moved out, and loaded in a single operation. (*Courier-Journal* and *Louisville Times*.)

Petroleum: Oil and Natural Gas

Oil was first produced in Rumania, in 1857. Soon afterward, in 1859, the first oil well in North America was drilled at Titusville, Pennsylvania. That event was epoch making. The well, 23m deep, produced 25 barrels[3] per day of a substance that spelled the doom of candles and whale-oil lamps. More than a hundred years later, in 1972, the United States was producing approximately ten million barrels of oil per day from about half a million wells, the deepest of which extended nearly 9km down into the crust. Of this huge production, 90 per cent is used as fuel; the remainder goes into lubricants, without which the fuel would turn very few wheels, and into thousands of products manufactured by the petrochemical industry.

[3] A barrel contains 42 gallons, and is the unit of volume generally used when oil is discussed.

Oil and gas are the two chief kinds of petroleum. We can define **petroleum** as *gaseous, liquid, or solid substances, occurring naturally and consisting chiefly of chemical compounds of carbon and hydrogen.* Oil and gas occur together and are searched for in the same way. We can therefore follow general practice and talk about oil pools, oil exploration, and the origin of oil with the understanding that we mean not only oil but gas as well. In discussing oil, we shall deal with its occurrences, its origin, the means used to find new occurrences, and oil production, in that order.

Occurrence: Oil Pools. The accumulated experience of a century of exploration, drilling, and producing has taught us much about where and how oil occurs. Oil possesses two important properties that affect its occurrence. It is fluid, and it is generally lighter than water. Oil is produced from pools (an **oil pool** is *an underground accumulation of oil or gas in a reservoir limited by geologic barriers*). The word gives a wrong impression because an oil pool is not a lake of oil. It is a body of rock in which oil occupies the pore spaces. *A group of pools, usually of similar type, or a single pool in an isolated position* constitutes an **oil field.** The pools in a field can be side by side or one on top of the other.

For oil or gas to accumulate in a pool, four essential requirements must be met. (1) There must be a reservoir rock to hold the oil, and this rock must be permeable so that the oil can percolate through it. (2) The reservoir rock must be overlain by a layer of impermeable roof rock, such as shale, to prevent upward escape of the oil, which is floating on ground water. (3) The reservoir rock and roof rock must form a trap that holds the oil and prevents it from moving any farther under the pressure of the water beneath it (Fig. 20.7). These requirements are much like those of an artesian-water system but with the essential difference that the artesian aquifer connects with the surface, whereas the oil pool does not. Although these three features—reservoir, roof, and trap—are essential, they do not guarantee a pool. In many places where they occur together, drilling has shown that no pool exists, generally because of lack of a source from which oil could enter the trap. So, to the three requirements for a pool, we must add another: (4) there must be source rock to provide oil.

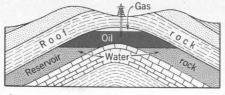

A

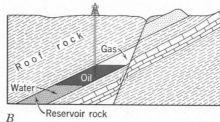

B

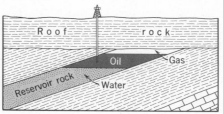

C

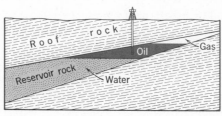

D

Figure 20.7 Four of the many kinds of oil traps. *A*, *B*, *C* are structural traps; *D* is a stratigraphic trap. Gas (white) overlies oil (black), which floats on ground water (blue), saturates reservoir rock, and is held down by roof of claystone. Oil fills only the pore spaces in rock.

Origin. Oil is believed to be a product of the decomposition of organic matter, both plant and animal. This belief arises principally because of two observations: (1) oil possesses optical properties known only in hydrocarbon compounds derived from organic matter, and (2) oil contains nitrogen and certain compounds (porphyrins) that scientists believe can only originate in living matter.

Oil is nearly always found in marine sedimentary rocks. Indeed, in places on the sea floor, particularly on the continental shelves and at the bases of the continental slopes, sampling has shown that fine-grained sediment now accumulating contains up to 8 per cent organic matter—chemically, good potential oil substance. Thus, on the basis of the principle of uniformity, we can make the general suggestion that the oil we now find in rocks originated as organic matter deposited with marine sediment.

However, the suggestion is only a general one. Analyzing the organic matter in modern sediments, chemists have found significant differences between its composition and that of petroleum. Therefore, in order to effect the change from organic matter to petroleum, a complex and little-understood series of changes must occur. The following simplified theory about how the steps occur is widely held and is supported by enough facts to be at least somewhere near the truth.

1. The raw material consists mainly of microscopic marine organisms, mostly plants, living in multitudes at and near the sea surface. Measurements show that the sea grows at least 31,500kg of organic matter per square kilometer per year, and the most productive inshore waters grow as much as six times this amount. The latter value represents more than could be harvested in a year from the richest farmland.

2. When the microscopic plants die, they fall to the bottom. Some of the dead plant matter is devoured by scavengers, but the rest becomes buried in the accumulating pile of sediment. There it is attacked and decomposed by bacteria, which split off and remove oxygen, nitrogen, and other elements, leaving residual carbon and hydrogen. Accumulating sediment that is rich in organic matter teems with bacteria.

3. Deep burial beneath further fine sediment destroys the bacteria and provides pressure, heat, and time for further chemical changes that convert the substance into droplets of liquid oil and minute bubbles of gas.

4. Gradual compaction of the enclosing sediments, under the pressure of their own increasing weight, reduces the space between the rock particles and squeezes out oil and gas substance into nearby layers of sand or sandstone, where open spaces are larger.

5. Aided by their buoyancy and perhaps by artesian water circulation, oil and gas migrate generally upward through the sand until they reach the surface and are lost, or until they are caught in a trap and form a pool.

We repeat that this statement is greatly oversimplified. A long and complex chemistry is involved in the conversion of the original organic constituents to crude petroleum. Also, chemical changes may occur in oil and gas even after they have migrated into their reservoirs. This may help explain why chemical differences exist between the oil in one pool and that in another.

The migration of oil needs more explanation. The sediment in which oil substance is accumulating today is rich in clay minerals, whereas most of the strata that constitute oil pools are sandstones consisting of quartz grains. It seems obvious, therefore, that oil forms in one kind of material and at some later time migrates to another. The migration process involves essentially the principles of movement of ground water. When, as we mentioned above, oil is squeezed out of the clay-rich sediment in which it originated and enters a body of sandstone somewhere above, it can migrate more easily than before. One reason is that sandstone, as explained in Chapter 9, is far more permeable than are clay-rich rocks. Another reason is that the force of molecular attraction between oil and quartz is less strong than that between water and quartz. Hence, because oil and water do not mix, water remains fastened to the quartz grains while oil occupies the central parts of the larger openings (like those shown in Fig. 9.2). Because it is lighter than water, the oil tends to glide upward past the quartz-held water. In this way it becomes segregated from the water; and when it encounters a trap it can form a pool.

Exploration. When U.S. commercial production began in western Pennsylvania in 1860, the first pools were discovered at places where oil and gas advertised their presence by seeping out at the surface. It soon became apparent that at all the producing wells the rocks were sedimentary and that at most wells the structure consisted of anticlines (Fig. 20.7). In 1861 the theory was advanced that oil had migrated upward into anticlines and lay trapped beneath their crests. In the 1880s this theory was tested and confirmed by drilling holes in certain anticlines where no oil had been found previously; the holes were successful wells.

Geologic Exploration. Proof of the anticline theory led to a period (1890 to 1925) in which geologists searched for and mapped anticlines from surface exposures. During this period traps other than anticlines were discovered (Fig. 20.7,*B–D*), and the search for pools widened rapidly.

The occurrence of buried traps, such as the one in Figure 20.7,*C*, showed that mapping of exposures would have to be supplemented by methods that would reveal structural and stratigraphic traps not apparent at the surface (Fig. 20.7,*D*). This was done first, and is still being done, by core drilling in a pattern beneath a suspected area. Study of the cores makes it possible to construct a graphic log from each core and to correlate the strata by physical character and contained fossils (Fig. 2.8) almost as well as though the rocks were exposed at the surface. The depths at which a recognizable bed is encountered by a series of drill holes give a strong clue to structure (Fig. 20.8).

Core drilling, however, is expensive. This fact stimulated the development of electric logging, which can be done by means of holes drilled by cheaper methods. Common salt and other dissolved substances are present in the ground water that saturates all deep-lying sedimentary rocks. As salt solution is a good conductor, the conductivity of a sedimentary layer at depth varies with water content. The conductivity at any desired depth is measured by devices lowered into the hole by a cable. The resulting electric logs are widely used in correlation of strata.

Geophysical Exploration. In the 1920s geophysical exploration began to be used widely. Twenty years later this kind of exploration was uncovering

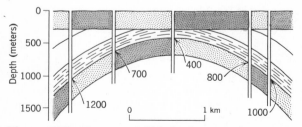

Figure 20.8 Five core-drill holes intersect the same stratum at different depths, showing that flat-lying layers at the surface conceal an anticline lying unconformably beneath it. Numbers show depth (in meters) at which the top of the sandstone stratum was encountered.

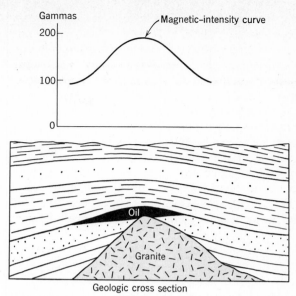

Figure 20.9 Curve constructed from a magnetometer survey showed an abrupt local increase in magnetic intensity. Later test drilling disclosed a buried hill of magnetic granite, with sedimentary rock, containing an oil pool, draped over its summit. (*Sharpe Instruments of Canada.*)

most of the new pools, simply because the majority of those discoverable by geologic methods alone had been found already. Geophysical exploration is designed to detect variations, within the rocks, of some physical property such as density, magnetism, or ability to transmit or reflect seismic waves. As most of these variations are related to structures, they help to indicate the presence of possible pools. If, for instance, a stratum of high density has been lifted up during the making of an anticline, the pull of gravity will be increased very slightly compared with that in the surrounding area. As the increase is not likely to be more than about one ten-millionth, the measuring instrument (a gravity meter or *gravimeter*) has to be extremely sensitive.

Again, distortions of Earth's magnetic field by rock bodies with varying magnetic properties are measured with *magnetometers*. Because igneous rocks, containing magnetite, are much more magnetic than sedimentary rocks, magnetometer surveys are most successful in areas in which igneous rocks project into overlying sedimentary rocks by intrusion, folding, or faulting, or as erosional hills beneath surfaces of unconformity (Fig. 20.9). The magnetometer can detect differences among sedimentary rocks also.

In the search for oil the most useful geophysical exploration method is seismic surveying. An artificial earthquake is created by exploding buried dynamite, and in the common method the waves reflected from the upper surfaces of strata having high wave-reflection potential are picked up at ground level by portable seismographs located a kilometer or so from the site of the explosion. Velocity of wave transmission being known, depths of the reflecting layers can be calculated from the travel time of the waves (Fig. 20.10). Special techniques have been developed for using seismic methods to locate pools beneath areas that are covered by water. Figure 20.11 illustrates one of the techniques.

Areal Pattern of Exploration. Exploration proceeds by elimination. Experience shows that, essentially, oil occurs only in sedimentary strata and that the thinner the pile of such strata the less likely they are to yield oil. Maps showing possibly productive areas are continually being made and revised by companies engaged in exploration. On the maps, areas of igneous and metamorphic rock are labeled "improbable"; areas of sedimentary strata that are very thin or that have unfavorable structures are likely to be put into the same class. With these, too,

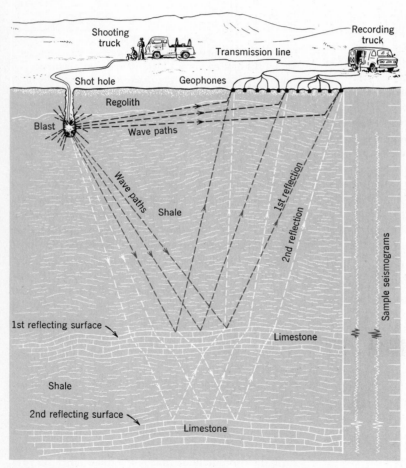

Figure 20.10   Determination of structure of rock layers by means of seismic exploration. Blast creates seismic waves that are reflected by interfaces between layers of limestone and shale. Analyses of reflections from interfaces reveal attitude of strata. (*After C. A. Heiland, Geophysical exploration. By permission of Prentice-Hall, Inc., Englewood Cliffs, N.J.*)

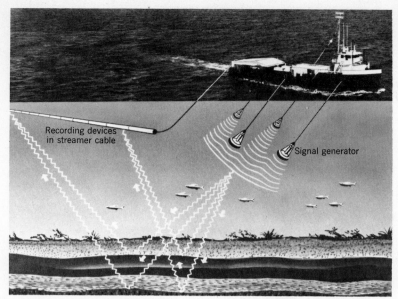

Figure 20.11 Seismic exploration at sea. Signal generators send out waves that travel through seawater but are reflected by layers of sediment beneath the sea floor. Reflected waves are recorded by detectors housed in the streamer cable, which may be as much as 2.5km in length. (*Courtesy of Continental Oil Company, and American Petroleum Institute.*)

go areas that, possessing favorable rocks and structure, have been drilled and yet have not yielded oil. Figure 20.12 is a map of this kind for North America. People who risk a million dollars or more in drilling a hole do not rely solely on such general data; they prepare large-scale maps of small areas before making decisions.

The map shows in outline the continental shelf, which adds almost one-third to the potentially productive area of the United States. Both geophysical exploration and drilling have shown that in at least some places the shelf is highly favorable territory. In the Gulf Coast area, considerable production has been developed on the shelf, and wells are being drilled as far as 150km offshore (Fig. 20.13), and beneath water more than 150m deep. Continental shelves have large possibilities for future production, as has been shown by the recent discoveries of large gas fields beneath the North Sea, near the coasts of The Netherlands and England.

Tar Sands and Oil Shales. Another source of petroleum consists of nonfluid hydrocarbons. They occur in the solid state and like the fluid hydrocarbons, fill the pores in sedimentary strata. By far

the largest occurrences in North America are the Athabaska tar sands in northern Alberta and the Green River oil shales in Colorado, Utah, and Wyoming. Together they have been estimated to represent potential petroleum reserves twice as great as the continent's presently known reserves of liquid petroleum.

Actually the "sands" in Alberta are sandstone containing asphalt ("tar") that occupies the pore spaces, having migrated into them long ago. The asphalt is too cold to flow. It must be mined and then heated to separate asphalt from quartz grains. The asphalt must then be refined to produce gasoline and other useful products. In Alberta, a large mining and heating plant is now digging out 100,000 tons of sand a day, from which it produces about 50,000 barrels of petroleum. The Athabaska tar sands cover an area of 75,000 square kilometers, and in places reach a thickness of 60m. Other, much smaller bodies of tar sand occur in parts of western United States.

The Green River oil shales contain as even larger volume of hydrocarbon derivatives than do the tar sands. They consist of strata deposited in very ancient lakes. The hydrocarbon material is unusual in

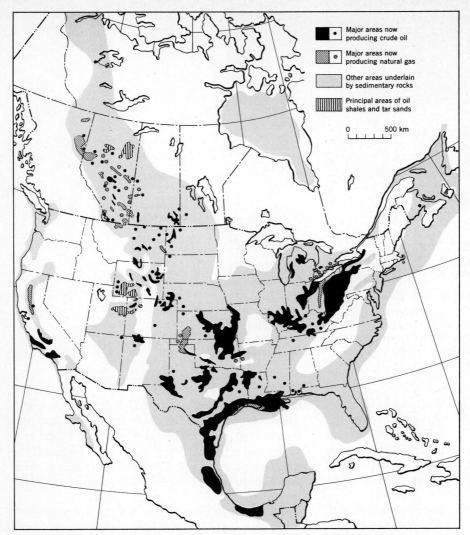

Figure 20.12 North America (exclusive of northern Canada and Alaska), showing areas of petroleum production, areas of tar sands and oil shales, and areas of sedimentary rock (possible future production). (*Data from American Petroleum Inst., Geol. Survey of Canada, and other sources.*)

being of freshwater origin and also in occurring not in a reservoir but in the clay-rich sediments in which it was formed originally. When oil shale is finely ground and heated under pressure, much of the organic matter is driven off as petroleum. Methods of mining and treating oil shales have been devised and are presently being used in Russia, Estonia, and China—one-third of the Chinese petroleum production in 1972 came from oil shale. In the United States, pilot plants have been operated successfully, but commercial production has never been attempted. Probably oil shale will become an important source of petroleum for the United States.

How Much Fossil Fuel?

Because of their "one-crop" availability, are supplies of fossil fuel adequate to meet future demands? If we use a barrel of oil as our unit of measurement, we can compare quantities of all fossil fuels directly.

Figure 20.13 Drilling rig on a platform in water 30m deep, 50km offshore near Galveston, Texas, 1965. Crew lives on platform, which has its own heliport (upper left). (*Shell Oil Company.*)

Thus approximately 0.22 tons of coal produce the same amount of heat energy as one barrel of oil; so the 8415×10^9 tons of coal stated earlier to be the world's recoverable coal reserves are equivalent to about 38,000 billion barrels of oil.

While the total amount of recoverable coal is known fairly accurately, much uncertainty surrounds estimates of the total amounts of oil and gas. The amounts and locations of oil in known pools can be readily computed: lumping oil and gas together by taking 6000 cubic feet of gas as equal to one barrel of oil, we arrive at about 600 billion barrels of oil for the world's known reserves. But this includes only oil pools that have been discovered. Unlike coal, for which the volume of strata in a basin of sediment can be accurately estimated, the volume of oil still to be discovered in a sedimentary basin can only be guessed at. Using the accumulated experience of a century of oil production, many experts now predict that the total oil and gas production of

the world, including all of that still to be found at sea, will be about 2500 billion barrels. Exactly where all this oil and gas will be found remains a vital question for geologists to answer. But the distribution of oil in known pools, and the places where most oil is produced today, are shown in Figure 20.14.

The use of energy is increasing so rapidly that demand in the world doubles approximately every ten years. By 1975, the amount of energy used each year will equal that derived from the burning of 45 billion barrels of oil. Approximately half of this energy will actually consist of oil and gas. It does not take much arithmetic to find that with a predicted consumption rate of 24 billion barrels in 1975, the world's oil and gas reserves cannot see mankind through another century of high-energy industry. Fortunately the 38,000 billion barrel equivalent in coal reserves can be called on and can be used, for many purposes, in much the same way as oil and gas. So too, can reserves of oil shale and tar sands. The total energy available in the world's tar sands is estimated at what could be obtained from 600 billion barrels of oil, while the energy in oil shales is estimated to be even greater—about 1000 billion barrels.

From the numbers presented, it seems likely that in the years ahead, the world will slowly have to change its patterns of use of fossil fuels. The importance of oil and gas will decline, presently unconventional sources such as oil shales will be exploited, and the use of coal will increase rapidly. At the same time, probably other sources of energy will be developed. We must hope that other sources of energy will be developed quickly, because widespread mining of coal, oil shale, and tar sand will create massive problems for the environment in regions where the mining must be carried out.

Other Sources of Energy

Two sources of energy other than fossil fuels have already been developed to some extent: hydroelectric energy and nuclear energy. Two others—tides, and Earth's internal heat—have been tested and developed on a limited basis. Neither source has yet been developed on a large scale in the United States, but the day may not be far in the future when both will become important.

Hydroelectric energy is recovered from the potential energy of water in streams and rivers as they

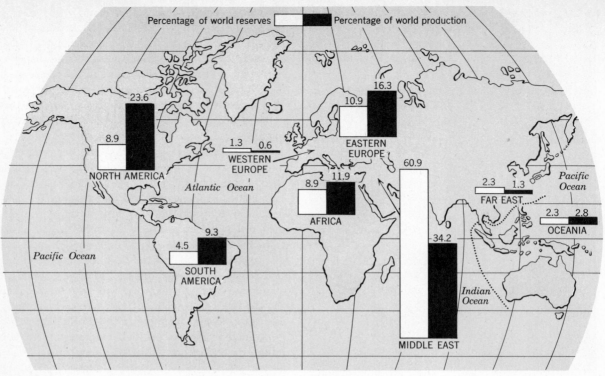

Figure 20.14 Distribution of the world's proved oil reserves (oil known to be in the ground) compared with amount of oil produced, by regions, in 1971. In that year North America possessed 8.9 per cent of the world's known oil but contributed 23.6 per cent of the world's production. (*American Petroleum Institute.*)

flow downward to the sea. Such energy already accounts for 4 per cent of the energy produced in the United States. But in the future, the total percentage of energy from hydroelectric sources will probably not rise much above this figure, because the number of streams that can be dammed for the installation of hydroelectric plants is limited.

Nuclear energy, unlike hydroelectric energy, is not limited in its potential. The use of nuclear energy in industry has already begun, and in time will play an important part in the world's economy. Nuclear energy really refers to the heat energy produced during radioactive decay. The same radioactive atoms that keep the Earth hot—uranium-238, uranium-235 and thorium-232 (Chap. 3)—can be mined and used to produce energy for mankind. Unfortunately, only the process that uses uranium-235, the least abundant of the three kinds of atoms, has been perfected so far. Supplies of uranium-235 are limited. But many experts predict that *breeder-reactors*,

capable of using the more abundant atoms of uranium-238 and thorium-232, will be developed in the years ahead. When this happens, problems of supply will virtually disappear, because both uranium-238 and thorium-232 are plentiful.

Tidal energy resembles hydroelectric energy in that it is recovered from flowing water. People have used tides to drive water wheels for many centuries; much of the flour used in Boston in the 17th and 18th centuries, was ground in a tidal mill. But today's interest in tides lies in their use as a source of water to drive electrical generators. Water retained behind a dam at high tide, and allowed to flow out again at low tide, can drive a generator in the same way that river water can. One important difference between hydroelectric power and tidal power is that rivers flow continually, while tides can be impounded only twice a day, so that electricity is produced spasmodically.

Along most of the world's coasts the differences

in height between high and low tide are too small—about 2m—to drive generators. In some restricted bays, however, tidal ranges exceed 15m, and here it is possible to recover tidal energy. In France, at the mouth of the River Rance, a tidal-power plant has been operating successfully for several years. Another plant has been built in the U.S.S.R. near Murmansk on the White Sea. But no tidal-power plants have been built in the United States. When they are, the most likely sites are in Passamaquoddy Bay in Maine, and in some of the bays and inlets of Alaska.

Geothermal energy, as Earth's heat energy is called, has been used for more than fifty years in Italy and Iceland and more recently in other parts of the world, including the United States. The steam produced in hot-spring areas (Chap. 9) can be used to power generators in the same way as steam produced in coal- and oil-fired boilers. To capture steam in hot-spring areas it is necessary to drill into the hot, underground reservoirs that feed the springs, bring the steam to the surface in pipes and feed it into a power plant (Fig. 20.15). The places where sources of heat are close enough to the surface to produce steam, and therefore where geothermal energy can be developed, are mostly in areas of current or recent volcanic activity. In the United States these areas include California, Nevada, Montana, Wyoming, New Mexico, Utah, and Alaska. In other countries, in addition to Italy and Iceland, large resources of geothermal energy exist in Japan, U.S.S.R., New Zealand, several countries in Central America, Ethiopia, and Kenya.

Hydroelectric, nuclear, tidal and geothermal energy will all contribute to our future energy needs, but the source that seems most likely to develop rapidly, and to have the brightest future, is nuclear energy. By its use, the decline predicted earlier for supplies of oil and gas, might be offset. But the development of nuclear energy will not solve all our energy problems. Nuclear energy is most conveniently used to produce electricity; it cannot, for example, be substituted for gasoline in automobiles. But nuclear energy is much less limited as to future supplies than is energy from fossil fuels. Probably, therefore, in the years ahead, uranium will slowly replace fossil fuels as a source for generating electricity. As a result, some experts predict that future patterns of energy production in the United States will change in the manner shown in Figure 20.16.

Sources of Material

Having discussed the principal sources of *energy*, we turn next to the mineral substances that provide *materials* from which things can be made. The number and diversity of such substances is so great it is almost impossible to make a simple classification covering all of them; almost every mineral substance known can be used for something. But certain sub-

Figure 20.15 "The Geysers," Sonoma County, California, where steam associated with hot springs is brought to the surface through deep wells and used to power a plant (extreme left) that produces electricity. White plumes of excess steam mark the site of geothermal wells. (*U.S. Geol. Survey.*)

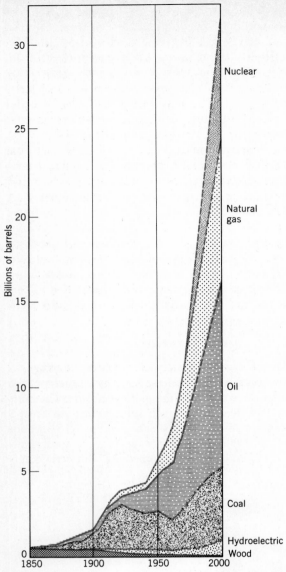

Figure 20.16 Changing pattern of energy sources in the United States.

In 1850, firewood supplied most needs, but as energy demands increased, coal, oil, and natural gas became increasingly important. In the years ahead the importance of nuclear energy is expected to grow significantly. Portions of curves that are dashed are projections. Sources of energy converted to equivalent numbers of barrels of oil, 1 barrel being taken as equal to 1.5 billion calories.

stances are so vital for industry that we can classify them on the basis of their most distinctive uses. This has been done in Table 20.1. The table is subdivided into two broad groups: (1) substances used for their metallic properties, such as iron, copper, and gold, and (2) nonmetallic substances.

The form in which a mineral substance is used may not be the form in which it is mined. Iron is used as a metal, but is mined as one of the oxide minerals, hematite or magnetite. Therefore mineral substances for industry are sought in deposits of certain preferred minerals from which the desired substances can be easily recovered. The more concentrated the preferred mineral is, the more valuable the mineral deposit. In some deposits, the desired minerals are so highly concentrated that even very rare substances, such as gold, or platinum, can be seen by the naked eye. But for all desired mineral substances there is a level of concentration below which the deposit cannot be economically worked. To distinguish between profitable and unprofitable mineral deposits, we use the word **ore,** meaning *an aggregate of minerals from which one or more materials can be extracted profitably.* It is not always possible to say exactly how much of a given mineral must be present in order to constitute an ore. For example, two deposits containing the same iron minerals may contain 60 per cent iron, but one will not be ore because the costs of mining and processing are so high that the price of the final product—metallic iron—is not competitive with iron from other deposits. Furthermore, as both costs and market prices fluctuate, a particular aggregate of minerals may be an ore at one time but not at another.

Along with ore minerals, from which the desired substances are extracted, are other minerals collectively termed **gangue** (pronounced *gang*). These are *the nonvaluable minerals of an ore.* Familiar minerals

Table 20.1

Principal mineral substances, grouped according to use.

(1) *Metals*
Iron, aluminum, copper, lead, zinc, nickel, silver, gold, tin, mercury, silver, chromium, maganese, molybdenum, uranium, and many others

(2) *Nonmetallic Substances*
(a) Used for chemicals: salt, sodium carbonate, sulfur, borax, fluorite.
(b) Used for fertilizers: phosphate, potash, sulfur, calcium carbonate, sodium nitrate
(c) Used for building: gypsum, calcite, clay, asbestos, sand, gravel, shale (for cement)
(d) Used for ceramics and abrasives: clay, feldspar, quartz, diamond, garnet, corundum

that commonly occur as gangue are quartz, feldspar, mica, calcite, and dolomite.

The ore problem has always been twofold: first, to find the ores (which altogether underlie an infinitesimally small proportion of Earth's land area), and second, to mine the ore and get rid of the gangue as cheaply as possible. Both getting rid of gangue and mining are technical problems; engineers have been so successful in solving many of them that some deposits now considered ore are only one-fifth as rich as were the lowest-grade ores 50 years ago.

Origin of Mineral Deposits

Table 20.1 subdivides mineral substances according to use, and as we have noted, the substances are recovered from desirable minerals that have been processed after being mined from deposits of restricted size. But Earth processes that form mineral deposits cannot be conveniently subdivided into metallic and nonmetallic categories. Concentrating processes operate without regard to our use of the minerals. Therefore we discuss the origin of ores without regard to their use. There are five principal ways by which minerals become sufficiently concentrated to form ores. They are:

1. by concentration within magma
2. by deposition from hot, watery solutions
3. by deposition from lake-water or seawater
4. by concentration due to weathering and ground water
5. by mechanical concentration in flowing water.

Concentration within Magma. When some kinds of ore containing sulfide minerals such as chalcopyrite and pyrrhotite are melted in a smelter, the molten sulfides melt, and being heavy sink to the bottom, whereas the molten silicates, being light, rise to the top as a sort of scum. Thus the ore melts into two liquids that do not mix, just as oil and water, instead of mixing, separate into two layers. This simple form of concentration occurs in nature when, for reasons not clearly understood, certain magmas separate into two liquids, of which one, rich in copper, nickel, or iron, sinks to the floor of the magma chamber. The world's greatest concentration of nickel ore, at Sudbury, Ontario, is believed to have formed in this fashion. Similarly the huge deposit of magnetite at Kiruna in northern Sweden,

site of one of Europe's most important iron mines, formed in this manner.

We noted in Chapter 5 that one form of magmatic differentiation occurs when dense minerals sink through a magma and accumulate on the floor of the chamber. Sometimes the dense minerals make desirable ores. Most of the world's chromium ores were formed in this manner by the accumulation of the mineral chromite. The largest chromite deposits are found in South Africa, Rhodesia, and the U.S.S.R. Similarly, vast deposits of ilmenite, a source of titanium, were formed by magmatic differentiation. Large deposits occur in the Adirondack Mountains.

Concentration by Deposition from Hot, Watery Solutions. Many of the most famous and valuable ores were formed when their essential minerals were deposited from hot, watery solutions. But despite their importance, the origins and compositions of the solutions remain problematical. Deposition occurs deep underground where we cannot see it happen, and when the deposit is finally uncovered by erosion, the solution is no longer present. Nevertheless, enough clues can be found so that the process of deposition is understood at least in a general way.

Composition of the Solutions. The principal ingredient of ore-forming solutions is water, although small amounts of carbon dioxide, methane, and sometimes other fluids are also present. This information comes from analyses made of tiny bubbles of ore fluid trapped inside crystals of quartz and other minerals of the ore (Fig. 20.17). The water is never pure and always contains, dissolved within it, salts such as sodium chloride, potassium chloride, calcium sulfate, and calcium chloride. The amount varies, but most solutions range from about the saltiness of seawater (3.5 per cent dissolved solids by weight) to about ten times as salty as seawater. The ore-forming solution is therefore a brine, and brines, unlike pure water, are capable of dissolving minute amounts of seemingly insoluble sulfide minerals such as chalcopyrite, galena, and sphalerite.

Origins of the Solutions. Traces of soluble salts exist in all rock. Water percolating through a sufficiently large volume of rock, therefore, can eventually become a brine. The percolating brine can also dissolve some of the widely dispersed copper, lead, zinc, and other minerals that are contained in the rock through which it passes. Some ore-forming

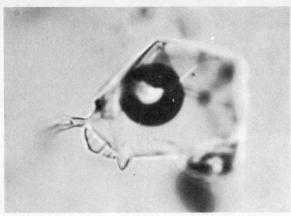

Figure 20.17 Seen through a microscope, tiny inclusion of ore-forming solution trapped inside a quartz crystal from Mayflower Mine, Utah. Dark, circular object is a gas bubble. Diameter of inclusion is four thousandths of a millimeter. (*J. T. Nash, U.S. Geol. Survey.*)

solutions seem to originate in this manner. Others form when seawater, trapped in the pores of buried sediments, is squeezed out by compaction. Still others form when granitic magma, and other magmas generated by wet partial-melting (Chap. 16), give off watery solutions as they cool. Ore-forming solutions having very similar compositions can apparently form in a number of ways. What seems to be important is not *where* the solution came from, but *what* made it precipitate the valuable minerals it carried.

Causes of Precipitation. If an ore-forming solution moves slowly upward, like ground water through a confined aquifer, changes happen very slowly. If valuable minerals are precipitated, they are spread out over great distances and will not be sufficiently concentrated to form ore. But if the solutions find an open fracture, a body of shattered rock, or any other place where rapid flow can occur, changes may happen rapidly. Among the changes that happen in such flowing solutions are sudden decrease of temperature and pressure, changes in composition caused by reaction with adjacent rocks, and dilution by mixing with ground water. Each change can cause dissolved minerals to be precipitated, and if precipitation is confined to a small space, a rich ore deposit can be formed.

Examples. As noted earlier, deposits formed by precipitation from hot, watery solutions are common. In most areas of metamorphic rock, for example, quartz veins occur (Fig. 20.18). Probably these were deposited by hot, salty solutions that slowly escaped upward as they responded to the pressure and heat involved in metamorphism.

In regions of volcanic activity, particularly where rhyolite and andesite occur, many veins containing valuable minerals are found. This is probably because volcanism heats solutions very near the surface, making them effective ore-formers. Formation of hot solutions by volcanism seems to occur even when volcanism takes place beneath the sea. The gold deposits at Kirkland Lake in Ontario, Kal-

Figure 20.18 Quartz vein cutting across fine-grained metamorphic rock. Such veins are formed when dissolved minerals are deposited from hot, watery solutions. Ruler is scaled in inches. West Cummington, Massachusetts. (*B. M. Shaub.*)

Figure 20.19 Circulating hot, watery solutions escaping from cooling magma, alter and bleach sedimentary rock. At the same time the composition of the solution changes, and ore minerals are precipitated. Sedimentary rock (black) in the Gaspé Copper Mine, Quebec, is limestone; ore minerals (microscopic grains within whitish altered areas) are chalcopyrite and sphalerite. (*Photo, J. Allcock.*)

goorlie in Australia, and Cripple Creek in Colorado, were formed in this fashion.

When a body of granitic magma cools near the surface, it is a source of heat just as a volcano is, and so can become a source of ore-forming solutions (Fig. 20.19). Many great ore bodies are associated with shallow intrusive rocks. The copper deposits at Butte, Montana, Bingham, Utah, and Bisbee, Arizona are examples.

Some deposits form solutions that apparently start as rainwater. The waters circulate through deep aquifers and become transformed to hot, salty, ore-forming solutions, then deposit ores as circulation brings them back close to the surface again. Examples are the zinc deposits near Pitcher, Oklahoma, the lead deposits in southeastern Missouri, and the uranium deposits of the Colorado Plateau.

Deposition from Lake Water and Seawater. Sodium chloride and other nonmetallic compounds precipitated in saline lakes, such as Great Salt Lake, are derived through tributary streams from the products of chemical weathering of the rocks within the lake's drainage area. This process of weathering—stream transport—precipitation in a water body, a prime example of natural concentration, has

been responsible for a wide variety of mineral deposits. Many salts of sodium (including common salt), potassium, and boron are recovered from saline lakes and playas (Chap. 10; Fig. 20.20). Enormous quantities of common salt, and of the calcium sulfates, gypsum and anhydrite, used in making plaster, are recovered from sedimentary layers precipitated in shallow, cut-off arms of ancient seas. Much of the production of salt in the United States comes from such layers in rocks of Silurian age, more than 400 million years old.

These and other nonmetallic substances have been concentrated by evaporation and are therefore evaporites. Metals and the important fertilizer-mineral apatite have also been concentrated from solution in the sea, but their abundance in average seawater is so slight that thick beds of iron ore, for example, could not have been concentrated solely by evaporation of ordinary seawater; some other process must have been responsible.

A good example is the Clinton iron ore, occurring as lenslike bodies in several sedimentary units, one of them locally more than 10m thick, extending from New York State southward for 1100km into Alabama. The steel industry of Birmingham, Alabama is based on the convenient proximity of this ore, coal, and limestone. Fossils in the ore strata show that the strata were deposited in a shallow sea during the Silurian Period. Ripple marks and mud cracks further indicate shallow water. Apparently these beds were parts of the thick pile of sediment that accumulated in the Appalachian geosyncline. The iron mineral is hematite, and the proportion of iron is 35 to 40 per cent.

It is believed that the iron was dissolved from iron-bearing minerals in mafic igneous rocks, carried by streams to the shallow sea (perhaps as a bicarbonate), and there precipitated as oxides. The sediment is a chemical sediment and the Clinton ore concentrated in this way amounts to billions of tons. The enormous iron-ore strata in the province of Lorraine in eastern France originated in a similar way. The Lake Superior iron ores (Fig. 20.21), a mainstay of the North American steel industry, and similar ores in Labrador, Venezuela, U.S.S.R., India, and Australia originated in a similar way. But ores of Lake Superior type differ from ores of Clinton- and Lorraine types in two important respects. First, whereas the Clinton ores contain fossils and other detritus, the Lake Superior ores are sediments of

Figure 20.20 Salts deposited by evaporation of ancient lake waters are mined near Boron, California. Boron salts (principally the mineral borax) were deposited about 15 million years ago. Evaporation of Rodgers Lake (white playa in background) is presently depositing sodium chloride. View looking to southwest; pit is about 100m deep and 1.5km across. (*Photo from U.S. Borax.*)

Figure 20.21 Susquehanna Location, a deep open-pit iron mine in Hibbing, Minnesota where Lake Superior-type sedimentary iron ore is mined. Strata overlying the ore are visible in the left-hand wall of the pit. Structure on upper right-hand edge of photograph is a head frame covering the entryway to an underground mine. (*Sawders-Cushing.*)

wholly chemical origin and are free of all detritus. Second, whereas the Clinton ores contain calcite as a gangue, the Lake Superior ores contain very fine bands of cherty quartz (Fig. 20.22). Because all the Lake Superior ores formed more than 2 billion years ago, we infer that the composition and properties of seawater may have been different from those of today.

The world's biggest deposits of manganese, in the U.S.S.R., originated by precipitation in the sea, as did the world's largest deposits of the phosphate mineral apatite. A very important chemical precipitate that is forming in the sea today consists of the nodular growths of manganese oxides found on the floors of the world's oceans (Fig. 12.12).

Concentration Caused by Weathering and Groundwater. Mineral deposits are concentrated in two general ways: secondary enrichment and residual concentration.

A. *Secondary Enrichment*. The related processes of sorting and concentration by ground water are chemical. They depend on the principle that change in the environment of a mineral may make it vulnerable to chemical attack. Many minerals found in ore bodies, although stable at the high temperatures at which they crystallized from solution, become unstable in the zone of aeration at Earth's surface. In areas where erosion has worn the surface down, even formerly deep-lying minerals are brought within the shallow zone in which atmospheric oxygen and

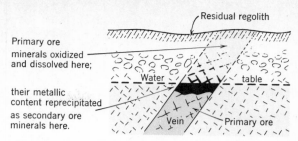

Figure 20.23 Descending ground water impoverishes ore above water table by removing soluble ore minerals; produces secondary enrichment below.

ground water can attack them. Thus, in the new environment of low temperature and low pressure, these minerals decay. Ground water dissolves pyrite, FeS_2, creating sulfuric acid, H_2SO_4, and residual iron oxides. The acid dissolves copper minerals, forming copper sulfate. Entering the zone of saturation, the solution loses oxygen and reacts with minerals there, depositing copper as sulfides. In a vein, the added sulfides enrich the ore already present or convert low-grade deposits into ore (Fig. 20.23). This process of addition is known as ***secondary enrichment,*** defined as *natural enrichment of an ore body by addition of mineral matter, generally from percolating solutions.*

Secondary enrichment has been important at the Utah Copper Mine (Fig. 21.2) at Bingham Canyon, Utah, the largest copper producer in the United States. The ore lies in the upper part of a body of intensely fractured intrusive igneous rock. The minerals containing the copper, gold, silver, and molybdenum, now found in the deposit, were carried upward by hot solutions from the magma underneath, and were deposited in the innumerable tiny fractures of the igneous rock. The upper part of the primary ore has been secondarily enriched through a zone more than 30m thick (Fig. 20.24). This enormous low-grade deposit is ore only because it does not demand underground mining and can be worked by low-cost surface methods. The sides of the canyon have been cut into giant steps 15m to 25m high, making a huge pit resembling a Roman amphitheater. The ore is blasted out and is loaded by power shovels into railroad cars. Waste is dumped into side canyons. The ore already mined exceeds 1.25 billion tons, and some ore containing as little as 0.4 per cent copper is being worked profitably.

Figure 20.22 Alternating layers of iron oxide (dark) and cherty quartz (light) formed by chemical precipitation in seawater, are found in all Lake Superior-type iron ores. Such ores are widely spread around the world. This sample is from the Hamersley Range, Western Australia, site of the world's richest iron ores. (*B. J. Skinner.*)

413

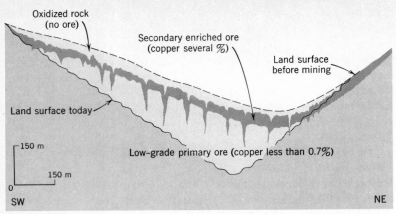

Oxidized rock
(no ore)

Secondary enriched ore
(copper several %)

Land surface
before mining

Land surface today

Low-grade primary ore (copper less than 0.7%)

150 m

150 m

0

SW

NE

Figure 20.24 Diagrammatic section across Utah Copper Mine (Fig. 21.2) at Bingham Canyon, Utah shows how weathering concentrated copper in an enriched zone related to a land surface.

B. *Residual Concentration.* In secondary enrichment, ground water adds new material to an existing body. Some ores, however, are developed by subtraction of old material. This is ***residual concentration,*** defined as *the natural concentration of a mineral substance by removal of a different substance with which it was associated.* An example is bauxite, a mixture of various hydrous aluminum oxide minerals and the common ore of aluminum (Fig. 20.25).

Bauxite is a product of chemical weathering. Aluminum was present as a constituent of original, primary silicate minerals in syenite, schist, or clay.

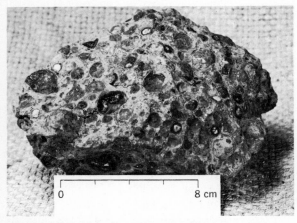

0   8 cm

Figure 20.25 Bauxite, the ore of aluminum, concentrated by deep, long-continued weathering of syenite. The rounded bodies are not pebbles but concretions, developed during weathering. Bauxite, Arkansas.

During weathering, silica was carried away in solution, gradually increasing the concentration of aluminum. We know this has happened because we can see bauxite grading downward into the underlying rock just as does any weathered regolith. From the locations of bauxite ores we know that the weathering has taken place beneath peneplains and other surfaces of low relief, with the water table close to the surface, with slow ground-water circulation, and always in tropical or subtropical climates. But the chemistry of the process is complex, and we still have much to learn about just what happens. Apparently ground water sometimes has just the right acidity to decompose the original silicate minerals and to carry silica away in solution, but not to remove the aluminum (and iron) oxides. Even though the rock originally contained no more than 1 per cent aluminum, the resulting bauxite may reach a concentration of 40 per cent aluminum.

Although some difference of opinion about the matter exists, it is likely that the richer iron ores of the districts in which ores of Lake Superior type occur, originate in a similar way. The primary ores contain about 30 per cent iron, but near the surface pockets as rich as 62 per cent iron are found. Apparently long-continued weathering has dissolved and removed the silica, leaving a residual concentration of almost pure hematite.

Mechanical Concentration in Flowing Water. The famous gold rush to California in 1849 resulted from the discovery that the sand and gravel in the bed of a small stream contained bits of gold. Indeed, many mining districts have been discovered by following trails of gold and other minerals upstream to their sources in bedrock. Because pure gold is heavy, and has a density of 19gm per cc., it is deposited by flowing water very quickly, while quartz, with a density of 2.65gm per cc., is washed away. Most silicate minerals are light by comparison with gold, and as a result gold becomes mechanically sorted out and concentrated in places where the velocity of water flow is least (Fig. 20.26). *A deposit of heavy minerals concentrated mechanically is a* **placer.** Besides gold, other heavy, durable metallic

minerals form placers. These include minerals that occur as pure metals, such as platinum and copper, as well as tinstone (cassiterite), and the nonmetallic minerals diamond, ruby, sapphire, and other gemstones.

Every phase of the conversion of gold in a fissure vein into placer gold has been traced. Chemical weathering of the exposed vein releases the gold, which then moves slowly downslope by mass-wasting (Fig. 20.27). In some places mass-wasting alone has concentrated gold or tinstone sufficiently to justify mining these metals.

More commonly, however, the mineral particles

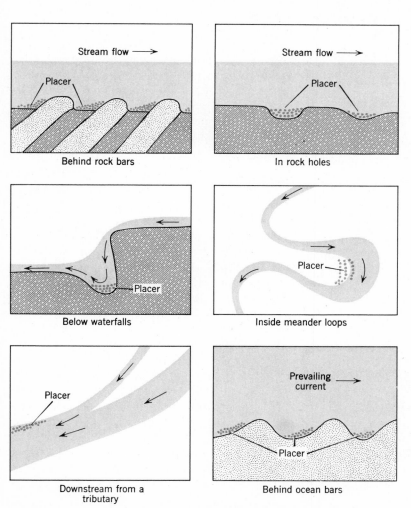

Figure 20.26 Typical placers occur where barriers allow faster-moving stream water to carry away suspended load of light-weight particles while trapping denser particles in bed load. Placers can form wherever water moves, but are most commonly associated with streams.

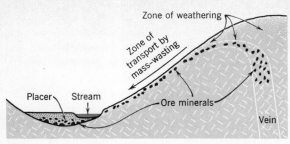

Figure 20.27 Ore minerals from weathered fissure veins creep down slopes and are redeposited by streams as placers.

get into a stream, which concentrates them more effectively than mass-wasting can do. Most placer gold occurs in grains the size of silt particles, the "gold dust" of placer miners. Some of it is coarser; pebble-sized fragments are *nuggets* (Fig. 20.28), of which the largest ever recorded weighed 70.9kg and at the current price of gold ($2.58 a gram) is worth $182,400. In following placers upstream, prospectors have learned that rounding and flattening (by pounding) increase downstream, just as does rounding of ordinary pebbles; when they find angular nuggets they know they are close to the bedrock source. Mining was done first by hand, simply by washing stream sediment around in a small pan of water. Later it was done by jetting water under high pressure against gravel banks and washing the sediment through troughs that caught the heavy gold behind cleats. Nowadays the more efficient method of dredging is used. The platinum placers of the Ural Mountains in Russia, the rich diamond placers of

the Kasai district in Zaire (formerly the Congo), and the hundreds of tin placers in Malaya and Indonesia are examples of other mechanical concentrations by streams.

Gold, diamonds, and several other minerals have been concentrated in beaches by surf.

Diamonds are being obtained in large quantities from gravelly beach placers, both above and below present sea level, along a 350-kilometer strip of the coast of South-West Africa. Weathered from deposits in the interior, the diamonds were transported by the Orange River to the coast and were spread southward by longshore drift. Later some beaches were uplifted, and others were submerged by rise of sea level.

Finally, gold is recovered in Mexico and in Australia from placers concentrated by wind. Quartz and other lightweight minerals have been deflated, leaving the heavier gold particles behind. Therefore, because mechanical concentration results from a variety of natural processes, the principles of erosion and deposition must be thoroughly understood by those who would discover new placers.

Future Supplies of Minerals

Mineral deposits are exploited in the least expensive way possible. For some deposits this means underground mines in which miners drill and blast away at narrow veins (Fig. 20.29); for others it means huge open pits from which the ore is removed by the truck- or trainload (Figs. 20.21, 21.2); for still other deposits, such as salt bodies, exploitation may mean

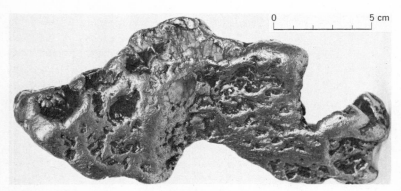

Figure 20.28 Gold nugget, found near Greenville, California, weighs 2550 grams. Its surface shows clearly the pounding it received during stream transport. (*George Switzer; collection U.S. National Museum.*)

Figure 20.29 Drilling copper ore in a New Mexico mine. (*American Smelting and Refining Co.*)

solution in hot water pumped down a drill hole and recovery of salt by evaporation from the resulting brine. All phases and kinds of mining have become increasingly efficient, so that once a deposit is discovered, it can quickly be worked. But how can we discover enough deposits? Minerals for industry, like energy to drive the industry, are being used at ever-increasing rates, and mineral deposits are only a one-crop resource.

At present, new mineral deposits are being found as fast as old ones are being mined out. But these finds are mostly in places such as South Africa, Australia, and Siberia, where intense prospecting has occurred only rather recently. In Europe, which has been intensively prospected for many years, new deposits are rarely found; indeed, in the areas conquered and occupied by the Romans, no new deposits of metals used by them have been discovered since they departed more than 1500 years ago. It has been suggested, therefore, that once all parts of the Earth's surface have been prospected intensively, mankind will have to look to kinds of deposits not presently exploited. This view is probably correct for metals such as mercury, silver, gold, and molybdenum. Because the amounts of such elements in the Earth's crust are very low to start with, the frequency with which ore deposits form is also low. For these scarcer materials, careful conservation of supplies will some day have to be practiced, substitutes may have to be found for some of their applications, and efficient recycling will become essential. For materials that are abundant in Earth's crust, such as iron, aluminum, common salt, and sulfur, supplies will probably always be sufficient to meet demand.

Probably the future of mineral supplies will be similar to the future of energy supplies. A sufficiency will be found, but present patterns of use will change; some materials will become more important, others less important, and yet other substances, not used today, will assume major importance. It is impossible to predict exactly how the changes will occur. But one prediction can safely be made: mining operations will get bigger, more materials will be used, and so pollution problems will become more pressing. Environments will come under even heavier pressure than that which endangers them today. In the next chapter we will see how man, and all the activities of his civilization, have become major factors in shaping the face of the Earth.

Summary

1. Ancient industry based largely on wood and muscle has given way to high-energy industry based mainly on fossil fuels and metals.

Fossil Fuels

2. Coal originated as plant matter in ancient swamps, and is both abundant and widely distributed.

3. Oil and gas probably originated as organic matter sedimented on sea floors and decomposed chemically. Later these fluids moved through reservoir rocks and were caught in geologic traps to form pools.

4. Oil pools are found by the use of both geologic and geophysical methods. Of the latter methods, seismic surveying is the one most commonly used.

Nuclear Energy

5. Nuclear energy is derived from atomic nuclei of unstable elements, chiefly uranium.

Mineral Deposits: Ores

6. The mineral industry is based mainly on local concentrations of useful minerals.

7. Most ores were formed by deposition of minerals from hot, watery solutions.

8. Many important iron ores were originally precipitated in ancient seas and probably later concentrated by removal of silica by circulating ground water.

9. Secondary enrichment by weathering and ground water has been important in further concentrating many copper deposits.

10. Gold, platinum, tinstone, diamonds, and copper have been mechanically concentrated to form placers.

Selected References

Averitt, Paul, 1969, Coal resources of the United States: U.S. Geol. Survey Bull. 1275, 20–44.

Bateman, A. M., 1951, The formation of mineral deposits: New York, John Wiley.

Bates, R. L., 1960, Geology of the industrial rocks and minerals: New York, Harper & Row.

Cloud, Preston (ed), 1969, Resources and man: San Francisco, W. H. Freeman.

Levorsen, A. I., 1967, Geology of petroleum, 2nd ed.: San Francisco, W. H. Freeman.

McDivitt, J. F., 1965, Minerals and man: Baltimore, Johns Hopkins Univ. Press.

Park, C. F., Jr., and MacDiarmid, R. A., 1970, Ore deposits, 2nd ed.: San Francisco, W. H. Freeman.

Skinner, B. J., 1969, Earth resources: Englewood Cliffs, N.J., Prentice-Hall.

Skinner, B. J., and Turekian, K. K., 1973, Man and the ocean: Englewood Cliffs, N.J., Prentice-Hall.

The Staff, 1965, Mineral facts and problems: U.S. Bureau of Mines Bull. 630.

Williamson, I. A., 1967, Coal mining geology: New York, Oxford Univ. Press.

Chapter 21

Man as a
Geologic Agent

Historical Background

Through much of this book we have described agents of geologic change at Earth's surface: streams, wind, ground water, waves along coasts, glaciers, and so forth. To those agents we must add man. Of course living things other than man create geologic changes too; witness the ability of ants and termites to move regolith, as we noted in Chapter 6. But we can logically single out man among all living species. Man is spread more widely over the face of the globe than any other large animal. He can rapidly bring about changes in composition of atmosphere, streams, lakes, and even the ocean. He can change the distribution of matter and the rates of many natural activities. By distorting the cyclic activity of Earth's external processes, man has become a potent geologic agent.

The distortions caused by man have not happened either suddenly or recently. They are the cumulative product of activities that have been in progress for a very long time. They are a product of gradual evolution just as the agent, man, is a product of gradual evolution. Over the years these man-created activities have increased in intensity, for two basic reasons. First, the number of people has increased spectacularly. Second, there has been an equally spectacular increase in the capability of people to use energy beyond the strength of their own muscles. A short review of the industrial history of people brings this out.

Before man appeared, the interactions between the biosphere and the inorganic spheres were in a condition of steady state. Through the time, probably more than three million years long, during which Stone-Age man was evolving, human influence on natural processes was almost nonexistent. Stone-Age people were comparatively few and their technical capabilities limited. Essentially they were just another species of animal, competing for survival. Through their greater intelligence, men gradually drew ahead of their competitors and became skilled and efficient hunters. Indeed they depended for their food supply on the killing of wild game. Although the matter is still uncertain, Stone-Age hunters may have been chiefly responsible for the extinction of many species of large mammals in Europe, North America, and elsewhere.

For this or other reasons, toward the close of the latest Glacial Age the traditional supply of animals dwindled, and in the southeastern Mediterranean region, people began to look around for supplementary food. They found it in plants, and before long they were growing and harvesting grain. Their domestic tools demonstrate that they soon learned how to clear land and till soil. Later, perhaps about 11,000 years ago, a different group of people in the Near East faced a similar problem, either because the climate became drier or because they themselves had over-hunted their territory. At any rate they began to domesticate goats and sheep so as to steady their meat supply. News of these two new methods of increasing the food supply got around, because before 10,000 years ago people east of the Mediterranean were combining both of them.

This seems to have been the path along which men turned from being roving hunters into being sedentary farmers and stock raisers, people who perforce had to build dwellings, from which they rarely strayed far. Besides resulting in a diet that included more pastry and fewer steaks, this comparatively sedentary way of life made use of the muscles of horses and cattle as well as of the muscles of people. With a given amount of human effort, domesticated animals could produce much more food. The production of food, likewise, was less dangerous than it had been in the old days of hunting.

The change from hunting to farming had two significant results. The first was extensive cutting-down of forests (chiefly in Eurasia at first) to create fields for growing crops. The way we learned when it started is interesting. Some of the pollen discharged by plants is carried by wind. It drops into small lakes and bogs and is preserved in their wet environments. Pollen grains are identifiable. They tell what vegetation was growing in the vicinity when the pollen was discharged. Examination of hundreds of core samples, from ancient lakes and bogs, their pollen layers dated by carbon-14, has shown that all over western Europe, beginning not long before 5000 years ago, forest trees were disappearing. They were being replaced by ragweed and other plants characteristic of deforested land. The plant cover of the world's lands was beginning to be changed by people. Historical and other data on an area of some 8 million square kilometers, embracing most or all of Central Europe, show that during the last millenium, forest land decreased from 80 per cent of the area in A.D. 900 to 24 per cent in A.D. 1900.

The second result was brought about by increase in the supply of artificially grown plant food. An acre of grain produces far more food energy than an acre of stock range, which is one of the reasons why a pound of beef is much more costly than a pound of bread. The growing of grain led to a big increase in human populations. Within a few thousand years the world's population grew from probably fewer than ten million people, mostly hunters, to nearly 100 million, mainly farmers. The increase has continued ever since, spurred on in large part by human ingenuity in devising ways to increase the production of food and other goods.

Among the many influences that helped increase populations still further was the Industrial Revolution, which began about A.D. 1760. That revolution consisted mainly of substituting the mass production of goods for hand fabrication by craftsmen. Thereby manufactured products became so abundant that more people could be clothed, housed, and fed, and the population curve climbed even more steeply. At the handicrafts stage, water power and simple machines had begun to be substituted for animal power. Starting from that base, mass production progressed into the age of steam. Fuel for steam-driven machines led to more cutting of forest trees—a new wave of cutting, for a new purpose.

Already in the 12th Century, hundreds of years before the age of steam, coal was being mined and burned. King John of England had made it a capital crime to burn coal because of the "noxious odors" that resulted. Later, nonetheless, the wood fuel used at first for raising steam gave way to coal, which possesses far greater heating power than wood does. Coal was more efficient; therefore it could support more people. But it had to be mined, and the mines developed undesirable side effects: they began to create significant changes in the lands.

The next great change brought about by the Industrial Revolution was the discovery of large quantities of oil and gas and the use of those fuels in newly invented engines powered by internal combustion. Those momentous events raised the curtain on an age of motors, which gradually replaced the already-existing age of steam. The motors, more efficient and more widely applicable than steam engines, made it possible for the populations of industrialized countries to increase still further. The use of power progressed so far that by mid-20th Century each human inhabitant of the United States had, working for him continuously, the equivalent of 14 horsepower.

After the age of motors had run for a good many decades, industrial power began to be derived from nuclear reactions, foreshadowing a future nuclear age. As the efficiency of production improved, populations grew to still greater size. One factor in this growth was geographic. Cheap fuels led to widespread systems of central heating and so to the expansion of population into high, cold latitudes. Population growth was further stimulated by inventions that improved medicines and medical care. The improvement was so great that the balance between births and deaths was sharply altered; many more individuals survived the hazards of childhood and so increased the sizes of populations. Now, more than 200 years after the Industrial Revolution began, the number of people that swarm over Earth's lands has increased from about 730 million to nearly 4000 million, an increase of more than 500 per cent.

As the curve in Figure 21.1 shows, Earth's human population has been increasing at a rate that is not linear but exponential. That is, the rate of increase is itself continually increasing. Stone-Age man, few in numbers and relying for his energy output only upon his own muscle, maintained a steady-state relationship with Earth's natural processes. He did no harm to his surroundings; in fact he was incapable of doing harm. In eating a meal he threw the waste on the ground, and in time the waste was transformed by decay into various substances that entered

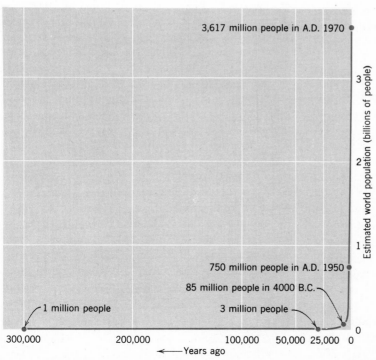

Figure 21.1 Growth of human population during the last 300,000 years.

the complex systems of biological activities. Stone-Age man was incapable of controlling, even to a small extent, the environment in which he lived. Indeed, much of the time he probably lived in fear of it.

Today a human population little short of 4 billion individuals is struggling, like its early ancestors, to survive. In the struggle it is competing within itself and also with other animals and with plants as well, other occupiers of the environments in which people live. Ever since the Stone Age a principal result of the struggle has been the destruction of other animals and of plants. The long list of animal species that have become extinct within the last 15,000 years or so testifies to the destruction. In the days of our Stone-Age ancestors, with their simple way of life, such alteration of the biosphere marked the limit of man's influence, which had little or no effect on atmosphere, water sphere, or lithosphere.

But with the beginning of farming over an ever-increasing area the natural vegetation began to be destroyed each year to permit the planting of crops. Destruction of the vegetation laid the regolith open to attack by running water and wind, causing increased erosion. Later, mining and construction of various kinds began to tear up protective vegetation or to cover the surface with concrete, asphalt, or buildings. The result is that erosion has increased still more.

Summary. Thus we see that across a span of time of perhaps 15,000 years, the human race has radically changed the part it plays in Earth's ongoing natural processes. Putting the cumulative change in capsule form, we can say it has involved four long-drawn-out events:

1. Extensive deforestation of Earth's surface.
2. An enormous increase in population.
3. An accelerating increase in technical capability.
4. (Only in industrialized countries) rapid urbanization whereby the majority of people live crowded into comparatively restricted areas, under generally artificial conditions. Such conditions necessitate the importation of food and water, and also demand the transport of people, over an average of about 10 miles per individual per day, between home and work.

Man as an Earth Mover

Now let us turn to specific ways in which human activities interfere with the steady state in natural processes, beginning with man's skill in moving earth—that is, regolith and rock (Fig. 21.2). The extent to which mankind has pushed up the old natural rates of erosion, merely by moving materials, seems hard to believe. Nonetheless the changes have been established by careful sampling. In a published report to a Maryland state commission, a geologist[1] remarked that the tonnage of sediment eroded from an acre of ground under construction in housing developments and highways can exceed 20,000 to 40,000 times the amount eroded naturally from an acre of woodland during the same length of time.

Rates of erosion of areas torn up by strip mining of coal probably are even greater. Although formerly most coal mining in the United States was underground, leaving the surface almost undisturbed, surface stripping, which began on a large scale in 1915, now accounts for more than half of U.S. coal production. Stripping is done with gigantic power shovels that cut horizontal shelves following the contours of hills. The shelves, as much as 50m wide, are floored with a jumble of broken rock waste, and are backed by walls of rock as much as 30m high (Figs. 21.3, 21.4). By 1972 the Appalachian region had become scarred with more than 20,000 linear miles of such cuts.[2] Thus far, little has been done to remedy the instability the cuts have created. Stripping destroys both vegetation and soil, lowers the water table, acidifies the ground water and pollutes streams, creates landslides, and promotes erosion, which leads to the deposition of sediment on valley floors.

The examples cited are extreme, but the broad effect of human activities on erosion rates is reflected clearly in thoughtful estimates of man-induced erosion on nationwide- and worldwide scales. In Table 21.1 we see the huge amounts of regolith and rock that are believed to be moved artificially each year. Farming accounts for the lion's share of the material moved, because the combined areas under cultivation far exceed the combined areas of all structures and all mines.

[1] M. G. Wolman (1964).
[2] Open-pit mining of this and other kinds has scarred one-half of one per cent of the area of the United States.

Figure 21.2 Long-continued removal of ore has made this hole, nearly 600m deep, in a mountain range in Utah, and is steadily enlarging it. The Bingham Canyon Copper Mine is one of Earth's biggest man-made excavations. (*Courtesy of Kennecott Copper Corporation.*)

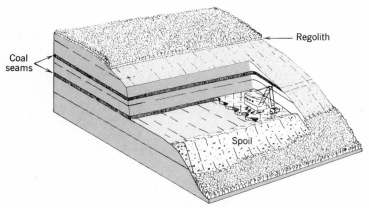

Figure 21.3 Shelf, wall, and spoil pile made by strip mining of coal. (*After a sketch published by R. D. Hill.*)

Figure 21.4 Strip mining of coal in West Virginia. The three shelves, one above the other like a flight of steps, reveal the positions of three seams of coal. Waste from the lowest and largest step is creeping and sliding down the hillsides, among the trees. (*U.S. Dept. of Agriculture—Soil Conservation Service.*)

Table 21.1

Regolith and Rock Moved Artificially, According to an Estimate Made in 1971. (S. Judson, Geol. Soc. America Abstrs., 1971, p. 615.) (Metric tons per year.)

Activity	USA	World
Mining and quarrying	5 billion (5×10^9)	30 billion (3×10^{10})
Construction (roads, buildings, dams, etc.)	10 billion (10^{10})	50 billion (5×10^{10})
Farming	1 trillion (10^{12})	6.5 trillion (6.5×10^{12})

Human Effects on Streams

The increased erosion caused by all these artificial activities should be reflected in increased loads of sediment carried by streams. And indeed it is. The rivers of the United States are estimated to be carrying to the ocean, now, 1 billion (10^9) metric tons of sediment and dissolved load annually. This is twice the amount they are believed to have been carrying before European man began to cultivate American soil.

This 100 per cent increase has occurred despite the fact that much has been done since about 1930 to reduce erosion on American farmland. At around that time a long succession of dry years coincided with the recent plowing of dry grassland to create farms. This combination disrupted the steady-state condition, mainly by wind erosion as described in Chapter 10. Many farms had to be abandoned, and distress was widespread. An educational campaign was mounted, and farmers not only in dry lands but throughout the nation were shown how to prevent and how to combat erosion by wind and by water. Undoubtedly the result has been a decrease in the sediment delivered from farmlands to streams, but undoubtedly also the decrease from that source has been offset, to an unknown extent, by the increasing pace of construction of roads, streets, and buildings, which likewise alters the steady state among natural processes.

Why is disruption of the steady state in this case a serious matter? The answer requires some explanation. Most of Earth's external processes involve whole chains of cause-and-effect activities. So we turn here from a cause (man's earth-moving capability) to one of its effects (on stream flow and stream transport of sediment), and under the latter heading continue the story, beginning with an answer to the question asked above.

Disruption of the steady state is serious here for this reason: In a climate sufficiently moist to permit the growth of forests, the leaves of trees shield the ground from the splashing, eroding impacts of raindrops. Beneath the surface, as we noted in Chapter 9, the regolith acts as a broad, thin sponge that absorbs water. This ground water percolates through the sponge at rates that are very slow compared with runoff over the surface, and little by little emerges in streams without having picked up any sediment at all. Fed with water at this slow rate, streams tend to fluctuate only moderately; they tend not to rise in high floods.

But when there are no trees and when regolith is covered only partly by vegetation, much of the fallen rain fails to enter the sponge below ground. Instead, it runs off along the surface, eroding regolith and picking up sediment as it runs (Fig. 21.5). Thus the water arrives at the nearest stream very quickly, swelling the stream and so creating a flood. The floodwater is turbid with sediment picked up on the hillslopes above. As the flood subsides, this load of mud is deposited somewhere downstream, often covering a meadowland with an unwanted layer of infertile silt.

So, an unbroken cover of vegetation on a watershed promotes the smooth operation of one part of the water cycle with minimum fluctuation, and is equally important for one part of the rock cycle because it promotes erosion and deposition of sediment at slow, steady rates.

It is interesting to watch the progress of a flood down a sizable river. A flood is merely a marked, temporary increase in the discharge of a stream. As we read in Chapter 8, with increased discharge the stream becomes deeper and wider, and also flows faster. The increased depth is achieved partly by erosion of the bed and partly by rise of the water surface. This rise is the most obvious aspect of a flood. A bulge, a sort of broad wave, travels down a river at a rate of a few miles per hour, and in some large rivers reaches 10m or more in height. This bulge is measured by gages fixed to the bank. They float, moving up or down in vertical slots that are graduated like a thermometer. Each gage marks the height of the flood as it passes by, so that from all the gages along a river the history of a flood can be reconstructed.

Figure 21.6 is not a picture of the bulge. It is a curve showing how discharge varies with time, increasing to a peak and then gradually tapering off. The white curve represents the flood when stream and watershed are in their natural states. The blue curve shows the flood after the watershed has been urbanized. For a given rainfall, the runoff from streets and roofs and through sewers is faster, and puts water into the stream more quickly; so the broken curve has a higher peak and steeper sides, and occurs farther to the left (that is, earlier in time) than the continuous curve. The difference between the two curves illustrates the special importance of

Figure 21.5 On this steep hillside in western Oregon, cutting of timber has resulted in rapid erosion, aggravated by the presence of caterpillar gouges and a temporary road. (*U.S. Dept. of Agriculture—Soil Conservation Service.*)

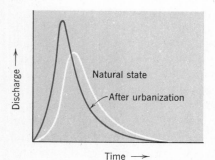

Figure 21.6 Graph showing how the variation of flood discharge of a stream, through time, is distorted by urbanization of the stream's watershed. (*After Leopold, U.S. Geol. Survey Circ. 554, 1968.*)

advance planning for land use (described in a later part of this chapter) in city areas that occupy stream valleys. In this connection Figure 8.7 is worth looking at again.

Floods: The Mississippi River System

What has just been said explains why new suburban communities sometimes experience flooding of streets and waterlogging of lawns and gardens despite the fact that in the years before construction began, flooding had been unknown. On a larger scale, it also affords one reason why overbank floods

Figure 21.7 Mississippi River in flood near Louisiana, Missouri. The river's channel is in the foreground, bounded by a levee crowned with trees. Beyond are embanked roads with rows of trees showing above flooded land. (*Corps of Engineers, U.S. Army.*)

The inset shows Mississippi River floodwater pouring through a break in a levee and over a railroad near Alexandria, Missouri in June 1947, adding to the water over the flood-plain. (*Wide World Photos.*)

along the Mississippi River and its principal tributary streams became both higher and more frequent during the 19th Century and the first half of the 20th. As we noted in Chapter 8, the Mississippi River watershed stretches from Pennsylvania to Montana and includes about 41 per cent of the area of the conterminous United States.

When spring rains add to the water from melting snow and frozen regolith, a large volume of water runs off into small streams, and so through larger streams into the channel of the Mississippi itself. The results—part of the steady-state condition among all the factors in the huge river system and not an event caused by man—are annual floods that occasionally go overbank.

This state of affairs was well known to the early French settlers in Louisiana, who saw the Mississippi spilling over its natural levees (Chap. 8), to convert the lower lands beyond into vast, shallow lakes (Fig. 21.7). To protect buildings and farmland the settlers, early in the 18th Century, started a program of heightening the natural levees by erecting artificial levees, dikes of earth with surface armor, on top of them. The program was continued and gradually expanded, so that the lower Mississippi is now confined within a system of artificial levees. Together these have a combined length of more than 2000 miles. Each levee is 300 feet wide from toe to toe and about 30 feet high.

The levee-heightening program was based on the

assumption that it was principally the lower river that must be considered if destructive floods were to be prevented. Gradually, however, it became evident that all the rest of the vast watershed must be considered also. The floods downstream began to be understood as partly the result of the hurrying of water across a field in Ohio or along a roadside ditch in Dakota, instead of letting it soak into the regolith and reaching the nearest stream by delayed action. The hurrying of little waters on the watershed made flood peaks both higher and narrower, as Figure 21.6 shows.

While the levees along the Mississippi were being built ever higher, these other things were happening over a large part of the watershed.

1. Deforestation.

2. Conversion of former woodland and grassland into cropland.

3. Draining of swamps and other wet areas on farms, the water flowing through tile drains instead of passing through the ground by percolation. (Between 1950 and 1970, 2 per cent of all the marshland in the United States was filled or drained.)

4. Overgrazing, thus impairing grassland sod and thereby partly destroying the sponge beneath the surface.

5. Construction of all kinds: buildings, paved streets, highways, and other structures, each with ditches and other conduits designed to lead rainwater off over the surface.

All these changes heightened and narrowed the peaks of floods and so made matters worse in downstream areas. Since the 1930s the "hurrying of runoff" across farmland has been reduced somewhat, by the use of measures such as (a) plowing *along* contours instead of up-and-down the slopes, (b) alternating strips of grass with strips of crops along the contours of slopes (Fig. 21.8), and (c) leaving a litter of straw, stalks, and other organic trash scattered over bare fields in order to impede surface water.

These measures are undoubted improvements. We do not know, however, to what extent their effects have been offset by those of construction, which is covering an ever-increasing proportion of the remaining natural surface with concrete, asphalt, and buildings.

To give a balanced picture of the Mississippi River flood situation, however, we must say that besides the system of levees there is now an elaborate set of channels that can be opened at critical places in the levees. These permit floodwater, at controlled points, to be diverted into lowlands where damage will be least. Essentially these points are analogous to a fuse in an electric circuit. Also, along some of

Figure 21.8 Crops, planted in horizontal strips, wind around the contours of hills in southern Wisconsin farmland. (*U.S. Dept. of Agriculture—Soil Conservation Service.*)

the Mississippi's tributaries, dams have been built. Some of the dams can be used to reduce flood heights by holding floodwater temporarily and letting it pass through gradually.

However, successful though they have been in confining floods, the levees have created an undesirable side effect. Because they extend down almost to the river's mouth, the stream they confine no longer spreads sideways when in flood, depositing its load of silt and adding new land to the delta. Instead, it deposits its load farther offshore in deeper water, and so creates land less rapidly and offsets erosion by waves less effectively. In this side effect is another example of the basic truth that when a complex system is tampered with, it reacts by moving toward a new condition of steady state among the altered factors.

Lakes

A lake is a trap that catches and holds the water brought to it by streams, and also the sediment and substances dissolved in the water. Because it is a trap, a lake is specially vulnerable to contamination. Within limits, a stream can clean itself by flushing, but a lake cannot. An example is Lake Baikal (Fig. 21.9) in southwestern Siberia, with twice the volume of Lake Superior and a depth of nearly two kilometers. Fed by 336 streams, surrounded by forested mountains, and populated by a unique assemblage of animals and plants, it holds great scientific interest.

Decades ago, large-scale lumbering began along some of the tributary streams. As one result, much sediment was discharged into the lake, along with sewage and great numbers of cut logs, many of which sank to the lake bottom. By 1960 industrial development had begun on the lake shore itself, and the lake began to receive effluent from pulp and paper plants in amounts greater than could be absorbed. In consequence the lake is becoming rapidly polluted.

Lake Erie, a smaller, much shallower, and less spectacular water body than Lake Baikal, records a longer history of pollution and is in a condition of greater deterioration. From 1910 to 1960 the human population of the Lake Erie basin increased from 3 million to 10 million, and industrialization proceeded rapidly. Detergents, as well as fertilizer

Figure 21.9 Lake Baikal, world's deepest lake and a source of commercial fishing, is endangered by industrial pollution of its water. (*M. Redkin, Sovfoto.*)

leached from agricultural areas poured, and are still pouring, into the lake. Their content of nitrogen and phosphorus has stimulated great growths of green algae in surface water where sunlight is greatest, thus screening sunlight from the bottom and inhibiting plant growth there. Further, dead algae drift down from above, decompose by oxidation, and so use up the oxygen in the bottom water. A result is suffocation of fish and other animals. The lake bottom is widely covered with slime, and in shallow places organic matter extends from the bottom to the surface. Fishing has been greatly reduced, as has recreational use of the lake. If these changes were to continue, the lake would become filled with organic matter and so be converted into a bog.

Others of the Great Lakes are in much earlier stages of the same process. Pollutants are pouring into them also, disturbing the natural, steady-state condition. An ambitious and costly plan to halt and reverse this deterioration, a pilot plan financed jointly by public agencies at federal, state, and municipal levels, began to be implemented in 1971. Within a pilot region it will divert all sewage and industrial effluent away from Lake Michigan, and instead discharge them into specially created lagoons for thorough biochemical treatment. The treated liquid will then be distributed through irrigation systems to fertilize soil that although presently unproductive, can be made productive with the fertilizer. The plan, if adopted widely, is believed to hold promise of cleaning up the Great Lakes.

However, such a cleanup will affect pollution of another kind—thermal pollution, which results from adding to rivers and lakes the warm waste water from nuclear power stations. This not only suppresses cold-water game fish but also promotes the growth of algae with the unpleasant consequences already mentioned.

An example of a different kind is the Caspian Sea, a lake without outlet. Eighty per cent of its water is fed to it through the Volga River. Historically the surface of the lake has fluctuated under the influence of changes in rainfall and in temperature. From 1847 to 1928 a warming and drying climate caused it to fall at an average rate of 6mm/year. Then human influence was felt. A program of building dams across the Volga was carried out, water was diverted from the river for irrigation, and from 1929 to 1965 the level of the Caspian fell 67mm/year. The area of the northern, shallow part of the lake has decreased greatly. In addition pollutants were contributed from the Volga, from industry around the lake shores, and oil drilling, with leaks and spills, out in the lake itself. Among other unwanted results was a drastic reduction in the fish population. This not only reduced the commercial catch but also, because there were fewer fish to eat the mosquito larvae, there were epidemics of malaria.

The recent history of the Caspian has led to consideration of a plan to reverse and divert into the Volga certain rivers, in northern Siberia, that presently flow into the Arctic Ocean. Such a plan would tamper still further with natural systems, and it would be difficult indeed to foresee all the consequences.

Salt-Water Intrusion

Contamination of Ground Water Near Coasts.

At this point we can add to the discussion, in Chapter 9, that deals with the withdrawal of ground water through wells. When withdrawal does not exceed recharge, the general condition of steady state is not affected significantly. But when industrial or agricultural withdrawal does exceed recharge, not only does the future supply become uncertain, but ground water near a coast can become contaminated with salt-water.

The ground water near a seacoast consists of salt water derived from the sea and fresh water derived from rainfall (Fig. 21.10,A). The fresh water, having lower density, floats on the salt water, and normal recharge from above gives it enough volume and weight to hold in place the salt water beneath it. The interface between the two water bodies is distinct, and normally its position changes little or not at all. Wells that penetrate the fresh but not the salt layer can safely withdraw water as long as withdrawal does not exceed recharge.

However, in some coastal regions concentrated industry or intensive agriculture that demands irrigation has led to the drilling of many wells, which as a group pump fresh water in excess of recharge. What happens then is shown in Figure 21.10,B. The water table is gradually lowered. As the layer of fresh water becomes thinner and exerts less pressure on the layer beneath, the interface between the layers rises. When the rising interface—the dividing surface—reaches the bottom of a well, that well begins to deliver salt water. A group of wells that are thus affected puts an end to irrigation or to industrial activities that demand fresh water.

This kind of intrusion of salt water into coastal wells has occurred commonly, especially in areas where population is dense and where sources of fresh water are limited. Areas of occurrence include Long Island and coastal New Jersey, southern Florida, coastal Texas, and Los Angeles, as well as areas on coasts of other continents. In the case of the municipal water supply for Miami, Florida, begun in 1896, salt-water intrusion twice forced abandonment of a field of wells and their replacement with new fields in different areas. It has been estimated that by the year 1995 the projected population of Miami's Dade County would require an amount of water that would have to be supplied by recharge over an area of 2800 square miles.

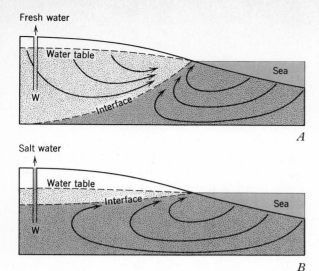

Figure 21.10 Salt-water intrusion beneath a coastal area.

A. Steady state in the ground-water system. Well (W) delivers fresh water.

B. Result of overpumping. Water table declines; salt water intrudes landward. Well delivers salt water.

Contamination of River Water. The fresh water of rivers becomes contaminated with salt from two principal sources, industrial and agricultural. Water is taken from a river or from the ground, used in various industrial processes that add salt to it, and the salty water is then poured into the river. In this way the salt content of some rivers has been increased by as much as 50 per cent.

In arid parts of some western states, fresh water for irrigating crop land is withdrawn from rivers that originate in high mountains. In dry land the regolith is apt to be salty, because salt is precipitated by evaporation of ground water derived from infrequent rains. Irrigation water sinks into the regolith, dissolves salt, and returns to the river by seepage. The added salt impairs the quality of the water for domestic and other uses. Since large-scale irrigation along the Colorado River began, the concentration of salt in the river water has almost doubled.

Artificially Induced Subsidence

Removal of Support. Human activities have long involved the moving of huge quantities of solids and fluids from one place to another, both on and below the ground. One of the results has been subsidence of the ground (Fig. 21.11). Sinking of the ground can occur either if support from below is withdrawn or if extra mass is added to it above. Examples of both are not hard to find.

Underground mining, especially of coal, comes to mind at once as an obvious way of creating big openings that remove support from the rock above.

Figure 21.11 This house slid into the hole created by the collapsing roof of a zinc mine in southeastern Oklahoma. (*U.S. Bureau of Mines.*)

In some mines, active or abandoned, roofs have simply fallen in. In others, collapse of the overlying rock occurs slowly, causing the ground to subside gradually. Old coal mines beneath the city of Scranton, Pennsylvania, represent the space left after removal of 200 million cubic yards of coal. Some of this space has been refilled with solid waste of various kinds, flushed down into it in places where heavy structures were to be placed above. Elsewhere the old tunnels and passages remain beneath the city, like the catacombs beneath the city of Rome.

Underlying western England, south of the city of Liverpool, the strata contain layers of salt, evaporated from ancient seawater. The thickest layers are as much as 180m thick and in places lie less than 60m below the surface. The salt has long been recovered by an extensive mining industry. Formerly it was worked in big underground mines, but nowadays the salt, dissolved in ground water to form a brine, is pumped out through wells and is recovered by evaporation. Many tens of millions of tons of salt have been recovered in this way. Removal of the salt by both mining and dissolution have been causing the ground to subside, locally at rates of more than 1m per year, destroying streets and buildings and

forming basins that fill with rainwater and create lakes, some of them deep.

Even where dissolution is not involved, the removal of fluids from below ground can cause subsidence of the ground itself. In an aquifer consisting of sediment or sedimentary rock, the weight of the rock material that overlies it is supported not only by the particles of sediment, but also partly by the water itself. In addition, the water helps keep the rock particles loosely packed. When more water is withdrawn from the aquifer than is recharged into it, part of the support that had been borne by water is removed. The weight of the overburden begins to pack the rock particles together more closely. Pore space in the aquifer is reduced permanently, and in some cases the ground above begins to subside.

In the central part of the San Joaquin Valley in California, long-continued pumping of water for irrigation so depleted an aquifer that within an area of 200km^2 the ground above it subsided, in places as much as 6m. Near Las Vegas, Nevada, which derives much of its water from pumped wells, the ground within an area 8km in diameter sank to form a conical depression (Fig. 21.12, *left*) that reflected the form of the cone of depression created in the

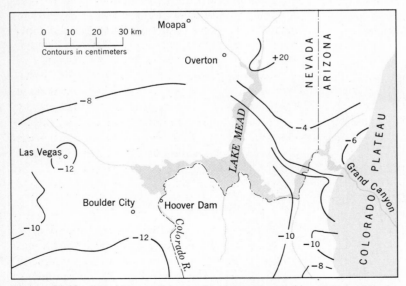

Figure 21.12 Subsidence of the surface beneath and around Lake Mead, Nevada-Arizona, during the period 1935–1950, shown by contours scaled in centimeters. Greatest subsidence was 17cm. (*Contours from U.S. Geol. Survey Circ. 346.*)
The small circular contour at Las Vegas, Nevada shows more localized subsidence caused by pumping of ground water from beneath that city.

water table underneath. Between 1935 and 1950 maximum subsidence was 36cm.

Withdrawal of ground water through more than 350 artesian wells resulted in sinking of the area of Mexico City, between 1891 and 1959, by as much as 7.5m. During 1953 the rate of subsidence reached 50cm/year. Subsequently the city reduced the trouble, partly by reducing the rate of withdrawal of water and partly by bringing in additional water from more distant sources.

Water is not the only fluid the removal of which can cause subsidence. Between 1937 and 1962 the oilfield area around Long Beach Harbor, south of Los Angeles, subsided to form a basin with a maximum depth of nearly 8m, as a direct result of pumping of oil, gas, and water from the underlying strata. Inasmuch as a part of the area of subsidence had lain only 2m to 3m above sea level before pumping began in 1936, and furthermore was intensively industrialized, this result was serious. Because large economic benefits were expected, $30 million were spent between 1958 and 1962 on an elaborate scheme in which nearly 600 million barrels of water were injected into the ground through more than 200 wells. As a result, by 1963 subsidence near the shoreline had been locally reversed, and the ground had risen by as much as 15 per cent of the earlier subsidence. It seems unlikely, however, that full recovery in any part of the subsided area will ever be attained.

Artificial Loading. In Figure 15.6 we saw the obvious effect of a vast, thick ice sheet that threw the crust beneath it out of isostatic equilibrium (Fig. 15.5; Chap. 17) and caused it to subside, becoming hundreds of meters lower. A much smaller load likewise causes subsidence, although on a smaller scale. A neat example is Lake Mead, the big reservoir held back of Hoover Dam, on the Colorado River some 25 miles southeast of Las Vegas, Nevada. Created in 1936, the lake is nearly 200km long, and at the dam is more than 150m deep. The weight of the enormous volume of new lake water (more than 40 billion tons) plus the additional weight of the newly deposited sediment caused the crust beneath it to sag down. The sagging was measured by a series of very precise measurements beginning in 1935 and repeated at intervals. The measurements soon showed that in subsiding, the crust was forming a roughly circular basin having a radius of more than 40km and with its center at the lake itself. Figure 21.12 shows the amount of subsidence that had occurred by the end of 15 years. If, tomorrow, the dam were opened and the lake drained, we could expect the crust to rise toward the form it had before 1936, much as it is still rising today, in the region formerly covered with ice, toward the form it possessed before the latest of the glacial ages.

Man-Made Earthquakes. Other artificially induced movements of Earth's crust have been caused by the forcible injection into it of liquid industrial wastes. On the outskirts of Denver, Colorado a hole was drilled to a depth of more than 3600m. Between 1962 and 1967 large amounts of liquid waste were injected into it intermittently. About 40 days after injection commenced, local shallow earthquakes began to be recorded. Through the entire 5-year period more than 1500 earthquakes occurred. During two time intervals when injection ceased, the number of recorded earthquakes decreased sharply. Apparently the injections triggered the earthquakes, probably by building up stresses that were relieved by movement along faults.

Tampering with Earth's crust in this way not only threatens earthquake hazards, but also, especially where noxious wastes are forced into the ground, threatens contamination of otherwise pure ground water. Where extensive artesian-water systems exist, like those mentioned in Chapter 9, contamination could occur at distances as great as hundreds of kilometers from the points of injection of waste. Deep injection of waste is no more than storage underground. It amounts, in essence, to getting the waste out of sight—in other words, sweeping it under the carpet. With few exceptions such waste remains with little or no change, a potential threat to the quality of ground-water supplies.

Man as a Producer of Waste

A tiny sample of city waste is pictured in Figure 1.1. The people of all the cities in the United States together produce, each year, enough solid refuse to form a layer 3m thick overlying 600 square kilometers of land. But around the New York metropolitan area (Fig. 21.13) there isn't enough land to receive its share of so great a covering of waste. Some of the refuse goes to sea. Each year nearly 5 million metric tons of it are dumped from barges into the

Figure 21.13 One day's normal production of trash by the people of New York City. (*New York Times photo.*)

sea off New York. Cities have always created waste. Archeologists have found evidence that through thousands of years, ancient cities gradually rose on their own debris, layer after layer, at rates averaging about 30cm per century. In later times many cities grew by spreading waste in low places, thus filling up swamps and lakes, narrowing river channels, and building seashores outward. Large parts of some cities consist of "made land" of this kind.

As city waste grows more bulky from year to year, there has ceased to be room for it in the cities themselves. It is deposited farther and farther away from urban centers and from smaller towns as well. Furthermore the waste includes increasing proportions of synthetic substances, such as certain plastics, that are either very resistant to chemical weathering or impervious to destruction by bacterial decay; the latter have no natural enemies and are not biodegradable. Quantities of plastic packing material, an industrial waste, have been found floating in the Sargasso Sea, a part of the Atlantic Ocean (Fig. 12.3) that lies between Florida and northwestern Africa. The plastic is a long way from its probable points

of origin, but it is neither eaten by organisms nor broken down in inorganic ways.

The bulk and character of city wastes are the joint result of increasing population, industrialization, and urbanization, and they pose serious problems. If burned, waste pollutes the air. If discharged into streams, lakes, or the ocean, it can pollute the water and also smother bottom-living life with blankets of slime. If buried, it can pollute ground water. In fact, waste in great quantity is a threat to public health and to the well-being of vast numbers of other organisms.

We say "in great quantity" because waste both solid and liquid plays a part in each of the natural cycles. All natural sediment is waste. As long as the steady state is not disturbed, as long as the input of waste into a cycle is moderate, the cycle absorbs it. But when the input of a substance is too great for the system to absorb, the steady state is disturbed and the substance becomes a pollutant. Indeed we define a ***pollutant*** as *any waste substance introduced into a natural system in greater amount than can be disposed of by the system.* Table 21.2 lists some of

Table 21.2
Some of the Wastes That Can Become Pollutants

Principal Sources	State S: solid L: liquid G: gas
Agriculture	
Residues from harvesting of crops	S
Residues from logging and pruning	S
Mineral fertilizers leached from fields	L
Animal wastes from feedlots and slaughter-houses	S,L
Salt dissolved by irrigation water and carried into streams	L
Mining and mineral processing	
Mines, blast furnaces, smelters, petro-chemical plants	L,S
Manufacturing and transportation	
Scrap metal, wood, paper, etc.	S
Cinders and ash; gases	S,G
Emissions from motors to atmosphere	G
Spills of oil, gasoline, etc., at wells, refineries, storage tanks, pipelines, ships, and filling stations	L
Wastes from burning of coal and petroleum	G,S
Salt placed on icy highways	L
Domestic	
Trash	S
Sewage	L
Waste from incinerators	G,S

the principal sources of wastes that become pollutants of air and water. Table 21.3 gives estimates of rates of production of solid wastes in a highly industrialized country. From it we learn that within the United States, every year, nearly 4 billion tons (nearly 20 tons per person) of rock, metal, paper, wood, and animal wastes are being poured onto land, into water, and into the air.

Table 21.3
Estimated Amounts of Solid Wastes Created Annually in the United States

Source of Waste	Estimated Amount (tons)
Agriculture	2,000,000,000
Mining and processing of minerals	1,700,000,000
Domestic	250,000,000
Junked cars	12,000,000
Total	3,962,000,000

Planning for Land Use

The example described in the foregoing section is one in which a particular use of land resulted in unforeseen side effects. It illustrates the need for careful planning based on the systematic collection of information about a land area before the area is put to a special use. The practice of setting aside tracts of land for specific uses is far from new. Small parks in the central parts of cities have long existed; on a wider scale the designation of land for large parks, forests, wilderness areas, and wildlife preserves and refuges, which cannot legally be put to other than specified uses, has been going on for scores and even hundreds of years. Still other tracts have been set aside for the creation of reservoirs and roadways.

Most of the assignments of land to such uses were made in past times when people were fewer and land was still comparatively abundant; hence those acts were generally haphazard and not parts of an overall plan. Such a plan might have recognized that valuable minerals might lie beneath an area thus assigned, or that a city with a rapid-transit system might some day need to encroach upon its boundaries. Now, however, open land has become scarce and new uses are coming into conflict with earlier uses. It has become clear that if improper or wasteful uses are to be avoided, decisions must be made only after careful long-term planning. Such planning is only just beginning to get under way.

Cities and Land Use. Although authorities recognize that planning for land use is needed wherever there are people, the need for planning is especially urgent in cities. Nearly 75 per cent of the population of the United States lives in cities. Even in countries that are only beginning to industrialize, a trend toward concentration of people in cities is evident. Wherever cities exist, most of them are growing. Because they contain people, they also contain structures: *buildings* for dwelling, for working, for learning, for worship, for recreation, and *other structures* such as highways, railroads, airports, cables and wires, conduits for fresh water and waste water, and so on almost endlessly. Besides these structures, cities also contain great numbers of vehicles. The fixed structures and the mobile people and vehicles are comparatively very close together, confined within restricted areas. Such crowding makes cities especially vulnerable to changes in their environment, by intensifying the interactions between people and environment. This happens no matter whether the changes are sudden and great or slow and gradual.

Sudden great changes include such events as earthquakes, landslides, and streamfloods. Falling walls, blocked streets, and washed-out foundations take a heavy toll of lives and property. The same natural events, having the same intensities but occurring in an open, little-inhabited country, would do but little damage to people and to artificial structures.

The slow, gradual changes in the environment of human populations result mainly from two groups of human activities. The first is depletion of substances by over-use (cutting of trees, lowering of water table). The second activity is the accumulation of wastes—solids, liquids, gases, and heat energy. The wastes accumulate on the surface, underground, in the atmosphere, and in streams, lakes, and the ocean, and cause many secondary effects, some of which are merely disagreeable while others are toxic.

These undesirable changes in environments are forcing public agencies in cities, states and provinces, and national governments to set up plans for action. Such planning, aimed mainly at controlling the uses to which various categories of land are put, has as its goal the return to a steady-state condition among the environmental factors. But this goal is a long way off; before an effective plan can be devised, a host of relevant facts concerning the environment of an area must be known. To obtain the facts, detailed surveys and analyses must be made by competent scientists, engineers, and other specialists. Geologic data constitute an important part of the needed facts; many geologists are currently employed in finding the facts on which planning for land use, including the wise use of natural resources, must be based. Many more will have to be enlisted before the environmental problems of human populations can be solved.

Postscript

We have touched on only a few of the ways in which man has interfered with natural processes, but we have chosen examples obviously pertinent to physi-cal geology. Others, at least equally important, especially concern the biosphere. They include such matters as pollution of the atmosphere, pollution of the ocean, and the artificial synthesis of substances that can be decomposed with difficulty or not at all by chemical weathering.

The examples we have given demonstrate that man *is* a geologic agent, and a considerable one. As human beings we are together responsible for the fact that today various natural systems are operating at rates different from those which prevailed before our early ancestors began to bend natural processes. We (that is, all mankind) have been taking substances we want from the lithosphere, biosphere, hydrosphere, and atmosphere. All such substances participate in one or more of Earth's external processes. We have changed, combined, or fabricated many such substances, physically or chemically. Then we have put residues or used-up fabrications back into one cycle or another, mostly by throwing them away, to become pollutants. We have left the cycles to cope with it as best they could. Many of the things we take involve destruction. Many of those we throw away cause pollution. In both cases the steady state is usually disrupted, often with very disagreeable consequences.

Both the destruction and the pollution were begun unwittingly. They have continued as deeply rooted habits, exercised thoughtlessly or carelessly. For long, each new generation of people accepted the Earth as they found it, without knowledge of steady states and assuming that things were normal. With very few exceptions they did not realize what they were doing. Then, just after the middle of the 20th Century, particularly in industrialized and urbanized countries, realization appeared and spread. It began to be evident that each of the natural cycles must be analyzed, not only qualitatively but also quantitatively, and in precise terms as well. The rapid growth of science has made such analysis possible, as a basis for change in existing practices.

Through our study of physical geology we fully realize that destructive human influence on Earth's ongoing cycles has been confined entirely to the very latest moment of geologic time—to an almost incredibly small part of Earth's history. Through almost all of the more than 3 billion years that have elapsed since the earliest fossils we know of today were living organisms, Earth's flow of energy and cycling of materials have been operating in mutual harmony. Significant human influence has existed during barely the last 15,000 years—only one two-hundred-thousandth part of that long time. If you should chisel about 1.5 inches of rock off the top of Mt. Everest, you would be lowering the altitude of that peak by a proportional amount, one two-hundred-thousandth.

Man's unique skill in technology derives from his great intelligence compared with that of other organisms. That same intelligence is also the basis of man's ability to perceive disturbances in Earth's cycles, measure them scientifically, and plan their elimination. In other words, the basic human factors that led to human tampering with Earth's environments are just as capable of intervening to restore steady states that have been thrown out of balance. Man is moving towards a future that is unknown, hoping for continuance of his species through long ages of time. It is this hope that may lead him to use his basic capacity to become, not an unsuccessful master, but a true son, of the Earth from which he has arisen.

Summary

1. Man is a geologic agent because he moves rock and regolith and changes the rates of many natural processes. By thus changing the steady state he distorts the processes.

2. When waste is introduced into a natural system in greater amount than can be disposed of by the system, it becomes a pollutant.

3. Distortion of natural processes is the result chiefly of increased population and technical capability, rapid urbanization, and extensive deforestation.

4. Man moves more Earth materials than are moved by any other single external process. Farming, construction, and mining are the chief activities by which the materials are moved.

5. Deforestation, swamp drainage, careless farming, and most kinds of construction tend to send runoff along the surface instead of through the ground. Among the results are increased erosion, increased flooding by streams, and lowered water tables.

6. Wastes that contain nitrogen and phosphorus, poured to excess into a lake (a natural trap), act to suppress oxidation and to inhibit many kinds of animal life in the lake.

7. When water, petroleum, and ores are removed from beneath the surface, or when a large artificial lake is created, the ground tends to subside.

8. Industrial wastes, salt from streets and highways, and irrigation water increase the salt content of ground water and water in reservoirs.

9. Excessive withdrawal of ground water in a coastal area can cause intrusion of salt water from the sea, and in any area can cause the ground to subside.

10. Injection of liquid wastes into the ground can cause pollution of ground water and can even cause earthquakes.

11. Disposal of solid waste by cities and towns involves pollution dangers of several kinds.

References

Committee on Geological Sciences (National Research Council), 1972, The Earth and human affairs: Canfield Press (Harper & Row).

Commoner, Barry, 1967, Science and survival: New York, Viking Press.

Detwyler, T. R. (ed.), 1971, Man's impact on environment: New York, McGraw-Hill.

Flawn, Peter, 1970, Environmental geology: New York, Harper & Row.

Goldman, M. I., 1972, The spoils of progress. Environmental pollution in the Soviet Union: Cambridge, Mass., M. I. T. Press.

Leopold, L. B., 1968, Hydrology for urban land planning—a guidebook on the hydrologic effects of urban land use: U.S. Geol. Survey Circ. 554.

McKenzie, G. D., and Utgard, R. O., 1972, Man and his physical environment. Readings in environmental geology: Minneapolis, Burgess.

Piper, A. M., 1969, Disposal of liquid wastes by injection underground—neither myth nor millenium: U.S. Geol. Survey Circ. 631.

Russell, W. M. S., 1969, Man, nature, and history: New York, Natural History Press.

Schneider, W. J., 1970, Hydrologic implications of solid-waste disposal: U.S. Geol. Survey Circ. 601-F.

Thomas, H. E., 1954, First fourteen years of Lake Mead: U.S. Geol. Survey Circ. 346.

Thomas, William, and others, 1956, Man's role in changing the face of the Earth: University of Chicago Press.

U.S. Department of the Interior, 1967, Surface mining and our environment: Washington, Government Printing Office.

U.S. Department of the Interior, 1968, Lake Erie report: Federal Water Pollution Control Administration, Great Lakes Region.

Appendices

Appendix A

Organization of Matter

Every object in the Universe is composed of matter. But can we say precisely what *matter* is? Ancient Greek philosophers first attempted the question and came close to a correct answer. They saw that matter occurs in three different states—solid, liquid, and gaseous—and that different forms of matter combine and react with each other, giving rise to new forms. Seeking some unifying principle, they examined properties such as color, hardness, taste, and odor of common materials, including rocks and minerals of many kinds. In a sense, some of those Greek philosophers were the first geologists, and the names they gave to some forms of matter are with us to this day. The modern word *copper* comes directly from the Greek word *cyprus,* and words such as *sapphire* and *magnetite* are little changed from the Greek originals.

Although the ancient philosophers learned much about matter, they were unable to resolve the question of what matter is. That question remained unanswered until the present era because no one was able to decide between two equally plausible possibilities, which can be stated in terms of questions that they themselves occasioned: can matter be endlessly subdivided into tiny pieces, all of which retain the properties of the whole; or is matter built up from submicroscopic particles of only a few kinds, so that the different properties of matter reflect different arrangements of the particles? As we learned in Chapter 4, although there is some truth in both lines of thinking, the second is essentially the correct one. Matter can be subdivided into atoms, and all atoms of a given chemical element have identical chemical properties. But atoms themselves are composed of still smaller subatomic particles, and these are the fundamental particles the Greek philosophers were seeking.

Structure of Atoms. Atoms are held together by electrical forces; we can therefore say that matter is basically electrical in character. Physicists have found that all atoms have a structure possessing a central portion called the *nucleus* which has a *positive electrical charge.* Spinning around the nucleus in defined paths, or orbits, are *negatively charged particles* called **electrons.** The positive charge of the nucleus exactly balances the negative charge of the orbiting electrons. Of all atoms, the simplest is hydrogen, its nucleus having a single positive charge. Only one orbiting electron, therefore, is required to provide electrical balance. The nucleus of the hydrogen atom gets its positive charge from a single particle called a proton. A **proton** is *a positively charged particle with a mass 1832 times greater than the mass of an electron.*

The second simplest atom is helium. Its nucleus has two positive charges, indicating that it contains two protons and therefore requiring two orbiting electrons to maintain electrical neutrality (Fig. A.1).

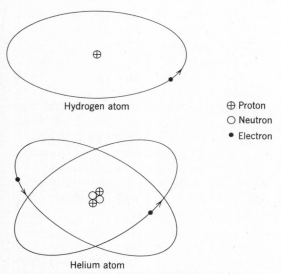

Hydrogen atom

⊕ Proton
○ Neutron
• Electron

Helium atom

Figure A.1 Hydrogen and helium atoms, shown schematically. Hydrogen is the simplest atom with one proton (positive) electrically balanced with one electron (negative). Helium has 2 electrons in orbit around a nucleus with 2 protons and 2 neutrons. The *atomic number* of helium is 2 (the number of protons); its *mass number* is 4 (the sum of protons and neutrons).

Table A.1
The Fundamental Particles of Matter

Name	Electric Charge	Relative Mass
Proton	+1	1832
Neutron	0	1833
Electron	−1	1

Sensitive measurements show, however, that the mass of a helium nucleus is equal to four protons. The increase in mass results from the presence of yet another type of particle in the nucleus. The newcomer is *an electrically neutral particle with a mass 1833 times greater than that of the electron,* called a **neutron.**

Protons, neutrons and electrons are the three principal sub-atomic particles that combine to form all atoms, and they are always the same, no matter in what atoms they occur. The way they fit together in a given atom determines the chemical properties of the atom. Protons, neutrons and electrons are therefore the fundamental particles of matter and, as the Greeks speculated, their different arrangements determine the different properties of matter.

Because protons and neutrons are so much heavier than electrons (Table A.1), 99.9 per cent of the mass of an atom resides in the nucleus. We are already aware that atoms are tiny particles: the radius of an atom is about 10^{-8} centimeters. The radius of the atomic nucleus, however, is very much smaller—about 10^{-12} centimeters. Therefore, as Figure A.1 reveals, an atom is mostly open space. It has a tiny but heavy nucleus, surrounded by a diffuse cloud of electrons that move in distant orbits.

We saw in Chapter 4 that elements are built systematically, starting with hydrogen, which has one proton and one electron, then helium, with two protons and two electrons, and so on. The number of protons in the nucleus of an atom was defined in Chapter 4 as the atomic number. All the known elements, with their atomic numbers and their abundance in Earth's crust, are listed in Table A.2.

Isotopes and Radioactivity. Although the number of neutrons that fit into the nucleus of an atom does not follow a simple pattern, nevertheless it is an important property of the atom because neutrons and protons together determine its mass. *The sum of the protons and neutrons in the nucleus of an atom* is called the **mass number.** Both atomic number and

Table A.2
Alphabetical List of the Elements (After K. K. Turekian, 1969)

Element	Symbol	Atomic Number	Crustal Abundance, Weight Per Cent	Element	Symbol	Atomic Number	Crustal Abundance, Weight Per Cent
Actinium	Ac	89	Man-made	Neptunium	Np	93	Man-made
Aluminum	Al	13	8.00	Nickel	Ni	28	0.0072
Americium	Am	95	Man-made	Niobium	Nb	41	0.0020
Antimony	Sb	51	0.00002	Nitrogen	N	7	0.0020
Argon	Ar	18	Not known	Nobelium	No	102	Man-made
Arsenic	As	33	0.00020	Osmium	Os	76	0.00000002
Astatine	At	85	Man-made	Oxygen[b]	O	8	45.2
Barium	Ba	56	0.0380	Palladium	Pd	46	0.0000003
Berkelium	Bk	97	Man-made	Phosphorus	P	15	0.1010
Beryllium	Be	4	0.00020	Platinum	Pt	78	0.0000005
Bismuth	Bi	83	0.0000004	Plutonium	Pu	94	Man-made
Boron	B	5	0.0007	Polonium	Po	84	Footnote[d]
Bromine	Br	35	0.00040	Potassium	K	19	1.68
Cadmium	Cd	48	0.000018	Praseodymium	Pr	59	0.0013
Calcium	Ca	20	5.06	Promethium	Pm	61	Man-made
Californium	Cf	98	Man-made	Protactinium	Pa	91	Footnote[d]
Carbon[a]	C	6	0.02	Radium	Ra	88	Footnote[d]
Cerium	Ce	58	0.0083	Radon	Rn	86	Footnote[d]
Cesium	Cs	55	0.00016	Rhenium	Re	75	0.00000004
Chlorine	Cl	17	0.0190	Rhodium[c]	Rh	45	0.00000001
Chromium	Cr	24	0.0096	Rubidium	Rb	37	0.0070
Cobalt	Co	27	0.0028	Ruthenium[c]	Ru	44	0.00000001
Copper	Cu	29	0.0058	Samarium	Sm	62	0.00077
Curium	Cm	96	Man-made	Scandium	Sc	21	0.0022
Dysprosium	Dy	66	0.00085	Selenium	Se	34	0.000005
Einsteinium	Es	99	Man-made	Silicon	Si	14	27.20
Erbium	Er	68	0.00036	Silver	Ag	47	0.000008
Europium	Eu	63	0.00022	Sodium	Na	11	2.32
Fermium	Fm	100	Man-made	Strontium	Sr	38	0.0450
Fluorine	F	9	0.0460	Sulfur	S	16	0.030
Francium	Fr	87	Man-made	Tantalum	Ta	73	0.00024
Gadolinium	Gd	64	0.00063	Technetium	Tc	43	Man-made
Gallium	Ga	31	0.0017	Tellurium[c]	Te	52	0.000001
Germanium	Ge	32	0.00013	Terbium	Tb	65	0.00010
Gold	Au	79	0.0000002	Thallium	Tl	81	0.000047
Hafnium	Hf	72	0.0004	Thorium	Th	90	0.00058
Helium	He	2	Not known	Thulium	Tm	69	0.000052
Holmium	Ho	67	0.00016	Tin	Sn	50	0.00015
Hydrogen[b]	H	1	0.14	Titanium	Ti	22	0.86
Indium	In	49	0.00002	Tungsten	W	74	0.00010
Iodine	I	53	0.00005	Uranium	U	92	0.00016
Iridium	Ir	77	0.00000002	Vanadium	V	23	0.0170
Iron	Fe	26	5.80	Xenon	Xe	54	Not known
Krypton	Kr	36	Not known	Ytterbium	Yb	70	0.00034
Lanthanum	La	57	0.0050	Yttrium	Y	39	0.0035
Lawrencium	Lw	103	Man-made	Zinc	Zn	30	0.0082
Lead	Pb	82	0.0010	Zirconium	Zr	40	0.0140
Lithium	Li	3	0.0020				
Lutetium	Lu	71	0.000080				
Magnesium	Mg	12	2.77				
Manganese	Mn	25	0.100				
Mendelevium	Md	101	Man-made				
Mercury	Hg	80	0.000002				
Molybdenum	Mo	42	0.00012				
Neodymium	Nd	60	0.0044				
Neon	Ne	10	Not known				

[a] Estimate from S. R. Taylor (1964).
[b] Analyses of crustal rocks do not usually include separate determinations for hydrogen and oxygen. Both combine in essentially constant proportions with other elements, so abundances can be calculated.
[c] Estimates are uncertain and have a very low reliability.
[d] Elements formed by radioactive decay of long-lived uranium and thorium. The daughter products are themselves radioactive but have such short half-lives that their crustal accumulations are too low to be measured accurately.

mass number are such important properties of atoms that they are recorded as subscripts and superscripts respectively, usually on the left-hand side of the chemical symbol for the element. For example, $^{40}_{19}K$ means an atom of potassium with mass number 40 and atomic number 19. The atom therefore contains 21 neutrons and 19 protons.

We learned in Chapter 4 that *atoms having the same atomic number but differing numbers of neutrons in the nucleus* are called **isotopes.** Isotopes of an

Table A.3
Naturally Occurring Elements Listed in Order of Atomic Numbers, Together with the Naturally Occurring Isotopes of Each Element, Listed in Order of Mass Numbers

Atomic number[a]	Name	Symbol	Mass Numbers[b] of Natural Isotopes
1	Hydrogen	H	1, 2, ③[c]
2	Helium	He	3, 4
3	Lithium	Li	6, 7
4	Beryllium	Be	9
5	Boron	B	10, 11
6	Carbon	C	12, 13, ⑭
7	Nitrogen	N	14, 15
8	Oxygen	O	16, 17, 18
9	Fluorine	F	19
10	Neon	Ne	20, 21, 22
11	Sodium	Na	23
12	Magnesium	Mg	24, 25, 26
13	Aluminum	Al	27
14	Silicon	Si	28, 29, 30
15	Phosphorus	P	31
16	Sulfur	S	32, 33, 34, 36
17	Chlorine	Cl	35, 37
18	Argon	A	36, 38, 40
19	Potassium	K	39, ④⓪, 41
20	Calcium	Ca	40, 42, 43, 44, 46, ④⑧
21	Scandium	Sc	45
22	Titanium	Ti	46, 47, 48, 49, 50
23	Vanadium	V	⑤⓪, 51
24	Chromium	Cr	50, 52, 53, 54
25	Manganese	Mn	55
26	Iron	Fe	54, 56, 57, 58
27	Cobalt	Co	59
28	Nickel	Ni	58, 60, 61, 62, 64
29	Copper	Cu	63, 65
30	Zinc	Zn	64, 66, 67, 68, 70
31	Gallium	Ga	69, 71
32	Germanium	Ge	70, 72, 73, 74, 76
33	Arsenic	As	75
34	Selenium	Se	74, 76, 77, 80, 82
35	Bromine	Br	79, 81
36	Krypton	Kr	78, 80, 82, 83, 84, 86
37	Rubidium	Rb	85, ⑧⑦
38	Strontium	Sr	84, 86, 87, 88
39	Yttrium	Y	89
40	Zirconium	Zr	90, 91, 92, 94, 96
41	Niobium	Nb	93
42	Molybdenum	Mo	92, 94, 95, 96, 97, 98, 100
44	Ruthenium	Ru	96, 98, 99, 100, 101, 102, 104
45	Rhodium	Rh	103

element therefore have identical atomic numbers but different mass numbers. All isotopes that are known to occur in the Earth are listed in Table A.3.

Hydrogen has three isotopes. Each has one proton and one electron. As shown in Figure A.2, the three isotopes of hydrogen are $_1^1H$, $_1^2H$, and $_1^3H$, known in scientific language as proteum, deuterium, and tritium, respectively. Every element has two or more isotopes, and for some elements as many as ten isotopes have been discovered. The number of iso-

Table A.3 (*Continued*)

Atomic number[a]	Name	Symbol	Mass Numbers[b] of Natural Isotopes
46	Palladium	Pd	102, 104, 105, 106, 108, 110
47	Silver	Ag	107, 109
48	Cadmium	Cd	106, 108, 110, 111, 112, 113, 114, 116
49	Indium	In	113, $\boxed{115}$
50	Tin	Sn	112, 114, 115, 116, 117, 118, 119, 120, 122, 124
51	Antimony	Sb	121, 123
52	Tellurium	Te	120, 122, 123, 124, 125, 126, 128, 130
53	Iodine	I	127
54	Xenon	Xe	124, 126, 128, 129, 130, 131, 132, 134, 136
55	Cesium	Cs	133
56	Barium	Ba	130, 132, 134, 135, 136, 137, 138
57	Lanthanum	La	$\boxed{138}$, 139
58	Cerium	Ce	136, 138, 140, $\boxed{142}$
59	Praseodymium	Pr	141
60	Neodymium	Nd	142, 143, $\boxed{144}$, 145, 146, 148, 150
62	Samarium	Sm	144, $\boxed{147}$, $\boxed{148}$, $\boxed{149}$, 150, 152, 154
63	Europium	Eu	151, 153
64	Gadolinium	Gd	$\boxed{152}$, 154, 155, 156, 157, 158, 160
65	Terbium	Tb	159
66	Dysprosium	Dy	156, 158, 160, 161, 162, 163, 164
67	Holmium	Ho	165
68	Erbium	Er	162, 166, 167, 168, 170
69	Thulium	Tm	169
70	Ytterbium	Yb	168, 170, 171, 172, 173, 174, 176
71	Lutetium	Lu	175, $\boxed{176}$
72	Hafnium	Hf	174, 176, 177, 178, 179, 180
73	Tantalum	Ta	180, 181
74	Tungsten	W	180, 182, 183, 184, 186
75	Rhenium	Re	185, $\boxed{187}$
76	Osmium	Os	184, 186, 187, 188, 189, 190, 192
77	Iridium	Ir	191, 193
78	Platinum	Pt	190, $\boxed{192}$, 194, 195, 196, 198
79	Gold	Au	197
80	Mercury	Hg	196, 198, 199, 200, 201, 202, 204
81	Thallium	Tl	203, 205
82	Lead	Pb	$\boxed{204}$, 206, 207, 208
83	Bismuth	Bi	209
84	Polonium	Po	$\boxed{210}$
86	Radon	Rn	$\boxed{222}$
88	Radium	Ra	$\boxed{226}$
90	Thorium	Th	$\boxed{232}$
91	Protactinium	Pa	$\boxed{231}$
92	Uranium	U	$\boxed{234}$, $\boxed{235}$, $\boxed{238}$

[a] Atomic number = number of protons.
[b] Mass number = protons + neutrons.
[c] $\boxed{}$ Indicates isotope is radioactive.

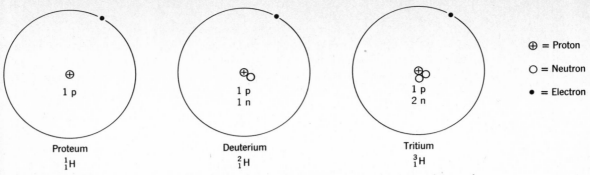

Figure A.2 Three isotopes of hydrogen, each of which has one proton in the nucleus and one electron in orbit. The isotopes differ in the number of neutrons each nucleus contains. All isotopes of an element have the same atomic number, but differ in their mass numbers.

topes that are possible for any given element is limited, however, because the mass numbers of elements can vary only within certain limits. Many combinations of protons and neutrons are unstable and break up spontaneously. *The decay process by which an unstable atomic nucleus spontaneously disintegrates,* emitting energy in the process, is called **radioactivity.**

The rates at which radioactive isotopes disintegrate are inherent properties of each isotope and are unaffected by changes in temperature or pressure. As we noted in Chapter 2, the constancy of decay rates is very important because decay rates of certain isotopes such as those of $^{14}_6C$ and $^{40}_{19}K$ are so slow that we can use them as accurate clocks for timing events in Earth's history that happened very long ago.

When a radioactive isotope decays, it may form either a new isotope of the same element or an isotope of a different element. The decay products are called *daughter products,* which may, in turn, be either stable or radioactive. If they are radioactive, then further disintegrations will eventually occur, and still newer daughter products will form.

When an atom disintegrates radioactively, some of the energy that holds the nucleus together is released. The disintegrating atom may also lose some of its sub-atomic particles. These may be either *alpha-particles* (α-particles) or *beta-particles* (β-particles). Alpha-particles are ^4_2He nuclei, stripped of their electrons so that their loss reduces the mass number by 4 and the atomic number by 2. Beta-particles are electrons expelled from the nucleus; their loss converts a neutron to a proton and thus increases the atomic number by 1 but leaves the mass number unchanged.

The results of radioactive emissions of α- and β-particles are illustrated in Figure A.3, which depicts a sequence of decays beginning with the radioactive isotope of uranium, $^{238}_{92}U$. The uranium atom emits an α-particle to form a daughter atom that is an isotope of thorium, $^{234}_{90}Th$. The thorium isotope then emits a β-particle to become $^{234}_{91}Pa$, and the new daughter product in turn emits a second β-particle to become $^{234}_{92}U$. The decay process continues through a total of fourteen radioactive daughter products until a stable atom, $^{206}_{82}Pb$, is formed. We infer from this decay scheme that the uranium content of the Earth must be slowly decreasing and, conversely, that the total lead content must be slowly increasing. Fortunately, the rate of decay of uranium is quite slow, and there is plenty left for us to mine. The rate of decay is also slow enough so that it has been possible to measure the way in which Earth's lead content has changed through its history, and as we saw in Chapter 2, the measurements allow us to infer a great deal about the age of the Earth.

When an atom decays radioactively, the principal way energy is released is as electromagnetic rays of very short wavelength, called *gamma-rays* (γ-rays), and as heat, which is generated whenever a nuclear disintegration occurs. As we noted in Chapter 3, the generation of heat by spontaneous nuclear decay has played a vitally important part in the history of Planet Earth. Radioactive decay of natural elements is believed to be the fuel that drives the planet's great internal processes.

Matter and Energy. We stated earlier that the nature of matter is essentially electrical. We also know that electricity is a form of energy (Chap. 3), so we can modify the earlier statement to say that

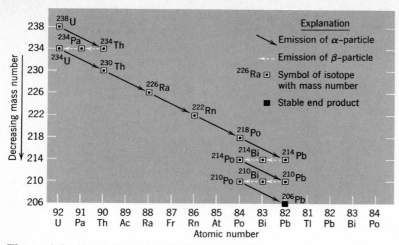

Figure A.3 Example of a radioactive-decay series. It commences with the most common isotope of uranium, $^{238}_{92}$U. The stable end-product is an isotope of lead, $^{206}_{82}$Pb, but in order to reach it a number of different isotopes and different elements are produced as intermediate steps.

matter is essentially energy. The first person to understand this important concept was Albert Einstein who, in 1905, showed that matter and energy are connected by the equation

$$E = mc^2$$

where E is energy, m is mass and c is the velocity of electromagnetic waves.

When we consider the structure of atoms, the equivalence of matter and energy become vitally important. For example, the nucleus of an atom of $^4_2$He contains 2 protons and 2 neutrons. Expressed in atomic mass units (AMU), a proton weighs 1.00758 AMU and a neutron 1.00893 AMU. The mass of $^4_2$He should therefore be $(2 \times 1.00758) + (2 \times 1.00893) = 4.03303$ AMU. When the helium nucleus is weighed, however, it is found to be only 4.00260 AMU—some of the mass has been lost. Explaining the mystery of the lost mass was one of the greatest triumphs of atomic physics. When the particles join together to form a helium nucleus, some of their mass is converted to energy, which appears partly as electromagnetic radiation and, because the new atomic nucleus has a high speed, partly as heat energy. The fusion of light particles to form heavier ones proceeds continuously in the Sun. Since nuclei of hydrogen, which are simply protons, are abundant in the Sun, astronomers believe that hydrogen fusion is the main source of solar energy. As a result of fusion, the Sun

radiates a continuous stream of electromagnetic waves into space, and is slowly converting more of its matter into energy. *Fusion* of light particles to form heavy ones does not happen naturally on Earth, because the high temperatures needed to trigger it do not occur, but it is the process that goes on in the hydrogen bomb.

A nucleus can be broken down into its constituent particles only by the addition of sufficient energy to replace that lost during the joining process. The missing mass of the particles in the nucleus acts, in a sense, as a sort of "glue" that binds the nucleus together. The energy given off when protons and neutrons combine to form an atomic nucleus is, therefore, called the *binding energy* of the nucleus. Binding energy increases as the mass number of an element increases, but as we can see in Figure A.4, the relation is not a simple one. The curve dips down for elements with very high mass numbers. This means that if we split a nucleus such as uranium into one or more elements with low mass numbers, we move up the binding-energy curve, and energy is released in the same way as it is when light nuclei fuse to form heavier ones. Splitting a heavy atom, $^{235}_{92}$U, into lighter atoms of barium and krypton produces the energy of the atom bomb (Fig. A.5). We speak of heavy-atom splitting as *fission*.

Energy released by atomic fission appears both as electromagnetic waves and as heat. If we refer to Figure A.3, in which the natural radioactive decay

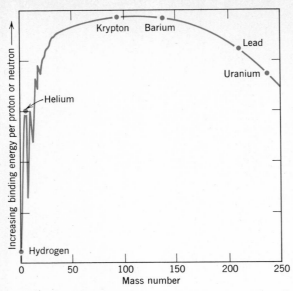

Figure A.4 Binding energy, the amount of energy produced by conversion of some of the mass of each proton and neutron combined in a nucleus, varies with the mass number of the nucleus (see text for further explanation). The higher an element is on the curve, the more energy will be given off when protons and neutrons combine to form its nucleus. The fusion of four hydrogen nucleii to form helium releases a vast amount of energy. This is the process that produces the Sun's energy and that indirectly drives most of Earth's external activities. The splitting of a heavy uranium atom into lighter krypton and barium atoms also releases energy. This is the process that occurs in the atom bomb. Lead sits higher on the curve than uranium. Natural radioactive decay of uranium to lead produces energy, and is one of the principal sources of energy for Earth's internal activities.

of $^{238}_{92}$U is depicted, we see that a stable daughter, $^{206}_{82}$Pb, results. Lead sits higher on the binding-energy curve than uranium; so energy is released during the process of natural decay of uranium. The fraction that is heat energy is an important source of Earth's internal heat energy.

Compounds. The chemical properties of an element (that is, the way elements combine together to form compounds) are determined by the orbiting electrons. Electrons are confined to specific orbits which are arranged at predetermined distances from the nucleus. Because the electrons in each orbit have a specific amount of energy characteristic for that

orbit, the orbit distances are commonly called *energy-level shells*. The maximum number of electrons that can occupy a given energy-level shell is fixed. Shell 1, closest to the nucleus, is small and can accommodate only 2 electrons; shell 2, however, can accommodate 8 electrons; shell 3, 18; and shell 4, 32.

When an energy-level shell is filled with electrons it is very stable, like an evenly loaded boat. To fill their energy-level shells and so reach a stable configuration, atoms share or transfer electrons among themselves. The movement of electrons naturally upsets the balance of electrical forces, for an atom that loses an electron has lost a negative electrical charge and therefore has a net positive charge, while one that gains an electron has a net negative charge. The sharing and transfer of electrons gives rise to electronic forces, or *bonds,* that bind atoms together into *compounds.* For example, as can be seen in Figure A.6, lithium has shell 1 filled, but has only one electron in shell 2. The lone outer electron is loosely held and easily transferred to an element such as fluorine, which has 7 electrons in shell 2, needing only one more to be completely filled. In this fashion both the lithium and fluorine finish with filled shells, and the resulting positive charge on the lithium (written Li^{+1}) and the negative charge on the fluorine (F^{-1}) bind the two atoms together. Lithium and fluorine form the compound lithium fluoride, which is written LiF to indicate that for every Li atom there is a counterbalancing F atom. Properties of compounds are quite different from the properties of their constituent elements. A single LiF pair is called a molecule of lithium fluoride. A *molecule* is *the smallest unit that retains all the properties of a compound.*

Some elements occur naturally with their energy-level shells completely filled. They occur as individual, electrically neutral atoms and show little or no tendency to react with other elements and form compounds. Because they are so unreactive, elements with normally filled outer shells are called *noble gases.* The noble-gas elements are helium, neon, argon, krypton, xenon, and radon. All elements other than the noble gases readily bond with other atoms and form compounds. Atoms form bonds with like- or unlike atoms, but regardless of the pairing, when they transfer electrons they finish with a net positive or negative charge, depending on whether they give up or receive the transferred electrons.

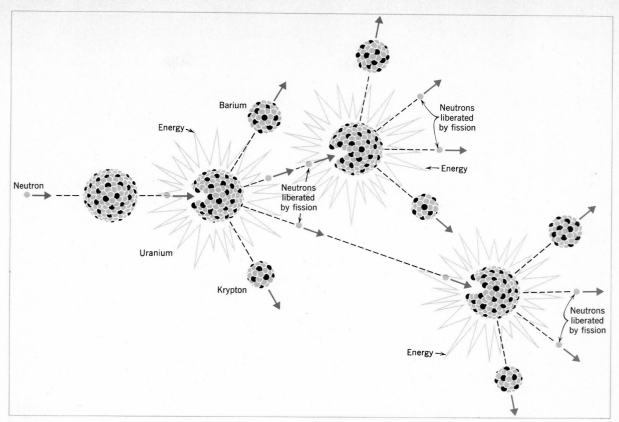

Figure A.5 Release of energy by a chain reaction in uranium. A neutron strikes the nucleus of a $^{235}_{92}$U atom, causing it to become unstable. It splits into lighter elements such as barium and krypton, releasing energy and more neutrons in the process. The newly produced neutrons cause more $^{235}_{92}$U atoms to disintegrate, producing a continuous chain reaction.

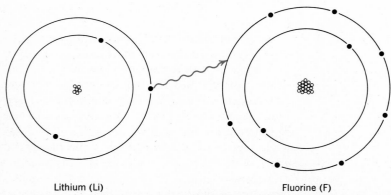

Lithium (Li)   Fluorine (F)

Figure A.6 To form the compound lithium fluoride, an atom of lithium, with one electron in its center shell, combines with an atom of fluorine with seven electrons in the outer shell. The lithium atom transfers its lone outer-shell electron to fill the fluorine atom's outer shell.

Appendix B

Minerals

Minerals that are abundant in rocks or common as ores number only a few dozens, and many can be identified without special equipment if sizable pieces are available. The techniques to be described consist of direct observations and simple tests. With these it is possible to recognize groups of rock-forming minerals and the common ore minerals. More exact determinations can be made with laboratory instruments beyond the scope of our introductory study.

Physical Properties

The simple physical properties useful in mineral identification are: appearance (expressed in luster, color, and form), the ways minerals break, the color of a mineral's powder, its hardness, specific gravity, and magnetic susceptibility.

Luster. *The quality and intensity of light reflected from a mineral* produce an effect known as **luster.** Two minerals with almost the same color can have totally different lusters. The more important are described as *metallic,* like that on a polished metal surface; *vitreous,* like that on glass; *resinous,* like that of yellow resin; *pearly,* like that of pearl; *greasy,* as if the surface were covered with a film of oil; and *adamantine,* having the brilliance of a diamond.

Mineral identification will soon become familiar procedure to those who approach it systematically. Minerals with metallic luster should be set aside for tests of streak and magnetic susceptibility. Most ore minerals have a metallic luster. Minerals with vitreous luster should be checked for hardness, inspected for kind of breakage surfaces, and given chemical tests.

Color. The color of a mineral is one of its striking properties, but unfortunately is not a very reliable means of identification. Color commonly results from impurities, which are present in only small amounts. Some minerals display various colors. Quartz, for example, can be clear and colorless, milky white, rose-colored, violet, and dark gray to black. Calcite, likewise, can be clear, milky white, pink, green, and gray. Among feldspars flesh-colored, cream-colored, pink, and light green characterize orthoclase and its relatives, whereas dead white and light blue typify plagioclases.

Form. As Chapter 4 explains, the form of a mineral depends partly on the crystal structure (a property unique to the mineral) and partly on conditions of growth (unrelated to the mineral).

We can quickly recognize the cubes of halite, pyrite, and galena, the 12-sided figures of garnet, and the 6-sided crystals of quartz. More commonly the particles are crowded together to form a mineral aggregate.

The form of asbestos and related minerals is *fibrous.* Many iron oxides are *earthy,* like crumbly soil. Sheet silicate minerals occur in thin sheets, a form termed *micaceous* because it is well developed in micas. Some ore minerals have grown in *forms resembling grapes closely bunched,* described as ***botryoidal.***

Ways in Which Minerals Break. Chapter 4 explains the relationship between the crystal structure and ways in which minerals break. Minerals with well-developed cleavage include micas, calcite, amphibole and pyroxene, halite, galena, and feldspars. Minerals lacking perceptible cleavage include garnet, which breaks along irregular fractures, and quartz and olivine, which fracture irregularly or display ***conchoidal fracture,*** *breakage resulting in smooth curved surfaces.* Some minerals break along splintery surfaces resembling those of wood.

Because of a distinctive habit of growth within the crystals the cleavage surfaces of plagioclase nearly always appear to be *striated.* Striations are a reliable means of distinguishing plagioclase from orthoclase. They are seen to best advantage with a hand lens when the cleavage surface reflects a bright light.

Streak. The ***streak*** is a *thin layer of powdered mineral made by rubbing a specimen on a nonglazed porcelain plate.* The powder diffuses light and gives a reliable color effect that is independent of the form and luster of the mineral specimen. Red streak characterizes hematite whether the specimen itself is red and earthy, like the streak, or black and metallic, like magnetite. Limonite streaks brown, and magnetite streaks black. Many minerals streak an undiagnostic white.

Hardness. *Relative resistance of a mineral to scratching,* or ***hardness,*** is one of the most noticeable ways in which minerals differ. A series of 10 minerals has been chosen as a standard scale, and any unknown specimen can be classified, 1 to 10, by trying it against the known specimens in the scale, which are arranged in order of decreasing hardness in Table 4.3. Fractional values of hardness are common. If a specimen scratches calcite but is distinctly scratched by fluorite, the approximate value of its hardness is 3.5.

In tests of hardness several precautions are necessary. A mineral softer than another may leave a mark that looks like a scratch, just as a soft pencil leaves its mark. A real scratch does not rub off. The physical structure of some minerals may make the hardness test difficult; if a specimen is powdery or in fine grains or if it breaks easily into splinters, an apparent scratch may be deceptive.

Specific Gravity. Density is not commonly measured directly. Instead, unit weights of different minerals are compared against the unit weight of a standard substance. In comparing unit weights, we commonly use water as a standard. The ***specific gravity*** of any substance is expressed as *a number stating the ratio of the weight of the substance to the weight of an equal volume of pure water.* Specific gravity can be approximated by comparing different minerals held in the hand. Metallic minerals such as galena feel "heavy" whereas nearly all others feel "light."

Magnetism. Of the common minerals only magnetite and pyrrhotite are strongly magnetic. They can be singled out at once by their strong attraction to a small magnet.

Chemical Properties

Only two chemical tests are commonly used in the beginning study of minerals: (1) taste test for halite and (2) acid test for calcite and dolomite. The salty taste of halite is distinctive. Carbonate minerals effervesce (make bubbles) in dilute hydrochloric acid. Calcite effervesces freely no matter what the size of the particles. Dolomite may not effervesce at all unless the specimen is powdered or the acid is heated. Dolomite powder effervesces slowly in cool dilute acid.

Caution. Hydrochloric acid is hard on teeth and has an unpleasant taste. Where many students share mineral specimens caution in use of acid is necessary. Acid should be applied in small drops and when the test is finished the specimen should be blotted dry. The next user of the specimen may decide to try the taste test.

Identification of Minerals. For convenience, the common rock-forming minerals are arranged alphabetically in Table B.1, and the common ore minerals in Table B.2. The chemical formulas in the second column of both tables are there for reference only, not to aid in identifications. The final column in Tables B.1 and B.2 list the physical properties most characteristic for each mineral, together with suggestions for distinguishing between similar-looking minerals. Most rock-forming minerals have white streak and vitreous luster and are either transparent or translucent. Ore minerals are almost all opaque, have metallic luster and high specific gravities. Most ore minerals will be encountered only in mineral deposits. A few, notably chalcopyrite, hematite, ilmenite, limonite, magnetite, pyrite, pyrrhotite, and rutile, may also be encountered as accessory minerals in common igneous, sedimentary or metamorphic rocks.

Gemstones are minerals with desirable properties of color and wearing ability. A list of the commonly observed gem minerals, together with the names of the gem varieties, is given in Table B.3.

Selected References

Dietrich, R. V., 1966, Mineral tables—hand specimen properties of 1500 minerals: Blacksburg, Va., Virginia Polytechnic Inst. Bull. 160.

Pearl, R. M., 1962, Successful mineral collecting and prospecting: New York, New American Library.

Vanders, Iris, and Kerr, P. F., 1967, Mineral recognition: New York, John Wiley.

Table B.1

Properties of the Rock-Forming Minerals (Ore Minerals, Some of Which Are Also Rock-Forming Minerals, Are Listed in Table B.2)

Mineral	Chemical composition	Form and habit	Cleavage	Hardness	Specific gravity	Other properties	Most distinctive properties
Amphiboles. (A complex family of minerals *Hornblende* is most common.)	$X_2Y_5Si_8O_{22}(OH)_2$ where X = Ca, Na, Y = Mg, Fe, Al	Long, six-sided crystals: also fibers and irregular grains	Two; intersecting at 56° and 124°.	5–6	2.9–3.8	Common in metamorphic and igneous rocks. *Hornblende* is dark green to black; *actinolite*, green; *tremolite*, white.	Cleavage, habit.
Andalusite	Al_2SiO_5	Long crystals, often square in cross-section.	Weak, parallel to length of crystal.	7.5	3.2	Found in metamorphic rocks. Often flesh colored.	Hardness, form.
Anhydrite	$CaSO_4$	Crystals are rare. Irregular grains or fibers.	Three, at right angles.	3	2.9	Alters to gypsum. Pearly luster, white or colorless.	Cleavage, hardness.
Apatite	$Ca_5(PO_4)_3(F, OH, Cl)$	Granular masses. Perfect six-sided crystals.	Poor. One direction.	5	3.2	Green, brown, blue or white. Common in many kinds of rocks in small amounts.	Hardness, form.
Aragonite	$CaCO_3$	Massive, or slender, needle-like crystals.	Poor. Two directions.	3.5	2.9	Colorless or white. Effervesces with dilute HCl.	Effervescence with acid. Poor cleavage distinguishes from calcite.
Asbestos			See Serpentine				
Augite			See Pyroxene				
Biotite			See Mica				
Calcite	$CaCO_3$	Tapering crystals and granular masses.	Three perfect; at oblique angles to give a rhomb-shaped fragment.	3	2.7	Colorless or white. Effervesces with dilute HCL.	Cleavage, effervescence with acid.
Chlorite	$(Mg, Fe)_5(Al,Fe)_2 Si_3O_{10}(OH)_8$	Flaky masses of minute scales.	One perfect; parallel to flakes.	2–2.5	2.6–2.9	Common in metamorphic rocks. Light to dark green. Greasy luster.	Cleavage—flakes not elastic, distinguishes from mica. Color.
Dolomite	$CaMg(CO_3)_2$	Crystals with rhomb-shaped faces. Granular masses.	Perfect in three directions as in calcite.	3.5	2.8	White or gray. Does not effervesce in cold, dilute HCl unless powdered. Pearly luster.	Cleavage. Lack of effervescence with acid.
Epidote	Complex silicate of Ca, Fe and Al.	Small elongate crystals. Fibrous.	One perfect, one poor.	6–7	3.4	Yellow-green to dark green. Common in metamorphic rocks.	Habit, color. Hardness distinguishes from chlorite.
Feldspars: Orthoclase	$KAlSi_3O_8$	Prism-shaped crystals, granular masses.	Two perfect, at right angles.	6	2.6	Common mineral. Flesh colored, pink, white or gray.	Color, cleavage.

Mineral	Chemical composition	Form and habit	Cleavage	Hardness	Specific gravity	Other properties	Most distinctive properties
Plagioclase	$NaAlSi_3O_8$ (albite) and $CaAl_2Si_2O_8$ (anorthite) and all compositions between.	Irregular grains, cleavable masses. Rarely as tabular crystals.	Two perfect, not quite at right angles.	6–6.5	2.6–2.7	White to dark gray. Cleavage planes may show fine parallel striations.	Cleavage, striations on cleavage planes.
Fluorite	CaF_2	Cubic crystals, granular masses.	Perfect in four directions.	4	3.2	Colorless, blue green. Always an accessory mineral.	Hardness, cleavage, does not effervesce with acid.
Garnets	$X_3Y_2 (SiO_4)_3$; X = Ca, Mg, Fe, Mn, Y = Al, Fe, Ti, Cr.	Perfect crystals with 12 or 24 sides. Granular masses.	None. Uneven fracture.	6.5–7.5	3.5–4.3	Common in metamorphic rocks. Red, brown, yellow green, black.	Crystals, hardness, no cleavage.
Graphite	C	Scaly masses.	One, perfect. Forms slippery flakes.	1–2	2.2	Metamorphic rocks. Black with metallic to dull luster.	Cleavage, color. Marks paper.
Gypsum	$CaSO_4 \cdot 2H_2O$	Elongate or tabular crystals. Fibrous and earthy masses.	One, perfect. Flakes bend but are not elastic.	2	2.3	Vitreous to pearly luster. Colorless.	Hardness, cleavage.
Halite	NaCl	Cubic crystals.	Perfect to give cubes.	2.5	2.2	Tastes salty. Colorless, blue.	Taste, cleavage.
Hornblende					See Amphibole		
Kaolinite	$Al_2Si_2O_5(OH)_4$	Soft, earthy masses. Submicroscopic crystals.	One, perfect.	2–2.5	2.6	White, yellowish. Plastic when wet; emits clay odor. Dull luster.	Feel, plasticity, odor.
Kyanite	Al_2SiO_5	Bladed crystals.	One perfect. One imperfect.	4.5 parallel to blade, 7 across blade	3.6	Blue, white, gray. Common in metamorphic rocks.	Variable hardness, distinguishes from sillimanite. Color.
Mica: Biotite	$K(Mg, Fe)_3AlSi_3O_{10}(OH)_2$	Irregular masses of flakes.	One, perfect.	2.5–3	2.8–3.2	Common in igneous and metamorphic rocks. Black, brown, dark green.	Cleavage, color.
Muscovite	$KAl_3Si_3O_{10}(OH)_2$	Thin flakes.	One, perfect.	2–2.5	2.7–3.1	Common in igneous and metamorphic rocks. Colorless, pale green or brown.	Cleavage, color. Flakes are elastic.
Olivine	$(Mg, Fe)_2SiO_4$	Small grains, granular masses.	None. Conchoidal fracture.	6.5–7	3.2–4.3	Igneous rocks. Olive green to yellow green.	Color, fracture, habit.
Orthoclase					See Feldspar		
Plagioclase					See Feldspar		

Mineral	Chemical composition	Form and habit	Cleavage	Hardness / Specific gravity		Other properties	Most distinctive properties
Pyroxene (A complex family of minerals. *Augite* is most common.)	$XY(SiO_3)_2$ $X = Y = Ca, Mg, Fe$	8-sided stubby crystals. Granular masses.	Two perfect, nearly at right angles.	5–6	3.2– 3.9	Igneous and metamorphic rocks. *Augite*, dark green to black; other varieties, white to green.	Cleavage.
Quartz	SiO_2	6-sided crystals, granular masses.	None. Conchoidal fracture.	7	2.6	Colorless, white, gray, but may have any color, depending on impurities. Vitreous to greasy luster.	Form, fracture, striations across crystal faces at right angles to long dimension.
Sillimanite	Al_2SiO_5	Long needle crystals, fibers.	One, perfect.	6–7	3.2	White, gray. Metamorphic rocks.	Hardness distinguishes from kyanite. Habit.
Serpentine (Fibrous variety is *asbestos*)	$Mg_3Si_2O_5(OH)_4$	Platy or fibrous.	Breaks irregularly, except in fibrous variety.	2.5– 5	2.2– 2.6	Light to dark green. Smooth, greasy feel.	Habit, hardness.
Talc	$Mg_3Si_4O_{10}(OH)_2$	Small scales, compact masses.	One, perfect.	1	2.6– 2.8	Feels slippery. Pearly luster. White to greenish.	Hardness, luster, feel, cleavage.
Tourmaline	Complex silicate of B, Al, Na, Ca, Fe, Li and Mg.	Elongate crystals, commonly with triangular cross section.	None.	7– 7.5	3– 3.3	Black, brown, red, pink, green, blue and yellow. An accessory mineral in many rocks.	Habit.
Wollastonite	$CaSiO_3$	Fibrous or bladed aggregates of crystals.	Two, perfect.	4.5– 5	2.8– 2.9	Colorless, white, yellowish. Metamorphic rocks. Soluble in HCl.	Habit. Solubility in HCl and hardness distinguish amphiboles, kyanite, sillimanite.

Table B.2

Properties of the Common Ore Minerals (Some Ore Minerals Are Also Rock-Forming Minerals)

Mineral	Chemical composition	Form and habit	Cleavage	Hardness / Specific gravity	Other properties	Most distinctive properties
Bornite	Cu_5FeS_4	Massive. Crystals very rare.	None. Uneven fracture.	3 / 5	Brownish bronze on fresh surface. Tarnishes purple, blue and black. Gray black streak.	Color, streak.
Chalcocite	Cu_2S	Massive. Crystals very rare.	None. Conchoidal fracture.	2.5 / 5.7	Steel-gray to black. Dark gray streak.	Streak.
Chalcopyrite	$CuFeS_2$	Massive or granular.	None. Uneven fracture.	3.5–4 / 4.2	Golden yellow to brassy yellow. Dark green to black streak.	Streak. Hardness distinguishes from pyrite.
Chromite	$FeCr_2O_4$	Massive or granular.	None. Uneven fracture.	5.5 / 4.6	Iron black to brownish black. Dark brown streak	Streak and lack of magnetism distinguishes from ilmenite and magnetite.
Copper	Cu	Massive, twisted leaves and wires.	None. Can be cut with a knife.	2.5–3 / 9	Copper color but commonly stained green.	Color, specific gravity, malleable.
Galena	PbS	Cubic crystals, coarse or fine grained granular masses.	Perfect in three directions at right angles.	2.5 / 7.6	Lead gray color. Gray to gray black streak.	Cleavage and streak.
Gold	Au	Small irregular grains.	None. Malleable.	2.5 / 19.3	Gold color. Can be flattened without breakage.	Color, specific gravity, malleability.
Hematite	Fe_2O_3	Massive, granular, micaceous	Uneven fracture.	5–6 / 5	Red-brown, gray to black. Red-brown streak.	Streak, hardness.
Ilmenite	$FeTiO_3$	Massive or irregular grains.	Uneven fracture.	5.5–6 / 4.7	Iron-black. Brown-red streak differing from hematite.	Streak distinguishes hematite. Lack of magnetism distinguishes magnetite.
Limonite (*Goethite* is most common.)	A complex mixture of minerals, mainly hydrous iron oxides.	Massive, coatings, botryoidal crusts, earthy masses.	None.	1–5.5 / 3.5–4	Yellow, brown, black. Yellow-brown streak.	Streak.
Magnetite	Fe_3O_4	Massive, granular. Crystals have octahedral shape.	None. Uneven fracture.	5.5–6.5 / 5	Black. Black streak. Strongly attracted to a magnet.	Streak, magnetism.
Pyrite ("Fool's gold")	FeS_2	Cubic crystals with striated faces. Massive.	None. Uneven fracture.	6–6.5 / 5.2	Pale brass-yellow, darker if tarnished. Greenish black streak.	Streak. Hardness distinguishes from chalcopyrite. Not malleable, which distinguishes from gold.
Pyrolusite	MnO_2	Crystals rare. Massive, coatings on fracture surfaces.	Crystals have a perfect cleavage. Massive breaks unevenly.	2–6.5 / 5	Dark gray, black on bluish black. Black streak.	Color, streak.

Table B.2 (*Continued*)

Mineral	Chemical composition	Form and habit	Cleavage	Hardness / Specific gravity		Other properties	Most distinctive properties
Pyrrhotite	FeS	Crystals rare. Massive or granular.	None. Conchoidal fracture.	4	4.6	Brownish-bronze. Black streak. Magnetic.	Color and hardness distinguish from pyrite, magnetism from chalcopyrite.
Rutile	TiO$_2$	Slender, prismatic crystals or granular masses.	Good in one direction. Conchoidal fracture in others.	6–6.5	4.2	Red-brown (common), black (rare). Brownish streak. Adamantine luster.	Luster, habit, hardness.
Sphalerite (zinc blende)	ZnS	Fine to coarse granular masses. Tetrahedron shaped crystals.	Perfect in six directions.	3.5–4	4	Yellow brown to black. White to yellow-brown streak. Resinous luster.	Cleavage, hardness, luster.
Uraninite	UO$_2$ to U$_3$O$_8$	Massive, with botryoidal forms. Rare crystals with cube shapes.	None. Uneven fracture.	5–6	6.5–10	Black to dark brown. Streak black to dark brown. Dull luster.	Luster and specific gravity distinguish from magnetite. Streak distinguishes from ilmenite and hematite.

Table B.3
Properties of Some Common Gemstones

Mineral and variety	Composition	Form and habit	Cleavage	Hardness / Specific gravity	Other properties	Most distinctive properties
Beryl: *Aquamarine* (blue) *Emerald* (green) *Golden beryl* (golden-yellow)	$Be_3Al_2Si_6O_{18}$	Six-sided, elongate crystals common.	Weak.	7.5– 8 / 2.75	Bluish green, green, yellow, white colorless. Common in pegmatites.	Form. Distinguished from apatite by its hardness.
Corundum: *Ruby* (red) *Sapphire* (blue)	Al_2O_3	Six-sided, barrel-shaped crystals.	None, but breaks easily across its crystal.	9 / 4	Brown, pink, red, blue, colorless, Common in metamorphic rocks. Star sapphire is opalescent with a six-sided light spot showing.	Hardness.
Diamond	C	Octahedron-shaped crystals.	Perfect, parallel to faces of octahedron.	10 / 3.5	Colorless, yellow, rarely red, orange, green, blue or black.	Hardness, cleavage.
Garnet: *Almandite* (red) *Grossularite* (green, cinnamon-brown) *Demantoid* (green)	A rock-forming mineral—See Table B.1					
Opal (A mineraloid)	$SiO_2 \cdot nH_2O$	Massive, thin coating. Amorphous.	None. Conchoidal fracture.	5–6 / 2– 2.2	Colorless, white, yellow, red, brown, green, gray, opalescent.	Hardness, color, form.
Quartz: (1) Coarse crystals *Amethyst* (violet) *Cairngorm* (brown) *Citrine* (yellow) *Rock crystal* (colorless) *Rose quartz* (pink) (2) Fine-grained *Agate* (banded, many colors) *Chalcedony* (brown, gray) *Heliotrope* (green) *Jasper* (red)	A rock-forming mineral—See Table B.1					
Topaz	$Al_2SiO_4(OH, F)_2$	Prism shaped crystals, granular masses.	One, perfect.	8 / 3.5	Colorless, yellow, blue, brown.	Hardness, form, color.
Tourmaline	A rock-forming mineral—See Table B.1					
Zircon	$ZrSiO_4$	Four-sided elongate crystals, square in cross-section.	None.	7.5 / 4.7	Brown, red, green, blue, black.	Habit, hardness.

Appendix C

Identification of Common Rocks

The three major classes of rocks (igneous, sedimentary, metamorphic) are defined and discussed in Chap. 5. In each class are many distinct kinds, each with a specific name. Fortunately, fewer than 30 in all make up the great bulk of the visible part of the Earth's crust. These common kinds must be learned well if we are to read correctly the history recorded in the crust. By studying representative specimens of all the important kinds, we can learn the properties some have in common, and the distinctive features by which each kind is identified.

Ideally, we should see each specimen in the field as part of an exposure in which the larger features and relations are clearly shown. The sedimentary layering of sandstone or limestone, and the intrusive relation of a dike tell us at once that the rock is sedimentary or igneous. Some hand specimens in a laboratory may not have clear indications of their general classification. But any systematic description of the common rocks lists them according to class, and we welcome any clue that may tell us at the start whether an unknown specimen is igneous, sedimentary, or metamorphic. A number of such clues are found with practice, and even without them a specimen can ordinarily be traced quickly to its class by a process of elimination.

Texture. We examine a rock specimen closely for the pattern of visible constituents, just as we inspect the weave (texture) in cloth. Many rocks have visible mineral grains, and if these can be made out with the unaided eye over the entire surface the rock has *granular texture, coarse* if the average grains are 5mm or more across, *medium* if the average is 1 to 5mm across, and *fine* if less than 1mm across. Some igneous rocks have *glassy* texture; the cooling from magma was too rapid for any grains to form.

Many igneous rocks are porphyritic, with distinct phenocrysts (Fig. 5.17); this proves igneous origin. Certain textures, then, tell us the general classification of some rock specimens as a first step toward their identification.

In examining texture, we are interested not only in the size of grains but also in their shapes and the way they fit together. If the grains are angular and dovetail one into another to fill all the space, they must have been formed by crystallization, and the rock is probably either igneous or metamorphic. If the grains are separated by irregular spaces filled with fine cementing material, probably they are fragments and the rock may be of either sedimentary or volcanic origin.

Mineral Composition. Many rocks are identified by their component minerals. The critical minerals are determined by their physical properties, as explained in Appendix B. Grains large enough for clear visibility are required for study without a microscope. A hand lens that magnifies 8 or 10 times is a useful aid in studying the mineral grains, even in coarse-textured rocks.

Other Properties. Some limestones look superficially like fine-grained igneous rocks but are much softer. Every rock specimen under study should be tested for hardness.

Some kinds of rocks show characteristic forms on fracture surfaces. Other tests are mentioned in the descriptions of specific rocks.

The principal group of rocks discussed in this Appendix is the igneous group. They offer more trouble to a beginning student than sedimentary or metamorphic rocks. The following discussion, and Table C.1, therefore present the necessary data for identifying common igneous rocks. Identification of sedimentary and metamorphic rocks can be effected by studying Chap. 5 and using Tables C.2 and C.3 respectively.

Table of Igneous Rocks

In an introductory study igneous rocks can be classified more systematically than others, and details required for satisfactory analysis of laboratory specimens are here separated from the general treatment in the body of the book. Table C.1 supplements Fig. 5.18, and the two can be used together with profit. Igneous rocks are grouped in the table according to (1) texture and (2) composition. Any classification for use with hand specimens must be general, and the number of names in the table is reduced to a minimum.

Nature does not draw sharp boundaries, and there are all conceivable gradations in texture and composition. The separating lines in the table, therefore, are somewhat fictitious. We can find a series of specimens that will bridge the gaps in composition between granite and granodiorite, granodiorite and diorite.

Minerals of igneous rocks are divided generally into a light-colored group (including the feldspars, quartz, and muscovite) and a dark-colored group (including biotite, pyroxene, hornblende, and olivine). Quartz grains in an igneous rock indicate a surplus of silica in the parent magma; therefore the presence or absence of quartz grains is a logical basis for drawing a line between granitic rocks and diorite. In the rocks that have no quartz the feldspars divide honors with the dark-colored minerals, and the boundary between diorite and gabbro is drawn where the dark minerals exceed 50 per cent of the total. This boundary is carried through between andesite and basalt on the basis of color. Inspecting thin edges of specimens in strong light is the best test; andesite transmits some of the light, basalt is opaque.

Kinds of Igneous Rocks

Granite. Feldspar and quartz are the chief minerals in granite. Some biotite usually is present, and many granites have scattered grains of hornblende. The dark minerals commonly are in nearly perfect crystals; this suggests that they formed first, while most of the mass was molten. The feldspar formed next, and the grains crowded against and hampered each other in growth. Quartz, the surplus silica, crystallized last and so is molded around the angular grains of the earlier minerals. This *interlocking arrangement of visible mineral grains characteristic of granite* is called **granular texture.**

Technically, the term granite is applied only to quartz-bearing rocks in which potassium feldspar is predominant, and the name granodiorite applies to

Table C.1
Aids in Identification of Common Igneous Rocks

To use the table, first determine the texture, and find which entries in Column 1 best describe it. Then determine if quartz is present, and if it is sparse or abundant. These options are listed in Column 2. Finally, determine which feldspars are present (Column 3), and what the remaining minerals are (Column 4). These four sets of observations uniquely determine the rock type, listed in Column 5. Column 6 is a review column in which the key identification features are mentioned together with suggestions on general rock features that may be helpful.

Texture	Minerals			Rock Name	Helpful Distinguishing Features
	Quartz	Feldspar	Other		
Coarse-grained. Grains uniform in size.	Abundant.	Abundant. Orthoclase exceeds plagioclase.	Muscovite and/or biotite common. Hornblende sometimes present.	**Granite**	Quartz and feldspar predominant. Light-colored rock, commonly pink, white, shades of gray. Make sure orthoclase exceeds plagioclase. Easily confused with granodiorite.
Coarse-grained. Grains uniform in size.	Abundant.	Abundant. Plagioclase exceeds orthoclase.	Muscovite and/or biotite common. Hornblende sometimes present.	**Grandiorite**	Quartz and feldspar predominant. Shades of gray.
Coarse-grained. Grains uniform in size.	Sparse or absent.	Abundant. Plagioclase. Orthoclase rare or absent.	Biotite and/or hornblende common. Pyroxene sometimes present.	**Diorite**	About equal amounts of light- and dark-colored minerals. A darker rock than grandiorite.
Coarse-grained. Grains uniform in size.	Sparse or absent.	Abundant. Orthoclase exceeds plagioclase.	Biotite, hornblende, nepheline may be present.	**Syenite**	Commonly pink or red. Distinguish from granite by quartz content.
Coarse-grained. Grains uniform in size.	Absent.	Common. Plagioclase only.	Pyroxene abundant. Olivine may be present.	**Gabbro**	Dark minerals exceed light. A dark-colored rock. Distinguish from peridotite and pyroxenite by common plagioclase.
Coarse-grained. Grains uniform in size.	Absent.	Rare or absent.	Pyroxene abundant. Olivine may be present.	**Pyroxenite**	A dark-colored rock consisting very largely of pyroxenes.
Coarse-grained. Grains uniform in size.	Absent.	Rare or absent.	Olivine abundant. Pyroxene common to abundant.	**Peridotite**	Dark-colored rock. Olivine is commonly a clear green and grains rounded.
Medium-grained. Grains uniform in size.	Rare or absent.	Abundant. Plagioclase only.	Pyroxene common. Olivine may be present.	**Diabase**	A common medium-grained, dark gray-colored rock. Look for pyroxene and plagioclase. Distinguish from basalt by grain size and lack of extrusive volcanic features. Often called trap rock.
Fine-grained. Grains uniform in size.	Abundant. Hard to see because of grain size.	Abundant. Orthoclase exceeds plagioclase.	Hornblende, biotite may be present.	**Rhyolite**	A light-colored volcanic rock. White, gray, red, purple. May contain some glass. Often shows signs of flowage.
Fine-grained. Grains uniform in size.	Sparse or absent.	Abundant. Plagioclase exceeds orthoclase.	Pyroxene, hornblende, biotite may be present.	**Andesite**	A dark-colored volcanic rock. Shades of gray, brown, green. Glass is not common.

Table C.1 (*Continued*)

Texture	Minerals			Rock Name	Helpful Distinguishing Features
	Quartz	Feldspar	Other		
Fine-grained. Grains uniform in size.	Absent.	Abundant. Plagioclase only.	Pyroxene common. Olivine often present.	**Basalt**	A common dark-colored volcanic rock. No quartz present. Often rings like a bell when struck with a hammer.
Glassy	—	—	—	**Obsidian**	A dense glass. May contain some vesicles.
Glassy	—	A few feldspar crystals may be present.	—	**Pumice**	A glassy froth.
Phenocrysts in a coarse- or medium-grained ground mass.	Determine overall composition of rock. If a granite composition, rock is a **granite porphyry,** if a gabbro composition, rock is a **gabbro porphyry.** Common varieties are granite, granodiorite and diorite porphyries. Phenocrysts commonly quartz, feldspar, hornblende.				
Phenocrysts in a fine-grained ground mass.	Determine overall composition of rock. Texture indicates a porphyritic volcanic rock. Most common varieties are **rhyolite porphyry** and **andesite porphyry.** Phenocrysts commonly quartz, feldspar, biotite.				

similar rocks in which plagioclase is the chief feldspar. Without special equipment, the differences in feldspars are not always easily recognized, and in a general study the term granite sometimes is extended to this whole group of rocks. We sometimes recognize the variation in mineral composition by speaking of the *granitic rocks*. They are widespread in all the continents.

A special kind of granite is granite **pegmatite,** or "giant granite," which *has abnormally large grains.* Quarries in pegmatite produce in commercial quantities large sheets of mica and minerals that yield the valuable elements lithium and beryllium.

Diorite. In *diorite* the chief mineral is feldspar, mainly plagioclase, though this may not be evident to the unaided eye. Generally the dark minerals are more abundant than in granite. Diorite forms many large masses, but it is not nearly so abundant as granitic rocks.

Gabbro. Dark diorite grades into *gabbro* as the dark minerals exceed 50 per cent of the rock and plagioclase becomes subordinate. The chief dark mineral in gabbro is pyroxene, commonly with some olivine. These minerals are heavier than feldspar, and gabbro is distinctly heavier than granite and average diorite.

Diabase is fine grained, intermediate in texture between gabbro and basalt.

Pyroxenite and Peridotite. As the dark minerals displace plagioclase entirely, we reach the extreme in composition from that of granite. A granular rock composed almost entirely of pyroxene is *pyroxenite*. If considerable olivine is present with the pyroxene, the rock is *peridotite*. Both these rocks are very dark and heavy and both are commonly associated with ores containing the metals nickel, platinum, and iron. In many masses the pyroxene and olivine have been altered to serpentine, probably by hot water and gases from underlying magma.

Porphyritic Rocks. Both coarse- and fine-grained igneous rocks commonly have prominent phenocrysts. If the larger grains make up less than about 25 per cent of the rock mass, we say that the rock is *porphyritic* and give it the name suited to its groundmass (*e.g.,* porphyritic granite, porphyritic diorite, porphyritic andesite). If the proportion of phenocrysts is more than 25 per cent, we call the rock a *porphyry* and combine this term with the name that is proper for the groundmass (*e.g.,* granite porphyry, diorite porphyry, rhyolite porphyry).

Rhyolite. A fine-grained rock with phenocrysts of quartz is *rhyolite.* The quartz indicates an excess of silica and therefore a close chemical kinship to granite. Rhyolites usually have phenocrysts of feldspar and biotite as well. Colors of the groundmass range from nearly white through shades of gray, yellow, red, or purple. Rhyolite commonly has irregular bands made by flowage of stiff magma shortly before it became solid.

Andesite. A fine-grained rock generally similar to rhyolite but lacking the quartz phenocrysts is *andesite.* Usually it has phenocrysts of feldspar and dark minerals. Common colors are shades of gray and green, but some andesites are very dark, even black. Freshly broken, thin edges of dark andesites transmit some light and appear almost white when held before a bright source of light. In this way they are distinguished from basalt, which is opaque even on thin edges. The lighter-colored andesites commonly have irregular banding similar to that of rhyolite.

Andesite is extremely abundant as a volcanic rock, especially around the margins of the Pacific Ocean. The name comes from the Andes of South America.

Basalt. *Basalt* is a fine-grained rock that appears dark even on freshly broken thin edges. Common colors are black, dark brown or green, and very dark gray. In the upper parts of lava flows the rock generally is *vesicular*—filled with small openings or *vesicles* made by escaping gases. In many flows these openings have been filled with calcite, quartz, or some other mineral deposited from solution.

Glassy Rocks. Quick chilling of magma forms natural glass. *Obsidian* is a highly lustrous glassy rock. Obsidian displays a conchoidal pattern when broken.

Clear natural glass is not unknown, but most obsidians appear dark, even black. Because many of them correspond in chemical composition to rhyolite and granite, they seem to contradict the rule that rocks with high content of silica are light-colored. But obsidian chipped to a thin edge appears white, even transparent. The dark coloring results from a small content of dark mineral matter distributed evenly in the glass.

Pumice is glass froth, full of cavities made by gases escaping through stiff, rapidly cooling magma. Because many of these cavities are small winding tubes, some of them sealed, pumice will float on water for a long time. As the thin walls of the cavities transmit light, pumice is almost white, though it may form the cap of a black sheet of obsidian.

Basalt glass forms in the outer parts of some basaltic flows. It is opaque, like ordinary basalt, but has glassy luster.

Extrusive Fragmental Rocks. *Volcanic ash,* the fine debris from explosive eruptions, becomes consolidated to form *tuff.* Commonly, the particles of magma are made into froth by the expanding gas, and therefore the flakes making up an ash deposit are in large part pumice, mixed with pieces of older bedrock blasted from the walls of the vent.

The particles of ash in some falls are very hot, and when they land, the temperature is high enough so the particles weld together. *A glassy or fine-grained rock formed by fusion of volcanic ash during deposition is **welded tuff.*** Some natural glasses are now known to be welded tuffs instead of chilled lava flows. Volcanic rocks in southeastern Arizona that strongly resemble flows of rhyolite are now recognized as welded tuffs.

With increase in size of particles, tuff grades into *volcanic breccia.*

Many volcanic tuffs and breccias are stratified and look much like sedimentary rocks. Successive layers of ash are spread by air currents and become solidified as distinct beds. Furthermore, many explosive eruptions are accompanied by heavy rains, and the ash is swept by the runoff to low ground where it is spread out in thin uniform beds. The loose volcanic debris on steep slopes becomes saturated, and masses of it move down as mudflows and debris. Some pyroclastic rocks, therefore, are hybrids in classification—partly igneous and partly sedimentary.

Kinds of Sedimentary Rock

A study of the common sedimentary rocks, with good specimens in a laboratory, is a good way to become familiar with the different kinds of sedimentary rock. To aid this study, a brief description of common sedimentary rock types is given below, and is summarized in Table C.2.

*A clastic sedimentary rock that contains numerous rounded pebbles or larger particles is **conglomerate.*** The pebbles, cobbles, and boulders have been more or less rounded during transport by streams or gla-

Table C.2
Aids in Identification of Sedimentary Rocks

Rock Name	Composition	Critical Tests
1. Clastic sedimentary rocks		
Conglomerate	Cemented particles, somewhat rounded, considerable percentage of pebble size	Larger particles more than 2mm in diameter; smaller particles and binding cement in interstices
Breccia	Fragments conspicuously angular, with binding cement	Large particles of pebble size or larger
Sandstone	Rounded fragments of sand size, 0.02 to 2mm; binding cement	Grains commonly quartz, but other rock materials qualify in general classification
Arkose	Important percentage of feldspar grains, sand size or larger	Essential that feldspar grains make 25 per cent or more of rock; some may be larger than sand size
Graywacke	Fragments of quartz, feldspar, rock fragments of any kind, with considerable clay	Poor assortment of several kinds of ingredients, with considerable clay in matrix
Siltstone	Chiefly silt particles, some clay	Surface is slightly gritty to feel
Claystone and *Shale*	Chiefly clay minerals	Surface has smooth feel, no grit apparent
2. Rocks of organic and chemical origin		
Limestone	Calcite; may be even grained and crystalline	Easily scratched with knife; effervesces in cold dilute hydrochloric acid
Dolostone	Dolomite; may be even grained and crystalline	Harder than limestone, softer than steel; requires scratching or powdering for effervescence in cold dilute hydrochloric acid

cier ice or in buffeting by waves along a shore (Fig. 5.22). They can consist of any kind of rock but most commonly of the kinds rich in the durable mineral quartz. Usually the spaces between pebbles contain sand cemented with silica, clay, limonite, or calcite.

Sedimentary breccia is *a clastic sedimentary rock that resembles conglomerate, but most of whose fragments are angular instead of rounded.* We find all graduations between conglomerate and breccia.

The term *breccia* should always be used with a modifier to indicate origin; for example, sedimentary breccia, or volcanic breccia.

A clastic sedimentary rock consisting of cemented sand grains is **sandstone.** With progressive change in size of grain, coarse sandstone grades into *siltstone.* In many rocks, grain sizes are mixed; so we speak of conglomeratic sandstone or sandy siltstone.

In sandstone the grains consist almost entirely of quartz. The cementing material varies, as in conglomerate; calcium carbonate is common, but silica makes a more durable rock. Color in sandstone, produced partly by the color of the grains, partly by that of the cementing material, varies within a wide range.

Arkose is *a variety of sandstone with a large proportion of feldspar grains.* A composition consisting of feldspar and quartz suggests granite, so that arkose might be mistaken for it. In arkose the grains do not interlock; they are rounded and separated by fine-grained cementing material.

Claystone (also called *shale*) is *a clastic sedimentary rock made of compacted clay and silt.* It is so fine grained that to the unaided eye it seems homogeneous. Claystone is soft and generally feels smooth and greasy, but some fine sand or coarse silt may make it feel gritty. Claystones generally split into thin layers or flakes parallel to the sedimentary layering. Rocks of similar composition but with thick, blocky layers are termed *mudstone.*

The color of claystones and mudstones ranges through shades of gray, green, red, and brown. Some layers that contain considerable carbon are black.

Limestone, which may be either *a clastic or chemical sedimentary rock, consisting chiefly of calcite,* has many impurities and varies greatly in appearance. Some limestones that are uniformly fine-grained probably were formed as chemical precipitates, aided more or less by tiny organisms. Some of the sediment on today's sea floors probably represents an early stage in the formation of fine-grained limestone. By contrast, many limestones are coarse-grained, either from crystallization of the calcium carbonate or because they are made largely of detrital shell fragments.

Dolostone, like limestone, is either *a clastic or a chemical sedimentary rock consisting chiefly of* the mineral *dolomite,* $CaMg(CO_3)_2$. Dolomite looks like calcite, which is why dolostone looks like limestone. But dolomite is slightly harder than calcite and only effervesces with acid on a scratched surface or in powdered form.

Kinds of Metamorphic Rock

Metamorphic rocks are classified on the basis of texture (Chap. 5). Where a particular mineral is very obvious, its name may be used as a prefix. Examples of the seven common metamorphic rocks are given below and identifying features are summarized in Table C.3.

Slate is *fine-grained metamorphic rock with a pronounced cleavage.* Cleavage planes separate slates into thin, flat plates, commonly cutting across original sedimentary layering. Although surfaces of the cleavage slabs have considerable luster, mineral grains can be seen only with very high magnification (Fig. 5.23). A common color is dark bluish gray, generally known as "slate color," but many slates are red, green or black.

Phyllite is *exceptionally lustrous rock representing a higher stage of metamorphism than slate.* The mica flakes responsible for the luster can be seen only with magnification, but the mineral grains are coarser than in slates and some phyllites have visible grains of garnet and other minerals (Fig. 5.25). The cleavage plates commonly are wrinkled or even sharply bent.

Schist, a *well-foliated metamorphic rock in which the component platy minerals are clearly visible,* represents a higher stage in metamorphism than phyllite. Mica schist is rich in mica (either biotite, muscovite, or both together). Chlorite schist and hornblende schist also are common. Quartz is abun-

Table C.3

Aids in Identification of Metamorphic Rocks

Rock Name	Distinguishing Characteristics
1. Foliated metamorphic rocks	
Slate	Cleaves into thin, plane plates that have considerable luster; commonly the sedimentary layers of parent rock make lines on plates; thin slabs ring when they are tapped sharply
Phyllite	Surfaces of plates highly lustrous; plates commonly wrinkled or sharply bent; grains of garnets and other minerals on some plates
Schist	Well foliated, with visible flaky or elongate minerals (mica, chlorite, hornblende); quartz a prominent ingredient; grains of garnet and other accessory minerals common; foliae may be wrinkled
Gneiss	Generally coarse-grained, with imperfect but conspicuous foliation; lenses and layers differ in mineral composition; feldspar, quartz, and mica are common ingredients
2. Nonfoliated metamorphic rocks	
Quartzite	Consists wholly of quartz sand cemented with quartz; outlines of sand grains show on broken surfaces; the breaks passing through the grains; wide range in shades of color
Marble	Wholly crystallized limestone or dolostone; grain varies from coarse to fine; responds to hydrochloric acid test, as do calcite and dolomite; accessory minerals have developed from impurities in original rock
Hornfels	Hard, massive, fine-grained rock, commonly with scattered grains or crystals of garnet, andalusite, staurolite, or other minerals that are common in zones of contact metamorphism

dant in all kinds of schists and many are studded with garnets and other metamorphic minerals.

Gneiss is *coarse-grained, foliated metamorphic rock, commonly with marked layering but with imperfect cleavage.* Many gneisses have a streaky, roughly banded appearance, caused by alternating layers that differ in mineral composition (Fig. 5.24). Feldspars, quartz, micas, amphiboles, and garnets are common minerals in gneisses. **Granite gneiss** is *distinctly banded rock with the mineral composition of granite.*

Marble is created by metamorphism of limestone, during which the calcite grains grow until the entire rock mass has become coarse grained. Dolomite marble is formed in the same way from dolostone. **Marble,** then, is merely *coarsely crystalline limestone or dolostone.* Impurities in the original rock may cause the growth of small amounts of pyroxene, amphibole, and other minerals, which in many marbles make striking patterns (Fig. 1.7 shows an example).

Some commercial "marbles" are actually non-metamorphosed limestones and dolostones that "take a good polish."

Quartzite is *a metamorphic rock developed from sandstone by introduction of silica into all spaces between the original grains of quartz.* When quartzite is broken, the fracture passes through the original quartz grains, not around them as in ordinary sandstone. Quartzites usually have no patterns of foliation. Quartzite and marble are examples of nonfoliated metamorphic rocks, in contrast to gneiss, schist, phyllite, and slate, which are all foliated.

Hornfels is a rock formed near intrusive igneous bodies where the invaded rock is greatly altered by high temperature. Shale and some other fine-grained rocks are thus changed to hornfels, *a very hard, nonfoliated metamorphic rock, commonly studded with small crystals of mica and garnet.*

Appendix D

Maps, Cross Sections, and Field Measurements

Uses of Maps

An important part of the accumulated information about the geology and morphology of the Earth's crust exists in the form of maps. Nearly everyone has used automobile road maps in planning a trip or in following an unmarked road. A road map of a state, province, or county does what all maps have done since their invention at some unknown time more than 5,000 years ago; it reduces the pattern of part of the Earth's surface to a size small enough to be seen as a whole. Maps are especially important for an understanding of geologic relations because a continent, a mountain chain, and a major river valley are of such large size that they cannot be viewed as a whole unless represented on a map.

Furthermore, although most geologic maps are supplemented by text material, they contain information that cannot be stated as fully and accurately in words or by any other means. Maps are the core of most geologic reports based on field study and are supplementary to many other kinds of geologic papers. This appendix is designed to outline basic information about maps and about common field measurements used in the construction of geologic maps.

A map can be made to express much information within a small space by the use of various kinds of symbols. Just as some aspects of physics and chemistry use the symbolic language of mathematics to express significant relationships, so many aspects of geology use the simple symbolic language of maps to depict relationships too large to be observed within a single view. Maps made or used by geologists generally depict either one or two of three sorts of things:

1. Hills, valleys, and other surface forms. Most maps of this kind are *geographic maps*.

2. Distribution and attitudes of bodies of rock or regolith. Maps showing such things are *geologic maps*.

3. Geographic features such as mountains, rivers, and seas, not as they are today but reconstructed as they are inferred to have been at some time in the past (Fig. 14.13). Maps of this kind are *paleogeographic maps*.

The first two express the results of direct observation and measurement and are frequently included on a single map. The third expresses concepts built up from whole groups of observations and necessarily is shown on special maps separate from those representing present-day features.

Base Maps

Every map is made for some special purpose. Road maps, charts for sea or air navigation, and geologic maps are examples of three special purposes. But whatever the purpose, all maps have two classes of data: *base* data and *special-purpose* data. As base data most geologic maps show a latitude-longitude grid, streams, and inhabited places; many also show roads and railroads. A geologist may take an existing *base map* containing such data and plot geologic information on it, or he may start with blank paper and plot on it both base and geologic data, a much slower process if the map is made accurately.

Two-Dimensional Base Maps. Many base maps used for plotting geologic data are two-dimensional; that is, they represent length and breadth but not height. A point can be located only in terms of its horizontal distance, in a particular direction, from some other point. Hence a base map always embodies the basic concepts of direction and distance. Two natural reference points on Earth are the North and South Poles. Using these two, the ancient Greeks established a grid by means of which any other point could be located. The grid we use now consists of lines of *longitude* (half circles joining the poles) and *latitude* (parallel circles concentric to the

poles) (Fig. D.1). The longitude lines (*meridians*) run exactly north-south, crossing the east-west *parallels* of latitude at right angles. The circumference of the Earth at its Equator and the somewhat smaller circumference through its two poles being known with fair accuracy, it is possible to define any point on the Earth in terms of direction and distance from either pole or from the point of intersection of any parallel with any meridian.

For convenience in reading, most maps are drawn so that the north direction is at the top or upper edge of the map. This is an arbitrary convention adopted mainly to save time. The north direction could just as well be placed elsewhere, provided its position is clearly indicated.

The accuracy with which distance is represented determines the accuracy of the map. *The proportion between a unit of distance on a map and the unit it represents on the Earth's surface* is the **scale** of the map. It is expressed as a simple proportion, such as 1 : 1,000,000. This ratio means that 1 foot, meter, or other unit on the map represents exactly 1,000,000 feet, meters, or other units on the Earth's surface; it works out to 1 inch equals about 16 miles and is approximately the scale of many of the road maps widely used by motorists. Scale is also expressed graphically by means of a numbered bar, as is done on most of the maps in this book. A map

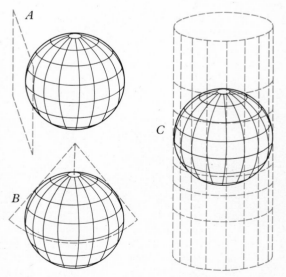

Figure D.1 The Earth's latitude-longitude grid can be projected onto a plane *A*, a cylinder *C*, or a cone *B* that theoretically can be cut and flattened out.

with a latitude-longitude grid needs no other indication of scale (except for convenience) because the lengths of a degree of longitude (varying from 69.17 statute miles at the Equator to 0 at the poles) and of latitude (varying from 68.70 statute miles at the Equator to 69.41 at the poles) are known.

Map Projections. The Earth's surface is nearly spherical, whereas nearly all maps other than globes are planes, usually sheets of paper. It is geometrically impossible to represent any part of a spherical surface on a plane surface without distortion (Fig. D.2). The latitude-longitude grid has to be *projected* from the curved surface to the flat one. This can be done in various ways, each of which has advantages, but all of which represent a sacrifice of accuracy in that the resulting scale on the flat map will vary from one part of the map to another. Examples of cylindrical projections are Figs. 12.3 and 12.13. The most famous of these is the Mercator projection; although it distorts the Polar Regions very greatly, compass directions drawn on it are straight lines. Because this is of enormous value in navigation, the Mercator projection is widely used in navigators' charts.

Figure 16.8,*A* is an example of conic projection. Some commonly used varieties are polyconic, in which not one cone, as in Fig. D.1,*B*, but several cones are employed, each one tangent to the globe at a different latitude. This device reduces distortion.

Figure 20.14 represents a plane projection, and Fig. 10.1 is a projection with certain areas, unimportant for the purpose of this map, eliminated. In a map of a very small area, such as Fig. 8.33, the distortion is of course slight, but it is there nevertheless.

Topographic Maps. A more complete kind of base map is three-dimensional; it represents not only length and breadth but also height. Therefore it shows **relief** (*the difference in altitude between the high and low parts of a land surface*) and also **topography,** defined as *the relief and form of the land. A map that shows topography* is a **topographic map.** Topographic maps can give the form of the land in various ways. The maps most commonly used by geologists show it by contour lines.

Contours. A **contour line** (often called simply a *contour*) is *a line passing through points having the same altitude above sea level.* If we start at a certain altitude on an irregular surface and walk in such a way as to go neither uphill nor downhill, we will trace out a path that corresponds to a contour line. Such a path will curve around hills, bend upstream in valleys, and swing outward around spurs. Viewed broadly, every contour must be a closed line, just as the shoreline of an island or of a continent returns upon itself, however long it may be. Even on maps of small areas, many contours are closed lines, such as those at or near the tops of hills. Many, however, do not close within a given map area; they extend to the edges of the map and join the contours on adjacent maps.

Imagine an island in the sea crowned by two prominent isolated hills, with much steeper slopes on one side than on the other and with an irregular shoreline. The shoreline is a contour line (the zero contour) because the surface of the water is horizontal. If the island is pictured as submerged until only the two isolated peaks project above the sea, and then raised above the sea 20 feet at a time, the successive new shorelines will form a series of contour lines separated by 20-foot contour intervals. (A **contour interval** is *the vertical distance between two successive contour lines,* and is commonly the same throughout any one map.) At first, two small islands will appear, each with its own shoreline, and the contours marking their shorelines will have the form of two closed lines. When the main mass of the island rises above the water, the remaining shorelines or contours will pass completely around the land mass. The final shoreline is represented by the zero

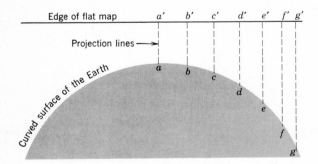

Figure D.2 Equally spaced points (*a, b, c,* etc.) along a line in any direction on the Earth's surface become unequally spaced when projected onto a plane. This is why all flat maps are distorted. Compare Fig. 4.8, in which points equally spaced in a plane are distorted by the Earth's curved surface.

contour, which now forms the lowest of a series of contours separated by vertical distances of 20 feet.

As the island is raised, the successive new shorelines are not displaced through so great a horizontal distance where the slope is steep as where it is more gradual. In other words, the water retreats through a shorter horizontal distance in falling from one level to the next along the steep slope than along the gentle slope. Therefore, when these successive shorelines are projected upon the flat surface of a map, they will be crowded where the slope is steep and farther apart where it is moderate. In order to facilitate reading the contours on a map, certain contours (usually every fifth line) are drawn with a wider line. Contours are numbered at convenient intervals for ready identification. The numbers are always multiples of the contour interval and are placed between broken ends of the contour they designate.

Because the contours that represent a depression without an outlet resemble those of an isolated hill, it is necessary to give them a distinctive appearance. Depression contours therefore are *hatched;* that is, they are marked on the downslope side with short transverse lines called *hachures.* An example is shown on one contour in Fig. D.4. The contour interval employed is the same as in other contours on the same map.

Idealized Example. Figures D.3 and D.4 show the relation between the surface of the land and the contour map representing it. Figure D.3, a perspective sketch, shows a stream valley between two hills, viewed from the south. In the foreground is the sea, with a bay sheltered by a curving spit. Terraces in which small streams have excavated gullies border the valley. The hill on the east has a rounded summit and sloping spurs. Some of the spurs are truncated at their lower ends by a wave-cut cliff, at the base of which is a beach. The hill on the west stands abruptly above the valley with a steep scarp and slopes gently westward, trenched by a few shallow gullies.

Each of the features on the map (Fig. D.4) is represented by contours directly beneath its position in the sketch.

Air Photographs

Figures 10.7 and 11.20 are photographs made from airplanes, with cameras pointing obliquely at the Earth's surface. Oblique air photographs enable us to see over a much larger area than could be seen from a single point on the ground and are therefore useful in making clear the broad rather than the detailed relations of various geologic features. The figures referred to above were selected for this very purpose. All oblique photographs are "pictorial" in that they show hills and valleys in perspective.

In contrast, Figs. 9.18 and 16.5 are photographs made with cameras pointing down vertically at the Earth's surface. Unlike oblique views, vertical photographs do not show perspective; they look "flat." But although distortion is always present (Fig. D.2) they show the pattern of the ground with distortion at a minimum. The vertical air photograph, therefore, is one kind of map. It is used widely as a base map on which geologic data are plotted in the field. Generally the scale of such a photograph must be 1 : 20,000 or larger, for it shows so much detail that on a smaller scale the various features can become blurred.

Used by itself, a vertical air photograph is a two-

Figure D.3 Perspective sketch of a landscape. (Modified from U. S. Geol. Survey.)

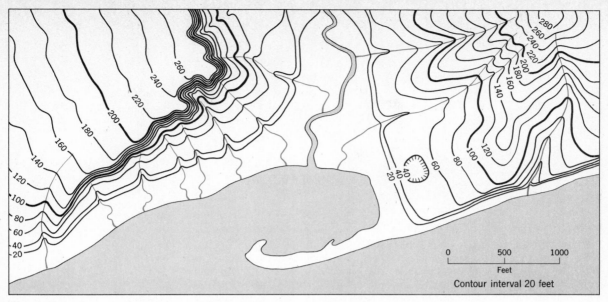

Figure D.4 Contour map of the area shown in Fig. D.3. (Modified from U. S. Geol. Survey.)

dimensional map. However, two photographs taken in sequence from a flying airplane, so that they overlap by a mile or two, can form a three-dimensional map with the vertical dimension exaggerated. The camera lens acts as a series of "eyes" with an interpupillary separation equal to the distance flown between successive pictures. This separation is so much greater than the distance between the two eyes of a human being that when overlapping photographs are viewed through a stereoscope, the relief of the ground surface becomes startlingly apparent. In fact we can see much more than we could by looking down from above at the same field of view. Not only is this a direct aid in plotting geologic data on the photograph, but also it forms the basis of a method of drawing an accurate contour map of the area photographed without the necessity of making a slow, laborious, and costly ground survey. Topographic contour maps are made largely by methods based on this principle. Finally, an individual air photograph can be used as a map, on which geologic features are then plotted.

Measurements Made in the Field

The most common kind of *geologic map* is *a map that shows the distribution, at the surface, of rocks of various kinds or of various ages.* Examples are Fig.

8.15, a sketch map of bodies of alluvium of successive ages, Fig. 15.20,*B*, a sketch of eroded and folded strata, and Fig. 16.8, sketch maps showing the distribution of basalt compared with that of other rocks.

Field Equipment. A first essential for making a geologic map in the field is a base map, preferably a topographic map with a contour interval no greater than 10 or 20 feet. Other probable necessities are a hammer, a steel tape for measuring thicknesses of strata, and a pocket transit (geologists' compass) known almost universally in the United States and Canada as a Brunton compass, or simply a brunton, after the name of its designer. This compact instrument (Fig. D.5) is a compass, a clinometer, and a sighting device used in reading compass directions and in hand leveling.

Strike and Dip. One of the commonest kinds of field measurement is determination of the attitude of a stratum. To represent the orientation of an inclined plane, we need to remember two principles of geometry: (1) the intersection of two planes defines a line; (2) in an inclined plane only one horizontal direction exists. The horizontal line formed by the intersection of the inclined plane with the horizontal plane can be visualized as the water line on a boat-launching ramp or by placing the edge

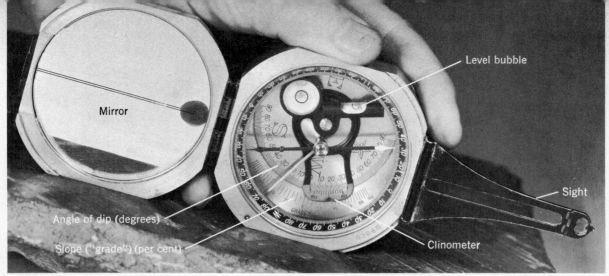

Figure D.5 Use a brunton (pocket transit) to measure angle of dip of a stratum. The sides of the brunton are plane, parallel surfaces. With one side placed on a sloping plane surface in the direction of dip, the level bubble of the clinometer is centered by means of a lever on the back of the case (not visible). The angle is read on one of the two arcs below. The inner semicircular arc, calibrated in degrees from 0° to 90°, shows a reading of 17.5°; the outer, shorter scale, calibrated in per cent, reads 31 per cent. Compass directions are read on the dark outer scale when the instrument is held face up and is leveled by use of the circular level bubble. The hinged mirror and sight aid in taking bearings on selected points, and in using the brunton as a hand level (Fig. D.8), with the clinometer set to read zero degrees.

of a carpenter's level in a horizontal position on a sloping plane (Fig. D.6). Instead of the level we can place one edge of a brunton, in the level position, against the inclined surface of a stratum.

The compass direction of the horizontal line in an inclined plane is the **strike** of the plane. In the United States and Canada, strike and other compass directions commonly are expressed as angles between 0° and 90° east or west of true north. A strike trending 20° east of north would be written N20E (Fig. D.7,*B*); one trending 72° west of north would be written N72W. In Europe and elsewhere strikes are measured as angles clockwise from true north (0°), through a full circle of 360°.

Once we know the strike we need only one more measurement to fix the orientation of the plane. That is the **dip,** *the angle in degrees between a horizontal plane and the inclined plane, measured down from horizontal in a plane perpendicular to the strike.* Dip is measured with a *clinometer,* usually a brunton in the position seen in Fig. D.5. Not only the angle but the direction of dip (always at right angles to the strike) must be noted.

When plotted on a map as symbols (Fig. D.7,*A*),

orientation measurements made at each locality graphically convey the significant structural features of an area (Fig. D.7,*C*). In a similar manner dip directions of cross strata can be plotted on a map to indicate directions of flow of ancient currents (Fig. 14.13).

Thicknesses of Strata. The perpendicular distance between the upper and lower surfaces of a stratum, the *thickness* of the stratum, can be measured directly with a steel tape or other scale or with a hand level (Fig. D.8,*A*) or altimeter. Thickness of an inclined stratum is usually calculated from simple trigonometric data (Fig. D.8,*B*).

Geologic Maps

Patterns Made by Strata. *Horizontal Strata.* Figure D.9,*A* is a geologic sketch map showing three units: two shale units with a unit of limestone between them. Each is a *formation* as defined in Chapter 14. The sketch map has been constructed on a topographic base representing two rounded hills with a stream between them. The two black lines

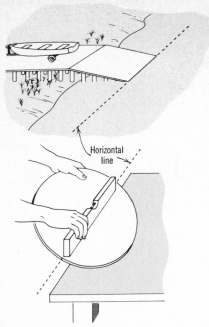

Figure D.6 The one horizontal direction in an inclined plane is illustrated by the water line against a boat-launching ramp, and by a carpenter's level held against an inclined board. The two horizontal lines shown are the directions of strike of the two inclined planes (see text).

represent the traces, along the land surface, of the interface of contact between the base of the limestone and underlying shale and of that between the top of the limestone and overlying shale. Symbols (Fig. D.10,A) at five places indicate that the strata exposed at those places were found to be horizontal; that is, their dip is zero.

The geologist who drew the black lines, or "contacts" determined the altitude of the lower one by hand level from a known point farther down the slope and then walked around each hill, following by eye, and then plotting on his map, the change in type of rock from shale below to limestone above. His circuit of each hill brought him back to his starting point, and the contact he had plotted maintained a constant altitude around both hills. The line representing it therefore extends between the same two contours around both hills and is parallel with the contours.

The geologist then repeated the process with the higher contact, thereby completing the map. By hand leveling (Fig. D.8,A) he could measure the vertical distance between base and top of the limestone and thereby find that the stratum is 14 feet thick. For the thicknesses of each of the two layers of shale, however, he could measure only minimum values, because neither the base of one of them nor the top of the other one is exposed within the area of the map.

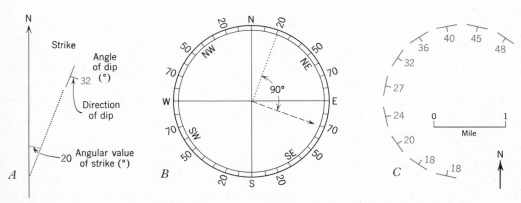

Figure D.7 A. Strike and dip are plotted on maps with the T-shaped symbol shown in blue. B. Strike and direction of dip shown in A are indicated by blue lines on the face of a compass. Strike is N20E, direction of dip S70E. Angle of dip, being a vertical angle, can not be shown. C. Strike and dip symbols, from measurements at several localities on the same stratum, plotted on a geologic map. Evidently the structure is a syncline plunging southeast, the northeast limb dipping more steeply than the southwest (Fig. 17-9).

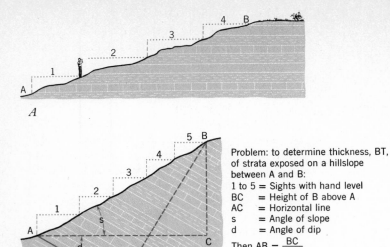

Problem: to determine thickness, BT, of strata exposed on a hillslope between A and B:

1 to 5 = Sights with hand level
BC = Height of B above A
AC = Horizontal line
s = Angle of slope
d = Angle of dip

Then $AB = \dfrac{BC}{\sin s}$

$BT = AB \sin (s + d)$

Figure D.8 Use of the hand level in measuring heights, and thicknesses of strata exposed in slopes. *A*. Thicknesses of horizontal strata are measured directly by hand level. The geologist determines accurately the height of his eyes above the ground, and multiplies this figure by the number of "sights" (1, 2, 3, etc.) he makes. If, instead of a hand level, he uses an altimeter, he computes differences between readings at critical points. *B*. Where strata are inclined, vertical distance between base and top is measured, up the slope, by hand level. Angle of dip is measured with a clinometer. Thickness is determined by the trigonometric computation given above.

The area, on a geologic map, shown as occupied by a particular rock unit is the **outcrop area** of that unit. An outcrop area, therefore, is the area of the Earth's surface in which some particular rock unit constitutes the highest part of the underlying rock, whether exposed at the surface or covered by regolith derived from it. Generally the aggregate area of actual exposure, free from overlying regolith, is far less than the outcrop area.

Vertical Strata. The lines representing contacts between strata whose dip is vertical appear as straight lines on a geologic map (Fig. D.9,*B*). They change direction only where the strike of an interface of contact changes, or where the contact is offset by a fault.

Inclined Strata. On a geologic map, the lines of contact between strata that are inclined cross the contours in directions that vary with angle of dip of the strata and with slope of the surface as repre-

sented by trend and spacing of contours. In Fig. D.9,*C* we note that the lines representing the top and base of the limestone layer swing west through broad arcs in each of the two hills and bend east in a chevronlike pattern as they cross the valley. Measurements of strike and dip at five localities show that all three strata are dipping east. If the dip were west instead of east, the pattern would be reversed, with arcs swinging east and the chevron pointing west.

Faults. On a geologic map faults (which are, of course, one kind of surface of contact) are represented by lines thicker than those used for all other kinds of contact. Where faults displace rock units they also displace other contacts between the units; these appear on a map as offsets (Fig. D.9,*D*). The details of faults and other structures are brought out on geologic maps by means of special symbols (Fig. D.10,*A*).

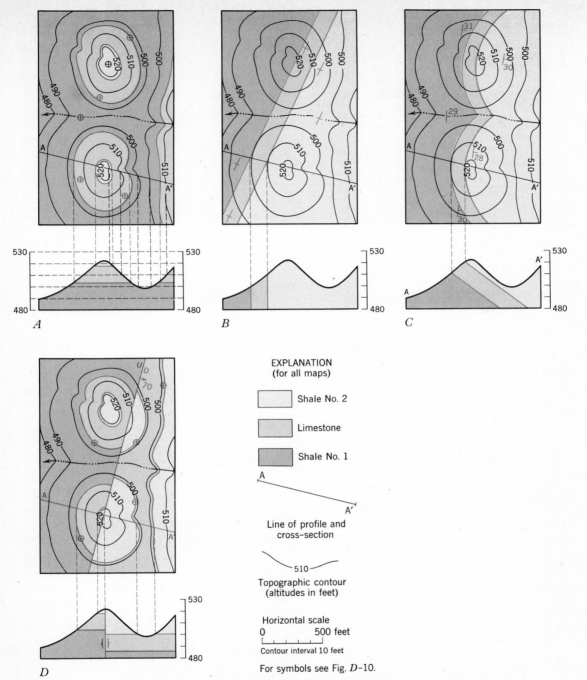

EXPLANATION
(for all maps)

Shale No. 2

Limestone

Shale No. 1

A — — — A'

Line of profile and
cross-section

—510—

Topographic contour
(altitudes in feet)

Horizontal scale

0   500 feet

Contour interval 10 feet

For symbols see Fig. D-10.

Figure D.9   Simple geologic sketch maps, each with topographic profile and geologic cross-section below it, showing relation between pattern on map and geometry of section in four situations: *A.* Strata horizontal. *B.* Strata vertical. *C.* Strata dipping east at 30° angle. *D.* Horizontal strata cut by a fault.

Symbol	Explanation
$\overline{}_{18}$	Strike and dip of strata
$\diagdown_{90}$	Strike of vertical strata; tops of strata are on side marked with angle of dip
⊕	Structure of horizontal strata; no strike, dip = 0
$\overline{}_{43}$	Strike and dip of foliation in metamorphic rocks
◆	Strike of vertical foliation
↑↓	Anticline; arrows show directions of dip away from axis
	Syncline; arrows show directions of dip toward axis
↗ 21	Anticline, showing direction and angle of plunge
↗ 15	Syncline, showing direction and angle of plunge
⊤⊤⊤⊤⊤⊤⊤	Normal fault; hachures on downthrown side
⊤⊤⊤⊤⊤⊤	Reverse fault; arrow shows direction of dip, hachures on downthrown side
50 U D	Dip of fault surface; D, downthrown side; U, upthrown side
	Directions of relative horizontal movement along a fault
▲▲▲	Low-angle thrust fault; barbs on upper block

A

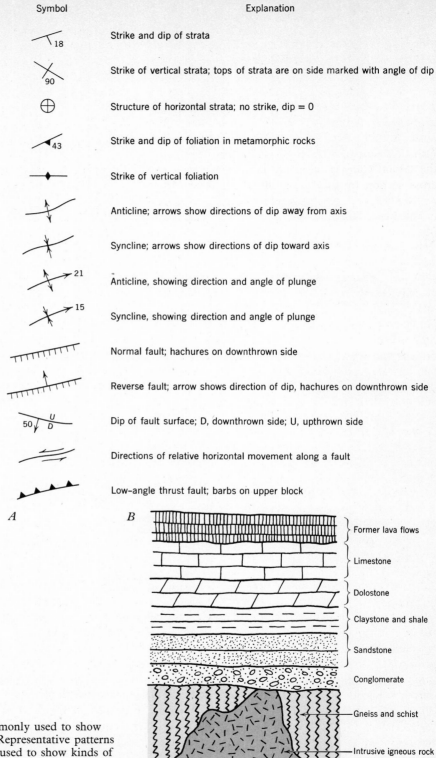

B

Former lava flows
Limestone
Dolostone
Claystone and shale
Sandstone
Conglomerate
Gneiss and schist
Intrusive igneous rock

Figure D.10 *A*. Symbols commonly used to show structure on geologic maps. *B*. Representative patterns commonly, but not universally, used to show kinds of rock in geologic cross-sections.

Geologic Cross Sections

Once a geologic map has been constructed, it can be used as the basis for constructing cross-sections along planes that extend downward below the surface, at right angles to the plane of the map. A *geologic cross section* is *a diagram showing the arrangement of rocks in a vertical plane*. It represents what would be revealed, as outcrops, in the vertical wall of a deep trench. In a natural trench such as the Grand Canyon we could construct a geologic cross section by measuring the exposed rocks directly, with an accuracy as great as would be possible in making a geologic map of a comparable area. Most sections, however, are made by geometric projection of the contacts already drawn on a map. The accuracy of a section constructed in this way varies with the kinds of rock bodies present. Where strata of originally wide extent have been deformed, surfaces of contact are generally parallel and can be projected downward with fair accuracy. But where rock bodies having irregular surfaces of contact are present and where exposures of the rocks are few, sections are less accurate and some are little more than speculative. Sections of this kind can be checked and corrected only by comparing them with subsurface data obtained from the records of drilling or geophysical exploration.

The method of constructing geologic cross sections is illustrated in Fig. D.9. A distinctive unit of limestone, shown in four geologic sketch maps, is horizontal in *A*, vertical in *B*, dipping at 30° in *C*, and cut by a fault in *D*. What would be the geometry of this unit as seen in a vertical plane?

Below the map we construct a grid of horizontal lines, spaced to represent the contours. Dashed vertical lines are drawn from the intersection of line AA′ with each contour on the map to corresponding horizontal lines on the grid. By connecting the points of intersection, we construct an east-west profile of the hill, a topographic profile. Next we draw lines from borders of the outcropping unit to intersect the profile on the grid. By connecting points on the profile that represent, respectively, the top and bottom of the unit, the cross section of the layer is completed.

Because the topography on all four base maps is the same, construction lines for making the topographic profile are shown for *A* only. On the identical profiles for *B*, *C*, and *D*, the lines dropped from the

lower and upper boundaries of the outcropping limestone unit give the points required for completing the sections.

Patterns commonly used in geologic cross sections to represent rocks of various kinds are illustrated in Fig. D.10,*B*.

Maps of Igneous and Metamorphic Rocks

Because sedimentary rocks are commonly stratified, with rather distinct upper and lower surfaces of contact, and because many of them are widely extensive, they present fewer mapping problems than do most igneous and metamorphic rocks. An exception is volcanic rocks that occur in widespread sheets like sedimentary strata and that may be interbedded with layers of volcanic ash. In some districts it is possible to map these units separately just as sedimentary formations are mapped. But in many volcanic fields, bodies of extrusive igneous rock and bodies of volcanic ash, erupted from a number of centers at irregular intervals, are mixed in great confusion. Such complexes can be mapped only as general assemblages, without regard to the various kinds of volcanic rock that compose them.

Intrusive igneous bodies exposed at the surface are highly varied in form and in their relation to associated rocks. Commonly they cut across older sedimentary units and their outcrop areas range widely in shape and in size. Those large enough to be shown clearly to the scale of the map are outlined and marked with distinctive colors or patterns.

Most metamorphic rocks have complex structure and are not easily divisible into distinctive units. Many such rocks therefore are represented on maps by a single color or pattern. In many mountain zones, metamorphic rocks intermixed with bodies of granite or other intrusive rocks are treated as a unit and are identified on many maps as "basement complex."

In some areas, on the other hand, metamorphic rocks are mapped in great detail. Various kinds of schists and gneisses are mapped individually, as formations; attitudes of foliation are recorded at many points by symbols (Fig. D.10,*A*), and faults are mapped, many of them through long distances. In such areas the various intensities of metamorphism that have affected the rocks are identified through critical minerals. With such identification at a large number of points, it is possible to draw a series of isograds. In this case an ***isograd*** is *a line on a map, connecting points of first occurrence of a given mineral in metamorphic rocks.* The trends and patterns of isograds give clues to the directions in which the former pressures were applied, and also suggest former temperature gradients.

Isopach Maps

Somewhat analogous to an isograd is an ***isopach,*** *a line on a map, that connects points of equal thickness of a rock unit.* Isopachs are usually shown on special maps of limited areas where thickness is particularly significant. For example, an isopach map might show the thickness of a surface cover of loess (Fig. 10.23), volcanic ash (Fig. 5.9), alluvium, or glacial drift. Again it might represent the thinning, in some direction, of a buried unit similar to the one appearing in cross section in Fig. 20.7,*D*. An isopach map of that unit probably would be useful in a local search for petroleum.

Selected References

Base Maps

Raisz, Erwin, 1948, General cartography, 2nd ed.: New York, McGraw-Hill.
Robinson, A. H., 1960, Elements of cartography, 2nd ed.: New York, John Wiley. (See also Lobeck and Tellington, below, p. 1–198.)

Air Photographs

American Society of Photogrammetry, 1952, Manual of photogrammetry, 2nd ed.: Washington, American Society of Photogrammetry.
American Society of Photogrammetry, 1960, Manual of photographic interpretation: Washington, American Society of Photogrammetry.
Lobeck, A. K., and Tellington, W. J., 1944, Military maps and air photographs: New York, McGraw-Hill, p. 199–250.
Miller, V. C., and Miller, C. F., 1961, Photogeology: New York, McGraw-Hill.

Geologic Maps and Cross Sections

Blyth, F. G. H., 1965, Geological maps and their interpretation: London, Edward Arnold.

Field Techniques

Compton, R. R., 1962, Manual of field geology: New York, John Wiley.

Appendix E

Units and Their Conversions

Multiples and Submultiples

Multiples and Submultiples	Prefixes
$1,000,000 = 10^6$	mega
$1,000 = 10^3$	kilo
$100 = 10^2$	hecto
$10 = 10$	deka
$0.1 = 10^{-1}$	deci
$0.01 = 10^{-2}$	centi
$0.001 = 10^{-3}$	milli
$0.000001 = 10^{-6}$	micro

Units of Measure

Linear Measure

1 mile (mi.)	= 5280 feet (ft)
1 foot (ft)	= 12 inches (in.)
1 kilometer (km)	= 1000 meters (m) = 10^3m
1 meter (m)	= 100 centimeters (cm) = 10^2cm
1 centimeter (cm)	= 0.01m = 10^{-2}m
1 millimeter (mm)	= 0.001m = 10^{-3}m
1 angstrom (Å)	= 0.0000000001m = 10^{-10}m
1 micron (μ)	= 0.001mm

Area Measure

1 square mile	= 640 acres
1 acre	= 43,560 square feet
1 acre	= 4,840 square yards
1 square meter	= 10,000 square centimeters (cm^2)
1 square kilometer	= 100 hectares

Volume and Cubic Measure

1 cubic foot	= 1,728 cubic inches
1 cubic yard	= 27 cubic feet
1 barrel (oil)	= 42 gallons
1 liter	= 0.001 cubic meter
	= 1 cubic decimeter
1 liter	= 1,000 milliliters
10 milliliters	= 1 centiliter
1 milliliter	= approximately 1 cubic centimeter (cc)
1 cubic meter (m^3)	= 1,000,000 cubic centimeters

Weights and Masses

1 short ton	= 2,000 pounds
1 long ton	= 2,240 pounds
1 pound (avoirdupois)	= 7,000 grains
1 ounce (avoirdupois)	= 437.5 grains
1 gram	= 15.432 grains
1,000 grams	= 1 kilogram
1,000 kilograms	= 1 metric ton

Units of Energy and Power

Energy

1 erg = the work done by a force of 1 dyne when its point of application moves through a distance of 1 centimeter in the direction of the force.

1 calorie (cal) = the amount of heat that will raise the temperature of 1 gram of water 1 degree Celsius with the water at 4 degrees Celsius.

1 erg = 9.48×10^{-11} British thermal unit (BTU)
1 erg = 7.367×10^{-8} foot-pounds
1 erg = 2.778×10^{-14} kilowatt-hours
1 kilowatt-hour = 3,413 BTU = 3.6×10^{13} ergs
1 BTU = 2.928×10^{-4} kilowatt-hours
   $= 1.0548 \times 10^{10}$ ergs
1 joule = 1×10^7 ergs
1 calorie = 3.9685×10^{-3} BTU
   $= 4.186 \times 10^7$ ergs

Power (= energy per unit time)
1 watt = 3.4129 BTU/hour
1 watt = 1.341×10^{-3} horsepower
1 watt = 1 joule per second
1 watt = 14.34 calories per minute

Conversions

English-Metric Conversions

1 inch	= 25.4 millimeters
1 foot	= 0.3048 meter
1 yard	= 0.9144 meter
1 mile	= 1.609 kilometers
1 square inch	= 6.4516 square centimeters
1 square foot	= 0.0929 square meter
1 square yard	= 0.836 square meter
1 acre	= 0.4047 hectare
1 square mile	= 2.590 square kilometers
1 cubic inch	= 16.39 cubic centimeters
1 cubic foot	= 0.0283 cubic meter
1 cubic yard	= 0.7646 cubic meter
1 cubic mile	= 4.168 cubic kilometers
1 gallon	= 3.784 liters
1 ounce	= 28.33 grams
1 pound	= 0.4536 kilograms

Metric-English Conversions

1 millimeter	= 0.0394 inch
1 meter	= 3.281 feet
1 meter	= 1.094 yards
1 kilometer	= 0.6214 mile
1 sq centimeter	= 0.155 sq inch
1 sq meter	= 10.764 sq feet
1 sq meter	= 1.196 sq yards
1 hectare	= 2.471 acres
1 sq kilometer	= 0.386 sq mile
1 cu centimeter	= 0.061 cu inch
1 cu meter	= 35.3 cu feet
1 cu meter	= 1.308 cu yards
1 liter	= 1.057 quarts
1 cu meter	= 264.2 gallons (U.S.)
1 cu kilometer	= 0.240 cu miles
1 gram	= 0.0353 ounce
1 kilogram	= 2.205 pounds

Additional Conversions

1 acre-foot = 1,233.46 cubic meters

Temperature

To change from Fahrenheit (F) to Celsius (C)

$$°C = \frac{(°F - 32°)}{1.8}$$

To change from Celsius (C) to Fahrenheit (F)

$$°F = (°C \times 1.8) + 32°$$

Glossary[1]

Abrasion (caused by external processes). The mechanical wear of rock on rock.

Abyssal plain. A large flat area of deep-sea floor having slopes less than about 1m per km.

Accretion. The process by which solid particles gather together to form a planet.

Adjusted stream system. See *Stream system, adjusted.*

Alluvial fill. A body of alluvium, occupying a stream valley, and conspicuously thicker than the depth of the stream.

Alluvium. Sediment deposited in land environments by streams.

Angle of repose. The steepest angle, measured from the horizontal, at which material remains stable.

Angular unconformity. Unconformity marked by angular divergence between older and younger rocks.

Anion. A positively charged ion.

Antecedent stream. A stream that has maintained its course across an area of the crust that was raised across its path by folding or faulting.

Anticline. An upfold in the form of an arch.

Aquifer. A body of permeable rock or regolith through which ground water moves.

[1]Some definitions are not included in the glossary: mineral names can be found in Tables B.1, B.2, and B.3; rock names are defined in Tables C.1, C.2 and C.3, and Figure 5.18; units of measurement are defined in Appendix E.

Arête. A jagged, knife-edge ridge created where two groups of glaciers have eaten into the ridge from both sides.

Artesian well. A well in which water rises above the aquifer.

Atmosphere. The gaseous envelope that surrounds the Earth.

Atom. The smallest individual particle that retains all the properties of a given chemical element.

Atomic number. The number of protons in the nucleus of an atom.

Axial plane. An imaginary plane that divides a fold as symmetrically as possible, and that passes through the axis.

Axis. The median line between the limbs, along the crest of an anticline or the trough of a syncline.

Barrier island. A long island of sand, lying offshore and parallel to the coast.

Base level. The limiting level below which a stream cannot erode the land.

Base level, local. The level of a lake or any other base level that stands above sea level.

Base level, ultimate. Sea level, projected inland as an imaginary surface beneath all streams.

Batholith. A very large intrusive igneous body, of irregular shape, that cuts across the layering of the intruded rocks.

Bay barrier. A ridge of sand or gravel that completely blocks the mouth of a bay.

Beach. The wave-washed sediment along a coast, extending throughout the surf zone.

Bed load. The coarse solid particles, within a body of flowing fluid, moving along or close above the bed.

Bedrock. The continuous solid rock of the continental crust.

Biosphere. The totality of Earth's living things.

Blowout. A deflation basin excavated in shifting sand or other easily eroded regolith.

Botryoidal. Forms resembling grapes closely bunched.

Bottomset layer. The gently sloping, fine, thin part of each layer in a delta.

Boulder train. A group of erratics spread out fanwise.

Caldera. A roughly circular, steep-walled basin several kilometers or more in diameter.

Caliche. A whitish accumulation of calcium carbonate developed in a soil profile. Also known as *calcrete*.

Cation. A negatively charged ion.

Cavern. A large, roofed-over cavity in any kind of rock.

Chemical elements. The most fundamental substances into which matter can be separated by chemical means.

Chemical sediment. Sediment formed by precipitation of minerals from solution in waters of the Earth.

Cirque. A steep-walled niche, shaped like a half bowl, in a mountain side, excavated mainly by frost wedging and glacial plucking.

Cirque glacier. A very small glacier that occupies a cirque.

Clastic sediment. See **Detritus.**

Cleavage (in minerals). The tendency of a mineral to break in preferred directions along plane surfaces.

Col. A gap in a mountain crest where the headwalls of opposing cirques intersect each other.

Colluvium. Sediment deposited by any process of mass-wasting or by overland flow.

Column. (1) A stalactite connected with a stalagmite. (2) Sometimes used as an abbreviation for **geologic column.**

Columnar joints. Joints that split igneous rocks into long prisms or columns.

Composite volcano. A volcano that emits both fragmental material and viscous lava and that builds up a steep conical mound. Also called a **stratovolcano.**

Compound. A combination of atoms of different elements.

Conchoidal fracture. Breakage resulting in smooth curved surfaces.

Concretion. A localized rock body having distinct boundaries, enclosed in sedimentary rock, and consisting of substances precipitated from solution, commonly around a nucleus.

Cone of depression. A conical depression in the water table immediately surrounding a well.

Conglomerate. A clastic sedimentary rock containing numerous rounded pebbles or larger particles.

Consequent stream. A stream whose pattern is determined solely by the direction of the slope of the land.

Contact (between rock units). A surface along which two rock units meet.

Contact metamorphism. Metamorphism in the vicinity of a pluton and resulting from the effects of the magma on its surrounding rocks.

Continental shelf. A submerged platform of variable width, that forms a fringe around a continent.

Continental shield. Mass of Precambrian rock around which a continent has grown.

Continental slope. A pronounced slope beyond the seaward margin of a continental shelf.

Contour. See **Contour line.**

Contour interval. The vertical distance between two successive contour lines.

Contour line. A line passing through points having the same altitude above sea level; often simply called a contour.

Convection current. The movement of material, within a closed system, as a result of thermal convection arising from the unequal distribution of heat.

Coral reef. A ridge of limestone built by colonial marine organisms.

Core (Earth's). The innermost zone of the Earth. A spherical mass of metallic iron and nickel, 3488km in diameter, consisting of a molten outer part approximately 2128km thick and a solid inner sphere with a radius of 1360km.

Correlation. Determination of equivalence, in geologic age and position, of the sequences of strata found in two or more different areas.

Creep. The imperceptibly slow downslope movement of regolith.

Crevasse. A deep, gaping crack in the surface of a glacier.

Cross section. See **Geologic cross section.**

Cross-strata. Strata that are inclined with respect to a thicker stratum within which they occur.

Crust (Earth's). The outer part of the lithosphere.

Crust, continental. The Earth's crust beneath continents, 20 to 40km thick, consisting of an upper part having the same elastic properties as sialic rocks and a lower part having the same elastic properties as mafic rocks.

Crust, oceanic. The Earth's crust beneath ocean basins, 10 to 15km thick, and consisting of material having the same elastic properties as mafic rocks.

Crystal. A solid compound composed of ordered, three-dimensional arrays of atoms or ions chemically bonded together.

Crystal form. Geometric solid that is bounded by symmetrically arranged plane surfaces.

Crystal structure. The geometric pattern that atoms assume in a mineral.

Curie point (of a mineral). A temperature above which all magnetism is destroyed.

Debris flow. The rapid downslope plastic flow of a mass of debris.

Declination. See **Magnetic declination.**

Decomposition (as a process of weathering). The chemical alteration of rock materials.

Deflation. The picking up and removal of loose rock particles by the wind.

Deflation armor. A surface layer of coarse particles concentrated chiefly by deflation.

Delta. A body of sediment deposited by a stream flowing into standing water. The name comes from the similarity of the plan view to the shape of the Greek letter Δ.

Dendritic pattern. A stream pattern characterized by irregular branching in many directions.

Density. The average weight per unit volume.

Density current. A localized current, within a body of water, caused by dense water sinking through less-dense water.

Detritus. The accumulated particles of broken rock and of skeletal remains of dead organisms.

Dike. A sheet of intrusive igneous rock cutting across the layering of pre-existing rock.

Dip (of a magnetic needle). See **Magnetic inclination.**

(of a stratum). The angle in degrees between a horizontal plane and an inclined plane, measured down from horizontal in a plane perpendicular to the strike. Dip is measured with a *clinometer*.

Discharge. The quantity of water that passes a given point in a given unit of time.

Disconformity. Unconformity that is not marked by angular divergence between two groups of strata in contact.

Disintegration (as a process of weathering). The mechanical loosening, during chemical weathering, of the individual mineral particles of bedrock.

Dissolved load. Matter dissolved in the water of a stream.

Divide. The line that separates adjacent drainage basins.

Drainage basin. The total area that contributes water to a stream.

Drift (glacial). The sediments deposited directly by glaciers (**till,** q. v.), or indirectly in glacial streams, lakes, and the sea (**stratified drift,** q. v.).

Dripstone. Calcite chemically precipitated from dripping water in an air-filled cavity.

Drumlin. A streamline hill consisting of drift, generally till, and elongated parallel with the direction of glacier movement.

Dune. A mound or ridge of sand deposited by the wind.

Dynamic equilibrium. See **Steady state.**

Earthquake focus. The point of first release of the energy that causes an earthquake.

Economy (of a stream). The input and consumption of energy within a stream or other system and the changes that result.

Electron. A negatively charged particle.

Elements. See **Chemical elements.**

End moraine. A ridgelike accumulation of drift, deposited by a glacier along its front margin.

Energy. The capacity to produce activity.

Epicenter. The point, on Earth's surface, that lies vertically above the focus of an earthquake.

Erosion. The complex group of related processes by which rock is broken down physically and chemically and its products removed.

Erratic (glacial). A transported fragment of rock that differs from the bedrock beneath it. The agent of transport was commonly glacier ice or floating ice.

Esker. A long narrow ridge, commonly sinuous, of stratified drift.

Evaporite. A nonclastic sedimentary rock whose constituent minerals were precipitated from water solution as a result of evaporation.

Exfoliation. The separation, during weathering, of successive shells from rock.

Exposure. A place where solid rock is exposed at Earth's surface.

Extrusive igneous rock. Rock formed by the cooling of magma poured out onto the Earth's surface.

Facies. A distinctive group of characteristics, within a rock unit, that differ as a group from those elsewhere in the same unit.

Fan. A fan-shaped body of alluvium built at the base of a steep slope. Known also as an *alluvial fan.*

Fault. A fracture along which the opposite sides have been displaced relative to each other. For kinds of faults see Fig. 15.11.

Fauna. The animals of any given region or age.

Faunal succession, law of. This law, discovered by William Smith, states that fossil faunas and floras succeed one another in a definite, recognizable order.

Ferromagnesian minerals. Silicate minerals in which iron and/or magnesium are the principal cations. Examples are the biotites, amphiboles, pyroxenes, and olivines.

Fiord. A glaciated trough partly submerged by the sea.

Fissure eruption. Extrusion of volcanic materials along an extensive fracture.

Floodplain. That part of any stream valley which is inundated during floods.

Flora. The plants of any given region or age.

Flowstone. Material chemically precipitated from flowing water in the open air or in an air-filled cavity.

Fold. A pronounced bend in layers of rock. For kinds of fold see Figure 15.19.

Foliation. A parallel or nearly parallel structure in metamorphic rock, caused by compositional layering.

Footwall. The surface of the block of rock below an inclined fault.

Foreset layer. The coarse, thick, steeply sloping part of each layer in a delta.

Formation. A succession of strata distinctive

enough to constitute a basic unit for mapping and description.

Fossil. The naturally preserved remains or traces of an animal or a plant.

Fossil fuel. A fuel that contains solar energy, locked up securely in chemical compounds by the plants or animals of former ages.

Frost heaving. The lifting of regolith or a small body of bedrock by freezing of contained water.

Frost wedging. The pushing apart of rock particles by water freezing to ice within rock or regolith.

Fumarole. A volcano that emits only gases.

Gangue. The nonvaluable minerals of an ore.

Geologic column. A composite diagram combining in a single column the succession of all known strata, fitted together on the basis of their fossils or of other evidence of relative age.

Geologic cross section. A diagram showing the arrangement of rocks in a vertical plane.

Geologic map. A map that shows the distribution, at the surface, of rocks of various kinds or of various ages.

Geology. The science of the Earth.

Geosyncline. A great trough that has received thick deposits of sediment during its slow subsidence through long geologic periods.

Geothermal gradient. The rate of increase of temperature downward in the Earth.

Geyser. An orifice that erupts steam and boiling water intermittently.

Glaciation. The alteration of a land surface by the massive movement over it of glacier ice.

Glacier. A body of ice, consisting mainly of recrystallized snow, flowing on a land surface.

Graben. A trench-like structure bounded by parallel normal faults.

Graded layer. A layer in which the particles grade upward from coarser to finer.

Gradient (of a stream). The slope measured along the stream, on the water surface or on the bottom.

Granular texture. Interlocking arrangement of visible mineral grains characteristic of granite.

Gravimeter (or **gravity meter**). A sensitive device for measuring the force of gravity at any locality.

Gravity (of Earth). The inward-acting force with which Earth tends to pull all objects toward its center.

Groin. A low wall, built on a beach, that crosses the shoreline at a right angle.

Ground moraine. Widespread thin drift with a smooth surface consisting of gently sloping knolls and shallow closed depressions.

Ground water. All the water contained in spaces within bedrock and regolith.

Guyot. A seamount with a conspicuously flat top well below sea level.

Half-life (in radioactive decay). The time required to reduce the number of parent atoms by one-half.

Hand specimen. A piece of rock that can be conveniently held in the hand for close study.

Hanging wall. The surface of the block of rock above an inclined fault.

Hardness. Relative resistance of a mineral to scratching.

Horn. A bare, pyramid-shaped peak left standing where glacial action in cirques has eaten into it from three or more sides.

Horst. An elevated elongate block bounded by parallel normal faults.

Humus. The decomposed residue of plant and animal tissues.

Hydrosphere. The Earth's discontinuous water envelope, including the oceans, lakes, streams, ground water, snow, and ice.

Ice-contact stratified drift. Stratified drift deposited in contact with its supporting ice.

Ice sheet. A broad glacier of irregular shape, generally blanketing a large land surface.

Igneous rock. Rock formed by the cooling and solidification of magma.

Intrusive igneous rock. Any igneous rock formed by cooling and solidification of magma below Earth's surface.

Ion. An atom that has excess positive or negative charges caused by electron transfers.

Island arcs. Great arcuate belts of andesitic and basaltic volcanic islands.

Isograd. A line on a map, connecting points of first occurrence of a given mineral in metamorphic rocks.

Isopach. A line on a map, that connects points of equal thickness of a rock unit.

Isostasy. The ideal property of flotational balance among segments of the lithosphere.

Isotopes. Atoms having the same atomic number but differing numbers of neutrons in the nucleus.

Joint. Fracture on which movement has not occurred in a direction parallel to the plane of the fracture.

Joint set. A widespread group of parallel joints.

Joint system. A combination of two or more intersecting sets of joints.

Kame. A body of ice-contact stratified drift in the form of a knoll or hummock.

Kame terrace. A terracelike body of ice-contact stratified drift along the side of a valley.

Karst topography. An assemblage of topographic forms consisting primarily of closely spaced sinks; strikingly developed in the Karst region of Yugoslavia.

Kettle (glacial). A basin in glacial drift, created by melting-out of a mass of underlying ice.

Laccolith. A lenticular intrusive igneous body above which the layers of invaded bedrock have been bent upward to form a dome.

Lagoon. A bay inshore from a line of barrier islands or from a coral reef.

Lava. Magma that has reached Earth's surface through a volcanic vent, and flows out as hot streams or sheets.

Leaching. The continued removal, by water, of soluble matter from regolith or bedrock.

Levee, natural. A broad, low ridge of fine alluvium built along the side of a stream channel by water that spreads out of the channel during floods.

Limbs. The sides of a fold.

Lithosphere. The outer zone of the solid Earth. The lithosphere includes the crust and grades compositionally into the mantle. The boundary between the mantle and the lithosphere is the low-velocity zone.

Load (of a stream). The material carried at a given time, by a stream, by a current of water, by the wind, or by a glacier.

Loess. Wind-deposited silt, usually accompanied by some clay and some fine sand.

Long profile (of a stream). A line that connects a series of points on the surface of a stream, along all or part of the length of the stream.

Longshore current. A current, within the surf zone, that flows parallel to the shore.

Luster. The quality and intensity of light reflected from a mineral.

M-discontinuity. See **Mohorovičić discontinuity.**

Magma. Molten silicate materials beneath the Earth's surface, including crystals derived from them and gases dissolved in them.

Magmatic differentiation by crystallization. The compositional changes that occur in magma by the separation of early-formed minerals from residual liquids.

Magmatic differentiation by partial melting. The process of forming magma with differing compositions by the incomplete melting of rocks.

Magnetic declination. The clockwise angle from true north assumed by a magnetic needle.

Magnetic field (Earth's). The magnetic lines of force surrounding the Earth.

Magnetic inclination (or **dip**). The angle with the horizontal assumed by a magnetic needle.

Mantle (Earth's). A zone of rocky matter, about 2800km thick, surrounding the Earth's core and covered by the thin lithosphere. The mantle occupies about 80 per cent of the total volume of the Earth.

Marine sediment. Sediment deposited in the sea.

Mascon (on the Moon). An anomalously high gravity pull.

Massive rock. Rock that is fairly uniform in appearance and lacks any breakage surfaces.

Mass number. The sum of the protons and neutrons in the nucleus of an atom.

Mass-wasting. The movement of regolith downslope by gravity without the aid of a stream, a glacier, or wind.

Mature soil. A soil that has a fully developed profile.

Meander. A looplike bend of a stream channel.

Metamorphic facies. Contrasting assemblages of minerals that reach equilibrium during metamorphism within a specific range of physical conditions.

Metamorphic rock. Rock formed within Earth's crust by transformation, in the solid state, of pre-existing rocks as a result of high temperature, high pressure, or both.

Metamorphism. The changes in mineral assemblage and rock texture, or both, that take place in the solid state within the Earth's crust as a result of high temperature and high pressure.

Meteorite. A small solid body formed during condensation of the planetary nebula.

Mid-ocean ridge. See *Oceanic ridge.*

Mineral assemblage. The variety and abundance of minerals present in a rock.

Minerals. All naturally occurring, crystalline, inorganic compounds.

Moho. See *Mohorovičić discontinuity.*

Mohorovičić discontinuity. The base of the crust. Also known as the *M-discontinuity,* and as the *moho.*

Molecule. The smallest unit that retains all the properties of a compound.

Monocline. A one-limbed flexure, on both sides of which the strata either are horizontal or dip uniformly at low angles.

Mountain chain. An elongate unit consisting of numerous ranges or groups of mountains, regardless of similarity in form or equivalence in age.

Mountain range. An elongate series of mountains belonging to a single geologic unit.

Mountain system. A group of ranges similar in general form, structure, and alignment, and presumably owing their origin to the same general causes.

Mud crack. Crack caused by shrinkage of wet mud as its surface becomes dry.

Mudflow. A debris flow in which the consistency of the flowing substance is that of mud.

Natural levee. See *Levee, natural.*

Neutron. An electrically neutral particle with a mass 1833 times greater than that of the electron.

Oceanic ridge (also called *mid-ocean ridge* and *oceanic rise*). A continuous rocky ridge on the ocean floor, many hundred to a few thousand kilometers wide, with a relief of 600m or more.

Oceanic rise. See *Oceanic ridge.*

Oil field. A group of oil pools, usually of similar type, or a single pool in an isolated position.

Oil pool. An underground accumulation of oil or gas in a reservoir limited by geologic barriers.

Ore. An aggregate of minerals from which one or more materials can be extracted profitably.

Organic compound. Chemical compound made from carbon and hydrogen, with or without other elements such as nitrogen and oxygen.

Organic matter. Mixtures of organic compounds, generally formed directly or indirectly from the activities of living organisms.

Outcrop area. The area, on a geologic map, shown as occupied by a particular rock unit.

Outwash. Stratified drift deposited by streams of meltwater as they flow away from a glacier.

Outwash plain. A body of outwash that forms a broad plain.

Overland flow. The movement of runoff in broad sheets or groups of small interconnecting rills.

Parallel strata. Strata whose individual layers are parallel.

Pedalfer. A soil in which much clay and iron have been added to the B horizon.

Pediment. A sloping surface, cut across bedrock, adjacent to the base of a highland in an arid climate.

Pedocal. A soil with calcium-rich upper horizons.

Peneplain. "Almost a plain." A land surface worn down to very low relief by streams and mass-wasting.

Perched water body. A water body that occupies a basin in impermeable material, perched in a position higher than the main water table.

Permafrost. Ground that is frozen perennially.

Permeability. Capacity for transmitting fluids.

Petroleum. Gaseous, liquid, or solid substances, occurring naturally and consisting chiefly of chemical compounds of carbon and hydrogen. Petroleum includes both oil and natural gas.

Phenocrysts. The isolated large crystals in porphyry.

Piedmont glacier. A glacier on a lowland at the base of a mountain, fed by one or more valley glaciers.

Placer. A deposit of heavy minerals concentrated mechanically. Common methods of mechanical concentration are sorting resulting from stream flow, wind action, and wave action.

Planetary nebula. A flattened rotating disk of gas.

Plateau basalt. Fissure basalt that forms widespread, nearly horizontal layers on continents.

Playa. A dry bed of a playa lake (q. v.); a nearly level area on the floor of an intermontane basin. Occasional floods deposit silt and clay that mantle a playa.

Playa lake. An ephemeral shallow lake in a desert basin.

Plucking (glacial). The lifting out and removal of fragments of bedrock by a glacier.

Plunge. The angle between a fold axis and the horizontal.

Plunging fold. A fold with an inclined axis.

Pluton. An intrusive igneous body, regardless of shape, size, or composition.

Pluvial lake. A lake that existed under a former climate, when rainfall on the surrounding region was greater than today's.

Pollutant. Any waste substance introduced into a natural system in greater amount than can be disposed of by the system.

Polymerization (applied to silicate minerals). The process of linking silica tetrahedra into larger groups.

Polymorph. A compound that occurs in more than one crystal form.

Porosity. The proportion, in per cent, of the total volume of a given body of bedrock or of regolith that consists of pore spaces (q. v.).

Porphyry. An igneous rock consisting of coarse mineral grains scattered through a mixture of fine mineral grains.

Potential energy. Stored energy waiting to be used.

Proton. A positively charged particle with a mass 1832 times greater than the mass of an electron.

Radioactivity. The decay process by which an unstable atomic nucleus spontaneously disintegrates.

Radiometric dating. Determination of absolute age of a mineral through measurement of the radioactivity of certain elements in the mineral.

Recharge. The addition of water to the zone of saturation (q. v.).

Rectangular pattern. A stream pattern characterized by right-angle bends in the stream.

Refraction. See **Wave refraction.**

Regolith. The blanket, consisting of loose, noncemented rock particles and mineral grains, that commonly overlies bedrock.

Rejuvenation. The development of youthful morphologic features in a land mass further advanced in the cycle of erosion. Regional uplift is the common cause of rejuvenation.

Relief. The difference in altitude between the high and low parts of a land surface.

Replacement. The process by which one mineral takes the place of another mineral. During replacement a fluid dissolves matter already present and at the same time deposits from solution an equal volume of a different substance.

Residual concentration. The natural concentration of a mineral substance by removal of a different substance with which it was associated.

Rock. Any naturally formed, firm and coherent aggregate or mass of mineral matter that constitutes part of Earth's crust.

Rock avalanche. The rapid downslope flow of a mass of dry rock particles.

Rock cleavage. The property by which a rock breaks into plate-like fragments along flat planes.

Rock cycle. The cyclic movement of rock material, in the course of which rock is created, destroyed, and altered through the operation of internal and external Earth processes.

Rockfall (and **debris fall**). The rapid descent of a mass of rock (debris), vertically from a cliff or by leaps down a slope.

Rock flour. Fine sand and silt produced by crushing and grinding in a glacier.

Rock glacier. A lobate, steep-fronted mass of coarse, angular regolith that extends out from the base of a cliff in a cold climate and that moves downslope with the aid of interstitial water and ice.

Rockslide (and **debris slide**). The rapid descent of a mass of rock (debris) by sliding down a slope.

Runoff. Water that flows over the lands.

Scale (of a map). The proportion between a unit of distance on a map and the unit it represents on the Earth's surface.

Sea-floor trench. A long, narrow, very deep basin in the sea floor.

Seamount. An isolated volcanic hill, more than 1000m high, rising above the deep-sea floor.

Secondary enrichment. Natural enrichment of an ore body by later addition of mineral matter, generally from percolating solutions.

Sediment. Regolith that is being or has been transported by any of Earth's external processes.

Sedimentary breccia. See **Conglomerate.** A clastic sedimentary rock that resembles conglomerate, but most of whose fragments are angular instead of rounded.

Sedimentary rock. Rock formed from sediment by cementation or by other processes that act at ordinary temperatures at or near Earth's surface.

Seismic belt. Large tract subject to frequent earthquake shocks.

Sheet erosion. The erosion performed by overland flow (q. v.).

Shield volcano. A volcano that emits fluid lava and builds up a broad, dome-shaped edifice, convex upward, with a surface slope of only a few degrees.

Sill. A sheet of intrusive igneous rock that is parallel to the layering of pre-existing rock.

Sink. A large solution cavity open to the sky, generally created by collapse of a cavern roof.

Slickensides. Striated or highly polished surfaces on hard rocks, abraded by movement along a fault.

Sliderock. Colluvium that constitutes a talus.

Slip face. The straight, lee slope of a dune.

Slump. The downward slipping of a coherent body of rock or regolith along a curved surface of rupture.

Snowfield. A wide cover, bank, or patch of snow, above the snowline, that persists throughout the summer season.

Snowline. The lower limit of perennial snow.

Soil. That part of the regolith which can support rooted plants.

Soil, mature. Soil that has a fully developed profile.

Soil profile. The succession of distinctive horizons in a soil, from the surface down to the unchanged parent material beneath.

Solar System. The system comprised of the Sun and the group of planetary bodies that revolve around it, held by gravitational attraction.

Solid solution. The substitution of one atom for another in a random fashion throughout a crystal structure.

Solifluction. Imperceptibly slow, downslope flow of water-saturated regolith.

Specific gravity. A number stating the ratio of the weight of the substance to the weight of an equal volume of pure water at $4°C$.

Spit. An elongate ridge of sand or gravel that projects from land and ends in open water.

Spring. A flow of ground water emerging naturally onto the surface.

Stalactite. An icicle-like form of dripstone and flowstone that hangs from the ceiling of a cavern.

Stalagmite. A blunt, icicle-like form of flowstone projecting upward from the floor of a cavern.

Steady state. A condition in which the rate of arrival of some materials equals the rate of escape of other materials.

Stock. A small body with the same characteristics as a batholith.

Stratification. Layered arrangement of the particles that constitute sediment or sedimentary rock.

Stratified drift. Drift that is sorted and stratified.

Stratigraphic superposition, principle of. The principle which states that in any sequence of strata, not later disturbed, the order in which they were deposited is from bottom to top.

Stratum (plural, **strata**). A distinct layer of sedimentary or igneous rock consisting of material that has been spread out upon the Earth's surface.

Streak. A thin layer of powdered mineral made by rubbing a specimen on a nonglazed porcelain plate.

Stream. A body of water carrying suspended and dissolved substances and flowing down a slope along a definite path, the stream's channel.

Stream capture. (Known also as **stream piracy.**) The diversion of a stream by the headward growth of another stream.

Stream flow. The flow of surface water between well-defined banks.

Stream system, adjusted. A stream system in which most of the streams occupy weak-rock positions.

Stream terrace. A bench along the side of a valley,

the upper surface of which was formerly the alluvial floor of the valley.

Striations (glacial). Scratches and grooves on bedrock surfaces, caused by grinding of rock against rock during movement of glacier ice.

Strike. The compass direction of the horizontal line in an inclined plane.

Stratovolcano. See **Composite volcano.**

Subsequent stream. A stream whose course has become adjusted so that it occupies belts of weak rock.

Superposed stream. A stream that was let down, or superposed, from overlying strata onto buried bedrock having composition or structure unlike that of the covering strata.

Surf. Wave activity between the line of breakers and the shore.

Surface ocean currents. Broad, slow drifts of surface water in an ocean.

Suspended load. The fine solid particles turbulently suspended within a body of flowing fluid.

Syncline. A downfold with a troughlike form.

Talus. An apron of rock waste that slopes outward from a cliff from which it is derived.

Tectonics. The study of Earth's broad structural features.

Texture (of a rock). The sizes and shapes of the individual particles in a rock, and the mutual relationship between them.

Thin section. A thin rock slice prepared for microscopic study.

Thrust fault. Low-angle reverse fault with dip generally less than 45°.

Till. Nonsorted glacial drift.

Tillite. Till converted to solid rock.

Tombolo. A ridge of sand or gravel that connects an island to the mainland or to another island.

Topographic map. A map that delineates surface forms.

Topset layer. Stream sediment that overlies the foreset layers in a delta.

Transform fault. A special class of strike-slip fault that links major structural features.

Trellis pattern. A rectangular stream pattern in which tributary streams are parallel and very long.

Tuff. An extrusive igneous rock formed by the agglomeration of small volcanic particles.

Turbidity current. A density current whose excess density results from suspended sediment.

Unconformity. A lack of continuity between two units of rock in contact, corresponding to a gap in the geologic record.

Uniformity, principle of. The principle that Earth's external and internal processes have been operating unchanged, and within the same range of rates, throughout most of Earth's history.

Valley glacier. A glacier that flows downward through a valley.

Valley train. A body of outwash that partly fills a valley.

Varve. A pair of sedimentary layers deposited during the cycle of the year.

Ventifact. A stone, the surface of which has been abraded by wind-blown sand.

Vesicle. Small opening, in extrusive igneous rock, made by escaping gas originally held in solution under high pressure while the parent magma was underground.

Viscosity. The internal property of a substance that offers resistance to flow.

Volcanic crater. A funnel-shaped depression from which gases, fragments of rock, and lava are ejected.

Volcanic mudflow. A mudflow whose chief constituent is volcanic ash.

Volcano. The vent from which molten igneous matter, solid rock debris, and gases are erupted.

Water cycle. The cyclic movement and change of state of water substance through evaporation, wind transport, stream flow, percolation, and related processes. (Also called **hydrologic cycle.**)

Water gap. A pass, in a ridge or mountain, through which a stream flows.

Water table. The upper surface of the zone of saturation.

Wave-built terrace. The body of wave-washed sediment that extends seaward from the line of breakers.

Wave-cut bench. A bench or platform cut across bedrock by surf.

Wave-cut cliff. A coastal cliff cut by surf.

Wave refraction. The process by which the direction of travel of a wave, moving in shallow water at an angle to the shoreline, is altered.

Weathering. The chemical alteration and mechanical breakdown of rock materials during exposure to air, moisture, and organic matter.

Weight (of a body). The force that gravity exerts on a body.

Welded tuff. A glassy or fine-grained rock formed by fusion of volcanic ash during deposition.

Zone of aeration. The zone in which the open spaces in regolith or bedrock are normally filled mainly with air.

Zone of saturation. The subsurface zone in which all openings are filled with water.

Index

Numbers of pages on which terms are defined are in *boldface*. Asterisks indicate illustrations.